LONG-TERM ECOLOGICAL RESEARCH NETWORK SERIES
LTER Publications Committee

The Ecology of Agricultural Landscapes

Long-Term Research on the Path to Sustainability

Edited by

STEPHEN K. HAMILTON, JULIE E. DOLL, AND
G. PHILIP ROBERTSON

OXFORD
UNIVERSITY PRESS

Oxford University Press is a department of the University of
Oxford. It furthers the University's objective of excellence in research,
scholarship, and education by publishing worldwide.

Oxford   New York
Auckland   Cape Town   Dar es Salaam   Hong Kong   Karachi
Kuala Lumpur   Madrid   Melbourne   Mexico City   Nairobi
New Delhi   Shanghai   Taipei   Toronto

With offices in
Argentina   Austria   Brazil   Chile   Czech Republic   France   Greece
Guatemala   Hungary   Italy   Japan   Poland   Portugal   Singapore
South Korea   Switzerland   Thailand   Turkey   Ukraine   Vietnam

Oxford is a registered trademark of Oxford University Press
in the UK and certain other countries.

Published in the United States of America by
Oxford University Press
198 Madison Avenue, New York, NY 10016

Library of Congress Cataloging-in-Publication Data
The ecology of agricultural landscapes : long-term research on the path to sustainability / edited by
Stephen K. Hamilton, Julie E. Doll, and G. Philip Robertson.
pages cm. — (Long-Term Ecological Research Network series)
Includes bibliographical references and index.
ISBN 978–0–19–977335–0 (alk. paper)
1. Agricultural ecology—Middle West.   2. Cropping systems—Middle West.   I. Hamilton, Stephen K.,
editor.   II. Series: Long-Term Ecological Research Network series.
S444.E25 2015
577.5'5—dc23
2014040737

9 8 7 6 5 4 3 2
Printed in the United States of America
on acid-free paper

Contents

Preface

Agricultural ecosystems and landscapes are managed to produce food, fuel, and fiber but also have the potential, when managed appropriately, to provide society a host of other benefits known as ecosystem services. Examples include climate change mitigation, clean drinking water, beneficial insect habitat, and various cultural amenities like outdoor recreation and green space. The delivery of these services depends on how agriculture is managed as fields and landscapes, and often involves trade-offs. Historically we have managed agricultural systems more for yield than for other ecosystem services, though evidence in this volume and elsewhere suggests that many of these other services can be promoted without sacrificing yield. And increasingly, we realize how issues of environmental quality that extend well beyond the farm challenge the sustainability of agriculture in the long term.

By taking a systems approach to the study of agricultural ecosystems, it is possible to understand how different parts of the systems interact to enhance or diminish different ecosystem services, and to then evaluate inherent trade-offs. Managing these trade-offs for different outcomes provides the opportunity to make farming more sustainable.

To take such an approach requires taking into account every key part of the agricultural ecosystem: living organisms (crops, weeds, insects, microbes, animals, and humans) as well as their nonliving, physical environment (water, air, minerals, and soil). Understanding how all these parts interact can help to better utilize biological resources to control pests, provide nitrogen to crops, mitigate climate change, and build soil fertility.

This book is a synthesis of over two decades of research in agricultural ecology at the W. K. Kellogg Biological Station Long-Term Ecological Research site (KBS LTER), located in southwest Michigan, U.S. Here, scientists study agricultural ecosystems amid the matrix of unmanaged, successional forests and fields in which they reside, as well as wetlands, streams, and lakes in the broader landscape. Sustained sampling over many years documents the effects of episodic events such as drought or pest outbreaks, and as well allows observation of ecosystem processes that respond slowly, such as changes in soil carbon and microbial communities. Experimentation allows us to identify the organisms and processes responsible for different outcomes, and suggests ways that different systems might be managed to optimize the delivery of the most valued ecosystem services.

A unique aspect of the KBS LTER is the synergistic collaboration of agricultural scientists, ecologists, and social scientists, providing cross-disciplinary exchanges of ideas to generate new knowledge and new avenues of investigation. In this book, we present the current state of our understanding of row-crop agriculture at this site, drawing on comprehensive research extending from field to landscape scales. We show, for example, how KBS LTER scientists traced the cycle of nitrogen through soils, plants, and microbes to reveal how leaching moves different amounts of nitrogen to ground and surface waters in different cropping systems; quantified how different farming practices can reduce greenhouse gas emissions without diminishing crop yields, and why farmers might adopt such practices; and revealed how plant diversity in the surrounding landscape enhances the number of beneficial predators of agricultural pests in farm fields. These are but a few examples of how KBS LTER research can and has informed the design and management of more sustainable farming systems.

Never has the need been greater for an ecosystem approach to agriculture. As our global population grows to over 9 billion in the next 30 years, with a concomitant demand for agricultural products, ever more pressure will be placed on our agricultural systems. Meanwhile, climate change is altering the ecological settings in which agriculture is practiced, demanding adaptation. Knowledge generated by long-term research such as that at KBS will help to address one of the grand challenges of our time: how to meet sustainably the growing world demand for agricultural products—in a way that minimizes environmental harm and enhances the delivery of a diverse array of ecosystem services.

Throughout this book, the authors identify knowledge gaps and suggest new directions for future research to bring us further down the sustainability path. Readers will find chapters that stand on their own, but when read together offer a comprehensive, synthetic portrait of the ecology of row-crop ecosystems and unmanaged lands in agricultural landscapes. We hope this volume enhances readers' understanding of the nexus between agriculture, people, and the environment, and stimulates new research and educational efforts.

Data collected as part of core KBS LTER research activities are maintained online, in a publicly available database. This includes most of the data used in the following chapters. The KBS LTER Data Catalog (http://lter.kbs.msu.edu/datatables) is also incorporated in the LTER Network Information System (https://portal.lternet.edu/nis/home.jsp).

Acknowledgments

The KBS LTER is one of 26 sites in the national LTER network that spans a broad diversity of climates, biomes, and degrees of human influence. The U.S. National Science Foundation funds the core activities at all LTER sites, enabling observations and experimentation over time scales not addressable in conventional research awards. We are indebted to the foresight and commitment of many at NSF who realized this vision and have acted to sustain it. We particularly thank former and current program directors James T. Callahan, Scott L. Collins, Henry L. Gholz, Nancy J. Huntly, Matthew D. Kane, and Saran Twombly.

Many colleagues have contributed substantially to the success of the KBS LTER. We are particularly indebted to former principal investigators who were not able to author chapters, including Michael J. Klug, Richard R. Harwood, and Christopher K. Vanderpool. George H. Lauff and Bernard D. Knezek were also instrumental in the successful launch of KBS LTER, and James J. Gallagher and Charles W. "Andy" Anderson helped develop our K–12 education program.

Critical to the success of this large, multidisciplinary, long-term project has been the dedicated staff that manages agronomic operations, sampling and data collection, and information management. Past and current project managers Katherine M. Klingensmith, Sandra J. Halstead, Andrew T. Corbin, and Stacey L. VanderWulp; agronomic managers James A. Bronson, Robert L. Beeley, Mark A. Halvorson, and Joseph T. Simmons; information managers John B. Gorentz, Timothy T. Bergsma, Lolita S. Krievs, Garrett R. Ponciroli, Sven Bohm, and Suzanne J. Sippel; and recent science coordinator Justin M. Kunkle all contributed immeasurably to the progress of science at KBS, as did the numerous technicians, graduate students, and postdocs who conducted most of the field and laboratory measurements. We are also indebted to Barbara G. Fox and Jane L. Schuette for their invaluable assistance in producing this book.

We, in addition, gratefully acknowledge the many funding agencies and foundations that have provided financial support for KBS LTER research, education, and outreach. In addition to NSF, core support has been provided by Michigan State University and MSU AgBioResearch, and we gratefully acknowledge the consistent and enthusiastic support of administrators at all levels of the university.

Finally, the editors are grateful to the contributing authors for their chapters, and for their patience with this project.

S. K. H.

J. E. D.

G. P. R.

Contributors

Bruno Basso
W.K. Kellogg Biological Station
Department of Geological Sciences
Michigan State University
East Lansing, MI 48824

Subir Biswas
Department of Electrical and Computer
   Engineering
Michigan State University
East Lansing, MI 48824

Huilan Chen
Guangzhou Chengfa Investment Fund
   Management Co. Ltd.
Guangzhou 510623, China

Adam S. Davis
Agricultural Research Service
U.S. Department of Agriculture
Urbana, IL 61801

Julie E. Doll
W.K. Kellogg Biological Station
Michigan State University
Hickory Corners, MI 49060

Sarah Emery
Department of Biology
University of Louisville
Louisville, KY 40292

Jordan Fox
Cerner Corporation
816 SW Country Hill Dr
Grain Valley, MO 64029

Stuart H. Gage
Department of Entomology
Michigan State University
East Lansing, MI 48824

Ilya Gelfand
Great Lakes Bioenergy Research
   Center &
W.K. Kellogg Biological
   Station
Michigan State University
Hickory Corners, MI 49060

A. Stuart Grandy
Department of Natural Resources and
   the Environment

University of New Hampshire
Durham, NH 03824

Katherine L. Gross
W.K. Kellogg Biological Station &
Department of Plant Biology
Michigan State University
Hickory Corners, MI 49060

Stephen K. Hamilton
W.K. Kellogg Biological Station
Department of Integrative Biology
Michigan State University
Hickory Corners, MI 49060

M. Christina Jolejole-Foreman
Department of Global Health and
    Population
Harvard University School of Public
    Health
Roxbury Crossing, MA 02120

Wooyeong Joo
Department of Integrative Biology
Michigan State University
East Lansing, MI 48824

Eric P. Kasten
Global Observatory for Ecosystem
    Services
Michigan State University
East Lansing, MI 48824

Alexandra Kravchenko
Department of Plant, Soil and
    Microbial Sciences
Michigan State University
East Lansing, MI 48824

Douglas A. Landis
Department of Entomology
Michigan State University
East Lansing, MI 48824

Frank Lupi
Department of Agricultural, Food,
    and Resource Economics &
Department of Fisheries and Wildlife
Michigan State University
East Lansing, MI 48824

Shan Ma
The Natural Capital Project
Stanford University
Stanford, CA 94305

Neville Millar
Great Lakes Bioenergy Research
    Center &
W.K. Kellogg Biological Station
Michigan State University
Hickory Corners, MI 49060

Sherri Morris
Biology Department
Bradley University
Peoria, IL 61625

Eldor A. Paul
Natural Resource Ecology
    Laboratory
Colorado State University
Fort Collins, CO 80525

Natalie Rector
Corn Marketing Program of Michigan
Lansing, MI 48906

Joe T. Ritchie
Department of Agricultural and
    Biological Engineering
University of Florida
Gainesville, FL 32611

G. Philip Robertson
W.K. Kellogg Biological Station &
Department of Plant, Soil and
    Microbial Sciences
Michigan State University
Hickory Corners, MI 49060

Todd M.P. Robinson
W.K. Kellogg Biological Station
Department of Plant Biology
Michigan State University
Hickory Corners, MI 49060

Gene R. Safir
Department of Plant Pathology
Michigan State University
East Lansing, MI 48824

Thomas M. Schmidt
Department of Ecology and
  Evolutionary Biology
University of Michigan
Ann Arbor, MI 48109

Richard G. Smith
Department of Natural Resources and
  the Environment
University of New Hampshire
Durham, NH 03824

Sieglinde S. Snapp
W.K. Kellogg Biological Station &
Department of Plant, Soil and
  Microbial Sciences
Michigan State University
Hickory Corners, MI 49060

Scott M. Swinton
Department of Agricultural, Food, and
  Resource Economics
Michigan State University
East Lansing, MI 48824

Clive Waldron
Department of Ecology and
  Evolutionary Biology
University of Michigan
Ann Arbor, MI 48109

Wei Zhang
Environment and Production
  Technology Division
International Food Policy Research
  Institute
Washington, DC 20006

THE ECOLOGY OF
AGRICULTURAL LANDSCAPES

1

Long-Term Ecological Research at the Kellogg Biological Station LTER Site

Conceptual and Experimental Framework

G. Philip Robertson and Stephen K. Hamilton

Over half of the land area of the contiguous United States is in agricultural production, with over half devoted to row crops such as corn (*Zea mays* L.), soybean (*Glycine max* L.), and wheat (*Triticum aestivum* L.) (NASS 2013). These cropping systems thus represent one of the most extensive and important ecosystem types in North America. The vast majority of this cropland is managed intensively with tillage, chemical fertilizers, and pesticides to achieve high yields. And with well-known environmental impacts on soils, watersheds, surface and coastal waters, and the atmosphere (Matson et al. 1997, Robertson et al. 2004, Robertson and Vitousek 2009, Tilman et al. 2011), the environmental consequences of agricultural intensification extend well beyond the boundaries of individual farm fields.

While the catalog of agriculture's harmful environmental impacts is extensive—ranging from biogeochemical pollution to diminished biodiversity to human health risks—many of the benefits are substantial, not the least of which is human well-being from the provision of food and other products. Less well appreciated are agriculture's contributions to a number of other ecosystem services (Swinton et al. 2007, Power 2010)—clean water, flood protection, climate regulation, disease and pest suppression, soil fertility, habitat conservation, and recreational and aesthetic amenities, among others—all of which benefit people.

Also underappreciated is the degree to which agricultural ecosystems are linked to one another and to unmanaged areas of the surrounding landscape such as surface waters, wetlands, woodlots, and abandoned fields undergoing ecological succession. Only recently have we learned the importance of many of these understudied linkages (Robertson et al. 2007). For example, while it has been long known how bacteria in streams and wetlands can transform excess nitrate (NO_3^-) that leaves farm fields into harmless dinitrogen (N_2) gas (Lowrance et al. 1984, Robertson and Groffman 2015), only recently have headwater streams and small wetlands in agricultural landscapes been shown to disproportionately improve water quality (Mulholland et al. 2009, Hamilton 2015, Chapter 11 in this volume). Likewise, relatively small areas of uncropped habitat can disproportionately support biodiversity services via the provision of refugia for pollinators and insect predators important to pest suppression (Gardiner et al. 2009).

How can row crops be managed to balance or reduce the negative impacts of agricultural production? The answer lies in knowing how to manage cropland for an array of ecosystem services, and that area of research remains largely unexplored. Of particular importance is an understanding of how different cropping systems vary in their impacts—environmental, economic, as well as social—and how they interact with unmanaged ecosystems. By fully comprehending the causes and consequences of these impacts and interactions we can identify (1) which components and interactions are important for delivering the services we value and (2) how this knowledge can be used to promote beneficial services and minimize the negative impacts of agriculture at different geographic scales.

Many processes and attributes that provide ecosystem services in agricultural landscapes take decades to occur or become visible. Thus, long-term observations are crucial for detecting change (Magnuson 1990, Scheffer et al. 2009). Some changes are gradual, such as trends in soil organic matter (Paul et al. 2015, Chapter 5 in this volume) and shifts in soil microbial communities (Schmidt and Waldron 2015, Chapter 6 in this volume). Others may be more rapid with clear immediate effects but still have long-term, perhaps subtle consequences. The appearance and persistence of exotic pests and their predators (Landis and Gage 2015, Chapter 8 in this volume) and the adoption of new genomic technologies (Snapp et al. 2015, Chapter 15 in this volume) might fit this description. And still other changes can be highly episodic, such as the outbreak of a pest that affects a dominant competitor or a 20-year drought that affects later plant populations via seed bank changes (Gross et al. 2015, Chapter 7 in this volume). Short-term observations might entirely miss episodic events or lack the temporal context to fully understand events with long-term consequences. Century-long experiments at Rothamsted in England (Jenkinson 1991) and at a few U.S. sites (Rasmussen et al. 1998) have illustrated the importance of long-term observations and experiments for understanding the impact of agriculture on many slowly changing ecosystem attributes (Robertson et al. 2008a).

Understanding the complexity of intensive field–crop ecosystems thus requires a long-term systems perspective: understanding (1) potential ecosystem services and how multiple services can be delivered in synergistic ways; (2) how local, interdependent communities in agricultural landscapes interact across landscapes and regions; and (3) how the component parts of agricultural ecosystems behave and interact over appropriate, often long time scales. Sustainable agriculture depends on this knowledge (Robertson and Harwood 2013). And the prospect of human-induced climate change coupled with increasing demands for agriculture to produce both food and fuel (Robertson et al. 2008b, Tilman et al. 2009) makes the need for long-term agricultural research guided by a systems perspective ever more imperative. This has been, and remains, a primary motivation underlying research at the Kellogg Biological Station Long-Term Ecological Research site (KBS LTER).

Here, we present the context and conceptual basis for the KBS LTER program, including descriptions of the principal long-term experiments, their rationale, and their regional setting. Data collected as part of core KBS LTER research activities are maintained online, in a publicly available database. This includes most of the data used in this and the following chapters. The KBS LTER Data Catalog (http://lter.kbs.msu.edu/datatables) is also incorporated in the LTER Network Information System (https://portal.lternet.edu/nis/home.jsp).

The KBS Long-Term Ecological Research Program

The KBS LTER program is part of a nationwide network of 26 LTER sites representing a diversity of biomes (Robertson et al. 2012). KBS is the only LTER site focused on row-crop agriculture and is located in the USDA's North Central Region in southwest Michigan (42° 24'N, 85° 23'W; 288-m elevation; Fig. 1.1). Since its inception in 1987, LTER research at KBS has sought to better understand the ecology of intensively managed field crops and the landscape in which they reside. The emphasis of our research has been on corn, soybean, wheat, and alfalfa (*Medicago sativa* L.) (Gage et al. 2015, Chapter 4 in this volume)—crops that dominate the North Central Region and have a huge impact on human and environmental welfare. And in anticipation of the importance of cellulosic bioenergy crops over the coming decades, we have also studied hybrid poplar (*Populus* sp.) since 1987 and more

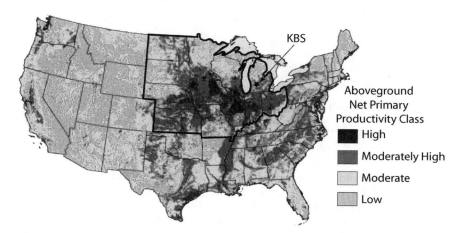

Figure 1.1. Location of the Kellogg Biological Station (KBS) in relation to estimates of U.S. net primary productivity. The area outlined in black is the USDA's North Central Region and includes the U.S. corn belt (Gage et al. 2015, this volume). Base map is modified from Nizeyimana et al. (2001).

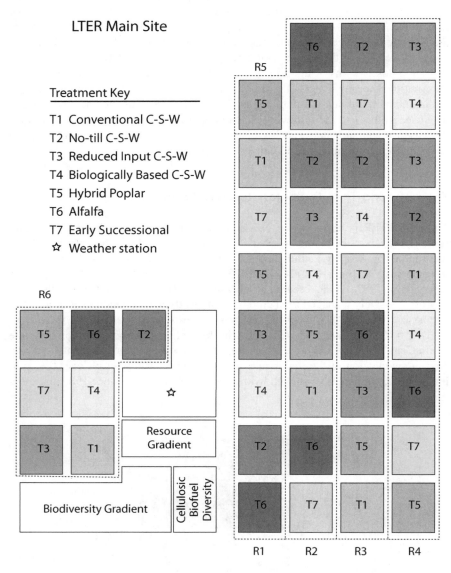

Figure 1.2. Experimental layout for seven systems of the KBS LTER Main Cropping System Experiment (MCSE) at the KBS LTER main site: four annual cropping systems (T1–T4), Alfalfa (T6) and hybrid Poplar (T5) perennial systems, and the Early Successional community (T7). All are replicated as 1-ha plots in six replicated blocks (R1–R6). C-S-W = corn-soybean-winter wheat rotation. Other MCSE systems are located as noted in Figure 1.3. See Table 1.1 and text for management details. Also shown are locations of several ancillary experiments.

recently switchgrass (*Panicum virgatum* L.), miscanthus (*Miscanthus × giganteus*), and mixed-species grassland communities. Our diverse agricultural ecosystems are compared to native forest and unmanaged successional communities close by.

Our original global hypothesis, still relevant today, is that agronomic management based on ecological knowledge can substitute for management based on chemical inputs without sacrificing the high yields necessary for human welfare. A corollary is that the delivery of other ecosystem services—including environmental benefits—can be concomitantly enhanced.

Many of our specific hypotheses have been addressed using the KBS LTER Main Cropping System Experiment (MCSE) established in 1988 to reflect the range of ecosystem types typical of field–crop landscapes in the upper Midwest. Model ecosystems replicated as 1-ha plots along a management intensity gradient include four annual cropping systems, three perennial crops, and unmanaged ecosystems ranging in successional stage from early to late (Figs. 1.2 and 1.3). The annual cropping systems are corn–soybean–winter wheat rotations ranging in management intensity from conventional to biologically based (the latter is a USDA-certified organic system without added compost or manure). Perennial crops include alfalfa, hybrid poplar trees, and conifers. Successional reference communities range in age from early succession (recently abandoned farmland) to late successional deciduous

Figure 1.3. Location of mid-successional and forested sites of the KBS LTER Main Cropping System Experiment (MCSE). Included are the Mown Grassland (never tilled) site (T8), and three Coniferous Forest (CF), Mid-successional (SF), and late-successional Deciduous Forest (DF) sites. See Figure 1.2 for LTER main site details and Table 1.1 for further description. Aerial photo background is from August 2011.

forest. Additional experiments have been added since 1988 to address additional long-term hypotheses as described later.

The Conceptual Basis for KBS LTER Research

Research at KBS LTER has steadily grown in scope and complexity since its initiation in 1988. It is now guided by a conceptual model (Fig. 1.4) that integrates both ecological and social perspectives and explicitly addresses questions about the ecosystem services delivered by agriculture. The model is derived from the press-pulse disturbance framework for social-ecological research developed by the national LTER community (Collins et al. 2011) and represents coupled natural and human systems, highlighting relationships between human socioeconomic systems and cropping systems and the landscapes in which they reside. This approach reflects the need to understand both human and natural elements and their interacting linkages. This need is especially acute in agricultural landscapes, where human decisions affect almost every aspect of ecosystem functioning and where the resulting ecological outcomes, in turn, strongly affect human well-being.

Farming for Services

Ecosystem services (Millennium Ecosystem Assessment 2005) provide a framework for examining the dependence of human welfare on ecosystems. Food, fiber, and fuel production are vital provisioning services supplied by

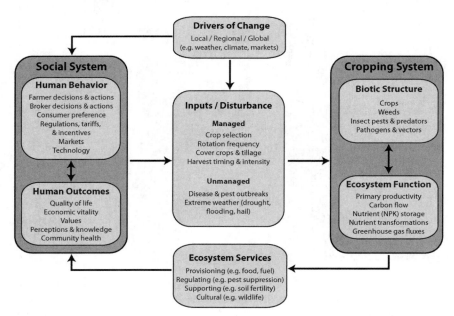

Figure 1.4. Conceptual model currently guiding KBS LTER research. Adapted from Collins et al. (2011).

agriculture, and increasingly, society is recognizing the potential for other services such as improved water quality, the protection and enhancement of biodiversity, climate stabilization via carbon sequestration and greenhouse gas abatement, and social amenities such as verdant landscapes and agrotourism (Robertson and Swinton 2005, Power 2010, Swinton et al. 2015a, Chapter 3 in this volume). Agriculture also produces disservices (Swinton et al. 2007): undesirable effects such as erosion, nitrate pollution (e.g., Syswerda et al. 2012), and emissions of greenhouse gases such as nitrous oxide (Gelfand and Robertson 2015, Chapter 12 in this volume). Mitigation services provided by alternative practices or other parts of the agricultural landscape can also be considered services provided by agriculture (Swinton et al. 2007). We refer in general to the implementation of agricultural practices that improve the delivery of ecosystem services as "farming for services" (Robertson et al. 2014).

Agriculture is typically subject to a complex set of drivers, including shifts in climate, commodity markets, human population and land use, and social and regulatory environments, as well as subject to new developments in agricultural technology such as genetically improved crop varieties and new tillage practices. Drivers of change that affect both human and natural systems occur on scales from local to landscape to global and operate under variable time scales. Conceptually, we view these drivers as disturbances to the biophysical or social systems (Fig. 1.4). They can be broadly classified into either "pulse" or "press" disturbances, depending on whether they occur as discrete events or as gradual changes over a more protracted period, respectively (Collins et al. 2011). They can be further grouped into those that are intentional management decisions vs. those that are unintentional and often unanticipated.

Intentional pulsed disturbances that affect field crops include tillage, planting, harvest, and fertilizer and pesticide applications; intentional presses include the gradual adoption of newly developed crop varieties and management technologies. Unintentional pulses include episodic weather events such as short-term droughts and late frosts as well as pest and disease outbreaks, whereas unintentional presses include climate change, increasing atmospheric carbon dioxide (CO_2) concentrations, and declining soil carbon stores. Pulses and presses may act alone or synergistically to affect how we farm, where we farm, and the profitability and sustainability of farming (Gage et al. 2015, Chapter 4 in this volume) as well as the short- and long-term impacts of agricultural activities on the environment at scales from local to global.

Most KBS LTER research to date has emphasized developing an ecosystem-level understanding of ecological structure and function—the right-hand portion of the model (Fig. 1.4). Biotic structure includes organisms and their adaptations, population and community assemblages, and the physical organization of different ecosystem habitats. Ecosystem function includes the processes carried out by organisms as mediated by the abiotic environment—for example, the cycling of carbon, nitrogen, and other nutrients, energy capture and flow, and hydrologic dynamics. Linkages between ecological structure and function largely define the mechanisms that support the delivery of ecosystem services.

Also important to consider is how factors beyond the field level affect the ability of row-crop ecosystems to deliver these services. Watershed position and landscape complexity can affect many aspects of ecological structure and function; examples include the movement of organisms, nutrients, and water between and among ecosystems, and the spatial patterns of soils and microclimates.

Organisms and Their Interactions

The main groups of organisms providing biological structure in cropping systems include (1) plants as they consume resources both above and below ground (Gross et al. 2015, Chapter 7 in this volume) and regulate the hydrologic cycle (Hamilton 2015, Chapter 11 in this volume); (2) microbes as they control organic matter turnover (Paul et al. 2015, Chapter 5 in this volume), nutrient availability (Millar and Robertson 2015, Chapter 9 in this volume; Snapp et al. 2015, Chapter 15 in this volume), and greenhouse gas fluxes (Schmidt and Waldron 2015, Chapter 6 in this volume; Gelfand and Robertson 2015, Chapter 12 in this volume); (3) insects and pathogens as they respond to changes in the plant community and affect plant productivity (Landis and Gage 2015, Chapter 8 in this volume); and (4) humans as they intentionally and unintentionally create biophysical and chemical disturbance (Swinton et al. 2015a, Chapter 3 in this volume). Each of these groups is a focal area of KBS LTER research and—together with research on watershed biogeochemistry (Hamilton 2015, Chapter 11 in this volume) and regionalization (Gage et al. 2015, Chapter 4 in this volume)—constitutes the core research areas of KBS LTER. Understanding the interactions and integration among these core areas is crucial for generating a comprehensive understanding of the drivers and dynamics of the coupled human–natural system we call agriculture.

The KBS LTER Experimental Setting

Factorial field experiments, wherein different experimental treatments are established in plots at a single geographic location, offer a powerful means for revealing the influence of individual factors or groups of factors on ecological interactions and agronomic performance. When treatments include a variety of cropping systems, the additional opportunity exists for identifying important interactions that can then be further untangled with nested, single-factor experiments. Furthermore, comparisons of cropping systems to unmanaged, reference plant communities at different stages of secondary succession allow us (1) to gauge the extent to which agriculture has produced long-term changes that may or may not be readily reversed and (2) to understand how noncrop habitats may provide resources for beneficial organisms and modify processes in a manner that might inform sustainable cropping system management.

Plot-scale experimentation provides the basis for sound statistical inference and its value cannot be overstated. But for many questions, the plots of such experiments can be too small to capture important interactions or processes, or the phenomena studied are significantly influenced by adjacent landscape elements. These include biodiversity questions when taxa are mobile or are influenced by other habitats in the landscape. For example, herbivorous insects and their predators as well as birds and other vertebrates typically respond to landscape structure at scales larger than can be accommodated in replicated field plots (Landis 1994, Landis and Marino 1999, Landis and Gage 2015, Chapter 8 in this volume). In some cases, noncrop habitats in the landscape can serve as metapopulation sources and sinks. Likewise, many biogeochemical questions depend on interactions that include landscape position and the presence and location of disproportionalities (Nowak et al. 2006), that is, hotspots of biogeochemical transformations such as high-phosphorus soils, or shallow streams and wetlands through which water flows on its way to larger rivers or lakes (Hamilton 2015, Chapter 11 in this volume).

Addressing these sorts of questions requires a landscape approach, rarely amenable to exact replication and instead more often dependent on regression and other inferential approaches (Robertson et al. 2007). Landscapes are often delineated hydrologically as watersheds or drainage basins, which are hierarchical by nature and can be grouped as needed to ask questions at larger scales. They can also be defined on the basis of other properties or processes—airsheds for questions related to nitrogen deposition or ozone impacts (e.g., Scheffe and Morris 1993), or foodsheds for questions related to the movement of nutrients and other materials related to food products (e.g., Peters et al. 2009).

An understanding of the roles of the social system (Fig. 1.4) requires expanding study boundaries to include pertinent drivers of change and human behaviors that respond to these drivers. In some cases, this might require regional surveys of farmers to understand the factors they weigh when making tillage or crop choices (e.g., Swinton et al. 2015b, Chapter 13 in this volume); in other cases this might require knowledge of the regional economy to understand land-use patterns and decisions (e.g., Feng and Babcock 2010). Where findings can be related back to the systems deployed in our field experiments, they will have the greatest power to contribute to our understanding of the interconnections between socioecological and biophysical realms in our conceptual model (Fig. 1.4).

The KBS LTER Main Cropping System Experiment

The KBS LTER Main Cropping System Experiment (MCSE) is an intensively studied factorial experiment that is the focus of much of the biophysical research at KBS LTER (Figs. 1.2 and 1.3). As mentioned earlier, it includes four annual and three perennial cropping systems plus four replicated reference communities in different stages of ecological succession, including an unmanaged late successional forest (Table 1.1). Seven systems were established and first sampled in 1989; the other four were already established and first sampled as noted below.

Each cropping system is intended to represent a model ecosystem relevant to agricultural landscapes of the region (Gage et al. 2015, Chapter 4 in this volume). They are not intended to represent all major crop × management combinations—to do this would require scores of additional experimental systems. Model systems are arranged along a gradient of decreasing chemical and management inputs. And differences that occur along this management intensity gradient can be understood,

Table 1.1. Description of the KBS LTER Main Cropping System Experiment (MCSE)[a]

Cropping System/ Community	Dominant Growth Form	Management
Annual Cropping Systems		
Conventional (T1)	Herbaceous annual	Prevailing norm for tilled corn–soybean–winter wheat (c–s–w) rotation; standard chemical inputs, chisel-plowed, no cover crops, no manure or compost
No-till (T2)	Herbaceous annual	Prevailing norm for no-till c–s–w rotation; standard chemical inputs, permanent no-till, no cover crops, no manure or compost
Reduced Input (T3)	Herbaceous annual	Biologically based c–s–w rotation managed to reduce synthetic chemical inputs; chisel-plowed, winter cover crop of red clover or annual rye, no manure or compost
Biologically Based (T4)	Herbaceous annual	Biologically based c–s–w rotation managed without synthetic chemical inputs; chisel-plowed, mechanical weed control, winter cover crop of red clover or annual rye, no manure or compost; USDA-certified organic
Perennial Cropping Systems		
Alfalfa (T6)	Herbaceous perennial	5- to 6-year rotation with winter wheat as a 1-year break crop
Poplar (T5)	Woody perennial	Hybrid poplar trees on a ca. 10-year harvest cycle, either replanted or coppiced after harvest
Coniferous Forest (CF)	Woody perennial	Planted conifers periodically thinned
Successional and Reference Communities		
Early Successional (T7)	Herbaceous perennial	Historically tilled cropland abandoned in 1988; unmanaged but for annual spring burn to control woody species
Mown Grassland (never tilled) (T8)	Herbaceous perennial	Cleared woodlot (late 1950s) never tilled, unmanaged but for annual fall mowing to control woody species
Mid-successional (SF)	Herbaceous annual + woody perennial	Historically tilled cropland abandoned ca. 1955; unmanaged, with regrowth in transition to forest
Deciduous Forest (DF)	Woody perennial	Late successional native forest never cleared (two sites) or logged once ca. 1900 (one site); unmanaged

[a]Codes that have been used throughout the project's history are given in parentheses.

predicted, simulated (Basso and Ritchie 2015, Chapter 10 in this volume), and extended to row-crop ecosystems in general.

The four annual KBS LTER cropping systems are corn–soybean–winter wheat rotations managed to reflect a gradient of synthetic chemical inputs:

• The Conventional system (T1) represents the management system practiced by most farmers in the region: standard varieties planted with conventional tillage and with chemical inputs at rates recommended by university and industry consultants. Crop varieties are chosen on the basis of yield performance in state variety trials (e.g., Thelen et al. 2011). Beginning in 2009 (for soybean) and 2011 (for corn), we have used varieties genetically modified for glyphosate resistance and (for corn) resistance to European corn borer (*Ostrinia nubilalis*) and root worm (*Diabrotica* spp.). Prior to this, we had used the same seed genetics in all cropping systems. Wheat varieties are in the soft red winter wheat class common in Michigan. Fertilizers (primarily nitrogen, phosphorus, and potassium) and agricultural lime (carbonate minerals that buffer soil acidity) are applied at rates recommended by Michigan State University (MSU) Extension following soil tests. No crops are irrigated. Herbicides and other pesticides are applied to all three crops as prescribed by integrated pest management (IPM) guidelines for Michigan (e.g., Difonzo and Warner 2010, Sprague and Everman 2011). Tillage for corn and soybean includes spring chisel plowing followed by secondary tillage to prepare the seed bed. Fall-planted winter wheat usually involves only secondary tillage. Crop residues are either harvested for animal bedding (wheat) or left on the field (corn, soybean).

- The No-till system (T2) is managed identically to the Conventional system except for tillage and herbicides. A no-till planter is used to drill seed directly into untilled soil through existing crop residue without primary or secondary tillage. When prescribed by IPM scouting, additional herbicide is used to control weeds that would otherwise be suppressed by tillage. The system has been managed without tillage since its establishment in 1989.

- The Reduced Input system (T3) differs from the Conventional system in the amounts of nitrogen fertilizer and pesticides applied, postplanting soil cultivation (prior to 2008), and winter plant cover. Crop varieties are identical to those in the Conventional system. During corn and soybean phases of the rotation, a winter cover crop is planted the preceding fall and plowed under prior to planting corn or soybean the following spring. A cover crop is not planted during wheat years because winter wheat is planted in the fall, immediately following soybean harvest. Nitrogen fertilizer is applied at reduced rates relative to the Conventional system: at 22% of the rate applied to Conventional corn and at 56% of the rate applied to Conventional wheat, for a full-rotation reduction to 33% of the Conventional system rate. Reduction in nitrogen inputs from Conventional management is expected to be made up through atmospheric N_2 fixation by legumes in the rotation: a winter cover crop of red clover (*Trifolium pratense* L.) follows wheat to precede corn, and soybean precedes wheat. A nonleguminous winter cover crop of fall-planted annual rye grass (*Lolium multiflorum* L.) follows corn to precede soybean.

The Reduced Input system thus has five species in the rotation—corn/ryegrass/soybean/winter wheat/red clover—so a crop is present at all times of the year during the entire 3-year rotation cycle. Crop varieties are the same as those used in the Conventional system, including genetically modified varieties since 2009.

Prior to 2008, weed control in corn and soybean phases of this rotation was provided by tillage and by applying herbicides at label rates only within rows (banding), so overall application rates were one-third of the amount applied in the Conventional system. Additional weed control was provided by mechanical means—rotary hoeing and between-row cultivation several times after planting. Since the use of glyphosate-resistant varieties was initiated in 2009, weed control for soybean currently relies on herbicide (glyphosate) as in the Conventional system. Weed control in wheat is provided mainly by narrow row spacing (19 cm [7.5 in.]) with no additional tillage; herbicide is only rarely applied to treat outbreak weed populations.

- The Biologically Based system (T4) is similar to the Reduced Input system except that neither nitrogen fertilizer nor pesticides are applied in this system and no genetically modified crop varieties are used. The system is entirely dependent on leguminous N_2 fixation for external nitrogen inputs, which supplements the 6–8 kg N ha^{-1} yr^{-1} received by all systems in rainfall (Hamilton 2015, Chapter 11 in this volume). Cover crops are as described for the Reduced Input system. Weed control is provided by tillage and by rotary hoeing and cultivation after planting. This system is certified organic by the USDA, but differs from conventional organic systems because it receives no manure or compost. This creates a system that is as reliant as possible on internal, biologically based nitrogen inputs.

In addition to four annual cropping systems, we have three perennial cropping systems, one herbaceous and two woody:

- Alfalfa (T6) represents a perennial herbaceous biomass system. Alfalfa is grown in a 6- to 8-year rotation with the duration defined by plant density: when the stand count declines below a recommended threshold, the stand is killed with herbicide and replanted. Because alfalfa reestablishment can be inhibited by autotoxicity, a break year is needed in the rotation and a small grain such as no-till oats or winter wheat is grown for one season in between alfalfa cycles. Alfalfa is commonly harvested three times per growing season for forage. Fertilizer (mainly phosphorus, potassium, and micronutrients such as boron and molybdenum) and lime applications follow MSU Extension recommendations following soil tests. Varieties are chosen on the basis of MSU yield trials.
- Poplar (T5) represents a short-rotation woody biomass production system. In 1989 hybrid poplar clones (*Populus* × *canadensis* Moench "Eugenei" ([*Populus deltoides* × *P. nigra*], also known as *Populus* × *euramericana* "Eugenei"), were planted as 15-cm stem cuttings on a 1 × 2 m row spacing, with nitrogen fertilizer applied only in the establishment year (123 kg N ha^{-1}). A cover crop of red fescue (*Festuca rubra* L.) was planted in 1990 for erosion control. Trees were allowed to grow for 10 years then harvested in February 1999 when they were dormant and frozen soil prevented undue soil disturbance. For the second rotation, trees were allowed to coppice (regrow from cut stems) and were harvested in the winter of 2008. After a

fallow break year during which new coppice growth, red fescue, and weeds were killed with glyphosate, in May 2009 trees were replanted as stem cuttings on a 1.5 × 2.4 m (5 ft × 8 ft) row spacing. For this third rotation, the variety *Populus nigra × P. maximowiczii* "NM6" was planted with no cover crop; weeds were controlled with herbicides applied in the first 2 years of establishment and fertilizer was applied once, in the third year of the rotation, at 156 kg N ha^{-1}.

- The Coniferous Forest (CF) includes three small long-rotation tree plantations established in 1965 and sampled as part of the MCSE beginning in 1993. One of the three sites is dominated (>10% of total biomass) by red pine (*Pinus resinosa* Aiton); a second is a mixture of Norway spruce (*Picea abies* [L.] Karst), red and white (*Pinus strobus* L.) pines, and now with significant black cherry (*Prunus serotina* Ehrh.) and large-tooth aspen (*Populus grandidentata* Michx.); and the third is dominated by white pine. The conifer stands have been periodically thinned and understory vegetation removed by prescribed burning as recommended by MSU Extension Forestry personnel.

Four successional ecosystems, either minimally managed or unmanaged, provide valuable reference communities for comparisons of specific processes and populations:

- Early Successional communities (T7) were allowed to establish naturally on land abandoned from row-crop agriculture in 1989 and have been left unmanaged but for annual spring burning (begun in 1997) to prevent tree colonization. Currently, the dominant plant species (>10% biomass) include Canada goldenrod (*Solidago canadensis* L.), red clover (*Trifolium pratense* L.), timothy grass (*Phleum pratense* L.), and Kentucky bluegrass (*Poa pratensis* L.).

- A Mown Grassland (never tilled) community (T8) that has never been in agriculture was established naturally following the removal of trees from a 10-ha woodlot in ca. 1959. The site has been mown annually in the fall since 1960 to inhibit tree colonization, with biomass left to decompose on site. At times between 1960 and 1984 the site may have received manure additions during winter months. Because the site has never been plowed, it retains an undisturbed, presettlement soil profile. KBS LTER sampling began in 1989. Plant community dominants (>10% biomass) include smooth brome grass (*Bromus inermis* Leyss.), tall oatgrass (*Arrhenatherum elatius* L.), and blackberry (*Rubus allegheniensis* Porter). Sampling occurs within four replicated 15 × 30 m plots randomly located within a portion of the field.

- Mid-successional communities (SF) occupy three sites that were abandoned from agriculture in the 1950s and 1960s (Burbank et al. 1992). Since that time they have been allowed to undergo succession, which is occurring at different rates across the replicates, possibly reflecting differences in soil fertility. One site (SF-1, abandoned in 1951) has limited overstory growth and is dominated (>10% biomass) by tall oatgrass, Canada goldenrod, quackgrass (*Elymus repens* L.), timothy grass, and Kentucky bluegrass.

Transition to forest is well under way in the remaining two sites, abandoned from agriculture in 1963 and 1964; overstory dominants reflect nearby late successional deciduous forests and understory dominants include the invasive shrubs oriental bittersweet (*Celastrus orbiculatus* Thunb.) and glossy buckthorn (*Rhamnus frangula* L.). KBS LTER sampling began in 1993.

- Late successional Deciduous Forest (DF) stands comprise the endpoint of the management intensity gradient. Soils of these three hardwood forest reference sites have never been plowed. Overstory dominants (>10% biomass) are the native trees red oak (*Quercus rubra* L.), pignut hickory (*Carya glabra* Mill.), and white oak (*Q. alba* L.); also present are black cherry (*Prunus serotina* Ehrh.), red maple (*Acer rubrum* L.), and sugar maple (*Acer saccharum* Marshall). Understory vegetation is patchy in nature and includes a variety of native forbs as well as some exotic species such as the shrubs honeysuckle (*Lonicera* spp. L.) and common buckthorn (*Rhamnus cathartica* L.), the woody vine oriental bittersweet (*Celastrus orbiculatus* Thunb), and the increasingly invasive forb garlic mustard (*Alliaria petiolata* M. Bieb.). Two of the three replicate sites have never been logged, while one was cut prior to 1900 and allowed to regrow. KBS LTER sampling started in 1993.

All MCSE systems and communities are replicated and most are within the same 60-ha experimental area, known as the LTER main site (Fig. 1.2); others, which for historical or size reasons could not be included in the main site layout, are on the same soil series within 1.5 km of the other plots (Fig. 1.3). Within the LTER main site are the four annual cropping systems, the Alfalfa and hybrid Poplar perennial cropping systems, and the Early Successional community. All are replicated as 1-ha plots in six blocks of a randomized complete block design (Fig. 1.2), for a total of 42 plots with blocks determined on the basis of an initial analysis of spatial variability in soils across the site (Robertson et al. 1997).

The Mown Grassland (never tilled) community is located about 200 m to the south of the LTER main site (Fig. 1.3); four replicated 15 × 30 m plots are located within a larger 1-ha area of the 10-ha former woodlot. The planted Coniferous Forests, the Mid-successional communities, and the late successional Deciduous Forests are each replicated three times in the landscape around the main site (Fig. 1.3). Within each replicated system, the sampling area is embedded within a larger area of similar vegetation and land-use history.

Plot sizes for MCSE systems in the main site (Fig. 1.3) are large (90 × 110 m = 1 ha) relative to plot sizes in most agronomic field experiments (Robertson et al. 2007). By adopting a 1-ha (2.5-acre) plot size, we encompass more of the spatial variability encountered in local landscapes (Robertson et al. 1997). This provides greater assurance that patterns discovered are relevant for more than a single landscape position and avoids statistical problems associated with spatial autocorrelation. Large plots also (1) allow the use of commercial-scale rather than plot-scale farm equipment, helping to ensure that agronomic practices are as similar as possible to those used by farmers; (2) help to ensure the integrity of long-term sampling by avoiding the danger of sampling the same locations multiple times years apart; and (3) avoid some of the scale effects associated with biodiversity questions for different taxa—for example, seed banks and noncrop plant diversity would not be well represented in 0.01-ha or smaller plots commonly studied in agricultural research, although even 1-ha plots are insufficient for research on more mobile taxa such as vertebrates and many arthropods.

In each MCSE replicate is a permanent set of five sampling stations near which most within-plot sampling is performed. Additionally, replicate plots typically host microplot experiments that focus on testing specific mechanistic hypotheses, such as N-addition plots to test the relationship between nutrient availability and plant diversity and predator-exclusion plots to examine the role of predators in controlling invasive insects. Some microplot experiments are permanent, such as annually tilled × N fertilized microplots within the Early Successional community (Gross et al. 2015, Chapter 7 in this volume); many have been shorter term.

Regular measurements for all 11 systems and communities in the MCSE include (1) plant species composition, above-ground net primary productivity, litter fall, and crop yield; (2) predaceous insects, in particular, coccinellids (ladybird beetles); (3) microbial biomass and abundance; (4) soil moisture, pH, bulk density, carbon, inorganic nitrogen, and nitrogen mineralization; (5) NO_3^- concentrations in low-tension lysimeters installed at a 1.2-m depth (Bt2/C horizon) in replicate plots of all systems; and (6) a number of weather variables measured at a weather station on the MCSE. Precipitation chemistry is monitored as part of the National Atmospheric Deposition Program/National Trends Network at another weather station 2 km away to avoid contamination by agricultural activities on site. Soil carbon is measured to 1-m depth at decadal intervals in all systems. The soil seed bank is sampled on a 6-year cycle.

Ancillary Experiments

In addition to the MCSE, several long- and shorter-term ancillary experiments address specific questions. In some cases these are located in subplots nested within the plots of the MCSE, and in others they are at independent locations. Here, we describe the most important.

The MCSE Scale-Up Experiment

The need to understand how findings from our 1-ha MCSE cropping systems scale up to commercially sized fields motivated the establishment of the MCSE Scale-Up Experiment (Fig. 1.5). Although larger than most agronomic research plots, the 1-ha MCSE plots may still suffer from artifacts related to plot size. For example, because plots are managed for research, agronomic operations may not be as influenced by labor issues as they might be on a commercial farm. The frequency and timing of operations such as mechanical weed control and planting date may affect weed densities and yields, and a commercial operator will have less flexibility for optimal scheduling due to labor constraints.

Additionally, our 1-ha plots are embedded in a matrix of other plots with different plant communities that could provide insect refugia or seed sources not typically

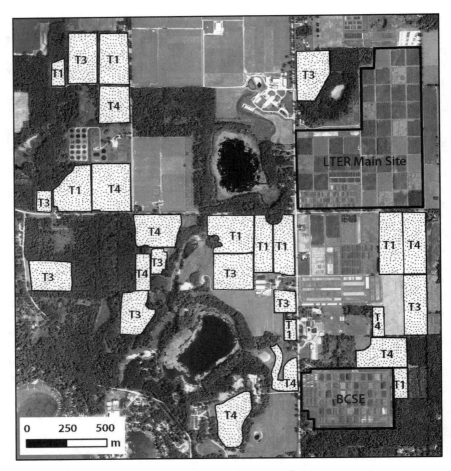

Figure 1.5. Main Cropping System Experiment (MCSE) Scale-up fields (n = 27) managed to address questions related to the scalability of results from the MCSE depicted in Figure 1.2. Management corresponds to the MCSE Conventional (T1), Reduced Input (T3), and Biologically Based systems (T4; see Table 1.1). Also shown is the location of the Bioenergy Cropping System Experiment (BCSE). Aerial photo background is from August 2011.

available in farm-scale fields. Farm-scale fields, on the other hand, will more often be bordered by larger successional areas or woodlots, important overwintering habitats for both insect herbivores and their natural enemies (Colunga-Garcia and Gage 1998, Landis and Gage 2015, Chapter 8 in this volume).

From the fall of 2006 through the fall of 2013, 27 fields managed by the Kellogg Farm were assigned to one of three MCSE annual cropping systems: Conventional, Reduced Input, or Biologically Based. Each was also assigned to one of three rotation entry points—corn, soybean, or wheat—and to one of three replicate blocks. This provides three replicate fields for each system × entry point combination (3 systems × 3 entry points × 3 replicates). Fields range in size from 1 to 7.5 ha, adjoin a number of different habitat types, and have a variety of perimeter complexities. Regular sampling activities include agronomic yields.

Biodiversity Gradient Experiment

The Biodiversity Gradient Experiment was established on the LTER main site (Fig. 1.2) in 2000 to investigate the effect of plant species diversity across a gradient ranging from bare ground to 1, 2, 3, 4, 6, and 10 species. Small plots (9 × 30 m) are within four randomized complete blocks and are managed much like the MCSE Biologically Based system (i.e., no synthetic chemical inputs). This experiment reveals how crop species and rotational complexity affect yield, weed competition, soil biogeochemical processes, microbial diversity, and other variables (Gross et al. 2015, Chapter 7 in this volume).

Resource Gradient Experiment

The Resource Gradient Experiment was established on the LTER main site (Fig. 1.2) in 2003 to investigate nitrogen and water constraints on crop yield. MCSE annual crops (either corn, soybean, or wheat) are nitrogen-fertilized at nine different rates and are either irrigated or rain-fed. Fertilizer rates differ by crop; for corn the range has been 0 to 292 kg N ha^{-1} and for wheat 0 to 180 kg N ha^{-1}. Soybeans are normally not fertilized (but were in 2012). Irrigation is sufficient to meet plant water needs as predicted by weather and SALUS, a crop growth model that calculates instantaneous water balance (Basso and Ritchie 2015, Chapter 10 in this volume). A linear move irrigation system applies water 0–3 times per week during the growing season depending on recent rainfall and crop water need. Crops are otherwise managed as for the MCSE No-till system. In addition to crop yield, greenhouse gas exchanges between soils and the atmosphere are measured in various treatments (Millar and Robertson 2015, Chapter 9 in this volume).

Living Field Lab Experiment

The Living Field Laboratory (LFL) was established on land just north of the MCSE in 1993 to investigate the benefits of leguminous cover crops and composted dairy manure in two integrated systems compared to a conventional and an organic agricultural system. "Integrated" refers to targeted, banded applications of herbicide, reduced tillage, and stringent accounting of nitrogen inputs using the pre-side-dress nitrate test (PSNT) or nitrogen analysis of composted dairy manure. During the past 15 years, a crop rotation of corn–corn–soybean–wheat has been compared to continuous corn where every entry point of the rotation was present each year. A number of soil and crop variables were measured at the LFL from 1993 to 2003 (Snapp et al. 2010); since 2006 the LFL has initiated new studies including a perennial wheat project (Snapp et al. 2015, Chapter 15 in this volume). The LFL was decommissioned in Fall 2014.

Bioenergy Cropping System Experiment

The Great Lakes Bioenergy Research Center's (GLBRC) Bioenergy Cropping System Experiment (BCSE) was established in 2008 south of the LTER main site (Fig. 1.5) to compare the productivity and environmental performance of alternative cellulosic biofuel cropping systems and to ask fundamental questions about their ecological functioning. Eight different cropping systems were established in a randomized complete block design (five replicate blocks of 30 × 40 m plots) that includes, in order of increasing plant diversity, continuous corn, a corn–soybean–canola rotation, switchgrass, miscanthus, hybrid poplar, mixed-species native grasses, successional vegetation, and native prairie. In 2012 the corn–soybean–canola system was terminated and two additional systems added: one a continuous corn + cover crop system and the other a corn–soybean + cover crop system. Regular measurements in the BCSE are similar to those made in the MCSE, but also include time domain reflectometry (TDR) soil water profiles and automated chamber measurements of soil-atmosphere greenhouse gas exchanges. An identical GLBRC-sponsored experiment on Mollisol soils is located in Arlington, Wisconsin.

In addition, larger biofuel scale-up fields of continuous corn, switchgrass, and restored prairie were established in 2009 on both existing cropland and on land that had been in the USDA Conservation Reserve Program (CRP) for 22 years. These KBS sites are about 10 km from the main biofuels experiment. The BCSE scale-up fields have eddy covariance flux towers to measure carbon dioxide and water exchange at the whole-ecosystem scale and are also sampled for yield and a variety of soil biogeochemical and insect diversity attributes.

Cellulosic Biofuel Diversity Experiment

The Cellulosic Biofuel Diversity Experiment is designed to test the long-term impact of plant diversity on the delivery of ecosystem services from cellulosic biofuel production systems. The experiment is located within the LTER main site (Fig. 1.2). Twelve different cropping systems vary in species composition and nitrogen input. Systems include continuous corn, corn–soybean, two varieties of switchgrass fertilized differently, a C_3 and C_4 grass plus legume mix, and four different prairie restorations with 6, 10, 18, or 30 different species at establishment. Replicate plots are 9 × 30 m replicated in four randomized blocks, established in 2008.

The Regional Setting

Climate, Soils, and Presettlement Vegetation

Climate at KBS is humid, continental, and temperate (Fig. 1.6). Annual precipitation averages 1005 mm yr^{-1}, with an average snowfall of ~1.3 m (1981–2010; NCDC 2013). Precipitation is lowest in winter (17% of total) and is otherwise evenly distributed among the other three seasons (25–30%). Potential evapotranspiration exceeds precipitation for about 4 months of the year (Crum et al. 1990; see

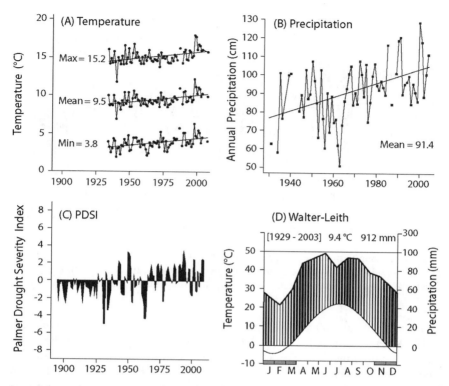

Figure 1.6. Long-term (1929–2008) trends for temperature and precipitation at KBS: (A) annual means of daily air temperatures showing maximum (upper line), minimum (bottom), and daily (24-hour) values (middle) in °C (means for the 80-year period are given to the left of each data series); (B) total annual precipitation (cm); (C) mean annual Palmer Drought Severity Index (PDSI); and (D) monthly mean air temperature and precipitation depicted as a Walter-Leith climate diagram. Negative PDSI indicates water deficit conditions for the region. Redrawn from Peters et al. (2013).

Hamilton 2015, Chapter 11 in this volume, Fig. 11.3). The mean annual temperature is 10.1°C, ranging from a monthly mean of –3.8°C in January to 22.9°C in July (1981–2010; NCDC 2013). Climate change models predict significant alterations in the amount of precipitation and its variability for the Midwest, in particular, the frequency and intensity of precipitation events (Easterling et al. 2000, Weltzin et al. 2003). At KBS, air temperature and precipitation have both shown increasing trends over the past several decades (Fig. 1.6), as has the incidence of large rain events. A warming trend is also apparent from the ice records of area lakes (Fig. 1.7).

The physiography of southwest Michigan is characteristic of a mature glacial outwash plain and moraine complex. The retreat of the Wisconsin glaciation, ~18,000 years ago in southwest Michigan, left a diverse depressional pattern of many kettle lakes and wetlands interspersed among undulating hills and outwash

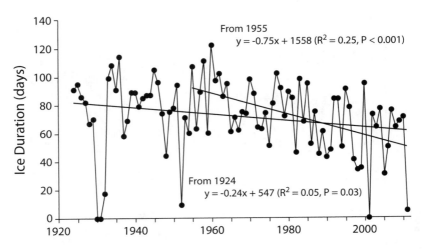

Figure 1.7. Long-term trends in ice duration on Gull Lake at KBS for the periods 1924–2011 and 1955–2011. Ice seasons potentially span two calendar years and therefore the x-axis depicts the year each winter began. From S.Hamilton (unpublished data).

channels. There are 200 lakes within 50 km of KBS, most of which originated as ice-block depressions in the outwash plains formed as the glacial ice melted. KBS is located within the Gull Creek/Gull Lake watershed (97 km²) and the Augusta Creek watershed (98 km²), both within the Kalamazoo River basin (5232 km²). At the watershed scale, most water movement occurs through groundwater aquifers, and water sources to all streams and most lakes and wetlands are dominated by groundwater inputs (Hamilton 2015, Chapter 11 in this volume).

Soils in the area thus developed on glacial till and outwash following the last glacial retreat. The predominant soils at and around KBS are Alfisols, developed under upland forest vegetation. MCSE soils are well-drained Alfisol loams of the Kalamazoo series (fine-loamy, mixed, mesic Typic Hapludalfs) co-mingled with well-drained loams of the Oshtemo series (coarse-loamy, mixed, mesic Typic Hapludalfs) (Mokma and Doolittle 1993, Crum and Collins 1995). Surface soil sand and clay contents average 43 and 17%, respectively (Robertson et al. 1997), and dominant silicate minerals include plagioclase, K-feldspar, quartz, and amphibole (Hamilton et al. 2007). Carbonate minerals (dolomite and calcite) are common in glacial drift and occur at depths below 1 m; they have been leached out of the upper soil profile at KBS (Kurzman 2006, Hamilton 2015, Chapter 11 in this volume), as is typical of glacial soils elsewhere in the Great Lakes region (Drees et al. 2001).

Pre-European settlement vegetation of the area consisted of a mixture of forests, oak savannas, and prairie grasslands (Gross and Emery 2007, Chapman and Brewer 2008). Southwest Michigan was part of the "prairie peninsula" (Transeau 1935) that appears to have developed during a prolonged dry period 4000–8000 years ago along the south and southeastern edge of Lake Michigan. Fires were likely frequent during this period and, beginning ca. 700 C.E., were actively promoted by local

Native Americans of the Mascouten and Potawatomi tribes in order to maintain game habitat (Legge et al. 1995).

Human Settlement and Agricultural Transitions

The current human landscape of SW Michigan (Fig. 1.8) is largely a product of its agricultural history, formed by demographic, social, and economic forces interacting within an ecological context of climate, soils, and natural vegetation. Rudy et al. (2008) provide a comprehensive and insightful account of the major periods of agricultural transitions within the region.

Southwest Michigan was inhabited beginning with glacial retreat ~16,000 B.C.E. By 8000 B.C.E. Paleo-Indians foraged for fish and game in the area, and evidence exists of at least one indigenous cultigen (a sunflower) by the start of the Early Woodland Period in 1000 B.C.E. By the Late Woodland Period (1200 C.E.), there was widespread incorporation of corn, bean, and squash cultivation around semi-permanent villages. By 1670 C.E., when Michigan's Lower Peninsula was depopulated by the Iroquois, the Potawatomi Indians had established large permanent villages with intercropped gardens of corn, pumpkin, squash, and beans in fields cleared of trees by girdling and fire. Following repopulation of the area in the early 1700s, the Potawatomi in the region's south and the Ottawa in the north cultivated corn and other vegetables as well as fruit trees, which supplemented the diets of as many as 10,000 Native Americans.

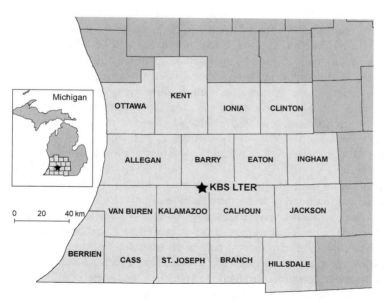

Figure 1.8. Southwest Michigan counties that comprise the regional setting for KBS LTER (Rudy et al. 2008).

By 1830, ~10,000 settlers of European descent had migrated to the area, driven by high land and low grain prices in the East, the opening of the Erie Canal in 1825, and the introduction of fruit trees. Most settlers were New Englanders (Gray 1996) and they first occupied the prairie and oak openings that had been farmed by the Potawatomi they displaced.

Rudy et al. (2008) describe six agricultural periods in southwest Michigan that define present-day agriculture. The first is the largely extensive development period that occurred prior to 1898 when land was cleared, drained, and farmed, mainly for the production of wheat for human consumption and hay for draft animals. The second is the 1899–1919 Golden Age of Agriculture marked by expanding international markets with their high grain prices and the introduction of new techniques for crop cultivation and animal breeding brought to farmers by the university-based Cooperative Extension Service.

The Agricultural Depression began ca. 1920 as a result of national overproduction following the return of European agriculture after World War I, and persisted throughout the Great Depression. It was exacerbated by mechanization and tractor-driven increases in productivity that at the same time opened to row-crop production pasturage that had been formerly used to feed draft animals. Agricultural Fordism (1941–1973) followed the Agricultural Depression and was marked by agriculture's increased need for capital goods such as tractors, hybrid seeds, and pesticides as well as by farm families' shift to a consumer orientation. Together, these trends encouraged agricultural intensification and, in particular, simple monoculture rotations and a near-singular focus on increasing productivity. The average farm size in this period grew from ~35 to ~60 ha (~90 to ~150 acres).

The period 1974–1989 found southwest Michigan farm operators squeezed between a continued downward trend in real prices for agricultural commodities and increasing production costs. The difference was relieved to some extent by government payments, which by the 1990s constituted >50% of net farm income. Southwest Michigan farmers also responded with strategies that included greater off-farm income and market opportunities for more diverse foods including organic and specialty crops. Even so, crop agriculture in the region was and remains grain dominated: in 2007 ~81% of cropland in the 17-county area was used to grown corn (47%), soybean (29%), and wheat (5%) (USDA 2009). Forage (11%; commonly alfalfa) and vegetables and orchards made up most of the remainder.

Globalization since 1990 marks Rudy et al.'s (2008) sixth major period, one in which rural agricultural dynamics are shifting rapidly. By the late 1900s, agrifood systems had an increasingly global scope with important local consequences. In southwest Michigan, as elsewhere, this has exacerbated tensions between agricultural production and environmental conservation as well as struggles over the social and environmental consequences of, in particular, exurban sprawl, agrichemical use, and industrial animal production. The local landscape now is a mixture of cultivated and successional fields, woodlots dominated by northern hardwood trees, private residences, and lakes and wetlands.

KBS LTER (MCSE) vs. Regional Crop Yields

KBS LTER crop yields are typical of rain-fed yields elsewhere in the North Central Region. For the 21-year period from 1989 to 2009, MCSE no-till soybean yields (2.6 ± 0.2 SD Mg ha^{-1} at standard 13% moisture; or 39 bu acre^{-1}) were similar to average Kalamazoo County yields (2.5 ± 0.1 Mg h^{-1}; 37 bu acre^{-1}), which were similar to soybean yields for the entire United States (2.8 ± 0.1 Mg ha^{-1}; 42 bu acre^{-1}) (NASS 2012a). No-till wheat yields at KBS LTER (3.7 ± 0.3 Mg ha^{-1} at standard 13% moisture; 55 bu acre^{-1}) were slightly higher than average Kalamazoo County yields (3.4 ± 0.3 Mg ha^{-1}; 51 bu acre^{-1}) and national yields (3.5 ± 0.1 Mg ha^{-1}; 52 bu acre^{-1}) for soft red wheat, which makes up ~25% of total U.S. wheat production and is the dominant class grown around KBS.

Corn yields are more variable, reflecting the greater sensitivity of corn yields to low rainfall periods and growing season heat waves, especially during pollination

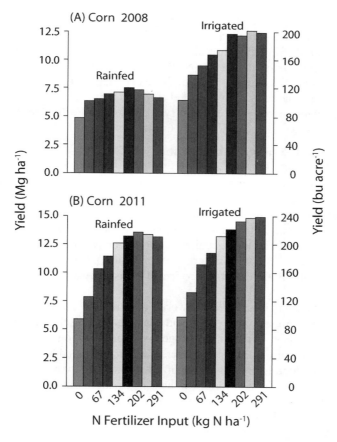

Figure 1.9. Corn yields in (A) 2008 and (B) 2011 in the KBS LTER Resource Gradient Experiment. For comparison, average U.S. corn yields in 2008 and 2011 (ERS 2013) were 9.7 and 9.2 Mg ha^{-1}, respectively (154 and 147 bu acre^{-1}).

(Hatfield et al. 2011). When rainfall is adequate, KBS LTER corn yields are ~12.5 Mg ha^{-1} at standard 15% moisture (199 bu acre^{-1}). In most years, however, yields are constrained by rainfall, as they were during a 2008 local drought when rain-fed corn yields in the Resource Gradient Experiment (Fig. 1.9) were only 7.5 Mg ha^{-1} (120 bu acre^{-1}), as compared to irrigated yields of 12.5 Mg ha^{-1} (199 bu acre^{-1}) and national average corn yields of 9.7 Mg ha^{-1} (155 bu acre^{-1}). In contrast, 2011 saw favorable precipitation; rain-fed corn yields were 13.4 Mg ha^{-1} (215 bu acre^{-1}) against a national average of 9.1 Mg ha^{-1} (147 bu acre^{-1}) and with less response to irrigation (Fig. 1.9).

Over all years during the 1989–2009 period, MCSE corn yields averaged 6.4 ± 0.7 Mg ha^{-1} (102 bu acre^{-1}). This is lower than county (7.4 ± 0.2 Mg ha^{-1}; 118 bu acre^{-1}) and national (8.4 ± 0.2 Mg ha^{-1}; 134 bu acre^{-1}) averages for the same period. For 4 of the 9 MCSE corn years in this period, yields were at or above county and national yields; for 3 years, corn yields were not significantly different from (but lower than) county and national yields; and for 2 years, corn yields were significantly lower than county and national averages. However, both county and national yields include those from irrigated acreage, which inflate yield comparisons relative to rain-fed MCSE yields. In Kalamazoo County about 38% of corn acreage is irrigated and, nationally, about 15% (NASS 2012b). Overall KBS LTER corn yields and variability are thus fairly typical of those experienced by rain-fed Kalamazoo County growers, and reflect how rain-fed corn will vary with the year-to-year variability in growing season rainfall that is typical for farms within the region.

Landscape and Regional Observations

As noted earlier, certain important ecosystem services that may not be evident at the field scale emerge at the scale of landscapes. Prominent examples include biodiversity-mediated services that require landscape-level habitat configurations (Gardiner et al. 2009) and recreational and aesthetic services that emerge from a landscape of varied vegetation and topography (Bolund and Hunhammar 1999, Swinton et al. 2015a, Chapter 3 in this volume). Likewise, the provision of high-quality water is an important service delivered by well-managed agricultural landscapes.

Experiments and observation networks designed to address landscape-level questions are by necessity specialized and do not lend themselves to a one-size-fits-all design (Robertson et al. 2007). Biogeochemical questions, for example, may require a diversity of flow paths and discrete watersheds to address (e.g., Hamilton et al. 2007). In contrast, questions about insect biodiversity may require a multi-county region that includes a variety of landscape patterns, crop rotations, or intensities (e.g., Landis et al. 2008, Landis and Gage 2015, Chapter 8 in this volume). And economic questions may require a social or market setting that encompasses scales from the regional (e.g., Jolejole 2009, Chen 2010, Ma et al. 2012) to the national (e.g., James et al. 2010) and international.

Consequently, there is no single landscape scale that is the focus for KBS LTER landscape-level research. Rather, our landscape research setting expands outward from MCSE sites to local fields (e.g., Gelfand et al. 2011); local watersheds (e.g., Hamilton 2015, Chapter 11 in this volume); southwest Michigan (e.g., Rudy et al. 2008); the state of Michigan (e.g., Ma et al. 2012); the Great Lakes states (e.g., Landis et al. 2008); and the U.S. Midwest (e.g., Grace et al. 2011, Gelfand et al. 2013), as dictated by the questions under investigation.

How large a landscape might KBS LTER research represent? Michigan is among the 12 states that produce most of the nation's corn, and is thus included in the USDA's designated North Central Region, part of which is known as the U.S. Corn Belt. Corn Belt states include Illinois, Indiana, Iowa, Kansas, Michigan, Minnesota, Missouri, Nebraska, North Dakota, Ohio, South Dakota, and Wisconsin. Though there are many caveats, KBS LTER research has been extended to the North Central Region by biogeochemical modeling used to forecast potential soil carbon sequestration (Fig. 1.10; Grace et al. 2006) and N_2O fluxes (Grace et al. 2011), as well as by crop modeling to develop regional crop stress indicators (Gage et al. 2015, Chapter 4 in this volume). Another, more robust approach to extend KBS LTER research findings would be to establish cooperative agricultural sites within the region at which coordinated experiments and observations might be conducted (Robertson et al. 2008a), similar in power to the many cross-site LTER syntheses now in the literature (Johnson et al. 2010). The nascent Long-Term Agricultural

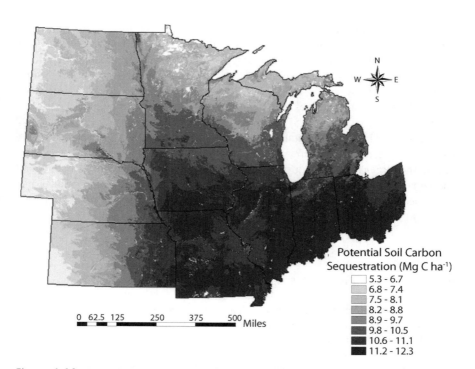

Figure 1.10. Potential soil carbon sequestration expected on adoption of no-till management in the USDA's North Central Region. Predicted values are modeled by the SOCRATES soil organic carbon model (Grace et al. 2006).

Research Network (Robertson et al. 2008a, Walbridge et al. 2011) may provide such opportunities in the future.

Summary

An ecological understanding of the row-crop ecosystem is necessary for designing agricultural systems and landscapes that depend less on exogenous inputs of chemicals and energy and more on internally provided resources for sustaining the production of food, fiber, and fuel, while also optimizing the delivery of other ecosystem services such as pest and disease suppression, nutrient acquisition and conservation, and water-quality protection. This understanding must be based on fundamental knowledge of interactions among major functional groups in agricultural ecosystems and landscapes—plants, microbes, arthropods, and humans—and how interactions change with management and natural disturbance to affect the provision of services.

A conceptual model that incorporates coupled natural and human systems is appropriate for asking many of the most relevant questions in agricultural ecology. The KBS LTER conceptual model (Fig. 1.4) provides a means for asking how the structure and function of row-crop ecosystems interact to deliver ecosystem services. The two linked realms of the model reflect how ecosystem services affect and are perceived by people, who might then directly or indirectly influence market and farmer decisions, public policy, and other actions that feed back to affect row-crop management—thus iteratively changing ecological interactions within the systems and subsequently the delivery of ecosystem services. Additionally, the model provides a framework to analyze the social and ecological consequences of external, unintentional drivers such as climate change.

The KBS LTER experimental approach is to intensively study interactions within model ecosystems, both cropped and unmanaged, and to then extend these findings to the larger landscape through knowledge of human interactions, both social and economic, and targeted observations made at the scale of commercial farms, watersheds, and broader landscapes. Models tested locally can extend insights still further to regional scales.

References

Basso, B., and J. T. Ritchie. 2015. Modeling crop growth and biogeochemical fluxes with SALUS. Pages 252–274 in S. K. Hamilton, J. E. Doll, and G. P. Robertson, editors. The ecology of agricultural Landscapes: long-term research on the path to sustainability. Oxford University Press, New York, New York, USA.

Bolund, P., and S. Hunhammar. 1999. Ecosystem services in urban areas. Ecological Economics 29:293–301.

Burbank, D. H., K. S. Pregitzer, and K. L. Gross. 1992. Vegetation of the W.K. Kellogg Biological Station. Michigan State University Agricultural Experiment Station Research Report 510, East Lansing, Michigan, USA.

Chapman, K. A., and R. Brewer. 2008. Prairie and savanna in southern Lower Michigan: history, classification, ecology. Michigan Botanist 47:1–48.

Chen, H. 2010. Ecosystem services from low input cropping systems and the public's willingness to pay for them. Thesis, Michigan State University, East Lansing, Michigan, USA.

Collins, S. L., S. R. Carpenter, S. M. Swinton, D. E. Orenstein, D. L. Childers, T. L. Gragson, N. B. Grimm, J. M. Grove, S. L. Harlan, J. P. Kaye, A. K. Knapp, G. P. Kofinas, J. J. Magnuson, W. H. McDowell, J. M. Melack, L. A. Ogden, G. P. Robertson, M. D. Smith, and A. C. Whitmer. 2011. An integrated conceptual framework for long-term social-ecological research. Frontiers in Ecology and the Environment 9:351–357.

Colunga-Garcia, M., and S. H. Gage. 1998. Arrival, establishment, and habitat use of the multicolored Asian lady beetle (Coleoptera: Coccinellidae) in a Michigan landscape. Environmental Entomology 27:1574–1580.

Crum, J. R., and H. P. Collins. 1995. KBS soils. Kellogg Biological Station Long-Term Ecological Research, Michigan State University, Hickory Corners, Michigan, USA. <http://lter.kbs.msu.edu/research/site-description-and-maps/soil-description>.

Crum, J. R., G. P. Robertson, and F. Nurenberger. 1990. Long-term climate trends and agricultural productivity in Southwestern Michigan. Pages 53–58 in D. Greenland and L. W. Swift, editors. Climate variability and ecosystem response. U.S. Department of Agriculture, U.S. Forest Service, Southeastern Forest Experiment Station, Asheville, North Carolina, USA.

DiFonzo, C., and F. Warner. 2010. Insect, nematode and disease control in Michigan field crops. Michigan State University Extension Bulletin E1582, East Lansing, Michigan, USA.

Drees, L. R., L. P. Wilding, and L. C. Nordt. 2001. Reconstruction of soil inorganic and organic carbon sequestration across broad geoclimatic regions. Pages 155–172 in R. Lal, editor. Soil carbon sequestration and the greenhouse effect. Special Publication 57, Soil Science Society of America, Madison, Wisconsin, USA.

Easterling, D. R., G. A. Meehl, C. Parmesan, S. A. Changnon, T. R. Karl, and L. O. Mearns. 2000. Climate extremes: observations, modeling, and impacts. Science 289:2068–2074.

ERS (Economic Research Service). 2013. Feed grains: Yearbook tables. Table 1. U.S. Department of Agriculture (USDA), Washington, DC, USA. <http://ers.usda.gov/data-products/feed-grains-database/feed-grains-yearbook-tables.aspx> Accessed February 6, 2013.

Feng, H., and B. A. Babcock. 2010. Impacts of ethanol on planted acreage in market equilibrium. American Journal of Agricultural Economics 92:789–802.

Gage, S. H., J. E. Doll, and G. R. Safir. 2015. A crop stress index to predict climatic effects on row-crop agriculture in the U.S. North Central Region. Pages 77–103 in S. K. Hamilton, J. E. Doll, and G. P. Robertson, editors. The ecology of agricultural Landscapes: long-term research on the path to sustainability. Oxford University Press, New York, New York, USA.

Gardiner, M. M., D. A. Landis, C. Gratton, C. D. DiFonzo, M. O'Neal, J. M. Chacon, M. T. Wayo, N. P. Schmidt, E. E. Mueller, and G. E. Heimpel. 2009. Landscape diversity enhances the biological control of an introduced crop pest in the north-central USA. Ecological Applications 19:143–154.

Gelfand, I., and G. P. Robertson. 2015. Mitigation of greenhouse gas emissions in agricultural ecosystems. Pages 310–339 in S. K. Hamilton, J. E. Doll, and G. P. Robertson, editors. The ecology of agricultural Landscapes: long-term research on the path to sustainability. Oxford University Press, New York, New York, USA.

Gelfand, I., R. Sahajpal, X. Zhang, C. R. Izaurralde, K. L. Gross, and G. P. Robertson. 2013. Sustainable bioenergy production from marginal lands in the US Midwest. Nature 493:514–517.

Gelfand, I., T. Zenone, P. Jasrotia, J. Chen, S. K. Hamilton, and G. P. Robertson. 2011. Carbon debt of Conservation Reserve Program (CRP) grasslands converted to bioenergy production. Proceedings of the National Academy of Sciences USA 108:13864–13869.

Grace, P. R., M. Colunga-Garcia, S. H. Gage, G. P. Robertson, and G. R. Safir. 2006. The potential impact of agricultural management and climate change on soil organic carbon of the North Central Region of the United States. Ecosystems 9:816–827.

Grace, P. R., G. P. Robertson, N. Millar, M. Colunga-Garcia, B. Basso, S. H. Gage, and J. Hoben. 2011. The contribution of maize cropping in the Midwest USA to global warming: a regional estimate. Agricultural Systems 104:292–296.

Gray, S. E. 1996. The Yankee West: community life on the Michigan frontier. The University of North Carolina Press, Chapel Hill, North Carolina, USA.

Gross, K. L., S. Emery, A. S. Davis, R. G. Smith, and T. M. P. Robinson. 2015. Plant community dynamics in agricultural and successional fields. Pages 158–187 in S. K. Hamilton, J. E. Doll, and G. P. Robertson, editors. The ecology of agricultural Landscapes: long-term research on the path to sustainability. Oxford University Press, New York, New York, USA.

Gross, K. L., and S. A. Emery. 2007. Succession and restoration in Michigan old-field communities. Pages 162–179 in V. Cramer and R. J. Hobbs, editors. Old fields: dynamics and restoration of abandoned farmland. Island Press, Washington, DC, USA.

Hamilton, S. K. 2015. Water quality and movement in agricultural landscapes. Pages 275–309 in S. K. Hamilton, J. E. Doll, and G. P. Robertson, editors. The ecology of agricultural Landscapes: long-term research on the path to sustainability. Oxford University Press, New York, New York, USA.

Hamilton, S. K., A. L. Kurzman, C. Arango, L. Jin, and G. P. Robertson. 2007. Evidence for carbon sequestration by agricultural liming. Global Biogeochemical Cycles 21:GB2021.

Hatfield, J. L., K. J. Boote, B. A. Kimball, L. H. Ziska, R. C. Izaurralde, D. Ort, A. M. Thomson, and D. Wolfe. 2011. Climate impacts on agriculture: implications for crop production. Agronomy Journal 103:351–370.

James, L. K., S. M. Swinton, and K. D. Thelen. 2010. Profitability analysis of cellulosic energy crops compared with corn. Agronomy Journal 102:675–687.

Jenkinson, D. S. 1991. The Rothamsted long-term experiments: Are they still of use? Agronomy Journal 83:2–10.

Johnson, J. C., R. R. Christian, J. W. Brunt, C. R. Hickman, and R. B. Waide. 2010. Evolution of collaboration within the US Long Term Ecological Research Network. BioScience 60:931–940.

Jolejole, M. C. B. 2009. *Trade-offs, incentives, and the supply of ecosystem services from cropland*. Thesis, Michigan State University, East Lansing, Michigan, USA.

Kurzman, A. L. 2006. Changes in major solute chemistry as water infiltrates soils: comparisons between managed agroecosystems and unmanaged vegetation. Dissertation, Michigan State University, East Lansing, Michigan, USA.

Landis, D. A. 1994. Arthropod sampling in agricultural landscapes: ecological considerations. Pages 15–31 in L. P. Pedigo and G. D. Buntin, editors. Handbook of sampling methods for arthropod pests in agriculture. CRC Press, Boca Raton, Florida, USA.

Landis, D. A., and S. H. Gage. 2015. Arthropod diversity and pest suppression in agricultural landscapes. Pages 188–212 in S. K. Hamilton, J. E. Doll, and G. P. Robertson, editors. The ecology of agricultural Landscapes: long-term research on the path to sustainability. Oxford University Press, New York, New York, USA.

Landis, D. A., M. M. Gardiner, W. van der Werf, and S. M. Swinton. 2008. Increasing corn for biofuel production reduces biocontrol services in agricultural landscapes. Proceedings of the National Academy of Sciences USA 105:20552–20557.

Landis, D. A., and P. Marino. 1999. Landscape structure and extra-field processes: impact on management of pests and beneficials. Pages 79–104 in J. Ruberson, editor. Handbook of pest management. Marcel Dekker, New York, New York, USA.

Legge, J. T., P. J. Hickman, P. J. Comer, M. R. Penskar, and M. C. Rabe. 1995. A floristic and natural features inventory of Fort Custer Training Center, Augusta, Michigan. Michigan Natural Features Inventory, Lansing, Michigan, USA.

Lowrance, R. R., R. L. Todd, J. Fail, O. Hendrickson, R. Leonard, and L. Asmussen. 1984. Riparian forests as nutrient filters in agricultural watersheds. BioScience 34:374–377.

Ma, S., S. M. Swinton, F. Lupi, and C. B. Jolejole-Foreman. 2012. Farmers' willingness to participate in payment-for-environmental-services programmes. Journal of Agricultural Economics 63:604–626.

Magnuson, J. J. 1990. Long-term ecological research and the invisible present. BioScience 40:495–500.

Matson, P. A., W. J. Parton, A. G. Power, and M. J. Swift. 1997. Agricultural intensification and ecosystem properties. Science 277:504–509.

Millar, N., and G. P. Robertson. 2015. Nitrogen transfers and transformations in row-crop ecosystems. Pages 213–251 in S. K. Hamilton, J. E. Doll, and G. P. Robertson, editors. The ecology of agricultural Landscapes: long-term research on the path to sustainability. Oxford University Press, New York, New York, USA.

Millennium Ecosystem Assessment. 2005. Ecosystems and human well-being: synthesis. Island Press, Washington, DC, USA.

Mokma, D. L., and J. A. Doolittle. 1993. Mapping some loamy alfisols in southwestern Michigan using ground-penetrating radar. Soil Survey Horizons 34:71–77.

Mulholland, P. J., R. O. Hall, D. J. Sobota, W. K. Dodds, S. E. G. Findlay, N. B. Grimm, S. K. Hamilton, W. H. McDowell, J. M. O'Brien, J. L. Tank, L. R. Ashkenas, L. W. Cooper, C. N. Dahm, S. V. Gregory, S. L. Johnson, J. L. Meyer, B. J. Peterson, G. C. Poole, H. M. Valett, J. R. Webster, C. P. Arango, J. J. Beaulieu, M. J. Bernot, A. J. Burgin, C. L. Crenshaw, A. M. Helton, L. T. Johnson, B. R. Niederlehner, J. D. Potter, R. W. Sheibley, and S. M. Thomas. 2009. Nitrate removal in stream ecosystems measured by N-15 addition experiments: denitrification. Limnology and Oceanography 54:666–680.

NASS (National Agricultural Statistics Service). 2012a. Statistics by subject. U.S. Department of Agriculture (USDA), Washington, DC, USA. <http://nass.usda.gov/Statistics_by_Subject/index> Accessed October 4, 2012.

NASS (National Agricultural Statistics Service). 2012b. Quick stats. U.S. Department of Agriculture (USDA), Washington, DC, USA. <http://quickstats.nass.usda.gov/> Accessed October 4, 2012.

NASS (National Agricultural Statistics Service). 2013. Crop production 2012 summary. U.S. Department of Agriculture (USDA), Washington, DC, USA. <http://usda.mannlib.cornell.edu/MannUsda/viewDocumentInfo.do?documentID=1047> Accessed February 6, 2013.

NCDC (National Climate Data Center). 2013. Summary of monthly normal 1981–2010. Gull Lake Biology Station, MI, USA. <http://www.ncdc.noaa.gov/cdo-web/search> Accessed January 17, 2013.

Nizeyimana, E. L., G. W. Petersen, M. L. Imhoff, H. R. Sinclair, S. W. Waltman, D. S. Reed-Margetan, E. R. Levine, and J. M. Russo. 2001. Assessing the impact of land conversion to urban use on soils with different productivity levels in the USA. Soil Science Society of America Journal 65:391–402.

Nowak, P., S. Bowen, and P. Cabot. 2006. Disproportionality as a framework for linking social and biophysical systems. Society & Natural Resources 19:153–173.

Paul, E. A., A. Kravchenko, A. S. Grandy, and S. Morris. 2015. Soil organic matter dynamics: controls and management for sustainable ecosystem functioning. Pages 104–134 in S. K. Hamilton, J. E. Doll, and G. P. Robertson, editors. The ecology of agricultural Landscapes: long-term research on the path to sustainability. Oxford University Press, New York, New York, USA.

Peters, C. J., N. L. Bills, J. L. Wilkins, and G. W. Fick. 2009. Foodshed analysis and its relevance to sustainability. Renewable Agriculture and Food Systems 24:1–7.

Peters, D. P. C., C. M. Laney, A. E. Lugo, S. L. Collins, C. Driscoll, P. M. Groffman, J. M. Grove, A. K. Knapp, T. K. Kratz, M. Ohman, R. B. Waide, and J. Yao. 2013. Long-term trends in ecological systems: a basis for understanding responses to global change. Technical Bulletin Number 1931, U.S. Department of Agriculture, Washington, DC.

Power, A. G. 2010. Ecosystem services and agriculture: tradeoffs and synergies. Philosophical Transactions of the Royal Society B: Biological Sciences 365:2959–2971.

Rasmussen, P. E., K. W. T. Goulding, J. R. Brown, P. R. Grace, H. H. Janzen, and M. Korschens. 1998. Long-term agroecosystem experiments: assessing agricultural sustainability and global change. Science 282:893–896.

Robertson, G. P., V. G. Allen, G. Boody, E. R. Boose, N. G. Creamer, L. E. Drinkwater, J. R. Gosz, L. Lynch, J. L. Havlin, L. E. Jackson, S. T. A. Pickett, L. Pitelka, A. Randall, A. S. Reed, T. R. Seastedt, R. B. Waide, and D. H. Wall. 2008a. Long-term agricultural research: a research, education, and extension imperative. BioScience 58:640–643.

Robertson, G. P., J. C. Broome, E. A. Chornesky, J. R. Frankenberger, P. Johnson, M. Lipson, J. A. Miranowski, E. D. Owens, D. Pimentel, and L. A. Thrupp. 2004. Rethinking the vision for environmental research in US agriculture. BioScience 54:61–65.

Robertson, G. P., L. W. Burger, C. L. Kling, R. Lowrance, and D. J. Mulla. 2007. New approaches to environmental management research at landscape and watershed scales. Pages 27–50 in M. Schnepf and C. Cox, editors. Managing agricultural landscapes for environmental quality. Soil and Water Conservation Society, Ankeny, Iowa, USA.

Robertson, G. P., S. L. Collins, D. R. Foster, N. Brokaw, H. W. Ducklow, T. L. Gragson, C. Gries, S. K. Hamilton, A. D. McGuire, J. C. Moore, E. H. Stanley, R. B. Waide, and M. W. Williams. 2012. Long-term ecological research in a human-dominated world. BioScience 62:342–353.

Robertson, G. P., V. H. Dale, O. C. Doering, S. P. Hamburg, J. M. Melillo, M. M. Wander, W. J. Parton, P. R. Adler, J. N. Barney, R. M. Cruse, C. S. Duke, P. M. Fearnside, R. F. Follett, H. K. Gibbs, J. Goldemberg, D. J. Miadenoff, D. Ojima, M. W. Palmer, A. Sharpley, L. Wallace, K. C. Weathers, J. A. Wiens, and W. W. Wilhelm. 2008b. Sustainable biofuels redux. Science 322:49.

Robertson, G. P., K. L. Gross, S. K. Hamilton, D. A. Landis, T. M. Schmidt, S. S. Snapp, and S. M. Swinton. 2014. Farming for services: An ecological approach to production agriculture. BioScience 64: 404-415. doi: 10.1093/biosci/biu037

Robertson, G. P., and P. M. Groffman. 2015. Nitrogen transformations. Pages 421–426 in E. A. Paul, editor. Soil microbiology, ecology, and biochemistry. Fourth edition. Academic Press, Burlington, Massachusetts, USA.

Robertson, G. P., and R. R. Harwood. 2013. Sustainable agriculture. Pages 111–118 in S. A. Levin, editor. Encyclopedia of Biodiversity. Second edition. Academic Press, New York, New York, USA.

Robertson, G. P., K. M. Klingensmith, M. J. Klug, E. A. Paul, J. R. Crum, and B. G. Ellis. 1997. Soil resources, microbial activity, and primary production across an agricultural ecosystem. Ecological Applications 7:158–170.

Robertson, G. P., and S. M. Swinton. 2005. Reconciling agricultural productivity and environmental integrity: a grand challenge for agriculture. Frontiers in Ecology and the Environment 3:38–46.

Robertson, G. P., and P. M. Vitousek. 2009. Nitrogen in agriculture: balancing the cost of an essential resource. Annual Review of Environment and Resources 34:97–125.

Rudy, A. P., C. K. Harris, B. J. Thomas, M. R. Worosz, S. C. Kaplan, and E. C. O'Donnell. 2008. The political ecology of Southwest Michigan Agriculture, 1837–2000. Pages 152–205 in C. L. Redman and D. R. Foster, editors. Agrarian landscapes in transition. Oxford University Press, New York, New York, USA.

Scheffe, R. D., and R. E. Morris. 1993. A review of the development and application of the urban airshed model. Atmospheric Environment 27:23–39.

Scheffer, M., J. Bascompte, W. A. Brock, V. Brovkin, S. R. Carpenter, V. Dakos, H. Held, E. H. van Nes, M. Rietkerk, and G. Sugihara. 2009. Early-warning signals for critical transitions. Nature 461:53–59.

Schmidt, T. M., and C. Waldron. 2015. Microbial diversity in agricultural soils and its relation to ecosystem function. Pages 135–157 in S. K. Hamilton, J. E. Doll, and G. P. Robertson, editors. The ecology of agricultural Landscapes: long-term research on the path to sustainability. Oxford University Press, New York, New York, USA.

Snapp, S. S., L. E. Gentry, and R. R. Harwood. 2010. Management intensity—not biodiversity—the driver of ecosystem services in a long-term row crop experiment. Agriculture, Ecosystems & Environment 138:242–248.

Snapp, S. S., R. G. Smith, and G. P. Robertson. 2015. Designing cropping systems for ecosystem services. Pages 378–408 in S. K. Hamilton, J. E. Doll, and G. P. Robertson, editors. The ecology of agricultural Landscapes: long-term research on the path to sustainability. Oxford University Press, New York, New York, USA.

Sprague, C., and W. Everman. 2011. Weed control guide for field crops. Michigan State University Extension Bulletin E0434, East Lansing, Michigan, USA.

Swinton, S. M., C. B. Jolejole-Foreman, F. Lupi, S. Ma, W. Zhang, and H. Chen. 2015a. The economic value of ecosystem services from agriculture. Pages 54–76 in S. K. Hamilton, J. E. Doll, and G. P. Robertson, editors. The ecology of agricultural Landscapes: long-term research on the path to sustainability. Oxford University Press, New York, New York, USA.

Swinton, S. M., F. Lupi, G. P. Robertson, and S. K. Hamilton. 2007. Ecosystem services and agriculture: cultivating agricultural ecosystems for diverse benefits. Ecological Economics 64:245–252.

Swinton, S. M., N. Rector, G. P. Robertson, C. B. Jolejole, and F. Lupi. 2015b. Farmer decisions about adopting environmentally beneficial practices. Pages 340–359 in S. K. Hamilton, J. E. Doll, and G. P. Robertson, editors. The ecology of agricultural Landscapes: long-term research on the path to sustainability. Oxford University Press, New York, New York, USA.

Syswerda, S. P., B. Basso, S. K. Hamilton, J. B. Tausig, and G. P. Robertson. 2012. Long-term nitrate loss along an agricultural intensity gradient in the Upper Midwest USA. Agriculture, Ecosystems and Environment 149:10–19.

Thelen, K., B. Widdicombe, and L. Williams. 2011. 2011 Michigan corn hybrids compared. Michigan State University Extension Bulletin E0431, East Lansing, Michigan, USA.

Tilman, D., C. Balzer, J. Hill, and B.L. Befort. 2011. Global food demand and the sustainable intensification of agriculture. Proceedings of the National Academy of Sciences USA 108:20260–20264.

Tilman, D., R. Socolow, J. A. Foley, J. Hill, E. Larson, L. Lynd, S. Pacala, J. Reilly, T. Searchinger, and C. Somerville. 2009. Beneficial biofuels—the food, energy, and environment trilemma. Science 325:270–271.

Transeau, E. 1935. The prairie peninsula. Ecology 16:423–427.

USDA (U.S. Department of Agriculture). 2009. 2007 Census of Agriculture: Michigan. USDA, Washington, DC, USA. <http://agcensus.usda.gov/Publications/2007/Full_Report/Volume_1,_Chapter_2_County_Level/Michigan/>

Walbridge, M. R., S. R. Shafer, C. Medley, G. Patterson, and M. Parker. 2011. A long-term agro-ecosystem research (LTAR) network for agriculture. Pages 26–30 in Proceedings of the Fourth Interagency Conference in the Watersheds: Observing, Studying, and Managing Change. U.S. Geological Scientific Investigations Report 2011–5169.

Weltzin, J. F., M. E. Loik, S. Schwinning, D. G. Williams, P. A. Fay, B. M. Haddad, J. Harte, T. E. Huxman, A. K. Knapp, G. Lin, W. T. Pockman, M. R. Shaw, E. E. Small, M. D. Smith, S. D. Smith, D. T. Tissue, and J. C. Zak. 2003. Assessing the response of terrestrial ecosystems to potential changes in precipitation. BioScience 53:941–952.

2

Farming for Ecosystem Services[*]

An Ecological Approach to Production Agriculture

G. Philip Robertson, Katherine L. Gross,
Stephen K. Hamilton, Douglas A. Landis,
Thomas M. Schmidt, Sieglinde S. Snapp, and
Scott M. Swinton

Row-crop agriculture is one of the most extensive and closely coupled natural–human systems and has extraordinary implications for human welfare and environmental well-being. The continued intensification of row-crop agriculture provides food for billions and, for at least the past 50 years, has slowed (but not stopped) the expansion of cropping onto lands valued for conservation and other environmental services. Nevertheless, intensification has also caused direct harm to the environment: The escape of reactive nitrogen and phosphorus from intensively managed fields pollutes surface and coastal waters and contaminates groundwater, pesticides kill nontarget organisms important to ecological communities and ecosystems sometimes far away, soil loss threatens waterways and long-term cropland fertility, accelerated carbon and nitrogen cycling contribute to climate destabilization, and irrigation depletes limited water resources.

The search for practices that attenuate, avoid, or even reverse these harms has produced a rich scientific literature and sporadic efforts to legislate solutions. That these harms persist and, indeed, are growing in the face of increased global demands for food and fuel underscores the challenge of identifying solutions that work in ways that are attractive to farmers and responsive to global markets. On one hand are farmers' needs for practices that ensure a sustained income in the face of market

[*] Co-published as Robertson et al. (2014) BioScience 64:404–415.

and consumer pressures to produce more for less; on the other are societal demands for a clean and healthful environment. Most growers are caught in the middle.

One avenue for addressing this conundrum is the potential for row-crop producers to farm for more than food, fuel, and fiber. Growing recognition that agriculture can provide ecosystem services other than yield (Swinton et al. 2007, Power 2010) opens a potential for society to pay for improvements in services provided by farming: a clean and well-regulated water supply, biodiversity, natural habitats for conservation and recreation, climate stabilization, and aesthetic and cultural amenities such as vibrant farmscapes.

Operationalizing such an enterprise, however, is far from straightforward: Farming for services requires knowledge of what services can be practically provided at what cost and how nonprovisioning services might be valued in the absence of markets. The costs of providing services are both direct (e.g., the cost of installing a streamside buffer strip) and indirect (e.g., the opportunity cost of sales lost by installing such a strip on otherwise productive cropland). Moreover, valuation includes not simply the monetary value of a provided service but also what society (consumers) might be willing to pay through mechanisms such as higher food prices or taxes.

Knowledge of the services themselves requires a fundamental understanding not only of the biophysical basis for the service but also of how different ecological processes interact to either synergize or offset the provisioning of different services: Farming is a systems enterprise with multiple moving parts and sometimes complex interactions. No-till practices, for example, can sequester soil carbon and reduce fossil fuel consumption but require more herbicide use and can increase the production of nitrous oxide (N_2O; van Kessel et al. 2013), a potent greenhouse gas. Understanding the basis for such trade-offs and synergies requires an ecological systems approach absent from most agricultural research.

Since 1988, we have pursued research to understand the fundamental processes that underpin the productivity and environmental performance of important row-crop systems of the upper U.S. Midwest. Our aim is to understand the key ecological interactions that constrain or enhance the performance of differently managed model cropping systems and, therefore, to provide insight into the provisioning of related services in a whole-systems context. Our global hypothesis is that ecological knowledge can substitute for most chemical inputs in intensively managed, highly productive, annual row crops. Together, long-term observations and experiments at both local and landscape scales uniquely inform our analysis.

Experimental Context: The Search for Services

The Main Cropping System Experiment (MCSE) of the Kellogg Biological Station (KBS), a member site of the U.S. Long Term Ecological Research (LTER) Network, was initiated in 1988 in southwest Michigan. The site is in the U.S. North Central Region, a 12-state region that is responsible for 80% of U.S. corn (*Zea mays*) and soybean (*Glycine max*) production and 50% of the U.S. wheat (*Triticum aestivum*) crop (NASS 2013a). The Great Lakes portion of the region is also an important dairy region, with alfalfa (*Medicago sativa*) being an important forage crop. Crop yields in Kalamazoo County, which surrounds the KBS LTER site, are similar to national average yields (NASS 2013a, b). The soils of the area are Typic Hapludalfs of moderate fertility, formed since the most recent glacial retreat ~18,000 years ago, and the climate is humid continental (1027 mm yr^{-1} average precipitation, 9.9°C mean annual temperature).

In 1988 we established a cropping-systems experiment along a management intensity gradient that, by 1992, included four annual and three perennial cropping systems plus four reference communities in different stages of ecological succession. The annual cropping systems are corn–soybean–winter wheat rotations managed in four different ways. One system is managed conventionally, on the basis of current cropping practices in the region, including tillage and, since 2009, genetically engineered soybean and corn. One is managed as a permanent No-till system, otherwise identical to the Conventional system. A third is managed as a Reduced Input system, with about one-third of the Conventional system's chemical inputs. In this system, winter cover crops provide additional nitrogen, and mechanical cultivation was used to control weeds until a 2009 shift to herbicide-resistant crops that allowed the use of the herbicide glyphosate for weed control in soybean and corn. A fourth system is managed biologically, with no synthetic chemicals (or manure) but with cover crops and mechanical cultivation as in the Reduced Input system. This system is U.S. Department of Agriculture–certified organic. The three perennial crops are continuous alfalfa, short-rotation hybrid poplar trees (*Populus* var.), and conifer stands planted in 1965.

The successional reference communities include (1) a set of Early Successional sites abandoned from cultivation in 1989 and undisturbed except for annual burning to exclude trees, (2) a set of Mown Grassland sites cleared from forest in 1960 and mown annually but never tilled, (3) a set of Mid-successional sites released from farming in the 1950s and 1960s that is now becoming forested, and (4) a set of late successional Eastern Deciduous Forest stands never cleared for agriculture. Complete descriptions of each system and community appear in Robertson and Hamilton (2015, Chapter 1 in this volume).

Delivering Ecosystem Services

We identify five major ecosystem services that our annual cropping systems could potentially provide: food and fuel, pest control, clean water, climate stabilization through greenhouse gas mitigation, and soil fertility. These services are provided to differing degrees in different systems and interact in sometimes unexpected ways. In many respects, however, their delivery comes in bundles that can be highly complementary.

Providing Food, Fuel, and Fiber

Without question, the most important ecosystem service of agriculture is the provision of food; fiber; and, more recently, fuel. To an ever-increasing extent, we are dependent on high yields from simplified, intensively managed row-crop ecosystems for this provisioning. But to what extent do high yields depend on current common management practices? The results from other long-term experiments (e.g., Drinkwater et al. 1998) suggest that more complex rotations using fewer inputs can provide similar or greater yields than those of conventional rotations. Our results suggest that simpler rotations of major grains can be managed to provide other ecosystem services as well.

Corn and soybean yields under Conventional management at the KBS LTER site are similar to the average yields for both the entire United States and Kalamazoo County; wheat yields are higher (Robertson and Hamilton 2015, Chapter 1 in this volume). In our Reduced Input system, corn and soybean yields slightly exceed those of our conventionally managed system, and wheat yields lag only slightly (Fig. 2.1). Indirect evidence points to nitrogen deficiency as the cause of the depressed wheat yields: Whereas corn follows a nitrogen-fixing winter cover crop and soybean fixes its own nitrogen, fall-planted wheat immediately follows the soybean harvest, which leaves relatively little nitrogen-rich residue for the wheat crop. This nitrogen deficit is especially apparent in the Biologically Based system, which lacks fertilizer nitrogen inputs: Wheat yields are ~60% of the yields under

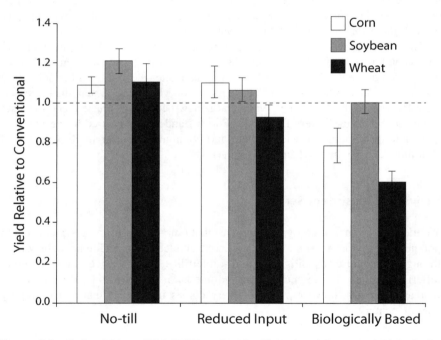

Figure 2.1. Grain yields at KBS LTER under No-till, Reduced Input, and Biologically Based management relative to Conventional management (dotted horizontal line) over the 23 year period of 1989–2012. Absolute yields for Conventional management are similar to county and U.S. national average yields. Error bars represent the standard error. Redrawn from Robertson et al. (2014).

Conventional management. This stands in contrast to soybean yields, for which the Biologically Based system is equivalent to the Conventional system (Fig. 2.1).

Rotational diversity clearly matters to the delivery of ecosystem services, including yield (Smith et al. 2008). A characteristic of intensive row-crop agriculture is its severe reduction of plant diversity of both crops and weeds. The conventional norm for most grain and other major commodity crops in the United States is weed-free monocultures or simple two-crop rotations. In the U.S. Midwest, corn is grown in a corn–soybean rotation on ~60% of corn acreage and in a continuous corn-only rotation on ~25% (Osteen et al. 2012). Simplified rotations date from the onset of highly mechanized agriculture in the 1940s. Until 1996 U.S. farm subsidies were linked to the area planted in selected crops (notably, wheat, corn, and other feed grains), which tended to encourage simplified rotations. Today, there are two federal programs that favor simpler rotations. The most important one is the 2007 legislative mandate to blend grain-based ethanol—made entirely from corn—into the national gasoline supply. This raises demand for corn and therefore its price, creating an incentive to increase its presence in crop rotations. The second is crop insurance subsidies that reduce farmer incentives to manage risk through crop diversity.

Simplified rotations and larger fields lead to simplified landscapes, because total cropland becomes constrained to two or three dominant species in ever-larger patches (Meehan et al. 2011, Wright and Wimberly 2013). Plant diversity is further constrained by increasingly effective weed control, with chemical technologies dating from the 1950s and genomic technologies dating from the 1990s. In 2011, 94% of U.S. soybean acreage and 70% of U.S. corn acreage were planted with herbicide-resistant varieties (Osteen et al. 2012).

Reduced plant diversity at both the field and the landscape scales can have negative consequences for many other taxa—most notably, arthropods; vertebrates; and, possibly, microbes and other soil organisms. The loss of these taxa can have important effects on community structure and dynamics—most notably on species extinctions and changes in trophic structure that can affect pest suppression—and on ecosystem processes, such as carbon flow and nitrogen cycling. To what extent might greater rotational complexity provide these important ecosystem services?

That continuous monocultures suffer a yield penalty that persists even in the presence of modern chemicals is well known. For millennia, agriculturalists have used multispecies rotations to improve yields by advancing soil fertility and suppressing pests and pathogens (Karlen et al. 1994, Bennett et al. 2012). Since the 1950s, monoculture penalties in grain crops have been largely ameliorated with chemical fertilizers and pesticides; the remaining penalties, which appear mainly from soil pathogens or other microbial factors (Bennett et al. 2012), are largely addressable with simple two-species rotations, such as corn and soybean.

To what extent might the restoration of rotational complexity in row crops substitute for today's use of external inputs? This is a fundamental question that underpins the success of low chemical input farming. As was noted above, the inclusion of legume cover crops plus mechanical weed control in our Reduced Input corn–soybean–wheat rotation alleviated the need for two-thirds of the synthetic nitrogen and herbicide inputs otherwise required for high yields (Fig. 2.1). Can rotational complexity substitute for the provision of all synthetic inputs? In our Biologically

Based system, only soybean, which provides its own nitrogen, matched the yields of crops managed with synthetic chemicals. In organic agriculture, manure or compost is generally required to achieve high yields in nonleguminous crops (e.g., Liebman et al. 2013). However, in another experiment at the KBS LTER site, designed specifically to address the impact of rotational diversity on yield in the absence of confounding management practices, Smith et al. (2008) found that a 3-year, six-species rotation of corn, soybean, and wheat, with three cover crops to provide nitrogen, could produce corn yields as high as the county average. In addition to yield, rotational complexity benefits other ecosystem services, as we will discuss below.

Providing Pest Protection through Biocontrol Services

Biodiversity at the landscape scale also affects the capacity of agriculture to deliver ecosystem services, especially those related to biocontrol and water quality. For example, ladybird beetles (Coleoptera: Coccinellidae) are important predators of aphids in field crops. In KBS LTER soybeans, ladybird beetles are responsible for most soybean aphid (*Aphis glycines*) control and are able to keep aphid populations below economic thresholds (Costamagna and Landis 2006); absent such control, soybean yields can be suppressed 40–60%. Coccinellid diversity is an important part of this control.

Because different coccinellid species use different habitats at different times for foraging or other purposes, such as overwintering, the diversity of habitats within a landscape becomes a key predictor of biocontrol efficacy (Fig. 2.2A). About a dozen coccinellid species with moderate to strong habitat preferences are present in the KBS landscape (Maredia et al. 1992a, Landis and Gage 2014). *Coleomegilla maculata*, for example, overwinters in woodlots and, prior to the summertime development of soybean aphid populations, depends on pollen from early flowering

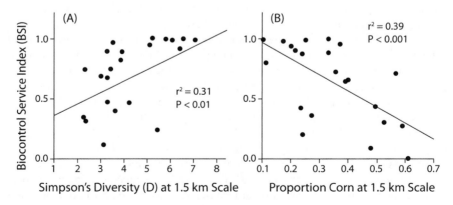

Figure 2.2. Biocontrol services from coccinellids as a function of landscape diversity (A) and the dominance of corn within 1.5 km of soybean fields (B). Panel (A) is redrawn from Gardiner et al. (2009) with permission from the Ecological Society of America; permission conveyed through Copyright Clearance Center, Inc. Panel (B) is redrawn from Landis et al. (2008).

plants such as Virginia springbeauty (*Claytonia virginica* L.) and the common dandelion (*Taraxacum officinale* F.H. Wigg.), and then on aphids in the winter wheat
and alfalfa crops (Colunga-Garcia 1996). Later in the season, after aphids have fed
on soybean, the Early Successional and Poplar communities support late-season
aphid infestations that are exploited by the coccinellids (Maredia et al. 1992b).

Landscape diversity can therefore be key for biocontrol services provided by
mobile predators. For coccinellids, the presence of heterogeneous habitats within
1.5 km of a soybean field is strongly correlated with soybean aphid suppression: Landscapes with greater proportions of the local area in corn and soybean
production have significantly less biocontrol (Fig. 2.2B; Gardiner et al. 2009).
Landis et al. (2008) estimated the value of hidden biocontrol in Michigan and three
adjacent states to be $239 million for 2007 on the basis of a $33 ha^{-1} increase in
profitability from higher production and lower pesticide costs among the soybean
farmers who used integrated pest management to control aphids.

Providing Clean Water

The quality of water draining from agricultural watersheds is a longstanding environmental problem. Sediment, phosphorus, and nitrate are important pollutants that
leave cropland and lead to compromised groundwater, surface freshwaters, and
marine ecosystems worldwide. In the United States, over 70% of the nitrogen and
phosphorus delivered to the Gulf of Mexico by the Mississippi River is derived
from agriculture (Alexander et al. 2008). Such deliveries create coastal hypoxic
zones worldwide (Diaz and Rosenberg 2008).

Must this necessarily be the case? Sediment and phosphorus loadings can be
reduced substantially with appropriate management practices: No-till and other
conservation tillage methods can often eliminate erosion and substantially reduce
the runoff that also carries phosphorus to surface waters, as can riparian plantings
along cropland waterways (Lowrance 1998). Nitrate mitigation is more problematic. Because nitrate is so mobile in soil, percolating water carries it to groundwater
reservoirs, where it resides for days to decades before it emerges in surface waters
and is then carried downstream (Hamilton 2012), eventually to coastal marine
systems.

Some of the transported nitrate can be captured by riparian communities
(Lowrance 1998) or can be processed streamside (Hedin et al. 1998) or in transit
(Beaulieu et al. 2011) to more reduced forms of nitrogen, including nitrogen gas.
If wetlands are in the flow path, a significant fraction can be immobilized in wet-
land sediments as organic nitrogen or can be denitrified into nitrogen gas, either by
heterotrophic or chemolithoautotrophic microbes (Whitmire and Hamilton 2005,
Burgin and Hamilton 2007). Restoring wetlands and the tortuosity of more natu-
ral channels can increase both streamside and within-stream processing of nitrate
(NRC 1995).

But, by far, the best approach to mitigating nitrate loss is avoiding it to begin
with—a major challenge in cropped ecosystems so dependent on large quantities
of plant-available nitrogen. The average nitrogen fertilizer rate for corn in the U.S.
Midwest is ~160 kg N ha^{-1} (ERS 2013), with only about 50% taken up by the crop,

on average (Robertson 1997). This contrasts with annual inputs of ~7 kg N ha⁻¹ delivered in precipitation at the KBS LTER site.

KBS LTER research has shown that crop management can substantially reduce long-term nitrate leaching. Over an 11-year period, beginning 6 years after establishment, the MCSE annual row-crop systems showed 2- to 3-fold differences in nitrate losses, ranging from average annual losses of 19 and 24 kg N ha⁻¹ in the Biologically Based and Reduced Input systems, respectively, and of 42 and 62 kg N ha⁻¹ in the No-till and Conventionally managed systems, respectively (Fig. 2.3; Syswerda et al. 2012). Even after accounting for yield differences (Fig. 2.1), leaching differences were substantial: 7.3 kg NO_3^-–N per megagram yield in the Reduced Input system, compared with 11.1 in the No-till and 17.9 in the Conventional systems.

What accounts for lower nitrate leaching rates? The better soil structure in No-till cropping systems allows water to leave more quickly (Strudley et al. 2008), which reduces equilibration with soil microsites where nitrate is formed. But a more important factor appears to be the presence of cover crops: Even with tillage, the Reduced Input and Biologically Based systems leached less nitrogen. Cover crops helped perennialize the crop year; that is, with the fields occupied by growing plants for a greater proportion of the year, more nitrate is scavenged from the soil profile and cycled through plant and microbial transformations (McSwiney et al. 2010). More soil water is also transpired, which reduces the opportunity for nitrate transport: Drainage in the Reduced Input and Biologically Based systems was only 50–70% of that in the Conventional and No-till systems (Fig. 2.3 inset). The rapidly growing systems with true perennial vegetation—the Poplar and Successional

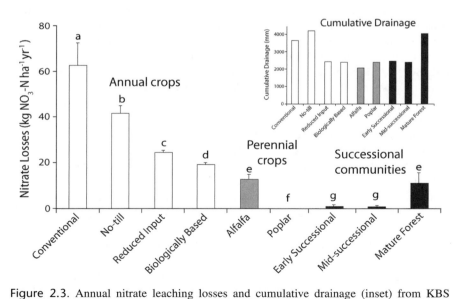

Figure 2.3. Annual nitrate leaching losses and cumulative drainage (inset) from KBS LTER cropping and successional systems between 1995 and 2006. Modified from Syswerda et al. (2012).

systems—had exceedingly small annual leaching rates of 0.1–1.1 kg N ha⁻¹, although that was, in part, due to very low or nonexistent rates of nitrogen fertilizer use. In a related experiment, a perennial cereal crop fertilized at agronomic levels leached 80% less nitrate than did its annual analog (Culman et al. 2013).

Providing Greenhouse Gas Mitigation

Agriculture is directly responsible for ~10–14% of total annual global anthropogenic greenhouse gas emissions (Smith et al. 2007). This is largely the result of nitrous oxide (N_2O) emitted from soil and manure and from methane (CH_4) emitted by ruminant animals and burned crop residues. Including the greenhouse gas costs of agricultural expansion, agronomic inputs, such as fertilizers and pesticides, and postharvest activities, such as food processing, transport, and refrigeration, bring agriculture's footprint to 26–36% of all anthropogenic greenhouse gas emissions (Barker et al. 2007). Mitigating some portion of this footprint could therefore significantly contribute to climate stabilization (Caldeira et al. 2004), as might the production of cellulosic biofuels if they were used to offset fossil fuel use (Robertson et al. 2008).

Global warming impact analyses can reveal the source of all significant greenhouse gas costs in any given cropping system and, therefore, the full potential for management to mitigate emissions. Such an analysis for KBS LTER cropping systems over a 20-year time frame (Fig. 2.4; Gelfand and Robertson 2014) shows how the overall costs can vary substantially with management. The Conventional annual cropping system had a net annual global warming impact (in CO_2 equivalents) of

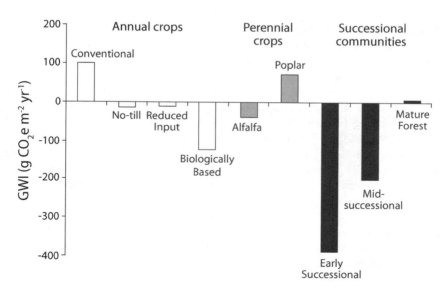

Figure 2.4. Net global warming impact (GWI) of cropped and unmanaged KBS LTER ecosystems. Annual crops include corn-soybean-wheat rotations. Redrawn from Robertson et al. (2014).

101 g CO_2e m^{-2}, whereas the No-till system exhibited net mitigation: -14 g CO_2e m^{-2}. The Early Successional system was the most mitigating, at -387 g CO_2e m^{-2}. Closer inspection reveals the basis for these differences: Although N_2O production and nitrogen fertilizer manufacture were the two greatest sources of global warming impact in the annual cropping systems, the soil carbon storage in the No-till system more than offset the CO_2e cost of no-till N_2O and fertilizer manufacture. And because the Biologically Based system sequestered carbon at an even greater rate and without the added cost of nitrogen fertilizer, the net mitigation was stronger still (Fig. 2.4; Gelfand and Robertson 2015, Chapter 12 in this volume).

Most of the substantial mitigation capacity of Early Successional fields is derived from their high rate of soil carbon storage, which will diminish over time. At the KBS LTER site, the carbon stored annually in Mid-successional soils was ~10% of that in Early Successional soils, and no net soil carbon storage occurred in the mature Deciduous Forest. As a result, the net CO_2e balance of the mature forest is close to 0 g CO_2e m^{-2}, with CH_4 oxidation offsetting most of the CO_2e cost of natural N_2O emissions (Fig. 2.4). Interesting, too, is the recovery of CH_4 oxidation during succession. Methane oxidation rates are typically decimated when natural vegetation is converted to agriculture (Del Grosso et al. 2000); that oxidation in the Mid-successional system is more than midway between that of the Early Successional system and that of the mature Deciduous Forest suggests an 80- to 100-year recovery phase. Recent evidence from the KBS LTER site suggests that methanotrophic bacterial diversity plays a role in CH_4 oxidation differences (Fig. 2.5; Levine et al. 2011).

In addition, if harvested biomass is used to produce energy that would otherwise be provided by fossil fuels, the net global warming impact of a system will be further reduced by avoided CO_2 emissions from the fossil fuels displaced by the biomass-derived energy. Sometimes—as with corn grain in conventional systems—the displacement is minor or even nonexistent because of the fossil fuel used to produce the biomass (Farrell et al. 2006) and the potential to incur carbon costs elsewhere by clearing land to replace that removed from food production (Searchinger et al. 2008). In contrast to the energy provided by corn grain is the energy provided by cellulosic biomass produced in the Early Successional system. Gelfand et al. (2013) calculated that harvesting successional vegetation for cellulosic biofuel could provide ~850 g CO_2e m^{-2} of greenhouse gas mitigation annually. Extrapolated yields to marginal lands across 10 U.S. Midwest states using finescale (0.4-ha) modeling yielded a potential climate benefit of ~44 MMT CO_2 yr^{-1}. However, such near-term benefits also depend on the methods used to establish the biofuel crop; killing the existing vegetation and replanting with purpose-grown feedstocks, such as switchgrass or miscanthus, can create substantial carbon debt (Fargione et al. 2008) that can take decades to repay (Gelfand et al. 2011); the debt is even greater if the replanted crop requires tillage (Ruan and Robertson 2013).

The provision of greenhouse gas mitigation is a service clearly within the capacity of modern cropping systems to provide. Various management practices have differing effects, sometimes in opposition (consider, e.g., no-till energy savings vs. the carbon cost of additional herbicides) and at other times synergistic (consider that leguminous cover crops in the Biologically Based system not only increased soil

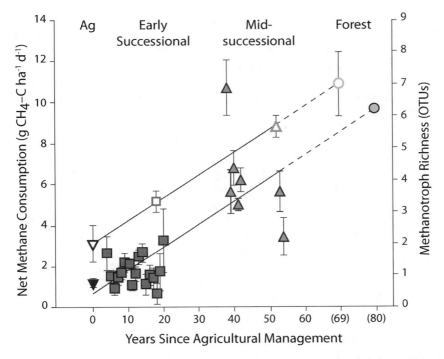

Figure 2.5. The increase in soil methanotroph diversity (open symbols) and atmospheric methane consumption (closed symbols) in ecological succession from row-crop fields (Ag, black) through early (dark gray) and mid-successional (medium grey) fields to mature forest (light gray) at KBS LTER. Redrawn from Levine et al. (2011).

carbon storage but also reduced the CO_2e costs of manufactured fertilizer nitrogen). Designing optimal systems is not difficult; there are many practice-based opportunities to diminish CO_2e sources or enhance CO_2e sinks and thereby help stabilize the climate.

Providing Soil Fertility, the Basis for Sustained Crop Production

Closely tied to other services, such as food production and greenhouse gas mitigation, is soil fertility. As a supporting service that underpins the provision of other services (MA 2005), soil fertility is under management control and is therefore a deliverable service; in its absence, fertility must be enhanced with greater quantities of external inputs, such as fertilizers, and the system is less able to withstand extreme events, such as drought. That said, soil fertility is not a panacea for reducing the environmental impacts of agricultural systems; for example, N_2O production was as high in our Biologically Based system as it was in the less fertile Conventional system (Robertson et al. 2000).

Soil fertility has many components. Physically, fertility is related to soil structure—porosity, aggregate stability, water-holding capacity, and erosivity.

Its chemical constituents include soil organic matter, pH, base saturation, cation exchange, and nutrient pools. Biologically, soil fertility is related to food web complexity, pest and pathogen suppression, and the delivery of mineralizable nutrients. Most of these components are interrelated, which frustrates attempts at a comprehensive definition of soil fertility or soil quality. At heart, however, soil fertility is the capacity of a soil to meet plant growth needs; all else equal, more fertile soils support higher rates of primary production.

Building soil fertility is closely tied to building soil organic matter: A century of work at Rothamsted and other long-term agricultural research sites (Rasmussen et al. 1998) has shown positive associations with most—if not all—of the indicators noted above. At the KBS LTER site, relative to the Conventional system, soil organic matter increased in the No-till, Reduced Input, and Biologically Based systems (Syswerda et al. 2011). A major reason for soil carbon gain in these systems is slower decomposition rates as a result of organic matter protection within soil aggregates, particularly within larger size classes. Grandy and Robertson (2007) found greater soil carbon accumulation in KBS LTER ecosystems with higher rates of large (2–8 mm) aggregate formation. The formation of large aggregates and carbon accumulation were greatest in the successional and mature forest systems, and lowest in the Conventional system; the Biologically Based, No-till, and perennial systems were intermediate. Aggregates in smaller size classes (<0.05–0.25 mm) expressed the opposite trend.

That the No-till system accumulated carbon and primarily in larger, more vulnerable aggregates is no surprise (West and Post 2002, Six et al. 2004); however, carbon and large aggregate accumulation in the heavily tilled Reduced Input and

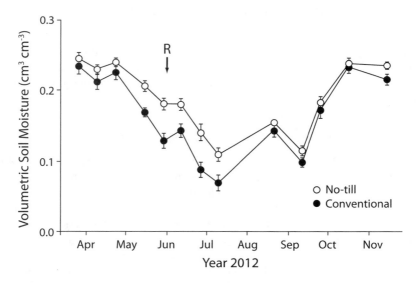

Figure 2.6. Seasonal variation in soil moisture in Conventional and No-till systems during the 2012 soybean growing season. The 6-week drought began after a June 3 rainfall (R on figure). Error bars denote the standard error (n = 6). Redrawn from Robertson et al. (2014).

Biologically Based systems was unexpected and likely related to the inclusion of leguminous cover crops in these rotations. Legumes may increase aggregate stability through greater polysaccharide production and different microbial communities (Haynes and Beare 1997).

That the No-till system better withstood the 2012 U.S. drought than did the other systems (1.9 ± 0.12 Mg ha^{-1} soybean grain in the No-till vs. 1.3 ± 0.05 Mg ha^{-1} soybean grain in the Conventional system) suggests a clear no-till benefit to soil fertility even when external inputs are high. Greater moisture stores in the better-structured no-till soils following the last significant rainfall before the drought (Fig. 2.6), equivalent to ~4 cm of stored water in the root zone, underscore the value of no-till agriculture to the 2012 soybean production. This enhanced water storage capacity may also help explain greater no-till productivity in more normal years; on average, yields in the No-till system were 9–21% higher than they were in the Conventional system (Fig. 2.1). In the Reduced Input system, soil fertility allowed competitive yields (Fig. 2.1) with only a fraction of the nitrogen and other inputs.

Valuing Ecosystem Services: The Social Component

The ability of row crops and agricultural landscapes to provide ecosystem services is only part of the farming for services equation. The other is farmers' willingness to implement practices that deliver additional services and, to the extent that adoption probably requires economic compensation, society's willingness to pay for these services.

The willingness of farmers to adopt new management practices that provide additional services depends on awareness, attitudes, available resources, and incentives (Swinton et al. 2015a, Chapter 3 in this volume). The current practices are largely the result of past practices; cultural norms; and the availability of technology, policies, and markets that support sustained profitability. Although environmental stewardship is a factor influencing many farmers' decisions, sustained profitability is usually the overriding concern.

Particularly for those services related to reducing the environmental impact of agriculture, farmers in Michigan—and presumably elsewhere—are more likely to adopt practices that provide direct, local benefits. These benefits might be monetary, such as higher profits or greater future land values, or nonmonetary, such as safer groundwater for family use. To learn how farmers weigh environmental benefits in their management decisions, Swinton et al. (2015a, Chapter 3 in this volume) conducted a series of six farmer focus groups in 2007 and a subsequent statewide survey of 1600 Michigan corn and soybean farms in 2008 (Jolejole 2009, Ma et al. 2012). When asked to consider six environmental benefits of reduced input agriculture and to rate their relative importance to themselves and to society, the participating farmers in both settings ranked benefits such as increased soil organic matter, soil conservation, and reduced nitrate leaching as significantly more important to themselves than to society (Fig. 2.7). In contrast, reduced global warming was ranked as more important to society than to the farmers. These attitudes conform to the economic distinction between private and public goods and

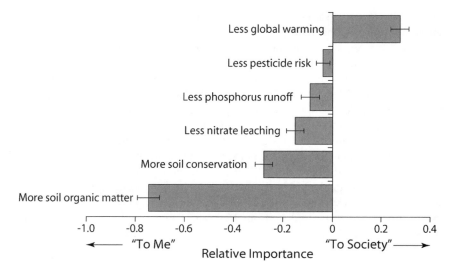

Figure 2.7. The relative importance to Michigan farmers and to society (as ranked by the farmers) of various environmental benefits potentially provided by agriculture. Redrawn from Robertson et al. (2014).

will strongly influence the farmers' willingness to accept payments for shouldering a perceived public burden.

The survey revealed that, of a variety of cropping practices known from KBS and other research to provide environmental benefits, two practices were currently used by over 80% of the participants (Swinton et al. 2015a, Chapter 3 in this volume). These included pest scouting prior to insecticide application and reduced tillage (e.g., chisel plowing). These practices saved labor or inputs or improved farmstead water quality without reducing expected crop revenue; they were therefore desirable with respect to both the environment and farm profitability.

A second group of three practices was viewed favorably by about half of the farmers: the addition of a small grain, such as wheat, to their standard corn–soybean rotation; incorporating rather than spreading manure; and no-till management, at least for specific crop years. In comparison to the first group of practices, these were perceived to have a greater risk of diminished revenues, higher costs, or greater labor demands during busy periods.

A third group of practices appealed to less than a third of the farmers: continuous no-till, banded application of fertilizer and pesticides at reduced rates, soil testing for nitrogen prior to nitrogen fertilization, and winter cover crops to substitute for most fertilizer nitrogen; these practices were seen as particularly high-risk. This result reveals that, for the rotation tested, although higher average yields (Fig. 2.1) and comparable profitability (Jolejole 2009) may be apparent under experimental conditions, they are not by themselves sufficient for the adoption of reduced input management.

A separate section of the survey elicited what levels of payment (if any) the farmer respondents would require to adopt environmentally beneficial cropping practices.

Farmers were presented with four increasingly demanding cropping systems with similarly increasing environmental benefits that ranged from a chisel-tilled corn–soybean rotation fertilized according to university recommendations (including a nitrate soil test) to a corn–soybean–wheat rotation with winter cover crops and reduced chemical inputs. The respondents were asked how much land they might enroll in each system for a predetermined payment level.

Three farmer traits—the belief that their production could benefit from nature, their years of prior experience, and the availability of suitable equipment— were collectively the best predictor of farmers' willingness to shift land into the more complex cropping systems associated with reduced chemical inputs (Ma et al. 2012). Not surprisingly, the simplest system attracted the most participation, regardless of farm size. However, among those willing to adopt the most environmentally beneficial system, farmers with over 200 ha were much more willing than were farmers with smaller farms to offer more acreage at higher payment levels. These larger farms are therefore most likely to be providers of environmental services at the lowest cost (Swinton et al. 2015b, Chapter 13 in this volume). This is probably related to economies of scale: Not only do larger farms have more land to enroll, but the additional fixed costs of no-till, banding, and cultivation equipment can be spread out over larger areas, therefore lowering capitalization barriers.

Clear from this research is that the provision of ecosystem services in agriculture will require incentives. Education is not the issue; most farmers are aware of the environmental benefits of alternative practices (except for greenhouse gas mitigation benefits). Indeed, those farmers who strongly valued environmental stewardship were willing to accept lower cost incentives to adopt the alternative practices (Ma et al. 2012). But almost all of the farmers—especially those with large farms— were willing to accept payments for services. This, then, raises the question: Are consumers willing to pay for such services? Regardless of the mechanisms whereby payments are made—direct payments to farmers through government or private programs, tax abatements, or higher prices to consumers from taxes on polluting inputs or tradable pollution credits (Lipper et al. 2009)—the cost of payments for ecosystem services must ultimately be borne by society.

The 2009 Michigan Environmental Survey (Chen 2010; Swinton et al. 2015b, Chapter 13 in this volume) provides insight on society's willingness to pay. The survey was returned by ~2400 households from every county in Michigan, stratified by population. The respondents were asked about their willingness to support a personal income tax increase to pay land managers to enroll in one of three stewardship programs that would, to varying degrees, reduce lake eutrophication and/or atmospheric greenhouse gas concentrations.

The responses to the survey showed substantial public willingness to finance policies that would pay farmers to adopt practices to abate lake eutrophication. In aggregate, the respondents were willing to pay $175 per household for a combined reduction of 170 eutrophic lakes and 0.5% lower greenhouse gas emissions. However, most of the households were unwilling to pay farmers for reduced greenhouse gas emissions alone. Over 60% of the households in 2009 were unconcerned about climate change. Of the 40% that were concerned, however, the mean household was willing to pay $141 per year for a 1% reduction per year in greenhouse gas emission levels.

On the supply side, then, the Michigan corn and soybean farmers were clearly willing to change their cropping practices to generate additional ecosystem services if they were paid to do so. The farmers would expand both the complexity of their farming practices and the acreage under these practices if they were given the opportunity and would thereby generate a supply of land managed to deliver additional ecosystem services. On the demand side, the state residents appeared to be willing to pay for reduced numbers of eutrophic lakes, and a significant fraction of the residents appeared to be willing to pay for reduced greenhouse gas emissions. How can we link buyers and sellers?

Important to both groups was how ecosystem services are characterized and bundled. It is difficult to measure the value of individual ecosystem services from agriculture. Management decisions affect multiple services simultaneously; farming is a systems-level enterprise with system-level responses (Robertson et al. 2004), such that ecosystem services come in bundles and should probably be marketed as such. Credit stacking in carbon and other payment for environmental services markets (e.g., Fox et al. 2011) cannot be avoided because of the varied objectives of the many willing governmental and nongovernmental payers. Moreover, credit stacking should probably be encouraged in order to take full advantage of co-benefits and fully exploit available synergies. Converting demand for additional ecosystem services into the area of land required to generate the desired change is a logical next step.

Although approaches to payment for ecosystem services deserve further research, they are but one among many policy tools available to meet the demand for additional ecosystem services. Exploring and testing alternative tools—especially in light of new precision-farming technologies—is an appropriate response to the evidence here that, at reasonable prices, farmers are willing to supply and consumers are willing to pay for a meaningful set of ecosystem services.

Where from Here?

Additional knowledge about row-crop ecosystems will reveal additional opportunities for providing services and delivering them more efficiently. One example might be to manage noncrop areas in agricultural landscapes to support natural enemies of crop pests (Landis et al. 2000). Another might be to more precisely estimate or meet crop nitrogen needs in order to avoid excess nitrogen fertilizer additions (Robertson and Vitousek 2009). And a third might be to manage the soil microbial community to restore the capacity to remove methane from the atmosphere (Levine et al. 2011). Understanding and evaluating the delivery of services in a systems context will allow the full suite of trade-offs and synergies to be considered.

Long-term agricultural research reveals ecological trends that build slowly and sometimes subtly. It also allows researchers to capture the expression of episodic events, such as weather extremes, pest outbreaks, and species introductions, and it permits the evaluation of biological change against slow but steady changes in climate; technology; markets; and public attitudes toward food, fuel, and the environment. In the coming decades, human population and income growth will drive agriculture to ever-higher intensities. Now is the time to guide this intensification in a way that enhances the delivery of ecosystem services that are not currently marketed. Delaying action will result in an environment further degraded and an agriculture further divorced from its biological roots, more vulnerable to climate extremes and pest outbreaks, and increasingly dependent on external energy and synthetic chemical inputs.

Systems-level research reveals how disparate parts of agricultural ecosystems interact in subtle, often surprising, and sometimes crucial ways. Connections among microbial community structure, the formation of soil organic matter, soil water-holding capacity, plant drought tolerance, and primary productivity and herbivory are difficult to detect in the absence of research in which multiple parts of the same system are studied simultaneously. And research that is too local and that fails to consider relationships among different cropped and noncropped habitats within the larger landscape will likewise fail to make apparent crucial opportunities for designing future cropping systems that are productive, resilient, and able to deliver a rich suite of ecosystem services. Systems-level research sufficiently reductionist to identify key organism-level interactions and processes will be increasingly valuable for delivering worthwhile opportunities.

Some of these opportunities will be more generalizable than others. They will all require adaptations to local environmental and economic conditions, and both policy and research must include the need for flexible solutions, especially as new genomic and other technologies enter the marketplace. Trade-offs and synergies must be recognized and evaluated (e.g., Syswerda and Robertson 2014) in order to design optimal systems for specific outcomes. Ultimately, modeling will be needed to help design specific solutions for specific locales.

Research from the KBS LTER site reveals a number of worthwhile opportunities for delivering services today. Almost all of those opportunities are interdependent. Some of these interdependencies are synergistic, suggesting multiple paths for farmer adoption; others are negative, suggesting the need for targeted incentives for particular services important to society. Identifying such interdependencies and how they respond to different management practices and environmental change is a need in cropping systems everywhere and has never been more urgent.

References

Alexander, R. B., R. A. Smith, G. E. Schwarz, E. W. Boyer, J. V. Nolan, and J. W. Brakebill. 2008. Differences in phosphorus and nitrogen delivery in the Gulf of Mexico from the Mississippi River Basin. Environmental Science and Technology 42:822–830.

Barker, T., I. Bashmakov, L. Bernstein, J. E. Bogner, P. R. Bosch, R. Dave, O. R. Davidson, B. S. Fisher, S. Gupta, K. Halsnæs, G. J. Geij, S. Kahn Riveiro, S. Kobayashi, M. D. Levine, D. L. Martino, O. Masera, B. Metz, L. A. Meyer, G. J. Nabuurs, A. Najam, N. Nakicenovic, H.-H. Rogner, J. Roy, J. Sathaye, R. Schock, P. Shukla, R. E. H. Sims, P. Smith, D. A. Tirpak, D. Urge-Vorsatz, and D. Zhou. 2007. Technical Summary. Pages 25–93 in B. Metz, O. R. Davidson, P. R. Bosch, R. Dave, and L. A. Meyer, editors. Climate change 2007: mitigation. Contribution of Working Group

III to the Fourth Assessment Report of the Intergovernmental Panel on Climate Change. Cambridge University Press, New York, New York, USA.

Beaulieu, J. J., J. L. Tank, S. K. Hamilton, W. M. Wollheim, R. O. Hall, Jr., P. J. Mulholland, B. J. Peterson, L. R. Ashkenas, L. W. Cooper, C. N. Dahm, W. K. Dodds, N. B. Grimm, S. L. Johnson, W. H. McDowell, G. C. Poole, H. M. Valett, C. P. Arango, M. J. Bernot, A. J. Burgin, C. L. Crenshaw, A. M. Helton, L. T. Johnson, J. M. O'Brien, J. D. Potter, R. W. Sheibley, D. J. Sobota, and S. M. Thomas. 2011. Nitrous oxide emission from denitrification in stream and river networks. Proceedings of the National Academy of Sciences USA 108:214–219.

Bennett, A. J., G. D. Bending, D. Chandler, S. Hilton, and P. Mills. 2012. Meeting the demand for crop production: the challenge of yield decline in crops grown in short rotations. Biological Reviews 97:52–71.

Burgin, A. J., and S. K. Hamilton. 2007. Have we overemphasized the role of dentirification in aquatic ecosystems? A review of nitrate removal pathways. Frontiers in Ecology and the Environment 5:89–96.

Caldeira, K., M. G. Morgan, D. Baldocchi, P. G. Brewer, C.-T. A. Chen, G.-J. Nabuurs, N. Nakicenovic, and G. P. Robertson. 2004. A portfolio of carbon management options. Pages 103–130 in C. B. Field and M. R. Raupach, editors. The global carbon cycle. Island Press, Washington, DC, USA.

Chen, H. 2010. Ecosystem services from low input cropping systems and the public's willingness to pay for them. Thesis, Michigan State University, East Lansing, Michigan, USA.

Colunga-Garcia, M. 1996. Interactions between landscape structure and ladybird beetles (Coleoptera: Coccinellidae) in field crop agroecosystems. Dissertation, Michigan State University, East Lansing, Michigan, USA.

Costamagna, A. C., and D. A. Landis. 2006. Predators exert top-down control of soybean aphid across a gradient of agricultural management systems. Ecological Applications 16:1619–1628.

Culman, S. W., S. S. Snapp, M. Ollenburger, B. Basso, and L. R. DeHaan. 2013. Soil and water quality rapidly responds to perennial grain Kernza wheatgrass. Agronomy Journal 105:735–744.

Del Grosso, S. J., W. J. Parton, A. R. Mosier, D. S. Ojima, C. S. Potter, W. Borken, R. Brumme, K. Butterbach-Bahl, P. M. Crill, K. E. Dobbie, and K. A. Smith. 2000. General CH_4 oxidation model and comparisons of CH_4 oxidation in natural and managed systems. Global Biogeochemical Cycles 14:999–1019.

Diaz, R. J., and R. Rosenberg. 2008. Spreading dead zones and consequences for marine ecosystems. Science 321:926–929.

Drinkwater, L. E., P. Wagoner, and M. Sarrantonio. 1998. Legume-based cropping systems have reduced carbon and nitrogen losses. Nature 396:262–265.

ERS (Economic Research Service). 2013. ARMS farm financial and crop production practices. <http://www.ers.usda.gov/data-products/arms-farm-financial-and-crop-production-practices.aspx#.UVCOT1d3dI1. U.S. Department of Agrilculture> Accessed March 15, 2013.

Fargione, J., J. Hill, D. Tilman, S. Polasky, and P. Hawthorne. 2008. Land clearing and the biofuel carbon debt. Science 319:1235–1237.

Farrell, A. E., R. J. Plevin, B. T. Turner, A. D. Jones, M. O'Hare, and D. M. Kammen. 2006. Ethanol can contribute to energy and environmental goals. Science 311:506–508.

Fox, J., R. C. Gardner, and T. Maki. 2011. Stacking opportunities and risks in environmental credit markets. Environmental Law Reporter 41:10121–10125.

Gardiner, M. M., D. A. Landis, C. Gratton, C. D. DiFonzo, M. O'Neal, J. M. Chacon, M. T. Wayo, N. P. Schmidt, E. E. Mueller, and G. E. Heimpel. 2009. Landscape diversity enhances the biological control of an introduced crop pest in the north-central USA. Ecological Applications 19:143–154.

Gelfand, I., and G. P. Robertson. 2015. Mitigation of greenhouse gas emissions in agricultural ecosystems. Pages 310–339 in S. K. Hamilton, J. E. Doll, and G. P. Robertson, editors. The ecology of agricultural Landscapes: long-term research on the path to sustainability. Oxford University Press, New York, New York, USA.

Gelfand, I., R. Sahajpal, X. Zhang, C. R. Izaurralde, K. L. Gross, and G. P. Robertson. 2013. Sustainable bioenergy production from marginal lands in the US Midwest. Nature 493:514–517.

Gelfand, I., T. Zenone, P. Jasrotia, J. Chen, S. K. Hamilton, and G. P. Robertson. 2011. Carbon debt of Conservation Reserve Program (CRP) grasslands converted to bioenergy production. Proceedings of the National Academy of Sciences USA 108:13864–13869.

Grandy, A. S., and G. P. Robertson. 2007. Land-use intensity effects on soil organic carbon accumulation rates and mechanisms. Ecosystems 10:58–73.

Hamilton, S. K. 2012. Biogeochemical time lags that may delay responses of streams to ecological restoration. Freshwater Biology 57 (Supp. 1):43–57.

Haynes, R. J., and M. H. Beare. 1997. Influence of six crop species on aggregate stability and some labile organic matter fractions. Soil Biology & Biochemistry 29:1647–1653.

Hedin, L. O., J. C. von Fischer, N. E. Ostrom, B. P. Kennedy, M. G. Brown, and G. P. Robertson. 1998. Thermodynamic constraints on nitrogen transformation and other biogeochemical processes at soil-stream interfaces. Ecology 79:684–703.

Jolejole, M. C. B. 2009. Trade-offs, incentives, and the supply of ecosystem services from cropland. Thesis, Michigan State University, East Lansing, Michigan, USA.

Karlen, D. L., G. E. Varvel, D. G. Bullock, and R. M. Cruse. 1994. Crop rotations for the 21st century. Advances in Agronomy 53:1–45.

Landis, D. A., and S. H. Gage. 2015. Arthropod diversity and pest suppression in agricultural landscapes. Pages 188–212 in S. K. Hamilton, J. E. Doll, and G. P. Robertson, editors. The ecology of agricultural Landscapes: long-term research on the path to sustainability. Oxford University Press, New York, New York, USA.

Landis, D. A., M. M. Gardiner, W. van der Werf, and S. M. Swinton. 2008. Increasing corn for biofuel production reduces biocontrol services in agricultural landscapes. Proceedings of the National Academy of Sciences USA 105:20552–20557.

Landis, D. A., S. D. Wratten, and G. M. Gurr. 2000. Habitat management to conserve natural enemies of arthropod pests in agriculture. Annual Review of Entomology 45:175–201.

Levine, U., T. K. Teal, G. P. Robertson, and T. M. Schmidt. 2011. Agriculture's impact on microbial diversity and associated fluxes of carbon dioxide and methane. The ISME Journal 5:1683–1691.

Liebman, M., M. J. Helmers, L. A. Schulte, and C. A. Chase. 2013. Using biodiversity to link agricultural productivity with environmental quality: results from three field experiments in Iowa. Renewable Agriculture and Food Systems 28:115–128.

Lipper, L., T. Sakuyama, R. Stringer, and D. Zilberman, editors. 2009. Payment for environmental services in agricultural landscapes: economic policies and poverty reduction in developing countries. Springer, Rome, Italy.

Lowrance, R. 1998. Riparian forest ecosystems as filters for nonpoint-source pollution. Pages 113–141 in M. L. Pace and P. M. Groffman, editors. Successes, limitations, and frontiers in ecosystem science. Springer-Verlag, New York, New York, USA.

MA (Millennium Ecosystem Assessment). 2005. Our human planet: summary for decision-makers. Island Press, Washington, DC, USA.

Ma, S., S. M. Swinton, F. Lupi, and C. B. Jolejole-Foreman. 2012. Farmers' willingness to participate in payment-for-environmental-services programmes. Journal of Agricultural Economics 63:604–626.

Maredia, K. M., S. H. Gage, D. A. Landis, and J. M. Scriber. 1992a. Habitat use patterns by the seven-spotted lady beetle (Coleoptera: Coccinellidae) in a diverse agricultural landscape. Biological Control 2:159–165.

Maredia, K. M., S. H. Gage, D. L. Landis, and T. M. Wirth. 1992b. Ecological observations on predatory Coccinellidae (Coleoptera) in southwestern Michigan. Great Lakes Entomologist 25:265–270.

McSwiney, C. P., S. S. Snapp, and L. E. Gentry. 2010. Use of N immobilization to tighten the N cycle in conventional agroecosystems. Ecological Applications 20:648–662.

Meehan, T. D., B. P. Werling, D. A. Landis, and C. Gratton. 2011. Agricultural landscape simplification and insecticide use in the Midwestern United States. Proceedings of the National Academy of Sciences USA 108:11500–11505.

NASS (National Agricultural Statistics Service). 2013a. Crop production 2012 summary. U.S. Department of Agriculture (USDA), Washington, DC. <http://usda.mannlib.cornell.edu/ MannUsda/viewDocumentInfo.do?documentID=1047> Accessed February 6, 2013.

NASS (National Agricultural Statistics Service). 2013b. Michigan agricultural statistics 2012–2013. U.S. Department of Agriculture (USDA), Washington, DC. <http://www. nass.usda.gov/Statistics_by_State/Michigan/index.asp> Accessed March 14, 2013.

NRC (National Research Council). 1995. Wetlands: characteristics and boundaries. National Academies Press, Washington, DC, USA.

Osteen, C., J. Gottlieb, and U. Vasavada. 2012. Agricultural resources and environmental indicators, 2012. EIB-98, U.S. Department of Agriculture, Economic Research Service.

Power, A. G. 2010. Ecosystem services and agriculture: tradeoffs and synergies. Philosophical Transactions of the Royal Society B: Biological Sciences 365:2959–2971.

Rasmussen, P. E., K. W. T. Goulding, J. R. Brown, P. R. Grace, H. H. Janzen, and M. Korschens. 1998. Long-term agroecosystem experiments: assessing agricultural sustainability and global change. Science 282:893–896.

Robertson, G. P. 1997. Nitrogen use efficiency in row crop agriculture: crop nitrogen use and soil nitrogen loss. Pages 347–365 in L. Jackson, editor. Ecology in agriculture. Academic Press, New York, New York, USA.

Robertson, G. P., J. C. Broome, E. A. Chornesky, J. R. Frankenberger, P. Johnson, M. Lipson, J. A. Miranowski, E. D. Owens, D. Pimentel, and L. A. Thrupp. 2004. Rethinking the vision for environmental research in US agriculture. BioScience 54:61–65.

Robertson, G. P., V. H. Dale, O. C. Doering, S. P. Hamburg, J. M. Melillo, M. M. Wander, W. J. Parton, P. R. Adler, J. N. Barney, R. M. Cruse, C. S. Duke, P. M. Fearnside, R. F. Follett, H. K. Gibbs, J. Goldemberg, D. J. Mladenoff, D. Ojima, M. W. Palmer, A. Sharpley, L. Wallace, K. C. Weathers, J. A. Wiens, and W. W. Wilhelm. 2008. Sustainable biofuels redux. Science 322:49–50.

Robertson, G. P., and S. K. Hamilton. 2015. Long-term ecological research at the Kellogg Biological Station LTER Site: Conceptual and experimental framework. Pages 1–32 in S. K. Hamilton, J. E. Doll, and G. P. Robertson, editors. The ecology of agricultural Landscapes: long-term research on the path to sustainability. Oxford University Press, New York, New York, USA.

Robertson, G. P., E. A. Paul, and R. R. Harwood. 2000. Greenhouse gases in intensive agriculture: contributions of individual gases to the radiative forcing of the atmosphere. Science 289:1922–1925.

Robertson, G. P., and P. M. Vitousek. 2009. Nitrogen in agriculture: balancing the cost of an essential resource. Annual Review of Environment and Resources 34:97–125.

Ruan, L., and G. P. Robertson. 2013. Initial nitrous oxide, carbon dioxide and methane costs of converting Conservation Reserve Program grassland to row crops under no-till vs. conventional tillage. Global Change Biology 19:2478–2489.

Searchinger, T., R. Heimlich, R. A. Houghton, F. Dong, A. Elobeid, J. Fabiosa, S. Tokgoz, D. Hayes, and T.-H. Yu. 2008. Use of U.S. croplands for biofuels increases greenhouse gases through emissions from land-use change. Science 319:1238–1240.

Six, J., S. M. Ogle, F. J. Breidt, R. T. Conant, A. R. Mosier, and K. Paustian. 2004. The potential to mitigate global warming with no-tillage management is only realized when practised in the long term. Global Change Biology 10:155–160.

Smith, P., D. Martino, Z. Cai, D. Gwary, H. Janzen, P. Kumar, B. McCarl, S. Ogle, F. O'Mara, C. Rice, B. Scholes, and O. Sirotenko. 2007. Agriculture. Pages 498–540 in B. Metz, O. R. Davidson, P. R. Bosch, R. Dave, and L. A. Meyer, editors. Climate change 2007: mitigation. Contribution of Working Group III to the Fourth Assessment Report of the Intergovernmental Panel on Climate Change. Cambridge University Press, Cambridge, United Kingdom and New York, New York, USA.

Smith, R. G., K. L. Gross, and G. P. Robertson. 2008. Effects of crop diversity on agroecosystem function: crop yield response Ecosystems 11:355–366.

Strudley, M. W., T. R. Green, and J. C. Ascough II. 2008. Tillage effects on soil hydraulic properties in space and time: State of the science. Soil & Tillage Research 99:4–48.

Swinton, S. M., C. B. Jolejole-Foreman, F. Lupi, S. Ma, W. Zhang, and H. Chen. 2015a. Economic value of ecosystem services from agriculture. Pages 54–76 in S. K. Hamilton, J. E. Doll, and G. P. Robertson, editors. The ecology of agricultural Landscapes: long-term research on the path to sustainability. Oxford University Press, New York, New York, USA.

Swinton, S. M., F. Lupi, G. P. Robertson, and S. K. Hamilton. 2007. Ecosystem services and agriculture: cultivating agricultural ecosystems for diverse benefits. Ecological Economics 64:245–252.

Swinton, S. M., N. Rector, G. P. Robertson, C. B. Jolejole-Foreman, and F. Lupi. 2015b. Farmer decisions about adopting environmentally beneficial practices. Pages 340–359 in S. K. Hamilton, J. E. Doll, and G. P. Robertson, editors. The ecology of agricultural Landscapes: long-term research on the path to sustainability. Oxford University Press, New York, New York, USA.

Syswerda, S. P., B. Basso, S. K. Hamilton, J. B. Tausig, and G. P. Robertson. 2012. Long-term nitrate loss along an agricultural intensity gradient in the Upper Midwest USA. Agriculture, Ecosystems & Environment 149:10–19.

Syswerda, S. P., A. T. Corbin, D. L. Mokma, A. N. Kravchenko, and G. P. Robertson. 2011. Agricultural management and soil carbon storage in surface vs. deep layers. Soil Science Society of America Journal 75:92–101.

Syswerda, S. P., and G. P. Robertson. 2014. Trade-offs and synergies in ecosystem services along a management intensity gradient in upper Midwest US cropping systems. Agriculture, Ecosystems & Environment 189:28–35.

van Kessel, C., R. Venterea, J. Six, M. A. Adviento-Borbe, B. Linquist, and K. J. van Groenigen. 2013. Climate, duration, and N placement determine N_2O emissions in reduced tillage systems: a meta-analysis. Global Change Biology 19:33–44.

West, T. O., and W. M. Post. 2002. Soil organic carbon sequestration rates by tillage and crop rotation: a global data analysis. Soil Science Society of America Journal 66:1930–1946.

Whitmire, S. L., and S. K. Hamilton. 2005. Rapid removal of nitrate and sulfate by freshwater wetland sediments. Journal of Environmental Quality 34:2062–2071.

Wright, C. K., and M. C. Wimberly. 2013. Recent land use change in the Western Corn Belt threatens grasslands and wetlands. Proceedings of the National Academy of Sciences USA 110:4134–4139.

3

Economic Value of Ecosystem Services from Agriculture

Scott M. Swinton, M. Christina Jolejole-Foreman, Frank Lupi, Shan Ma, Wei Zhang, and Huilan Chen

If ecosystem services describe the benefits that people get from nature, then those services must have value. What are they worth? Can we use values to choose desirable farming systems?

The idea of "value" in the sense of worth can be understood in two very different ways (Heal 2000). *Intrinsic value* refers to inherent worth. *Economic value* refers to relative scarcity. The diamond–water paradox elucidates the difference between the two (Heal 2000). Clean water, which is essential for human life, has great intrinsic value, yet its price is often very low. Diamonds have negligible intrinsic value yet they fetch very high prices. Prices express economic values based on supply and demand. *The amount of a good or service that producers will supply depends on the cost of producing it and the price offered for its purchase. The amount that consumers demand depends on how well they like it and its price of sale.* Economic methods for estimating the values of nonmarketed ecosystem services seek to capture these underlying market relationships.

Although the food, fiber, and bioenergy products from agroecosystems tend to be the only agricultural ecosystem services whose economic values are directly measured by market prices, research from the Kellogg Biological Station Long-Term Ecological Research (KBS LTER) project is beginning to provide estimates of the economic value of the nonmarketed ecosystem services from agriculture. In this chapter, we first introduce the principles for economic valuation of nonmarketed services, then offer a typology of valuation methods, and then review four KBS LTER–related studies we have conducted that estimate the value of nonmarketed ecosystem services.

Principles for Economic Valuation

A challenge to making sound measurements of nonmarket economic values is to capture the kinds of relationships that exist in markets. Markets are settings where people make choices about buying and selling. Market prices have three key traits: first, they are determined "at the margin." Put differently, prices are linked to quantities, so what a consumer is willing to pay depends on how much that person has already consumed. The price that a consumer and producer agree on is based on how badly the consumer wishes to buy a little more and what it would cost the producer to make that little more (above what each already has bought and produced, respectively). Second, there are limits to what choices are feasible. Consumer purchases are limited by budgets, and producer sales are limited by the productive resources and technology at hand. Third, both producers and consumers have substitutes and complements available to them. They tend to choose the most feasible alternative (not necessarily an extrapolation of current practice). So a farmer whose melon vines bear few fruit due to poor pollination may opt to rent honeybee hives rather than invest in habitat restoration for native pollinators. Some celebrated attempts at placing economic values on ecosystem services have extrapolated values to levels that violate these principles (Costanza et al. 1997), resulting in estimates that have been criticized for not being economically credible (Pearce 1998, Bockstael et al. 2000).

Economic valuation of ecosystem services uses methods that attempt to capture the effects of relevant markets. Those markets may be real or imagined. The relevant market for an ecosystem service varies with the scale over which people experience that ecosystem service. The nutrient cycling service of a soil microbial community may be fully captured at the farm field scale, whereas the climate regulation service rendered by the same microbial community (e.g., uptake of atmospheric methane) is realized only at the scale of global climate. For ecosystem services from agriculture, this scale effect means that farmers may care about certain services that directly benefit the farm, while viewing others as external to their management decisions (see focus group results in Swinton et al. 2015, Chapter 13 in this volume).

Depending on how consumers and producers experience an ecosystem service, there are many different methods to estimate its value (Freeman 2003, Shiferaw et al. 2005). The methods used for agricultural ecosystem services focus on values that people obtain from *use* of the services. Nonuse values—from existence of an ecosystem, the opportunity to pass it on intact to the next generation, or the possibility of discovering unknown benefits from it—are assumed to matter little in agricultural ecosystems. Research on economic valuation of agricultural ecosystem services in KBS LTER–related cropping systems can be divided into two strands. Revealed preference methods capture values revealed by existing markets. Production input, profitability trade-off, and stated preference methods estimate the value of changes to the status quo, such as changing current farmer cropping systems to include ecologically recommended practices.

Revealed Preference Estimates to Value Landscape-Level Ecosystem Services

The economic value of landscape-level ecosystem services such as wildlife habitat and recreation can be inferred from existing markets for land and recreational services. Revealed preference methods use information on expenditures and market prices to deduce the implied willingness to pay for environmental benefits. For example, the value of recreational fishing and hunting services can be inferred from what people spend to travel to fishing and hunting sites and from the characteristics of the sites they visit. Another revealed preference approach, hedonic valuation, uses price data and product characteristics to infer the component value of those characteristics by statistical regression methods. Just as real estate analysts use hedonic valuation to estimate the value that a second bathroom adds to a house, environmental economists can use the same method to estimate the value that adjacent forest adds to a farm field. Both travel cost and hedonic land price analyses have been used by KBS LTER economists (e.g., Knoche and Lupi 2007, Ma and Swinton 2011).

Among its many roles, agricultural land provides valuable habitat for wildlife. Hunting is one major ecosystem service experienced by 12.5 million adult Americans in 2006 (U.S. Department of the Interior and U.S. Department of Commerce 2006). Of those, ~750,000 hunted in Michigan. The value of Michigan's agricultural land as wildlife habitat is captured by a travel cost analysis of deer hunting in the state. Knoche and Lupi (2007) used data on the cost of hunting trips to calculate how much hunters are willing to pay for various attributes of the hunting experience and found that hunters were effectively paying $39 per acre for access to 10% of the private agricultural land in the southern Lower Peninsula of Michigan. This represents 7% of the per-acre market value of farm products in the area in 2004, a significant value (part of which is already captured by farmers who offer hunting leases for their lands). By providing a varied landscape with abundant food, agriculture enhances the habitat for deer. Knoche and Lupi (2007) estimated that in a nonagricultural landscape that supported only half as many deer, the annual value to hunters would decline by $15 million.

Hedonic valuation of ecosystem services through land prices can capture the values of a range of services, as compared to the single value from the travel cost of hunting trips. The price of a land parcel represents a bundle of attributes embodied in that parcel. Hedonic analysis applies statistical regression of land prices to different attributes of the parcel to infer the values of specific property traits (Palmquist 1989, Palmquist and Danielson 1989). In an attempt to measure the value of land-based ecosystem services in the KBS vicinity, Ma and Swinton (2011) estimated a hedonic model of land prices in four counties surrounding KBS—Allegan, Barry, Kalamazoo, and Eaton (Fig. 3.1). This work included variables describing traits of both the natural and built environments and subdivided these into traits that affect both the production value of the land and its consumption value (e.g., residential and recreational attributes). To capture the effect of surrounding ecosystems, the study analyzed spatial data on the proportions of land cover in a 1.5-km radius around property parcels.

The study inferred ecosystem service values from the influence that particular landscape features had on agricultural land prices in southwestern Michigan (Ma and Swinton, 2011). Three lessons stood out. First, recreational and production-supporting services tend to make the largest contributions to land values. On-site lakes and woodlands as well as nearby rivers, wetlands, and conservation lands enhance values, while on-site streams (which can flood) detract from value. Second, certain ecosystem services are likely to be only partially capitalized into land prices, either because landowners are unaware of their value or because the benefits are dispersed to areas external to the parcel. Examples include beneficial soil microbial communities and habitat for pollinators and natural enemies of agricultural pests. Third, it appears that land prices do not reflect benefits that are largely realized outside the parcel, such as greenhouse gas mitigation or habitat for large wildlife.

The contribution of the surrounding landscape to agricultural land prices is particularly meaningful in light of expanding research into the provision of ecosystem services at the landscape scale that involves multiple property owners. To date, much of the research on how landscape structure affects ecosystem services to agriculture has focused on arthropod-mediated ecosystem services, such as natural biocontrol of insect pests (Costamagna and Landis 2006, Gardiner et al. 2009, Landis and Gage 2015, Chapter 8 in this volume) and pollination (Kremen and Ricketts 2000, Kremen et al. 2002, Steffan-Dewenter et al. 2002, Ricketts et al. 2004, Ricketts et al. 2008). However, recreational and production-related ecosystem services at the landscape level are also significant, based on Ma and Swinton's (2011) research using rural property prices. For example, when the proportion of wetlands rose 1% in the 1.5-km radius of surrounding land, land parcel prices rose by 3%, suggesting that land markets place value on certain ecosystem services of wetlands (perhaps

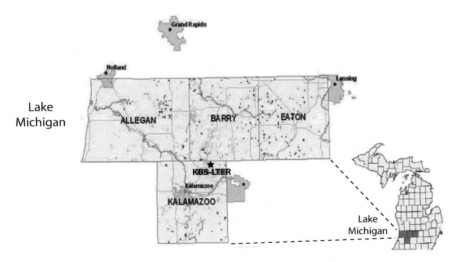

Figure 3.1. Property parcel locations (small black polygons) used in the hedonic study of ecosystem service values embodied in southwestern Michigan land prices. Counties and major cities are also shown. Redrawn from Ma (2010).

flood buffering capability, wildlife habitat, open space). A similar finding was that a 1% rise in surrounding conservation lands (e.g., natural preserves, parks, public forests) raised parcel prices by 2%. Value for purposes of recreation and crop irrigation likely explains the finding that parcel prices rose by 6% for each kilometer closer to a river (Ma and Swinton 2011). That some of these sources of landscape-level value result from land management choices—set-aside of conservation land and preservation of wetlands—points to the potential for landowners to improve land values by coordinating management across parcels within a given landscape, a direction deserving future research.

Valuing Ecosystem Services from Improved Cropping Practices

While land prices embody the local value of certain ecosystem services that emerge from landscape composition, other methods are needed to measure environmental values of specific farm management practices.

Developing estimates of economic values due to specific management practices is a two-stage process. Applying the framework of Collins et al. (2011), the first stage is to measure the changes in ecosystem service flows resulting from a change in crop management practices. The change must be measured from some baseline, such as the conventional corn-soybean-wheat management system of the KBS LTER Main Cropping Systems Experiment (MCSE) (Table 3.1) (Robertson and Hamilton 2014, Chapter 1, this volume). The second stage converts those changed service flows into economic values. As economic values, these are based on real or hypothetical markets that measure how much the people who gain would be willing to pay to obtain the changed service flows, or how much the people who lose would accept in order to be equally well off as they were before the change (Polasky and Segerson 2009). So, for example, if farmers reduce fertilizer use that prevents a lake from becoming eutrophic, economic value would be measured on the demand side by how much the lake users are willing to pay for maintaining its uses and on the supply side by how much farmers would be willing to accept as compensation for any income lost due to reduced fertilizer use. Where markets for ecosystem services or their near substitutes exist, prices may reflect an economic equilibrium where the value to those who gained from a specific change in ecosystem service is in balance with the compensation to those who lost by making the necessary management changes. Where markets do not exist, aspects of markets can be simulated to infer economic values.

The MCSE results point to several ecosystem services that alternative management of cropping systems can provide: nutrient cycling (biological in lieu of synthetic chemical fertilizer supplements), crop pest regulation (via natural biocontrol), climate regulation (via reduced greenhouse gas emissions), and water-quality regulation (via reduced nutrient leaching to groundwater and runoff to surface waters). Details can be found in other chapters in this volume (Paul et al. 2015, Chapter 5 in this volume; Landis and Gage 2015, Chapter 8 in this volume; Hamilton 2015, Chapter 11 in this volume; and Gelfand and Robertson 2015, Chapter 12 in this volume).

Table 3.1. Description of the KBS LTER Main Cropping System Experiment (MCSE) examined in economic valuation.[a]

Cropping System	Dominant Growth Form	Management
Annual Cropping Systems		
Conventional (T1)	Herbaceous annual	Prevailing norm for tilled corn–soybean–winter wheat (c–s–w) rotation; standard chemical inputs, chisel-plowed, no cover crops, no manure or compost
No-till (T2)	Herbaceous annual	Prevailing norm for no-till c–s–w rotation; standard chemical inputs, permanent no-till, no cover crops, no manure or compost
Reduced Input (T3)	Herbaceous annual	Biologically based c–s–w rotation managed to reduce synthetic chemical inputs; chisel-plowed, winter cover crop of red clover or annual rye, no manure or compost
Biologically Based (T4)	Herbaceous annual	Biologically based c–s–w rotation managed without synthetic chemical inputs; chisel-plowed, mechanical weed control, winter cover crop of red clover or annual rye, no manure or compost; certified organic
Perennial Cropping Systems		
Alfalfa (T6)	Herbaceous perennial	5- to 6-year rotation with winter wheat as a 1-year break crop
Poplar (T5)	Woody perennial	Hybrid poplar trees on a ca. 10-year harvest cycle, either replanted or coppiced after harvest

[a]Codes that have been used throughout the project's history are given in parentheses. Systems T1–T7 are replicated within the LTER main site. For further details, see Robertson and Hamilton (2015, Chapter 1 in this volume).

Inferring Value of Pest Regulation by Natural Enemies Using the Production Input Method

When an ecosystem service can substitute for an existing marketed input or when the service contributes to measurable marketed output, the economic value of changes in the level of the service can readily be inferred using information from the related input and/or crop (output) markets (Freeman 2003, Drechsel et al. 2005). An example of a widespread application of this method is the calculation of the fertilizer replacement value to measure the value of biological nutrient cycling in cereal–legume systems (Bundy et al. 1993).

Recognizing an opportunity to apply experimental results, KBS LTER researchers used the factor input method to estimate the value of a loss in natural pest biocontrol due to changed crop cover. Field research had revealed that more corn area in the landscape reduces natural biocontrol of the soybean aphid (Gardiner et al. 2009). When U.S. Midwest corn acreage jumped 19% in 2007 in response to soaring prices, Landis et al. (2008) realized that this change reduced habitat for natural enemies of the soybean aphid. They calculated the lost value of natural biocontrol services based on predicted soybean yield loss and associated increased insecticide costs. They estimated impacts both on farmers who follow integrated pest management (IPM) to guide insecticide sprays and on the roughly 1% of farmers who rely entirely on natural biocontrol. Depending on whether the soybean price was the 1996–2007 median or the higher post-2007 level, the reduced value of biocontrol services to soybeans due to the 19% increase in corn acreage was estimated to be $18–25 ha^{-1} for IPM farmers and $110–199 ha^{-1} for natural biocontrol farmers (Landis et al. 2008).

Another KBS LTER approach to measuring the economic value of the soybean aphid natural enemy complex used a bioeconomic model that predicted densities of the pest and its predators (as opposed to habitat features in the landscape). Zhang and Swinton (2009) developed a dynamic optimal control model for soybean aphid IPM that incorporates the economic value of natural enemy survival when making profit-maximizing decisions on insecticide applications to the aphid. They found that natural enemies are particularly valuable for suppressing low populations of soybean aphid, preventing them from multiplying to the point of causing significant crop damage. A single, typical natural enemy (comparable to the multicolored Asian lady beetle, *Harmonia axyridis*) per soybean plant is worth $32.60 ha^{-1} when 5 to 30 aphids are present per plant during its early flowering stage (Fig. 3.2). However, above 30 aphids, the value of a single natural enemy would fall to just $4.20 ha^{-1}, because at higher aphid populations, a single natural enemy could not control the infestation so insecticides would be needed to maximize profit (in spite of the collateral or nontarget damage to natural enemy populations) (Zhang and Swinton 2012). Based on evidence that soybean aphid density averaged 21 per plant in

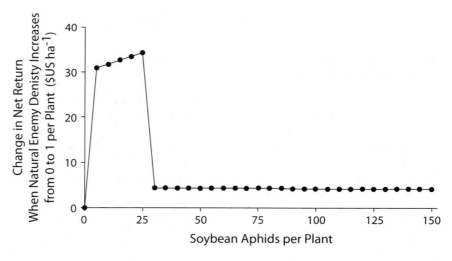

Figure 3.2. Value of one natural enemy per soybean plant (as compared to none) as a function of the abundance of soybean aphids. The aphids are assumed to be in reproductive stage R1, the natural enemy is assumed to have a daily predation rate of 35 aphids, and the initial soybean yield potential is assumed to be 2.69 Mg ha^{-1}. Redrawn from Zhang and Swinton (2012) with permission from Elsevier.

2005, the authors estimated that the value of going from no natural enemies to one per soybean plant is worth $32.60 ha^{-1}, amounting to some $84 million during 2005 for the U.S. states of Iowa, Illinois, Michigan, Minnesota, and Wisconsin. This estimate is based solely on natural enemy contributions to soybean profitability, so it ignores other sources of value to society, such as the health and environmental benefits of reduced insecticide use. Because the model uses the factor input valuation method, it is sensitive to assumptions about the aphid predation rate per natural enemy, as well as the price and pest-free yield of soybean.

Estimating Economic Supply and Demand for Ecosystem Services When Markets Are Missing

The economic valuation of climate-regulating and water-quality–regulating ecosystem services is more complicated because these services lack the direct market links of agricultural pest regulation services. Markets designed for climate- and water-quality–regulating services have been piloted since the late 1990s. However, localized water-quality nutrient trading has not been established successfully in the United States for a variety of reasons (Hoag and Hughes-Popp 1997), many of which also apply to carbon trading markets.

Developing economic values for these regulating services calls for understanding the cost to producers of supplying them and the benefits to consumers of enjoying them. Valuation of both costs and benefits can be measured by studying trade-offs (Polasky and Segerson 2009). On the supply side, how much would farmers need to be compensated to adopt practices that reduce net greenhouse gas emissions and/ or reduce waterborne nutrient losses from topsoil? On the demand side, how much would consumers be willing to pay to be equally well off with improved climate and water quality vs. degraded environmental quality?

Farmers earn their livelihoods from farming, so income matters a great deal. But farmers generally make management decisions based on the welfare of their households using more complex criteria than simply profit maximization. When eliciting from farmers how they weigh trade-offs between increased production costs and increased ecosystem services, these other objectives are automatically factored in. So-called stated preference methods of economic valuation that are described below can be used to capture this complex decision-making process.

But when analyzing ecological experimental data—such as that from the MCSE—making the simplifying assumption that farmers aim to maximize profit enables a trade-off analysis of private profitability compared to the public benefits from ecosystem services. A common feature of many ecosystem services is that they generate benefits beyond the boundary of the farm. These benefits may not be factored into the farmer's decisions, especially if generating them entails added costs. Given that agroecosystems generate a multiplicity of products and services, trade-off analysis enables comparing these outputs, typically using profitability as a numéraire for comparison (Wossink and Swinton 2007).

Trade-off Analysis of Profitability vs. Ecosystem Service Provision

Trade-off analysis can illustrate the relationship between profitability and ecosystem services such as greenhouse gas fluxes or nitrate leaching for the MCSE systems (Antle and Capalbo 2002). When graphed in two or three dimensions, the method provides a visual illustration of trade-off vs. win-win outcomes for farmers and the public. It also permits an indirect way to calculate the cost to the farmer of increasing output of nonmarketed ecosystem services. Farmers face two kinds of monetary costs: direct costs and opportunity costs. Direct costs are subtracted from revenues to calculate profitability. Opportunity costs are measured indirectly, as the difference in earnings between the most profitable system and an alternative. Trade-off analysis can measure the opportunity cost of reduced profitability in exchange for increased supply of ecosystem services.

Trade-off analysis can also be used to evaluate the efficiency of providing targeted outcomes. By comparing profitability and ecosystem service outcomes for the full set of MCSE systems, it identifies some systems that do not excel in any outcome. Such systems are termed "inefficient" because other ones (either alone or in combination) could provide the same or higher levels of all desired outcomes.

A caveat for trade-off analysis is that it only captures the supply side of economic value, focusing on the marginal cost to the farmer of providing more of an ecosystem service. In our research on the MCSE, we use budgets with static prices, so the analysis implicitly assumes that any shift to an alternative cropping system would be sufficiently limited in scale that it would not generate market price feedbacks.

The KBS LTER trade-off analyses begin with partial enterprise budgets for the MCSE systems. Annualized partial enterprise budgets were calculated by Jolejole et al. (2009) using standard enterprise budgeting techniques (Boehlje and Eidman 1984), with a focus on only those costs that vary across systems, as per the CIMMYT (1988) methodology for analysis of agronomic data. For systems involving perennial crops, net present values were calculated over the crop lifetime and converted to an annualized value using a standard financial annuity formula (Weston and Copeland 1986). The resulting profitability measure is the gross margin, which represents revenue above costs that vary. Gross margins capture the differences among MCSE cropping systems, although they do not account for other kinds of costs that are unchanging across MCSE systems but tend to vary substantially from farm to farm (e.g., land rental, compensation of family labor). Mean values for global warming impact (GWI) were compiled for all MCSE systems by Robertson et al. (2000) and Syswerda et al. (2011) and for nitrate leaching by Syswerda et al. (2012).

Global warming impact results show that among the six MCSE cropping systems evaluated during 1993–2007, Poplar had the lowest overall impact (–105 g CO_2e m^{-2} yr^{-1} or –1.05 Mg ha^{-1} yr^{-1}, where the negative sign connotes net CO_2 uptake from the atmosphere), but it was also one of the least profitable cropping systems ($73 acre^{-1} or $179 ha^{-1}), though more profitable than Alfalfa. In profitability, the No-till system dominated at standard prices ($139 acre^{-1} or $345 ha^{-1}), while the Biologically Based certified organic system dominated at organic prices ($185 acre^{-1}or $458 ha^{-1}) (Fig. 3.3).

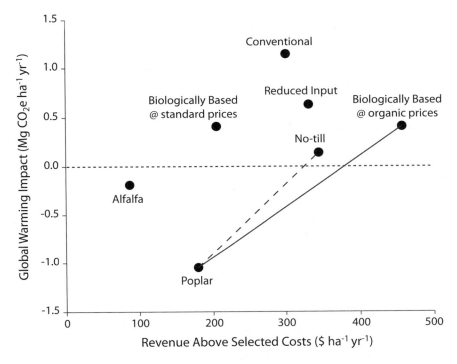

Figure 3.3. Trade-offs between global warming impact (in CO_2 equivalents, CO_2e) and revenue above selected costs for different Main Cropping System Experiment systems, 1993–2007 (except for Alfalfa, 1989–2004 and Poplar, 1989–1998). See text for explanation of lines. Adapted and updated from Jolejole (2009) and Robertson et al (2000).

Two important findings emerge. First, four of the MCSE systems are relatively inefficient (Conventional, Reduced Input, Poplar, and Biologically Based at standard nonorganic prices) in that other systems alone or in combination could provide better outcomes for both profitability and GWI. As illustrated in Table 3.2, at nonorganic prices, the No-till system offered greater profitability and lower GWI than the other three annual cropping systems (Conventional, Reduced Input, and Biologically Based). So switching from any of those systems to No-till would improve one or both outcomes. Likewise, Poplar dominates Alfalfa in both dimensions, so both profitability and GWI outcomes could be improved by switching land from Alfalfa to Poplar. The dashed line in Figure 3.3 illustrates the efficient frontier connecting the points that are efficient in the sense that at nonorganic prices, no other system excels in terms of both profitability and GWI. At organic prices, No-till is not dominated by Biologically Based alone (No-till has lower GWI), nor is it dominated by Poplar alone (No-till has greater profitability), but it is dominated by a combination of the two. So at organic prices, shifting land from No-till to a combination of the efficient systems (roughly three-quarters Biologically Based and one-quarter Poplar) could improve both profitability and GWI relative to No-till alone.

Table 3.2. Efficiency gains and trade-offs from cropping system changes.[a]

Change in Cropping System	Change in Global Warming Impact (Mg CO_2e ha^{-1} yr^{-1})	Change in Profitability ($ ha^{-1} yr^{-1})
Efficiency gains		
Conventional to No-till	−1.00	45
Conventional to Reduced Input	−0.51	31
Reduced Input to No-till	−0.49	14
Alfalfa to Poplar	−0.85	93
Trade-offs along efficient frontier		
No-till to Poplar	−1.19	−166
Biologically Based to Poplar	−1.46	−279

[a]Based on MCSE systems assuming nonorganic grain prices. Changes in GWI and profitability represent mean outcomes. Negative GWI indicates greenhouse gas mitigation, a positive outcome.
Source: Profitability data for 1993–2007 adapted and updated from Jolejole (2009). Global warming impact (GWI) data for 1991–1999 from Robertson et al. (2000).

The second important finding is the very high implied marginal cost per metric ton of reducing GWI by moving along each of the efficient frontiers. The marginal cost is the amount of crop net income given up per metric ton of GWI gain by moving from one efficient system to another along the efficient frontier. Arithmetically, it is the change in revenue above selected costs divided by the corresponding change in GWI. The dashed line between Poplar and No-till shows that the implied cost of reducing GWI by shifting land from the No-till corn–soybean–wheat rotation to Poplar is $140 Mg^{-1} CO_2e ha^{-1} yr^{-1}. As illustrated by the solid line from Poplar to Biologically Based, the higher value of certified organic production raises the unit cost of reducing global warming by switching between these systems to $191 Mg^{-1} CO_2e ha^{-1} yr^{-1}. These implied costs exceed the traded prices of carbon credits on international exchanges in the early 2000s by an order of magnitude. The implication is that other methods can abate CO_2e emissions at far lower cost (e.g., improving efficiency of coal-fired power plants or reducing N fertilizer use). Indeed, if substantial cropland were shifted out of grain crops into poplar, market prices of grain crops would rise and those of poplar would fall, making the implied marginal cost of shifting even greater than shown here.

Supply of Crop Land to Boost Ecosystem Service Provision: Application of Stated Preferences to Capture Farm Heterogeneity

Commercial farm conditions vary in terms of land quality, equipment availability, managerial ability, and farmer attitudes. Ecological experiments like the MCSE intentionally hold all these factors constant, limiting the scope of outcomes that can be explored in a trade-off analysis of experimental results. Farmers, however, vary in their resources, priorities, and perceptions of the costs and benefits of farming activities, and this variability generates a number of trade-off outcomes that can be compared. Jolejole (2009) and Ma et al. (2012) captured all three of these aspects of heterogeneity directly in their analyses of the 2008 Crop Management and Environmental Stewardship Survey of Michigan corn and soybean farmers.

The survey was motivated, in part, by a desire to understand why most Michigan field crop farmers choose different cropping systems than the corn–soybean–wheat rotations of the MCSE. In Michigan, corn and soybean are widely grown, often in a two-crop rotation, but wheat is only sometimes included in the rotation. In Michigan during 2006–2010, mean planted areas for corn and soybean were 970,000 and 790,000 ha, compared to 250,000 ha for wheat (NASS 2011). No-till crop farming has expanded greatly during the lifetime of the KBS LTER. By 2006, 48% of soybean land was farmed without tillage in Michigan, which reflected the national trend of 45%. Rates for corn and wheat were less than half this level (Horowitz et al. 2010), and only a fraction of the no-till area was in permanent no-till, as in the MCSE. Cover crops, which the MCSE uses to furnish nitrogen, augment soil organic matter, and prevent soil erosion, were planted on less than 20% of U.S. commercial family farms in 2001, with rates slightly lower on grain farms (Lambert et al. 2006). The same study found that fewer than 30% of farmers conducted soil tests before planting corn and soybean crops.

The purpose of the survey was to understand why Michigan corn and soybean farmers farm as they do, and how they perceive conservation practices like growing wheat, planting cover crops, and reducing fertilizer rates. The survey used contingent valuation methods to elicit whether farmers would be willing to adopt some of these practices in exchange for payments.

The survey questionnaire asked respondents to answer questions regarding four proposed cropping systems. The systems proposed to farmers were loosely based on MCSE Reduced Input and Biologically Based systems, but the first two proposed systems omitted wheat because it is less commonly grown in the region than the other two crops. The proposed systems were:

A. a chisel-tilled corn–soybean rotation fertilized according to university recommendations based on soil testing, including a pre-sidedress nitrate test for corn;
B. same as system A with winter cover crops added;
C. same as system B with winter wheat added to the rotation after soybean; and
D. same as system C but with fertilizer and pesticides reduced by one-third by banding applications over crop rows.

In order to elicit their willingness to change practices in exchange for payments, respondents were asked the following question:

"If a program run by the government or a nongovernmental organization would pay you $X per acre each year for 5 years for using cropping system (Y), how many acres of land would you enroll in this program?"

The question presented them with a predetermined payment level ($X), and the question was repeated with different payment levels for each of the four systems. If respondents answered that they would not participate, they were asked if they would be willing to consider participating in exchange for a higher payment. The questionnaire was sent out in 16 different versions that varied three experimental factors: (1) the payment levels offered, (2) whether the payment came from the government or a nongovernmental organization, and (3) whether the sequence of cropping systems went from least complex (A) to most complex (D) or vice versa. The sample of 3000 Michigan corn and soybean farms was stratified by farm size into four levels: under 100 acres, 101–500 acres, 501–1000 acres, and over 1000 acres. The sampling and mailing lists were managed by the National Agricultural Statistics Service Michigan field office. Usable responses were received from 1688 farms, representing a response rate of 56% (Jolejole 2009).

The econometric analysis of farmer willingness to change was divided into two steps: willingness to consider participation in the program (probit statistical model) and, for those willing to participate, the number of acres they would enroll (tobit regression) (Ma et al. 2012). The determinants of farmers' willingness to adopt these alternative systems differed sharply between the two levels of analysis.

Farmer conservation attitudes, prior experience, and equipment availability largely drove their willingness to consider participating in the hypothetical program to shift land into the proposed cropping systems in exchange for a payment (Ma et al. 2012). Respondents who agreed with the statement "nature provides services that improve my crop production" were 5% more likely to consider the program. Likewise, farmers with prior experience in federal agricultural programs that pay farmers for environmental stewardship practices were more likely to consider this program (though farmers involved in a state environmental assurance program were not). Not surprisingly, farmers who were already doing similar practices (such as no-till or planting wheat) were more inclined to consider proposed practices that were similar. This effect may be linked to the fact that farmers who owned the necessary equipment (e.g., band applicator for fertilizer or pesticides) were more prone to consider participating with the relevant practice than those who did not.

For those willing to consider participating in the program that would pay them for changed cropping practices, how much land they would enroll depended chiefly on benefit–cost and feasibility criteria (Ma et al. 2012). Most important was the size of the payment offered. For the obvious feasibility reason, farmers with more total cropland area would offer to enroll more land in the program. On the other hand, farmers using moldboard plows tended to enroll smaller acreages in the program.

The supply of land that farmers were willing to dedicate to specific cropping systems over a range of different subsidy payments represents their perceived underlying costs and benefits from adopting those practices. Some of these were the direct costs and opportunity costs discussed previously with the trade-off analysis of MCSE systems. But the land area offered for conservation practices also reflects the attitudes and preferences of individual farmers. The importance of attitudes was illustrated by increased program enrollment of farmers who expressed the belief that nature benefits the farm and who had previously participated in environmental programs. So the supply of land that respondents would devote to low-input cropping practices reflects not just farm heterogeneity due to differences in land and equipment, but also farmer heterogeneity due to differences in attitudes and management ability.

Two general patterns emerged from the supply response analysis. First, farmers were willing to supply much more land for System A, the simplest cropping system, than for the three systems that involved cover crops or more complicated management (Ma et al. 2012). In economic terms, this greater price elasticity of supply meant that for a given increase in payment, farmers would offer to devote more land to System A than to the other cropping systems. Second, farmers with over 500 acres (202 ha) were much more willing than operators of smaller farms to respond to higher payments by offering more acreage, especially for System D (Jolejole 2009). It was evident that these larger farms are the low-cost suppliers of environmental services. So payment-for-environmental-services policies that aim for cost-effective gains will likely achieve most of their impact from larger farms.

For measuring the economic value of *individual* ecosystem services from agriculture, the analysis was indeterminate for an important reason: management decisions affect multiple ecosystem services simultaneously. Put colloquially, ecosystem services come in bundles. A given agroecosystem generates ecosystem services in relatively fixed proportions (Antle and Capalbo 2002, Wossink and Swinton 2007). There exists no sound method for allocating costs among the different system outputs without an understanding of consumer demand for them.

Consumer Demand for Land-Based Ecosystem Services

How do consumers value the kinds of ecosystem services that farmers can help to produce? The answer to that question can inform the demand side of economic valuation for these services.

Few consumers perceive ecosystem services as scientists do. Ecosystem services like climate regulation, water quality regulation, nutrient cycling, and pest population regulation are meaningful to ecologists, but opaque to the general public. As a first step before designing a consumer survey, Chen (2010) developed a graphical model of how agricultural practices generate intermediate environmental changes that lead to the ecosystem services experienced by the general public (Fig. 3.4). To highlight one set of relationships in the figure, crop fertilization and tillage and their effects on nutrient cycling may carry little meaning for most citizens. But when lakes become eutrophic as a result of excess nutrients, the meaning to recreational users is very clear.

Based on the literature and pretests of the questionnaire, KBS LTER economists focused on two high-profile endpoints: the proportion of eutrophic lakes and percentage of progress toward international goals for abatement of climate change. The survey population were all residents of the state of Michigan. The 2009 Michigan Environmental Survey went to 6000 Michigan households stratified by population in each county to cover the full geographic extent of the state; the final response rate was 41% (Chen 2010). Respondents were first presented with information about climate change and eutrophication of lakes, along with the links between land management practices and changes in those outcomes. Householders were

Agricultural
Management Factor
Intermediate
Environmental Changes
Off-farm ES Changes
We Want To Measure

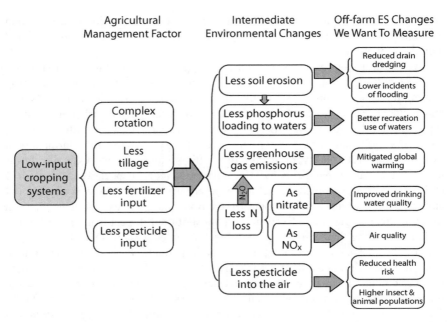

Figure 3.4. Conceptual model of linkages from agricultural management factors to off-farm ecosystem service (ES) changes. Modified from Chen (2010).

then presented with three land stewardship programs, with each proposing to make different changes in (1) the number of lakes with excess nutrient levels and (2) the percentage change in greenhouse gas emissions that scientists estimate is needed to slow global warming. For each of three programs, respondents were asked:

"Would you vote for program (Y) if it increased income taxes and your share of the increased tax was $X per year?"

The questionnaire was mailed in 14 versions, varying the tax rate ($X), the levels of eutrophic lakes and greenhouse gas abatement, and whether the recipients of the program payments were described as farmers or land managers.

The survey found significant public willingness to finance policies that would pay land managers for changed practices to mitigate lake eutrophication, but less support for financing mitigation of global warming (Chen 2010). The overall mean marginal willingness to pay of Michigan residents was $175 per household per year to reduce the number of eutrophic lakes by 170 and to reduce greenhouse gas emissions by 0.52% of their 2000 levels. They did not care whether the funds for changed land management went to farmers or other land managers.

Support for cleaner lakes was clear-cut. Respondents were willing to pay $0.45 per eutrophic lake per household per year, or $1.7 million annually per eutrophic lake, based on the 3.8 million households in Michigan.

Most households were unwilling to pay for reduced greenhouse gas emissions. This finding was due, in part, to the fact that 60% of households were unconcerned about climate change. But the unwillingness to pay may also have resulted from the smallness of the potential emission reductions—just 0 to 1.2%—based on the crop systems proposed in the 2008 Crop Management and Environmental Stewardship Survey. Ordinarily, economists expect that people will pay more to buy more. But statistical tests showed that willingness to pay was unaffected by the level of proposed reduction in greenhouse gas emissions. Hence, overall mean willingness to pay for reduced emissions was zero. Among the 40% of households that were concerned about climate change, however, the mean household would pay $141 per year for a 1% reduction, compared to year 2000 greenhouse gas emission levels. Scaling up to the 1.52 million Michigan households that were concerned about climate change, this would amount to $214 million annually for a 1% reduction in greenhouse gas emissions, compared to the year 2000 level.

Linking Demand and Supply for Ecosystem Services from Agricultural Stewardship

KBS LTER research into how people judge the economic value of ecosystem services from agriculture in the region around KBS has reached a critical stage for establishing value at the equilibrium between supply and demand. On the supply side, the results clearly document the willingness of corn and soybean farmers in Michigan to change their cropping practices so as to generate more ecosystem services if paid to do so. Farmers would expand both the complexity of management practices and the acreage under improved stewardship in response to rising payment levels—generating a supply of land under management for enhanced ecosystem services. On the demand side, state residents are willing to pay for reduced numbers of eutrophic lakes and—some residents at least—for reduced greenhouse gas emissions.

Bringing together the supply and demand sides presents two major challenges. First, the supply units need to be converted from all land area under a given practice to land area *under changed management that provides additional ecosystem services*. Resident taxpayers expect to buy *increases* in ecosystem services, not simply to express gratitude to environmentally minded farmers who were already providing those services. When the land area under stewardship practices offered by farmers is meticulously recalculated to ensure that it refers to practices that bring *additional* ecosystem services, the proportion of enrolled farms that newly adopt each practice may be small—our study revealed just 7% for chisel plow. But for other cropping practices, it may be quite large: 89% for pre-sidedress nitrate test, 96% for cover crop, 70% for adding wheat to a corn–soybean rotation, and 72% for reducing nitrogen fertilization by one-third (Ma 2011).

Second, a bridge is needed to connect the units of supply of ecosystem services with those of demand. Under the hypothetical payment-for-environmental-services program included here (which was modeled on existing U.S. farm programs such as the Environmental Quality Incentives Program, Conservation Reserve Enhancement Program, and Conservation Stewardship Program), farmers were offered payments to supply land under changed practices, not to supply changes in specific ecosystem services. Residents expressed a willingness to pay for enhanced levels of specific services that improve their experiences with lakes and climate. Because changed farming practices generate bundles of changed ecosystem service levels, the most practical way forward is to convert demand for changed ecosystem service levels into the area of land under changed management that would be required to generate jointly the desired levels of change.

Preliminary results using converted supply units, which incorporate stewardship as described above, indicate that what residents are willing to pay for enhanced ecosystem services would cover the costs to farmers of providing them. Moreover, the likely cost of such a program could be covered at the same level as the federal direct subsidy payments made to Michigan field crop farmers during the 2007–2012 period (Ma 2011).

Cautions and Emerging Opportunities from Economic Valuation of Agroecosystem Services

Ecosystem services to and from agriculture are valuable—both intrinsically and economically. KBS LTER–related research has estimated economic values using methods that range from simple to complex. The simplest methods require the most limiting assumptions. For example, trade-off analysis based on budgeting of experimental results assumes that farmer objectives are few and known (such as profitability and specific environmental outcomes), that prices of agricultural inputs and products are constant and known at levels from the period of study, and that the experimental biophysical setting and management practices are highly representative and known at observed levels. The most complex methods have fewer limiting assumptions, because they explore more fully the heterogeneity and dynamics of human behavioral interactions with ecosystems (e.g., as depicted in the conceptual model in Figure 1.4 in Robertson and Hamilton 2015, Chapter 1 in this volume).

Table 3.3 lists some of the ecosystem services whose economic values have been estimated in KBS LTER research. Each service has a range of estimates and a set of assumptions that arise from the valuation method (Champ et al. 2003, Freeman 2003). All are limited, too, by the time and place of the underlying data, because economic systems—like ecological ones—are subject to complex feedbacks. Hence, extrapolation of values to other settings calls for an understanding of system dynamics, methodological assumptions, and data limitations (Spash and Vatn 2006, Wilson and Hoehn 2006).

Economic valuation of ecosystem services can highlight the potential appeal of changes in agricultural management that deliver enhanced ecosystem services—specifically those supporting and regulating ecosystem services that lack markets. Two broad avenues exist for facilitating this: technological innovation and policy design.

Technological innovation can offer alternative ways to provide such nonmarketed ecosystem services as reduced greenhouse gas emissions and improved water quality. Ecological and economic knowledge from KBS LTER research has direct technological application and possibilities are emerging for manipulating agroecosystem components for newly understood benefits. One example is to inoculate soil

Table 3.3. Values of ecosystem services in agricultural systems: assumptions and indicative estimates.

Ecosystem Service	Proxy Variable(s)	Valuation Method	Key Assumptions	Units	Value(s)	Source
Pest regulation (soy aphid)	Corn area in 1.5-km radius	Production function and input cost	Crop yield response known to proxy variable. Prices 2005–2008	$ ha^{-1} yr^{-1}	$20–39 (IPM[a]) $70–264 (no insecticides)	(Landis et al. 2008)
Pest regulation (soy aphid)	Natural enemies and aphids per soy plant	Production function and input cost	Crop yield response known to proxy variable	$ ha^{-1} yr^{-1} for 1st NE[b] plant^{-1}	$4–33	(Zhang 2012)
Recreation (swim, fish)	Rivers near agricultural land parcels	Hedonic analysis of land prices	Regression model fully specified; Data: Agric. land in southwest Michigan 2003–2007	% change in price km^{-1} distance	3–9% of agric. land price	(Ma and Swinton 2011)
Recreation (hunting)	Hunting access to 10% of agric. land in southern Michigan	Travel cost	Full accounting of travel costs to hunt; Data: Hunter mail survey, Michigan 2003	$ trip^{-1} yr^{-1}	$1.90–2.20	(Knoche and Lupi 2007)
Flood regulation, native habitat	Wetland % area in 1.5-km radius of agric. land parcels	Hedonic analysis of land prices	Regression model fully specified; Data: Agric. land in southwest Michigan 2003–2007	% price change per % change in wetland area	2–4% of agric. land price	(Ma and Swinton 2011)
Water-quality regulation	Eutrophic lake number	Contingent valuation	Respondents fully understand scenario; Regression model fully specified; Data: Survey Michigan residents, 2009	$ hhd^{-1} lake^{-1} yr^{-1} kept noneutrophic	0.45[c]	(Chen 2010)
Water-quality regulation; Climate regulation; Nutrient cycling	Changed crop mgmt. practices (one of four crop systems) with additionality	Contingent valuation	Respondents fully understand scenario; Regression model fully specified; Data: Survey Michigan corn/soy farmers, 2008, and Mich. residents, 2009	$ ha^{-1} yr^{-1}	$30–67	(Ma 2011)

[a] Integrated Pest Management (IPM; includes insecticides when pest density exceeds economic threshold).

[b] Natural Enemy (NE; roughly equivalent to multicolored Asian lady beetle, *Harmonia axyridis*).

[c] This marginal value at the median was jointly determined with $141 hhd^{-1} yr^{-1} among the 40% of respondent households that were concerned about global warming for a 1% reduction in greenhouse gas emissions from the level in 2000.

Note: hhd = household.

with methane-consuming microbial communities to stimulate greater removal of atmospheric methane, a potent greenhouse gas (Levine et al. 2011). Another possibility is to manage noncrop areas in agricultural landscapes to support the natural enemies of agricultural pests, thereby reducing the need for chemical pest control (Landis et al. 2000). The viability of this strategy hinges on the opportunity cost of not growing crops (Zhang et al. 2010), which in turn depends on land productivity and crop prices. A third possibility is improvement of perennial grain crops such as wheat, so that grain production can be maintained while cycling nutrients and sequestering carbon in deep perennial root systems. This could enhance soil organic matter and reduce greenhouse gas emissions (DeHaan et al. 2005), though great strides remain for perennial grains to become competitive with current systems (Weir 2012). And a fourth is to manage nitrogen fertilizer more precisely to reduce emissions of the greenhouse gas nitrous oxide, using better estimators of fertilizer need, precision or on-the-go fertilizer application, or other emerging technologies (Liu et al. 2006, Millar et al. 2010).

Policy design is the second key to enhanced ecosystem services from agroecosystems. The contingent valuation survey research summarized above suggests that the public (at least in Michigan) is willing to pay what it would cost for farmers to adopt practices that improve water quality and climate. This evidence that the public is open to such a policy justifies research into the design of programs. Viable programs must tackle thorny problems, such as (1) how to monitor invisible management changes like lower fertilizer rates, and (2) how to balance the cost-effective provision of *additional* services desired by taxpayers with fairness in rewarding farmers (some who were practicing good stewardship without payment). Developing viable programs requires a sound understanding of why farmers adopt changed practices (Swinton et al. 2015, Chapter 13 in this volume). One way forward may be to target certain high-benefit practices. For example, reducing fertilizer application reduces nitrate and phosphorus movement to water as well as nitrous oxide emissions to the atmosphere, so a payment motivated by demand-side desire for fewer eutrophic lakes may generate reduce greenhouse gas emissions at no added cost (Reeling and Gramig 2012).

Summary

Evidence from economic research related to the KBS LTER indicates the potential for crop farming to be managed for higher levels of nonfood services. For example, crop management systems affect greenhouse gas fluxes and water-borne nutrients, which affect climate- and water-quality–regulating services. Land use and cover affect the abundance of natural enemies of crop pests, affecting the biocontrol regulating service.

Economic values of these and other ecosystem services have been calculated using both supply-side and demand-side approaches. KBS LTER research has used land prices and recreational hunting travel costs to estimate the values of recreational and provisioning services. The costs of supplying enhanced services by modifying crop management have been estimated using three methods. Where clear links exist between a specific ecosystem service and changes in input costs or crop yields, marketed products (such as reduced loss of grain yield from natural enemy predation of crop pests), agricultural input costs, and product prices have been used to estimate values of the ecosystem service. Where experimental data exist on how cropping practices link to multiple ecosystem services, trade-off analyses offer a limited way to rule out systems that are inefficient at generating desired services, as in the example provided here from the MCSE.

Understanding the economic value of complex changes in agroecological systems at large scales calls for a third method based on eliciting information from the people who would incur the costs and benefits of those changes. Data from farmers, such as those from contingent valuation surveys, can capture the costs of adopting modified cropping practices in a way that reflects the true heterogeneity of farm resources and people. Estimates of economic value become possible by linking such supply-side data on cost to provide ecosystem services with demand-side data on how much members of the public would willingly pay for those services. Evidence shows that the public is willing to pay for many ecosystem services at rates that many farmers find acceptable, but a challenge is to find practical ways to design an efficient and fair payment system for farmers supplying those services.

References

Antle, J. M., and S. M. Capalbo. 2002. Agriculture as a managed ecosystem: policy implications. Journal of Agricultural and Resource Economics 27:1–15.

Bockstael, N. E., A. M. Freeman, R. J. Kopp, P. R. Portney, and V. K. Smith. 2000. On measuring economic values for nature. Environmental Science & Technology 34:1384–1389.

Boehlje, M. D., and V. R. Eidman. 1984. Farm management. Wiley, New York, New York, USA.

Bundy, L. G., T. W. Andraski, and R. P. Wolkowski. 1993. Nitrogen credits in soybean-corn crop sequences on three soils. Agronomy Journal 85:1061–1067.

Champ, P. A., K. J. Boyle, and T. C. Brown. 2003. Primer on nonmarket valuation. Springer, Dordrecht, Netherlands.

Chen, H. 2010. Ecosystem services from low input cropping systems and the public's willingness to pay for them. Thesis, Michigan State University, East Lansing, Michigan, USA.

CIMMYT (Centro Internacional de Mejoramiento de Maíz y Trigo). 1988. From agronomic data to farmer recommendations: an economics training manual. Completely revised edition. CIMMYT, Mexico City, Mexico.

Collins, S. L., S. R. Carpenter, S. M. Swinton, D. E. Orenstein, D. L. Childers, T. L. Gragson, N. B. Grimm, J. M. Grove, S. L. Harlan, J. P. Kaye, A. K. Knapp, G. P. Kofinas, J. J. Magnuson, W. H. McDowell, J. M. Melack, L. A. Ogden, G. P. Robertson, M. D. Smith, and A. C. Whitmer. 2011. An integrated conceptual framework for long-term social-ecological research. Frontiers in Ecology and the Environment 9:351–357.

Costamagna, A. C., and D. A. Landis. 2006. Predators exert top-down control of soybean aphid across a gradient of agricultural management systems. Ecological Applications 16:1619–1628.

Costanza, R., R. d'Arge, R. de Groot, S. Farber, M. Grasso, B. Hannon, K. Limburg, S. Naeem, R. V. O'Neill, J. Paruelo, R. G. Raskin, P. Sutton, and M. van den Belt. 1997. The value of the world's ecosystem services and natural capital. Nature 387:253–260.

DeHaan, L. R., D. L. Van Tassel, and S. T. Cox. 2005. Perennial grain crops: a synthesis of ecology and plant breeding. Renewable Agriculture and Food Systems 20:5–14.

Drechsel, P., M. Giordano, and T. Enters. 2005. Valuing soil fertility change: selected methods and case studies. Pages 199–221 in B. Shiferaw, H. A. Freeman, and S. M. Swinton, editors. Natural resource management in agriculture: methods for assessing economic and environmental impacts. CABI Publishing, Wallingford, UK.

Freeman, A. M., III. 2003. The measurement of environmental and resource values: theory and methods. Second edition. Resources for the Future, Washington, DC, USA.

Gardiner, M. M., D. A. Landis, C. Gratton, C. D. DiFonzo, M. O'Neal, J. M. Chacon, M. T. Wayo, N. P. Schmidt, E. E. Mueller, and G. E. Heimpel. 2009. Landscape diversity enhances the biological control of an introduced crop pest in the north-central USA. Ecological Applications 19:143–154.

Gelfand, I., and G. P. Robertson. 2015. Mitigation of greenhouse gas emissions in agricultural ecosystems. Pages 310–339 in S. K. Hamilton, J. E. Doll, and G. P. Robertson, editors. The ecology of agricultural Landscapes: long-term research on the path to sustainability. Oxford University Press, New York, New York, USA.

Hamilton, S. K. 2015. Water quality and movement in agricultural landscapes. Pages 275–309 in S. K. Hamilton, J. E. Doll, and G. P. Robertson, editors. The ecology of agricultural Landscapes: long-term research on the path to sustainability. Oxford University Press, New York, New York, USA.

Heal, G. 2000. Nature and the marketplace: capturing the value of ecosystem services. Island Press, Washington, DC, USA.

Hoag, D. L., and J. S. Hughes-Popp. 1997. Theory and practice of pollution credit trading in water quality management. Review of Agricultural Economics 19:252–262.

Horowitz, J., R. Ebel, and K. Ueda. 2010. "No-till" farming is a growing practice. Economic Information Bulletin Number 70, U.S. Department of Agriculture, Economic Research Service, Washington, DC, USA.

Jolejole, C. B., S. M. Swinton, G. P. Robertson, and S. P. Syswerda. 2009. Profitability and environmental stewardship for row crop production: Are there trade-offs? International Association of Agricultural Economists, 2009 Triennial Conference, Beijing, China.

Jolejole, M. C. B. 2009. Trade-offs, incentives, and the supply of ecosystem services from cropland. Thesis, Michigan State University, East Lansing, Michigan, USA.

Knoche, S., and F. Lupi. 2007. Valuing deer hunting ecosystem services from farm landscapes. Ecological Economics 64:313–320.

Kremen, C., and T. Ricketts. 2000. Global perspectives on pollination disruptions. Conservation Biology 14:1226–1228.

Kremen, C., N. M. Williams, and R. W. Thorp. 2002. Crop pollination from native bees at risk from agricultural intensification. Proceedings of the National Academy of Sciences USA 99:16812–16816.

Lambert, D. H., P. Sullivan, R. Claassen, and L. Foreman. 2006. Conservation-compatible practices and programs: Who participates? U.S. Department of Agriculture, Economic Research Service, Washington, DC, USA.

Landis, D. A., and S. H. Gage. 2015. Arthropod diversity and pest suppression in agricultural landscapes. Pages 188–212 in S. K. Hamilton, J. E. Doll, and G. P. Robertson, editors. The ecology of agricultural Landscapes: long-term research on the path to sustainability. Oxford University Press, New York, New York, USA.

Landis, D. A., M. M. Gardiner, W. van der Werf, and S. M. Swinton. 2008. Increasing corn for biofuel production reduces biocontrol services in agricultural landscapes. Proceedings of the National Academy of Sciences USA 105:20552–20557.

Landis, D. A., S. D. Wratten, and G. M. Gurr. 2000. Habitat management to conserve natural enemies of arthropod pests in agriculture. Annual Review of Entomology 45:175–201.

Levine, U., T. K. Teal, G. P. Robertson, and T. M. Schmidt. 2011. Agriculture's impact on microbial diversity and associated fluxes of carbon dioxide and methane. The ISME Journal 5:1683–1691.

Liu, Y., S. M. Swinton, and N. R. Miller. 2006. Is site-specific yield response consistent over time? Does it pay? American Journal of Agricultural Economics 88:471–483.

Ma, S. 2010. Hedonic valuation of ecosystem services using agricultural land prices. Thesis, Michigan State University, East Lansing, Michigan, USA.

Ma, S. 2011. Supply and demand for ecosystem services from cropland in Michigan. Dissertation, Michigan State University, East Lansing, Michigan, USA.

Ma, S., and S. Swinton. 2011. Valuation of ecosystem services from rural landscapes using agricultural land prices. Ecological Economics 70:1649–1659.

Ma, S., S. M. Swinton, F. Lupi, and C. B. Jolejole-Foreman. 2012. Farmers' willingness to participate in Payment-for-Environmental-Services programmes. Journal of Agricultural Economics 63:604–626.

Millar, N., G. P. Robertson, P. R. Grace, R. J. Gehl, and J. P. Hoben. 2010. Nitrogen fertilizer management for nitrous oxide (N_2O) mitigation in intensive corn (Maize) production: an emissions reduction protocol for US Midwest agriculture. Mitigation and Adaptation Strategies for Global Change 15:185–204.

NASS (National Agricultural Statistics Service). 2011. Michigan agricultural statistics 2010–2011. U.S. Department of Agriculture, Michigan Field Office, Lansing, Michigan, USA.

Palmquist, R. B. 1989. Land as a differentiated factor of production: a hedonic model and its implication for welfare measurement. Land Economics 65:23–28.

Palmquist, R. B., and L. E. Danielson. 1989. A hedonic study of the effects of erosion control and drainage on farmland values. American Journal of Agricultural Economics 71:55–62.

Paul, E. A., A. Kravchenko, A. S. Grandy, and S. Morris. 2015. Soil organic matter dynamics: controls and management for ecosystem functioning. Pages 104–134 in S. K. Hamilton, J. E. Doll, and G. P. Robertson, editors. The ecology of agricultural Landscapes: long-term research on the path to sustainability. Oxford University Press, New York, New York, USA.

Pearce, D. 1998. Auditing the earth. Environment 40:23–28.

Polasky, S., and K. Segerson. 2009. Integrating ecology and economics in the study of ecosystem services: some lessons learned. Annual Review of Resource Economics 1:409–434.

Reeling, C. J., and B. M. Gramig. 2012. A novel framework for analysis of cross-media environmental effects from agricultural conservation practices. Agriculture, Ecosystems & Environment 146:44–51.

Ricketts, T. H., G. C. Daily, P. R. Ehrlich, and C. D. Michener. 2004. Economic value of tropical forest to coffee production. Proceedings of the National Academy of Sciences USA 101:12579–12582.

Ricketts, T. H., J. Regetz, I. Steffan-Dewenter, S. A. Cunningham, C. Kremen, A. Bobdanski, B. Gemmill-Herren, S. S. Greenleaf, A. M. Klein, M. M. Mayfield, L. A. Morandin, S. G. Potts, and B. F. Viana. 2008. Landscape effects on crop pollination services: Are there general patterns? Ecology Letters 11:499–515.

Robertson, G. P., and S. K. Hamilton. 2015. Long-term ecological research at the Kellogg Biological Station LTER Site: conceptual and experimental framework. Pages 1–32 in S. K. Hamilton, J. E. Doll, and G. P. Robertson, editors. The ecology of agricultural

Landscapes: long-term research on the path to sustainability. Oxford University Press, New York, New York, USA.

Robertson, G. P., E. A. Paul, and R. R. Harwood. 2000. Greenhouse gases in intensive agriculture: contributions of individual gases to the radiative forcing of the atmosphere. Science 289:1922–1925.

Shiferaw, B., H. A. Freeman, and S. Navrud. 2005. Valuation methods and approaches for assessing natural resource management impacts. Pages 19–51 in B. Shiferaw, H. A. Freeman, and S. M. Swinton, editors. Natural resource management in agriculture: methods for assessing economic and environmental impacts. CABI Publishing, Wallingford, UK.

Spash, C. L., and A. Vatn. 2006. Transferring environmental value estimates: issues and alternatives. Ecological Economics 60:379–388.

Steffan-Dewenter, I., U. Munzenberg, C. Burger, C. Thies, and T. Tscharntke. 2002. Scale-dependent effects of landscape context on three pollinator guilds. Ecology 83:1421–1432.

Swinton, S. M., N. Rector, G. P. Robertson, C. Jolejole-Foreman, and F. Lupi. 2015. Farmer decisions about adopting environmentally beneficial practices. Pages 340–359 in S. K. Hamilton, J. E. Doll, and G. P. Robertson, editors. The ecology of agricultural Landscapes: long-term research on the path to sustainability. Oxford University Press, New York, New York, USA.

Syswerda, S. P., B. Basso, S. K. Hamilton, J. B. Tausig, and G. P. Robertson. 2012. Long-term nitrate loss along an agricultural intensity gradient in the Upper Midwest USA. Agriculture, Ecosystems & Environment 149:10–19.

Syswerda, S. P., A. T. Corbin, D. L. Mokma, A. N. Kravchenko, and G. P. Robertson. 2011. Agricultural management and soil carbon storage in surface vs. deep layers. Soil Science Society of America Journal 75:92–101.

U.S. Department of the Interior, Fish and Wildlife Service, and U.S. Department of Commerce, U.S. Census Bureau. 2006. National survey of fishing, hunting, and wildlife-associated recreation. <http://www.census.gov/prod/2008pubs/fhw06-nat.pdf>.

Weir, A. 2012. Evaluating the economic feasibility of environmentally beneficial agricultural technologies compared to conventional technologies. Thesis, Michigan State University, East Lansing, Michigan, USA.

Weston, J. F., and T. E. Copeland. 1986. Managerial finance. Dryden Press, New York, New York, USA.

Wilson, M. A., and J. P. Hoehn. 2006. Valuing environmental goods and services using benefit transfer: the state-of-the art and science. Ecological Economics 60:335–342.

Wossink, A., and S. Swinton. 2007. Jointness in production and farmers' willingness to supply non-marketed ecosystem services. Ecological Economics 64:297–304.

Zhang, W., and S. M. Swinton. 2009. Incorporating natural enemies in an economic threshold for dynamically optimal pest management. Ecological Modelling 220:1315–1324.

Zhang, W., W. van der Werf, and S. M. Swinton. 2010. Spatially optimal habitat management for enhancing natural control of an invasive agricultural pest: soybean aphid. Resource and Energy Economics 32:551–565.

Zhang, W., and S. M. Swinton. 2012. Optimal control of soybean aphid in the presence of natural enemies and the implied value of their ecosystem services. Journal of Environmental Management 96:7–16.

4

A Crop Stress Index to Predict Climatic Effects on Row-Crop Agriculture in the U.S. North Central Region

Stuart H. Gage, Julie E. Doll, and Gene R. Safir

Corn (*Zea mays* L.) and soybean (*Glycine max* L.) are major U.S. grain crops with 38.6 million and 31.0 million ha (95.4 and 76.5 million acres) planted in 2013, respectively (NASS 2014a). These two crops support both local demand and international agricultural exports and greatly contribute to the U.S. economy. The value of production for corn and soybean in the United States has increased markedly in recent years and for 2012 was estimated at $77.4 billion and $43.2 billion, respectively (NASS 2014b). Corn accounted for 41% of the value of production of all U.S. field crops in 2012 (NASS 2014b). The United States is the world's largest exporter of corn, exporting 46.6 Tg (1.8 billion bushels) from the 2010 harvest of 317 Tg (12.5 billion bushels) (ERS 2011a). The U.S. North Central Region (NCR) (Fig. 4.1), often called the Corn Belt due to the success of its dominant crop (Hudson 1994), contains some of the most productive cropland in the world, growing over 80% of U.S. corn and soybean. The agricultural importance of this region will only continue to grow as demand for food, fuel, and fiber increases across the globe.

Natural vegetation in the NCR ranges from eastern deciduous forest in the north and east to tallgrass prairie in the central portion to shortgrass prairie in the west (Bailey 1998). Today, much of the NCR has been transformed to croplands (Fig. 4.1). Of its 228 million ha (563 million acres), 39.8% are row crops; corn and soybean account for 34% and 29%, respectively, of this area (NASS 2009a). The Kellogg Biological Station Long-term Ecological Research Site (KBS LTER) lies in the middle of the latitudinal range of the Corn Belt (Robertson and Hamilton 2015, Chapter 1 in this volume) and thus represents the climate stresses typical for row crops such as corn and soybean. The geographic position of KBS provides an

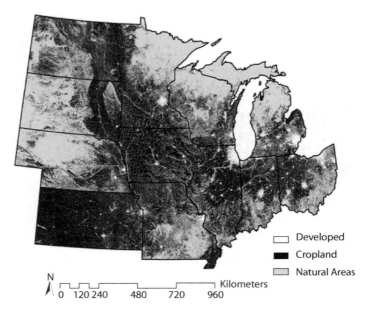

Figure 4.1. Land use or cover in the North Central Region (NCR) of the United States. Created from the USDA NASS Cropland Data Layers (NASS 2009b). States in the NCR include Illinois, Indiana, Iowa, Kansas, Michigan, Minnesota, Missouri, Nebraska, North Dakota, Ohio, South Dakota, and Wisconsin.

opportunity to investigate trends and patterns of crop production ranging from local to regional scales within the NCR.

As in any other ecosystem, production in cropping systems is determined by abiotic and biotic factors. Climate, soils, and nutrient availability can be abiotic constraints to producing a profitable crop. Biotic constraints include the yield potential of a given crop cultivar plus potential losses to weeds, insect herbivores, and plant pathogens. Management practices such as tillage, irrigation, fertilizers, and pesticides are attempts to attain crop yield potentials by overcoming limitations of climate, soil fertility, and pests and to favor agriculture in settings where it might otherwise be less productive. These practices have economic consequences for both growers and consumers, and often come at an environmental cost (Matson et al. 1997, Tilman et al. 2002, Robertson et al. 2004).

Climate regulates crop growth and productivity. Heat and precipitation are two important climatic drivers that operate at multiple spatial scales and are the principal physical determinants of crop yield. In this chapter, we present a simple Crop Stress Index (CSI), based on data sets of temperature and precipitation, which captures the main effect of climate on crop yields. We also compare the relationship between CSI and crop yields at county and regional levels. We end the chapter by addressing the implications of climate change projections for row-crop production in the NCR. In an era of incipient climate change impacts (IPCC 2013, Walsh et al. 2014) that will affect the NCR in ways not yet fully understood, it is crucial to have an understanding of how temperature and precipitation patterns have historically influenced crop productivity. Because trends cannot be understood in the absence of historical context, we begin the chapter with a brief historical background of the NCR, including the rise of corn and soybean as important crops, the industrialization of agriculture, and the advent of agricultural ecology as a scientific discipline.

A Short Agricultural History of the North Central Region

Land Transformation

The western movement of U.S. agriculture during the nineteenth century was facilitated by three primary factors. First, some 30 million immigrants came to the United States between 1815 and 1914, mainly from Germany, Italy, Ireland, and Austria-Hungary. Second, the cleared forest lands in much of the eastern United States had low productivity and were unable to produce sufficient yields for the increasing human population. And third, vast tracts of potential agricultural land—stretching from the Ohio Valley to the Rocky Mountains—were identified by expeditions such as those of Lewis and Clark in the early 1800s (DeVoto 1953). These areas were made accessible by mid-century transportation advances that included water routes such as the Erie Canal, which opened the Great Lakes portion of the NCR to westward expansion, and railroads that provided access to southern regions of the NCR.

The increasing availability of arable land to newcomers provided a huge incentive to establish row-crop agriculture in the NCR. Wheat (*Triticum aestivum* L.) was a dominant early crop because it was readily adaptable to the region's soils and climate. As the vast bison (*Bison bison*) herds in the western part of the NCR were harvested and replaced by domestic cattle and hogs, corn was grown to fatten grazed animals before transport to the emerging livestock markets in Chicago. Corn–livestock agriculture was well established in the southern states of the NCR by 1850 (Hudson 1994).

Corn–livestock agriculture spread slowly from southern states northward and by 1880 reached well into Michigan's southern peninsula, southern Wisconsin, and southern Minnesota (Hudson 1994). In the year 1890, the 41 packing houses in Chicago slaughtered 13 million head of livestock, accounting for 50% of the U.S. urban wholesale meat business (Hudson 1994). By the 1920s, the center of the NCR corn-growing region had shifted northward with the availability of high-yielding corn varieties adapted to more northern latitudes (Hudson 1994). And northern Ohio, Indiana, Illinois, and Iowa had rich soils well suited for growing corn closer to the Chicago livestock markets.

As agriculture expanded into the northern wooded portion of the NCR, trees were cleared to open farmland and provide other parts of the Midwest with lumber for buildings and furniture. Thousands of small sawmills were erected on waterways of Michigan and Wisconsin to process logs from vast inland stands of white pine and hardwood. In eastern Michigan, for example, the Saginaw River alone supported over 100 sawmills along a 56-km (35-mile) reach (Kilar 1990).

After working the land during the growing season, Michigan farmers often were employed to cut and haul timber to riverbanks, so the spring runoff could transport logs to Lake Michigan for subsequent shipment to the thriving Chicago market (Cronon 1991). The interplay between forest and agricultural resources played an important historical role in the growth and sustainability of the emerging farmsteads in the northern NCR: the availability of timber provided an initial source of income and made it possible to construct farm buildings such as barns, silos, and houses.

The Advent of Hybrid Corn and the Rise of Soybean

Open pollinated corn (i.e., corn that is naturally pollinated and produces fertile seeds) reached a yield plateau by 1900 in Illinois (Hudson 1994). Efforts to increase yield further by selecting seed from the best corn plants led to corn yields of only 4.4 Mg ha^{-1} (70 bushels per acre) in 1920 under ideal growing conditions. The discovery of hybrid vigor in corn led to single cross hybrids by 1934 that produced consistently higher yields (Weaver 1946). By 1940 hybrid corn was widely adopted and resulted in more than just increased yields (Hudson 1994): corn hybrids also hastened the shift to mechanization on account of greater stalk strength and increased demands for nitrogen and other nutrients that could more conveniently be supplied with fertilizers than with leguminous cover crops or manure (Robertson and Vitousek 2009). Fertilizers made yields profitable on even poor-quality soils. The widespread use of hybrids enabled more corn production west of the Mississippi River, enhanced by government-supported irrigation subsidies. In addition, northern corn production increased because hybrids performed well during a short growing season with long day length. These advances stimulated greater production and caused major overproduction of corn, depressing its market value.

After World War I, consumers' food preferences changed, causing an increased demand for vegetable oils rather than lard or "pig fat" (Hudson 1994). This cultural shift—combined with overproduction of corn and hogs—reduced demand and value and stimulated government subsidies for corn and hog farmers to reduce production. Disincentives for corn led the way for soybean, introduced to the United States by Benjamin Franklin, to become a new cash crop for the Corn Belt region, quickly replacing oats as a rotation crop (Hudson 1994). Soybean futures trading began in Chicago in 1936. Soybean oil meal was used by livestock and poultry producers as well as pet food manufacturers because of its high protein content (Hudson 1994). Other uses were also explored—for example, in 1936 Henry Ford Farms established 4900 ha (12,108 acres) of soybean in Michigan to explore soybeans for use in both food and industry, including soy-based plastics that could be used in cars. Ford had developed a laboratory to discover industrial uses for farm products ("chemurgy") then used mainly for food, with the aim of making farming more profitable (Lewis 1976).

Soybean adoption in the NCR was helped by the fact that farmers could plant soybeans at about the same time as corn with only minor adjustments in farming equipment. By the 1940s—even though no one in the United States had grown the crop commercially before 1920—soybean had become established as the second most profitable cash crop and was grown on the best lands in the NCR. Today, NCR soybean acreage is about the same as corn acreage, although the proportion can be affected by fluctuating corn ethanol prices (Feng and Babcock 2010).

The Industrialization of North Central Region Agriculture

By 1950 corn and soybean farming underwent post–World War II industrialization. Larger equipment was being used to plant, till, spray, and harvest crops. In addition, new varieties were developed to suit large-scale corn and soybean production, and new fertilizers and application methods were deployed. The advent of the insecticide DDT in the 1940s heralded the chemical era that grew exponentially during the 1960s and 1970s as the use of biocides expanded (Van Den Bosch 1978). Insecticides targeted insects such as rootworms (*Diabrotica* spp.), stalk worms including the European corn borer (*Ostrinia nubilalis* Hübner), and various insect defoliators. Herbicides also came into widespread use, with multiple types and formulations developed to combat both grass and broad-leaf weeds. And new types of fungicides were developed to attack plant diseases like root rot, mildew, and foliar pathogens.

Government subsidies for irrigation and advances in irrigation technology allowed cropland to expand west of the Mississippi. The Ogallala Aquifer, which underlies much of the Great Plains region, was tapped to supply water to row crops. Over 170,000 wells were drilled to enable corn production on land that without irrigation was ecologically suitable only for wheat. By 1977, 1.4 million ha (3.5 million acres) were irrigated by center pivot irrigation systems in Nebraska, Kansas, Texas, and Colorado. This enabled the development of both corn feedlots for cattle and decentralized, regional slaughtering facilities in these states (Hudson 1994), and led to the demise of the Chicago stockyards, most of which closed by 1970.

Increased international demand for corn also drove up production. This demand in the mid-1970s was met, in part, by abolishment of the Soil Bank—a government subsidy program initiated in 1956 that took agricultural land out of production to reduce crop surplus. As a result, millions of acres of land were put into corn production. For example, nearly 163,000 ha (402,721 acres) of corn were newly planted on drained wetlands in Michigan's Saginaw Bay watershed between 1969 and 1978.

Currently, corn is the largest U.S. crop in both volume and value. Iowa, Illinois, Nebraska, and Minnesota account for more than 50% of U.S. corn production. Other major corn-producing states include Indiana, Wisconsin, South Dakota, Michigan, Missouri, Kansas, Ohio, and Kentucky. Today's U.S. corn crop has three principal uses: animal feed, ethanol, and direct human consumption. Of the 13.13 billion bushels of corn used during the marketing year of September 2010 through August 2011, uses included: 38% for ethanol production; 38% for livestock feed and residual; 14% for export; 8% for high-fructose corn syrup, glucose, dextrose, and starch; and ~2% for cereals and other products (ERS 2011a). By 2015 about half of the 2009-equivalent corn crop is expected to be used for ethanol, with important environmental implications (Robertson et al. 2011).

Insect and Disease Influence on Corn and Soybean Production

The intensification and mechanization of grain production can increase crop susceptibility to plant pathogens and insect infestations because it provides extensive resource opportunities for these organisms. Major disease outbreaks are rare but can have significant impacts on corn production in the NCR. In 1970, for example, the Southern Corn Leaf Blight—caused by the fungus *Helminthosporium maydis* (Nisikado & Miyake)—resulted in a 10% reduction in national corn production, the largest decrease in corn yield due to a pathogen in U.S. history. The outbreak was limited to 1 year and was attributed mainly to an uncommon combination of favorable environmental conditions for the fungus (Tatum 1971).

Soybean production in the United States has been largely unaffected by disease, although a recent arrival introduces the potential for significant harm. Asian Soybean Rust, caused by the fungus *Phakopsora pachyrhizi* (Syd. & P. Syd. 1914), was present for many years in Asia before spreading to Africa and South America (Miles et al. 2003). This fungus was found in 2004 at a research farm in Louisiana (Schneider et al. 2005). Isard et al. (2005) produced a soybean rust aerobiology predictive model that guided soybean rust scouting operations after its initial discovery. Predictions of soybean yield decline are as high as 80% in the absence of effective management strategies, such as fungicide sprays now common in Brazil. Major research efforts are under way to develop more effective fungicides and disease-resistant soybean varieties. Although its impact in the United States thus far has been limited, this disease remains a significant threat to soybean production.

Insect pests have also affected corn and soybean yields. The European corn borer has plagued corn producers, who have waged major pesticide assaults against the insect. The borer can also be managed by rotating crops to break the insect's life cycle, and by cutting cornstalks close to the soil surface at harvest to remove overwintering habitat. Corn varieties with genes inserted to produce the bacterial toxin *Bacillus* thuringiensis (Bt), fatal to the larvae of moths and other lepidopterans, provide protection for 65% of U.S. corn acreage (ERS 2011b) at a disputed environmental cost not fully resolved (Rosi-Marshall et al. 2007, Beachy et al. 2008, Parrott 2008, Jensen et al. 2010, Tank et al. 2010).

The western corn rootworm (*Diabrotica virgifera virgifera* LeConte) is another serious corn pest responsible for severe root damage and subsequent yield loss when large populations are present. Insecticides applied to soil can reduce infestations but are costly. Although crop rotation can also be effective at reducing infestations, the rootworm has adapted to the normal rotation of corn and soybean in the NCR (Levine et al. 2002), so rotating into soybean is no longer an effective control technique. Bt genes have also been inserted into root tissue to combat rootworm, offering another pest management option for farmers. While such genetically modified crops have provided benefits to both agriculture and the environment, weed herbicide resistance is an important emerging problem (NRC 2010a) and questions remain regarding the long-term ecological effects of using genetically modified crops (NRC 2008, NRC 2010a).

In soybean, insect pests have been relatively rare until recently. In 2000 the soybean aphid (*Aphis glycines* Matsumura) invaded the NCR and has required widespread insecticide application (Landis et al. 2008). Major efforts are under way to understand biological regulation of this pest (Landis and Gage 2015, Chapter 8 in this volume). Other soil-inhabiting organisms such as the soybean cyst nematode also can reduce soybean yields and increase the cost of soybean production (Kaitany et al. 2000) by requiring management such as crop rotation, resistant varieties, nematicide application, or other cultural practices.

The Emergence of Agricultural Ecology

Many plant pathogens and insect pests flourish in large-scale crop production systems and are favored by the reduced use of crop rotations (Oerke 2007). Crop surveillance is a key factor in the early detection and control of such outbreaks. Although more farmers are beginning to apply principles of ecosystem management, such as scouting, to their cropping systems, there is limited coordination of such activities at regional scales (Isard et al. 2005).

In the 1970s, after years of attempting to manage agricultural insect pests with an arsenal of chemical inputs such as the insecticide DDT, the environmental risks of pesticide use were discovered. This, along with publications by Carson (1962) and others (Pimentel 1971, Van Den Bosch 1978), stimulated the need for a shift from indiscriminate pesticide use and the development of new approaches to pest control.

In response, insect ecologists developed the paradigm of Integrated Pest Management (IPM)—based on the greater use of ecological understanding and biological methods for pest regulation, such as intensive scouting prior to chemical use and using natural enemies of pests for biocontrol (Radcliffe et al. 2008). Integrated Pest Management challenged industrial agriculture's assumption that reliance on chemicals was the only effective means to manage and maintain production and economic capacity (Pimentel 1981). The controversy over chemical vs. biological pest management raged during the 1970s, at a time when corn and soybean production was vastly increasing on account of export demand. In the 1980s, the concepts of Sustainable Agriculture (Robertson and Harwood 2013) and Integrated Farming Systems emerged. Finally an ecosystem perspective became a lens through which to examine agriculture, and agricultural ecology had begun to mature as a scientific discipline (Tivy 1990, Soule and Piper 1992, Altieri 1987).

Climate and Crop Yields in the North Central Region

Patterns of Corn and Soybean Yield

Figure 4.2A shows the county-level percentage of NCR land area classified as suitable for agricultural production (i.e., arable land) without regard to water availability. Soil water-holding capacity (Fig. 4.2B) provides additional information about crop production potential.

(A) Arable Land

(B) Water Holding Capacity

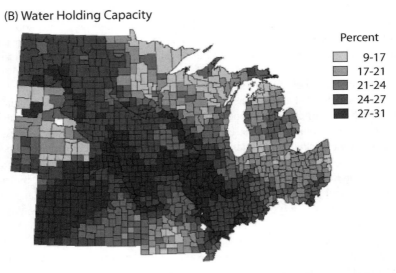

Figure 4.2. Distribution of (A) percentage of arable land and (B) soil water holding capacity by county in the NCR. Water holding capacity is the amount of water that can be held in a soil between its field capacity (drained upper limit) and the wilting point of the vegetation, expressed as the volumetric water content (% of total soil volume). This is a function of soil texture and is expressed for the upper 250 cm of the soil profile. Data from NRCS (1991).

Statewide 30-year patterns of rain-fed crop yields for corn and soybean range from 3–7 Mg ha^{-1} (48–112 bu acre^{-1}) for corn and 1.4–2.6 Mg ha^{-1} (21–39 bu acre^{-1}) for soybean across the NCR (Fig. 4.3). Corn and soybean yields for this time period (1971–2001) were highest in Iowa and lowest in North Dakota. The upward trends in corn and soybean yields for the NCR since 1971 (Fig. 4.4) reflect improvements

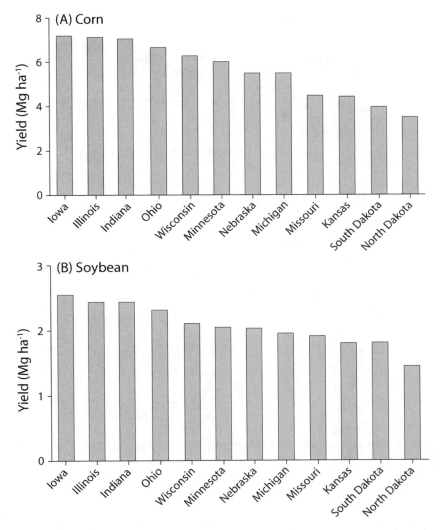

Figure 4.3. Mean non-irrigated (rainfed) grain yields (Mg ha⁻¹) for 1971–2001 for (A) corn and (B) soybean crops by state in the NCR. The regional crop dataset was compiled by the NCR Climate and Crop Committee (USDA and Cooperating States) from NASS (2011).

both in varieties (genetic improvement) and agronomic management—especially in the use of nitrogen fertilizers (Robertson and Vitousek 2009)—as well as favorable climate conditions (Twine and Kucharik 2009). Annual variability in yield largely reflects variation in climate. Although regional variability exists, the spatial pattern of corn and soybean yields provides an outline of the Corn Belt: a band of higher yields stretching from central Ohio through Indiana, Illinois, Iowa, and southern Wisconsin and Minnesota (Fig. 4.5).

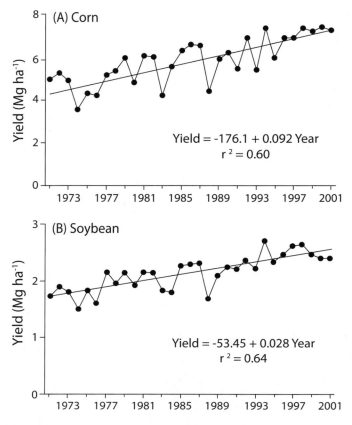

Figure 4.4. Trends in mean grain yields (Mg ha⁻¹) for non-irrigated (A) corn and (B) soybean in the NCR (1971–2001). See Fig. 4.3 legend for data source.

Crop Stress Indices

One of the stresses affecting corn and soybean growth and productivity—and plants in general—is a moisture deficit. Moisture deficits can cause physiological stress in plants and, when coupled with the additional stress of increasing temperatures, can lead to significant crop loss. Several indices have been derived to relate weather to crop stress and crop loss. Jackson et al. (1988) reexamined the theoretical basis of the crop water stress index and showed that measures of canopy temperature, wind speed, crop canopy resistance, evapotranspiration, solar radiation, and other factors are important considerations. In a review of drought indices, Heim (2002) noted that the Palmer Drought Index (PDI), despite some deficiencies, is still the most widely used index. However, the PDI requires data on precipitation, evapotranspiration, soil moisture loss and recharge, and runoff (Heim 2002), a range of variables for which few regions in the world have data.

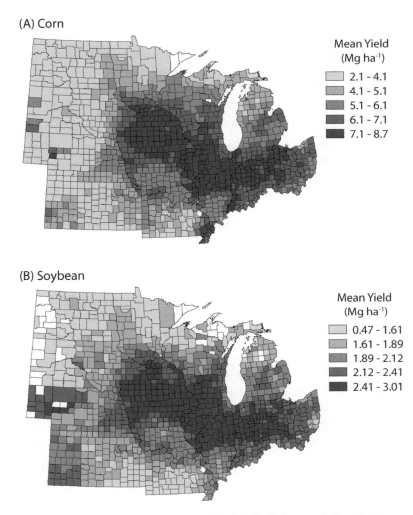

Figure 4.5. Distribution of mean grain yields (Mg ha^{-1}) for non-irrigated (A) corn and (B) soybean by county in the NCR (1971–2001). See Fig. 4.3 legend for data source.

The Crop Stress Index (CSI) is a simpler index to use and requires less data than the PDI. The CSI is estimated by the equation

$$\text{CSI} = D_{10}/(P + 1)$$

where D_{10} is degree-day, or the average daily air temperature as degrees in excess of 10°C ([(maximum temperature + minimum temperature)/2]–10°C) summed over a month-long period, and P is the amount of precipitation (mm) accumulated during the same month. Commonly, 10°C is used as the base temperature to calculate agricultural growing degree-days. The CSI has been correlated with wheat yields and grasshopper populations in Saskatchewan, a cotton pest in Australia, and regional drought in the NCR (Gage and Mukerji 1977, Hamilton and Gage 1986, Gage 2003). The CSI is relatively easy to compute at any scale—from local to global— since it requires only daily maximum and minimum temperature and precipitation data, which are generally available.

Patterns of Heat, Moisture, and Crop Stress

Coupling NCR databases of temperature, precipitation, and crop yield offers a powerful means for examining the relationship between yield and climate. The analyses that follow cover a 30-year period (1971–2001) for 1053 of the 1055 counties in the NCR where data were available, and incorporate more than 11 million daily climate records and over 35,000 rain-fed crop yield records. Figure 4.6 represents the structural organization of datasets used to compute the Crop Stress Index (CSI) from monthly records. The annual crop dataset shown in the lower panel of Fig. 4.6 illustrates the extraction of the rain-fed component of the crop database. The resulting database used in the analysis is the integration of the monthly climate and rain-fed crop datasets.

May through July degree-day accumulation, shown as means over the 30-year period, ranged from 334 in the northern NCR to 1258 degree-days in the south, a 3.8-fold difference (Fig. 4.7A). Precipitation (May through July) ranged from 177 mm in the western NCR to a high of 383 mm in the south central NCR, a 2.2-fold difference (Fig. 4.7B). Of interest is that the patterns of heat and precipitation are not consistent across the region. Accumulated degree-days increase from north to south, while precipitation tends to increase toward the central portion of the NCR, with highest accumulation in Iowa and lowest accumulation in western states. The CSI integrates these two patterns, and likewise varies across the region, as illustrated in Fig. 4.8. Monthly growing season patterns in the CSI (Fig. 4.8) show that sustained stress is concentrated in the southwestern part of the region in May, June, and July, but that May stress is also high in the northwest (Fig. 4.8A). In June, crop stress is high in the south central NCR (Fig. 4.8B), whereas in July it shifts primarily to the western NCR (Fig. 4.8C). The 3-month sum of average CSIs for each county shows that the western and southern parts of the NCR have the highest probability of crop stress (Fig. 4.8D), which is why most of the corn and soybean fields in these areas are irrigated.

A monthly distribution of the CSI illustrates the dynamics of stress over the 30-year period (Fig. 4.9). Although climate may affect crop growth and yield throughout the entire growing season, plants are most susceptible to stress, as measured by yield loss, during the critical months of May–July in the NCR (Fig. 4.9). The potential for crop stress differs over this period and tends to be greatest in July, followed by June and May The most intense crop stress occurs in late July when grain has already set and growth is beginning to slow; however, stress early in growing season can reduce crop yield quickly because young plants are more susceptible to moisture deficiency as they have not fully tapped into the belowground moisture. This was the case during the 1988 drought in the Corn Belt region (Gage 2003).

Regional Data Processing

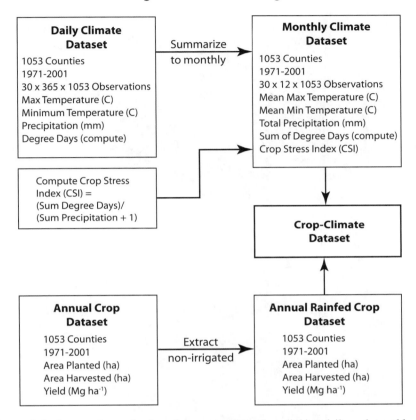

Figure 4.6. Structural organization of the regional database linking daily and monthly climate variables with annual yields of corn and soybean crops. Growing season degree-days are based on the average daily temperature (°C) - 10 °C. The climate data set is from the National Weather Service Cooperative Observer Network. Extrapolations to the 1053 of 1055 counties in the NCR where data were available were computed by the Midwestern Regional Climate Center (MRCC) and compiled by the North Central Regional Climate and Crop Committee (USDA and Cooperating States). See Fig. 4.3 legend for regional crop dataset source.

Crop Stress and Yields

Over the 30-year period of this analysis, technological advances—including new varieties, nutrient subsidies, soil management, and pest control—have improved yields of both corn and soybean. Regional yields of both crops trend upward during this period and have similar degrees of fit ($r^2 = 0.60$ for corn and $r^2 = 0.64$ for soybean, Fig. 4.4). Regional yields of both crops decline, however, with increasing crop stress (Fig. 4.10). The steeper negative slope for corn compared to soybean suggests that corn has a greater sensitivity to climate stress than soybean. This is consistent with the differential responses of these crops to heat stress: high

(A) Growing Degree-Days

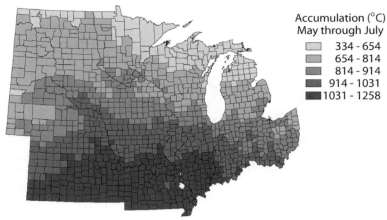

(B) Precipitation

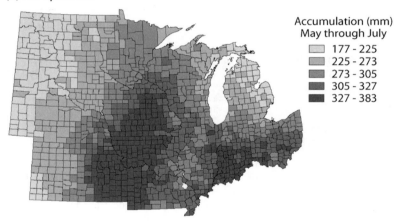

Figure 4.7. Distribution of accumulated (A) growing season degree-days and (B) precipitation (mm) by county in the NCR from May through July (1971–2001). See Fig. 4.6 for data sources.

temperatures during flowering and pollination depress yields in both, but because corn has a shorter reproductive period—on the order of only 1 week—corn is especially sensitive to short-term heat waves (Hatfield et al. 2011).

Incorporating both year and climate into multiple regression analyses significantly improves the prediction of corn ($r^2 = 0.80$) and soybean ($r^2 = 0.79$) yields in the NCR:

$$\text{Corn yield} = -135 + 0.0717 \text{ year} - 0.0955 \text{ CSI}$$

$$\text{Soybean yield} = -43.2 + 0.0230 \text{ year} - 0.0238 \text{ CSI}$$

where year = calendar year and CSI = Crop Stress Index.

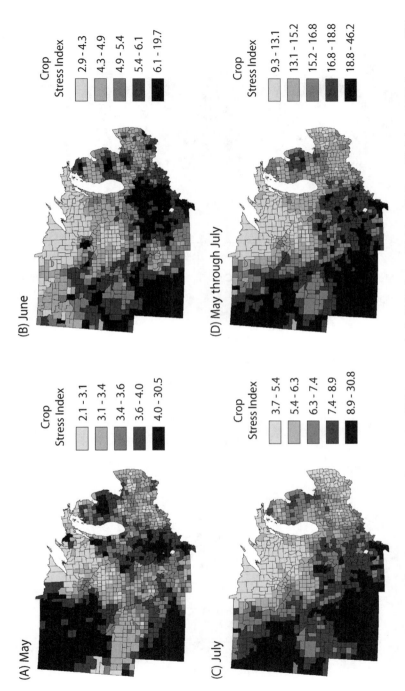

Figure 4.8. Distribution of the mean Crop Stress index (CSI) by county in the NCR during (A) May, (B) June, (C) July, and (D) the mean CSI sum for May—July from 1971–2001. See Fig. 4.6 for data sources.

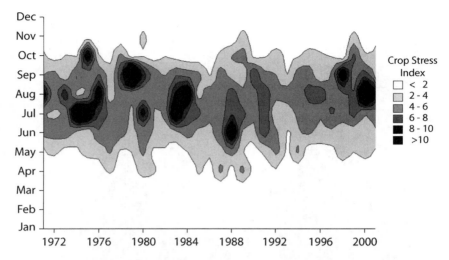

Figure 4.9. Month by year distribution of the Crop Stress Index (CSI) for the NCR (1971–2001). See Fig. 4.6 for data sources.

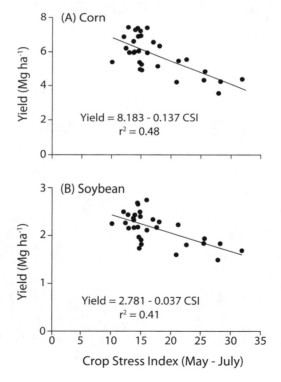

Figure 4.10. Average annual grain yields of non-irrigated (A) corn and (B) soybean in the NCR as a function of the May–July Crop Stress Index (1971–2001). See Figs. 4.3 and 4.6 for data sources.

The same approach can be used to examine the combined effect of year and climate on crop yield at the local scale. For example, Fig. 4.11 shows yield increases in corn and soybean in Kalamazoo County, Michigan—the location of the KBS LTER—during the 1971–2001 period, with slopes of 0.094 ($r^2 = 0.47$) and 0.042 ($r^2 = 0.51$), respectively. This yield trend also reflects technological advances, as it did at the regional level (Fig. 4.4). However, incorporating both year and climate does not improve the prediction of yield for Kalamazoo County (Fig. 4.12) as much as it did for the NCR (Fig. 4.10). At the county level, the CSI explains only 26% of the yield variance for corn (vs. 48% for the region) and 11% of the yield variance for soybeans (vs. 41% for the region).

The CSIs for Kalamazoo County during this time frame were not as extreme as they were in the NCR; in fact, locally only 1 year out of 30 had a CSI value greater than 25 (Fig. 4.12), whereas in the NCR, the CSI was greater than 25 in 5 out of the 30 years examined (Fig. 4.10). This is likely because Kalamazoo County is in the northern part of the NCR where temperatures are cooler and rainfall is greater than in the western

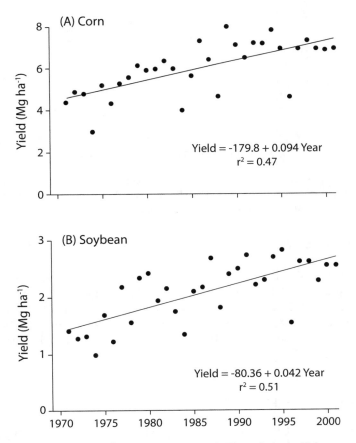

Figure 4.11. Trends in grain yields of (A) corn and (B) soybean in Kalamazoo County, Michigan (1971–2001). See Figs. 4.3 and 4.6 for data sources.

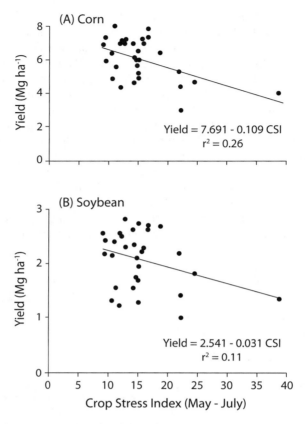

Figure 4.12. Average annual grain yields of (A) corn and (B) soybean in Kalamazoo County, Michigan, as a function of the Crop Stress Index (May–July) in Kalamazoo County, Michigan (1971–2001). See Figs. 4.3 and 4.6 for data sources.

portion. For example, in Kansas, the CSI was greater than 25 in 7 of the 30 years examined.

Also worth noting is that the slope of increasing corn yield over this period for Kalamazoo County (0.094 Mg ha^{-1} yr^{-1}; Fig. 4.11) is almost identical to the slope for the NCR (0.092 Mg ha^{-1} yr^{-1}; Fig. 4.4), showing how closely the KBS area tracked regional trends. Interestingly, the slope of soybean yields for Kalamazoo County over this same period (0.042 Mg ha^{-1} yr^{-1}) is greater than that for the NCR (0.028 Mg ha^{-1} yr^{-1}).

Climate Change Implications

Agriculture in the NCR will be greatly affected by climate change, with important consequences for crop stress. The earth's average global surface temperature rose 0.85°C from 1880 to 2012, with each of the last three decades being successively warmer than any decade since 1850 (IPCC 2013). In the Midwest, the largest temperature increases have occurred at night and in the winter (Pryor et al. 2014), and the length of the frost-free season has increased by 9 to 10 days across the NCR, leading to a longer growing season (Walsh et al. 2014). Since 1991, precipitation has increased across most of the NCR by 8 to 9%, as has the frequency of heavy downpours, especially in the Midwest and northern portion of the Great Plains region (Walsh et al. 2014). Heat waves have also increased, with three times the long-term average number of intense heat waves in 2011 and 2012 across the United States (Walsh et al. 2014).

Temperature trends in Michigan reflect global patterns with a cooling period from 1930 through 1980 followed by a warming trend beginning in the early 1980s (Andresen 2012). The warming has been concentrated in the winter months and mostly reflected in higher minimum temperatures. Mean precipitation has generally increased since the late 1930s—but with dry conditions in the late 1950s and early 1960s—as has the number of days with measureable precipitation, which is associated with more cloudiness (Andresen 2012). The decreasing amount and duration of ice cover on the Great Lakes have significant implications for Michigan's climate as the Great Lakes tend to moderate the local climate downwind of the lakes (Andresen 2012).

Climate Projections

For the next two decades, warming of 0.3 to 0.7°C in global mean surface air temperature is projected under a range of greenhouse gas scenarios, and even if greenhouse gas emissions are stopped, changes in the climate will continue for many centuries (IPCC 2013). By the end of the twenty-first century, projections of warming range from 0.3°C to 4.8°C depending on greenhouse gas emissions and other drivers, with projections showing a 1.5°C increase as likely under most emission scenarios (IPCC 2013). The highest rates of warming are expected to be over land, with more warm days and nights and fewer cold days and nights. Increases in the frequency of heat waves and heavy precipitation events are likely, with the contrast in precipitation between wet and dry regions of the globe increasing (IPCC 2013).

Climate change projections for specific regions are generally much more uncertain than global projections, and regionally focused studies are not yet available for every location (NRC 2010b). However, models consistently agree with projections of warmer annual temperatures for the NCR for both summer and winter months (IPCC 2013, Walsh et al. 2014), and the trend of longer frost-free and growing seasons is projected to continue for the region (Walsh et al. 2014). There is greater uncertainty in precipitation projections. For the northern half of North America, projections show increases in mean annual precipitation over winter and spring months, but models do not agree on summer or fall precipitation changes (IPCC 2013, Walsh et al. 2014), making it difficult to predict regional effects on agriculture. There is, however, high certainty that heavy precipitation events will increase in frequency and intensity, and the number of consecutive dry days is projected to increase (Walsh et al. 2014).

Implications for Row-Crop Production

Changes in climate greatly impact row-crop agriculture: temperature, precipitation amount and distribution pattern, cloud cover, and carbon dioxide (CO_2) levels affect plant growth, field practices, pests, and plant diseases, sometimes in conflicting ways (Tubiello et al. 2007, Hatfield et al. 2011, Hatfield et al. 2014, NRC 2010c). Over the last forty years, agricultural production in the United States has been affected by more climate disruptions, a trend that is expected to continue (Hatfield et al. 2014). The magnitude of the impacts on crop yields depends on location, the agricultural system, and the degree of warming (NRC 2010b) as well as the availability of water to the crop. We have used the CSI as a metric to assess the effects of changes in temperature and precipitation on corn and soybean yields in the NCR. While other factors will also affect crop yield, including rising atmospheric CO_2 levels, weed pressure, herbicide efficacy, and the spread of pests and diseases (Tubiello et al. 2007, Hatfield et al. 2011), there are too few data to develop an index that incorporates these effects.

A recent global analysis showed that from 1980 to 2008, corn and wheat yields were suppressed 3.8% and 5.5%, respectively, in many important agricultural countries because of increasing temperatures (Lobell et al. 2011). The United States was a notable exception, showing no detectable yield loss due to climate change. In fact, for the years 1982–2002, increased corn and soybean yields in the central and eastern United States were attributed to favorable climate conditions: a combination of more precipitation, longer growing seasons, and decreased summer average temperatures (Twine and Kucharik 2009). These favorable climate trends may have contributed 20–25% to the observed U.S. yield increases over this period. Kucharik and Serbin (2008) examined the variability of past temperature and precipitation county-level trends and their effect on crop yields in Wisconsin during 1976–2006. Yield trends were suppressed 5–10% in counties that had warmer summer temperatures. This negative impact was, however, counterbalanced by increases in precipitation that favored crop yields.

Our work shows that, in the 30-year period we analyzed, most periods of crop stress in the NCR have been short-term events in May through July (Fig. 4.9) that were characterized by high temperatures and below-average precipitation. For every unit increase in the CSI, yields decreased 0.14 Mg ha^{-1} for corn and 0.04 Mg ha^{-1} for soybean (Fig. 4.10). During this period, there were few years with back-to-back severe crop stress events (Fig. 4.4). Further back in the region's recorded climate history, however, severe continuous crop stress events helped create the 1930's Dust Bowl and depressed yields to near zero for several years in succession, illustrating the vulnerability of agricultural systems to prolonged and repeated climatic stress.

To date, U.S. agriculture has been effective at adapting to climate change (Hatfield et al. 2014), and under local temperature increases of up to 2°C adaptation has the potential to offset projected crop yield declines in North America (IPCC 2014a). At temperature increases of 4°C or more, however, the effectiveness of adaptation will be reduced and large risks to food security at global and regional scales are likely (IPCC 2014a,b). Climate disruptions are anticipated to have an increasingly negative impact on most U.S. crops by mid-century (Hatfield et al.

2014). Increased temperatures result in higher rates of soil water evaporation and crop transpiration, which could lead to an increase in soil water deficits (Hatfield et al. 2011). If climate change leads to longer periods of warmer and drier weather, producing high yields without irrigation will be increasingly challenging. While increased levels of CO_2 improve water-use efficiency for some plants (Hatfield et al. 2008), the benefit will be tempered by heat-related stresses that increase water demand (NRC 2010c). Moreover, NCR aquifers that provide irrigation water are already under stress because of unsustainable withdrawal rates (Kromm and White 1992) and are increasingly showing contamination by nitrate and pesticides. The IPCC projects an overall net negative impact of climate change on freshwater ecosystems (IPCC 2014a).

Recent studies have suggested that the effect of future warming on grain crops may be worse than previously recognized (Hatfield et al. 2011). For example, Kurcharik and Serbin (2008) found that for each degree of future warming, with no change in precipitation, corn yields could decrease by 13% and soybean by 16%. The authors note that while warmer and drier conditions during spring planting and fall harvest could help boost yields in some NCR states, higher summer temperatures will likely temper yield benefits. Likewise, Schlenker and Roberts (2009) estimated decreases in U.S. crop yields for corn, soybean, and cotton ranging from 30 to 46% by the end of the century under a slow warming scenario and 63 to 82% under a rapid warming scenario. By mid-century under conditions of increased temperatures and precipitation extremes, U.S. crop yields and farm profits are expected to decline while annual variation in crop production increases (Hatfield et al. 2014).

Agriculture will be affected both by changes in absolute values of temperature and precipitation and by increased climatic variation (Hatfield et al. 2011, 2014), and adaptive measures will be needed. Adaptation is not new to agriculture, and the analyses presented in this chapter highlight the extent to which agriculture has already adapted to trends and variability in temperature and precipitation. The challenge for the future is to adapt to more rapid and extreme climatic changes in the face of other environmental and social pressures and stresses (Easterling 2011, Hatfield et al. 2014), including increasing demands for agriculture to provide biomass for ethanol production (Robertson et al. 2008). Adaptive measures for agriculture include (1) relying on natural resources and inputs (e.g., water, energy, land); (2) technological innovation (e.g., breeding and genetic modification, water and soil conservation, pest management); (3) human ingenuity (e.g., relocating crop and animal production areas, improved agronomic practices); and (4) information and knowledge (e.g., environmental monitoring systems, risk management) (Easterling 2011). Although these measures have been effective in increasing crop yields to their current levels, it is not clear if further adaptation of agronomic practices and technologies—alone or in combination—will meet the challenge (Easterling 2011).

Uncertainties in climate projections for the NCR, coupled with varying climate trends at local levels (e.g., Kucharik and Serbin 2008), make adapting and planning for the future difficult. Agronomists are given the challenge of making cropping systems more resilient to climatic change (Hatfield et al. 2011) and using an ecosystem approach to agriculture (Robertson and Hamilton 2015, Chapter 1 in this volume) may help. Strategies such as cover and companion crop integration and no-till farming, which are included in the alternative crop management systems at the KBS LTER, can reduce the need for chemical subsidies and help manage crops under heat and/or water stress (Snapp et al. 2015, Chapter 15 in this volume). Increasing the organic matter content of soil increases soil water retention and reduces the need for water subsides, which will likely be required as the climate warms. Designing landscapes to optimize natural regulation of crop pests can reduce both crop loss and pesticide use (Landis and Gage 2015, Chapter 8 in this volume). Not only would such practices help agricultural ecosystems become more resilient and adaptable to climate change, they also have the potential to mitigate future climate change by sequestering carbon and by reducing the footprint of agronomic chemical use (Paul et al. 2015, Chapter 5 in this volume; Gelfand and Robertson 2015, Chapter 12 in this volume).

Because corn is more sensitive to heat stress than soybean (see Fig. 4.10), an immediate concern is whether the NCR will be able to sustain corn production under climate projections of increased heat and less water. How can farming in the NCR adapt to such projections? Formulating an answer to that question requires a long-term perspective based on the integration of both climate and crop production, such as in the development of the CSI—because if you cannot measure it, you cannot manage it. Interpreting and using indices such as the CSI will only become more important as an unprecedented global population places even greater demands on agricultural ecosystems for food, fuel, and fiber.

Summary

Since its conversion from prairie and forest following European settlement in the 1800s, the North Central Region (NCR) of the United States has become one of the most important crop-producing areas in the world. Government subsidies, economic forces, industrialization, and consumer preferences combined to shape the current agricultural landscape of the NCR. Temperature and precipitation patterns help to drive yield trends at regional and local scales. A simple Crop Stress Index (CSI) based on temperature and precipitation records across the region shows the strong climatic influence on yield of rain-fed corn and soybean over 1971–2001; each unit increase in the CSI results in a yield penalty of 0.14 and 0.04 Mg ha^{-1} for corn and soybean, respectively.

Overall yield trends for Kalamazoo County over 1971–2001, the location of KBS, are similar to those for the NCR, but crops in Kalamazoo County were less affected by climatic variability than crops in drier areas in the NCR. At both local and regional scales, few stress events spanned multiple years during this period. With projected changes in temperature and precipitation from human-induced climate change, crop stress is likely to increase, and with serious potential consequences. Climate uncertainties make adaptive measures both crucial and challenging to implement. A regional understanding of agriculture coupled with an ecosystem-level approach is needed to determine how interacting and ever-changing socioeconomic, climatic, and ecological forces will impact agriculture in the region.

References

Altieri, M. A. 1987. Agroecology: the scientific basis of alternative agriculture. Westview Press, Boulder, Colorado, USA.

Andresen, J. A. 2012. Historical climate trends in Michigan and the Great Lakes region. Pages 17–34 in T. Dietz and D. Bidwell, editors. Climate change in the Great Lakes region. Michigan State University Press, East Lansing, Michigan, USA.

Bailey, R. G. 1998. Ecoregions: the ecosystem geography of oceans and continents. Springer, New York, New York, USA.

Beachy, R. N., N. V. Federoff, R. B. Goldberg, and A. McHughen. 2008. The burden of proof: a response to Rosi-Marshall et al. Proceedings of the National Academy of Sciences USA 105:E9.

Carson, R. 1962. Silent spring. Houghton Mifflin, Boston, Massachusetts, USA.

Cronon, W. 1991. Nature's metropolis: Chicago and the Great West. W.W. Norton, New York, New York, USA.

DeVoto, B., editor. 1953. The journals of Lewis and Clark. Houghton Mifflin, Boston, Massachusetts, USA.

Easterling, W. E. 2011. Guidelines for adapting agriculture to climate change. Pages 269–286 in D. Hillel and C. Rosenzweig, editors. Handbook of climate change and agroecosystems: impacts, adaptation, and mitigation. Imperial College Press, London, UK.

ERS (Economic Research Service). 2011b. Adoption of genetically engineered crops in the U.S. U.S. Department of Agriculture, Washington, DC, USA. <http://www.ers.usda.gov/data/BiotechCrops/> Accessed September 15, 2011.

ERS (Economic Research Service). 2011a. Feed grains database: Yearbook tables. Tables 4 and 31. U.S. Department of Agriculture, Washington, DC, USA. <http://www.ers.usda.gov/Data/FeedGrains/FeedYearbook.aspx> Accessed October 13, 2011.

Feng, H., and B. A. Babcock. 2010. Impacts of ethanol on planted acreage in market equilibrium. American Journal of Agricultural Economics 92:789–802.

Gage, S. H. 2003. Climate variability in the North Central Region: characterizing drought severity patterns. Pages 56–73 in D. Greenland, D. Goodin, and R. Smith, editors. Climate variability and ecosystem response at long-term ecological research sites. Oxford University Press, New York, New York, USA.

Gage, S. H., and M. K. Mukerji. 1977. A perspective of grasshopper population distribution in Saskatchewan and interrelationship with weather. Environmental Entomology 6:469–479.

Gelfand, I., and G. P. Robertson. 2015. Mitigation of greenhouse gas emissions in agricultural ecosystems. Pages 310–339 in S. K. Hamilton, J. E. Doll, and G. P. Robertson, editors. The ecology of agricultural Landscapes: long-term research on the path to sustainability. Oxford University Press, New York, New York, USA.

Hamilton, J. G., and S. H. Gage. 1986. Outbreaks of the cotton tipworm, *Crocidosema plebejana,* related to weather in southeast Queensland, Australia. Environmental Entomology 15:1078–1082.

Hatfield, J., K. Boote, D. Fay, L. Hahn, C. Izaurralde, B. A. Kimball, T. Mader, J. Morgan, D. Ort, W. Polley, A. Thomson, and D. Wolfe. 2008. The effects of climate change on agriculture, land resources, water resources, and biodiversity in the United States. Pages 21–74 in P. Backlund, A. Janetos, D. Schimel, J. Hatfield, K. Boote, P. Fay, L. Hahn, C. Izaurralde, B. A. Kimball, T. Mader, J. Morgan, D. Ort, W. Polley, A. Thomson, D. Wofle, M. G. Ryan, S. R. Archer, R. Birdsey, C. Dahm, L. Heath, J. Hicke, D. Hollinger, T. Huxman, G. Okin, R. Oren, J. Randerson, W. Schlesinger, D. Lettenmaier, D. Major, L. Poff, S. Running, L. Hansen, D. Inouye, B. P. Kelly,

L. Meyerson, B. Peterson, and R. Shaw, editors. Synthesis and assessment Product 4.3. U. S. Department of Agriculture, Washington, DC, USA.

Hatfield, J. L., K. J. Boote, B. A. Kimball, L. H. Ziska, R. C. Izaurralde, D. Ort, A. M. Thomson, and D. Wolfe. 2011. Climate impacts on agriculture: implications for crop production. Agronomy Journal 103:351–370.

Hatfield, J., G. Takle, R. Grotjahn, P. Holden, R. C. Izaurralde, T. Mader, E. Marshall, and D. Liverman. 2014. Chapter 6: Agriculture. Pages 150–174 in M. Melillo, Terese (T. C.) Richmond, and G. W. Yohe, editors. Climate Change Impacts in the United States: The Third National Climate Assessment. U. S. Global Change Research Program. doi:10.7930/J02Z13FR.Heim, R. 2002. A review of twentieth-century drought indices used in the United States. Bulletin of the American Meteorological Society 83:1149–1165.

Hudson, J. C. 1994. Making the Corn Belt: a geographic history of middle-western agriculture. Indiana University Press, Bloomington, Indiana, USA.

IPCC (Intergovernmental Panel on Climate Change). 2013. Climate change 2013: the physical science basis. Contribution of Working Group I to the Fifth Assessment Report of the Intergovernmental Panel on Climate Change. T. F. Stocker, D. Qin, G.-K. Plattner, M. Tignor, S. K. Allen, J. Boschung, A. Nauels, Y. Xia, V. Bex and P. M. Midgley, editors. Cambridge University Press, Cambridge, UK. 1535 pages.

IPCC (Intergovernmental Panel on Climate Change). 2014a. Climate change 2014: impacts, adaptation, and vulnerability. Part B: Regional aspects. Contribution of Working Group II to the Fifth Assessment Report of the Intergovernmental Panel on Climate Change. V. R. Barros, C. B. Field, D. J. Dokken, M. D. Mastrandrea, K. J. Mach, T. E. Bilir, M. Chatterjee, K. L. Ebi, Y. O. Estrada, R. C. Genova, B. Girma, E. S. Kissel, A. N. Levy, S. MacCracken, P. R. Mastrandrea, and L. L. White, editors. Cambridge University Press, Cambridge, UK. 1132 pages.

IPCC (Intergovernmental Panel on Climate Change). 2014b. Climate change 2014: impacts, adaptation, and vulnerability. Part A: Global and sectoral aspects. Contribution of Working Group II to the Fifth Assessment Report of the Intergovernmental Panel on Climate Change. C. B. Field, V. R. Barros, D. J. Dokken, K. J. Mach, M. D. Mastrandrea, T. E. Bilir, M. Chatterjee, K. L. Ebi, Y. O. Estrada, R. C. Genova, B. Girma, E. S. Kissel, A. N. Levy, S. MacCracken, P. R. Mastrandrea, and L. L. White, editors. Cambridge University Press, Cambridge, UK. 688 pages.

Isard, S. A., S. H. Gage, P. Comtois, and J. M. Russo. 2005. Principle of the atmospheric pathway for invasive species applied to soybean rust. BioScience 55:851–861.

Jackson, R. D., W. P. Kustas, and B. J. Choudhury. 1988. A reexamination of the crop water stress index. Irrigation Science 9:309–317.

Jensen, P. D., G. P. Dively, C. M. Swan, and W. O. Lamp. 2010. Exposure and nontarget effects of transgenic Bt corn debris in streams. Environmental Entomology 39:707–714.

Kaitany, R., H. Melakeberhan, G. W. Bird, and G. Safir. 2000. Association of Phytophthora sojae with Heterodera glycines and nutrient stressed soybeans. Nematropica 30:193–199.

Kilar, J. 1990. Michigan's lumber towns: lumbermen and laborers in Saginaw, Bay City, and Muskegon, 1870–1905. Wayne State University Press, Detroit, Michigan, USA.

Kromm, D. E., and S. E. White, editors. 1992. Groundwater exploitation in the High Plains. University Press of Kansas, Lawrence, Kansas, USA.

Kucharik, C. J., and S. P. Serbin. 2008. Impacts of recent climate change on Wisconsin corn and soybean yield trends. Environmental Research Letters 3:034003.

Landis, D. A., and S. H. Gage. 2015. Arthropod diversity and pest suppression in agricultural landscapes. Pages 188–212 in S. K. Hamilton, J. E. Doll, and G. P. Robertson, editors. The ecology of agricultural Landscapes: long-term research on the path to sustainability. Oxford University Press, New York, New York, USA.

Landis, D. A., M. M. Gardiner, W. van der Werf, and S. M. Swinton. 2008. Increasing corn for biofuel production reduces biocontrol services in agricultural landscapes. Proceedings of the National Academy of Sciences USA 105:20552–20557.

Levine, E., J. L. Spencer, D. Onstad, and M. E. Gray. 2002. Adaptation of the western corn rootworm, Diabrotica vergifera Le Conte (Coleoptera: Chrusomelidae), to crop rotation: evolution of a new strain in response to a cultural control practice. American Entomologist 48:94–107.

Lewis, D. L. 1976. The public image of Henry Ford: an American folk hero and his company. Wayne State University, Detroit, Michigan, USA.

Lobell, D. B., W. Schlenker, and J. Costa-Roberts. 2011. Climate trends and global crop production since 1980. Science 333:616–620.

Matson, P. A., W. J. Parton, A. G. Power, and M. J. Swift. 1997. Agricultural intensification and ecosystem properties. Science 277:504–509.

Miles, M. R., R. D. Frederick, and G. L. Hartman. 2003. Soybean rust: Is the U.S. soybean crop at risk? APSnet Features. <http://www.apsnet.org/publications/apsnetfeatures/Pages/SoybeanRust.aspx> Accessed January 24, 2012.

NASS (National Agricultural Statistics Service). 2009a. 2007 Census of agriculture. Volume 1, Part 51, Geographic area series. United States summary and state data. U.S. Department of Agriculture, Washington, DC, USA.

NASS (National Agricultural Statistics Service). 2009b. Cropland data layer. 2009 edition [Geospatial Data Presentation Form: raster digital data]. U.S. Department of Agriculture, NASS Marketing and Information Services, Washington, DC, USA.

NASS (National Agricultural Statistics Service). 2011. Statistics by subject. U.S. Department of Agriculture, Washington, DC, USA. <http://www.nass.usda.gov/Data_and_Statistics/> Accessed on September 15, 2011.

NASS (National Agricultural Statistics Service). 2014a. Statistics by subject. U.S. Department of Agriculture, Washington, DC, USA. <http://www.nass.usda.gov/Data_and_Statistics/> Accessed on February 2, 2014.

NASS (National Agricultural Statistics Service). 2014b. Crop values annual summary. U.S. Department of Agriculture, Washington, DC, USA. <http://usda.mannlib.cornell.edu/MannUsda/viewDocumentInfo.do?documentID=1050> Accessed on February 2, 2014.

NRC (National Research Council). 2008. Genetically engineered organisms, wildlife, and habitat: a workshop summary. National Academies Press, Washington, DC, USA.

NRC (National Research Council). 2010a. The impact of genetically engineered crops on farm sustainability in the United States. National Academies Press, Washington, DC, USA.

NRC (National Research Council). 2010b. America's climate choices: Advancing the science of climate change. National Academies Press, Washington, DC, USA.

NRC (National Research Council). 2010c. America's climate choices: Adapting to the impacts of climate change. National Academies Press, Washington, DC, USA.

NRCS (Natural Resource Conservation Service). 1991. State soil geographic (STATSGO) data base: data use information. Miscellaneous Publication Number 1492, U. S. Department of Agriculture, Washington, DC, USA.

Oerke, E. C. 2007. Crop losses to animal pests, plant pathogens and weeds. Pages 116–120 in D. Pimentel, editor. Encyclopedia of pest management. Volume 2. CRC Press, Boca Raton, Florida, USA.

Parrott, W. 2008. Study of Bt impact on caddisflies overstates its conclusions: Response to Rosi-Marshall et al. Proceedings of the National Academy of Sciences USA 105:E10.

Paul, E. A., A. Kravchenko, A. S. Grandy, and S. Morris. 2015. Soil organic matter dynamics: controls and management for ecosystem functioning. Pages 104–134 in

S. K. Hamilton, J. E. Doll, and G. P. Robertson, editors. The ecology of agricultural Landscapes: long-term research on the path to sustainability. Oxford University Press, New York, New York, USA.

Pimentel, D. 1971. Evolutionary and environmental impact of pesticides. BioScience 21:109.

Pimentel, D., editor. 1981. Handbook of pest management, Volumes 1 and 2. CRC Press, Boca Raton, Florida, USA.

Pryor, S. C., D. Scavia, C. Downer, M. Gaden, L. Iverson, R. Nordstrom, J. Patz, and G. P. Robertson. 2014. Chapter 18: Midwest. Pages 418–440 in J. M. Melillo, Terese (T.C.) Richmond, and G. W. Yohe, editors. Climate change impacts in the United States: the Third National Climate Assessment. U.S. Global Change Research Program. doi:10.7930/J0J1012N.

Radcliffe, E. B., W. D. Hutchinson, and R. E. Cancelado, editors. 2008. Integrated pest management: concepts, tactics, strategies and case studies. Cambridge University Press, Cambridge, UK.

Robertson, G. P., J. C. Broome, E. A. Chornesky, J. R. Frankenberger, P. Johnson, M. Lipson, J. A. Miranowski, E. D. Owens, D. Pimentel, and L. A. Thrupp. 2004. Rethinking the vision for environmental research in US agriculture. BioScience 54:61–65.

Robertson, G. P., V. H. Dale, O. C. Doering, S. P. Hamburg, J. M. Melillo, M. M. Wander, W. J. Parton, P. R. Adler, J. N. Barney, R. M. Cruse, C. S. Duke, P. M. Fearnside, R. F. Follett, H. K. Gibbs, J. Goldemberg, D. J. Miadenoff, D. Ojima, M. W. Palmer, A. Sharpley, L. Wallace, K. C. Weathers, J. A. Wiens, and W. W. Wilhelm. 2008. Sustainable biofuels redux. Science 322:49.

Robertson, G. P., and S. K. Hamilton. 2015. Long-term ecological research at the Kellogg Biological Station LTER Site: Conceptual and experimental framework. Pages 1–32 in S. K. Hamilton, J. E. Doll, and G. P. Robertson, editors. The ecology of agricultural Landscapes: long-term research on the path to sustainability. Oxford University Press, New York, New York, USA.

Robertson, G. P., S. K. Hamilton, S. J. Del Grosso, and W. J. Parton. 2011. The biogeochemistry of bioenergy landscapes: carbon, nitrogen, and water considerations. Ecological Applications 21:1055–1067.

Robertson, G. P., and R. R. Harwood. 2013. Sustainable agriculture. Pages 111–118 in S. A. Levin, editor. Encyclopedia of biodiversity. Second edition. Volume 1. Academic Press, Waltham, Massachusetts, USA.

Robertson, G. P., and P. M. Vitousek. 2009. Nitrogen in agriculture: balancing the cost of an essential resource. Annual Review of Environment and Resources 34:97–125.

Rosi-Marshall, E. J., J. L. Tank, T. V. Royer, M. R. Whiles, M. Evans-White, C. Chambers, N. A. Griffiths, J. Pokelsek, and M. L. Stephen. 2007. Toxins in transgenic crop byproducts may affect headwater stream ecosystems. Proceedings of the National Academy of Sciences USA 104:16204–16208.

Schlenker, W., and M. J. Roberts. 2009. Nonlinear temperature effects indicate severe damages to U.S. crop yields under climate change. Proceedings of the National Academy of Sciences USA 106:15594–15598.

Schneider, R. W., C. A. Hollier, H. K. Whitam, M. E. Palm, J. M. McKemy, J. R. Hernández, L. Levy, and R. DeVries-Paterson. 2005. First report of soybean rust caused by *Phakopsora pachyrhizi* in the continental United States. Plant Disease 89:774.

Snapp, S. S., R. G. Smith, and G. P. Robertson. 2015. Designing cropping systems for ecosystem services. Pages 387–408 in S. K. Hamilton, J. E. Doll, and G. P. Robertson, editors. The ecology of agricultural Landscapes: long-term research on the path to sustainability. Oxford University Press, New York, New York, USA.

Soule, J. D., and J. K. Piper. 1992. Farming in nature's image: an ecological approach to agriculture. Island Press, Washington, DC, USA.

Tank, J. L., E. J. Rosi-Marshall, T. V. Royer, M. R. Whiles, N. A. Griffiths, T. C. Frauendorf, and D. J. Treering. 2010. Occurrence of maize detritus and a transgenic insecticidal protein (Cry1Ab) within the stream network of an agricultural landscape. Proceedings of the National Academy of Sciences USA 107:17645–17650.

Tatum, L. A. 1971. The southern corn leaf blight epidemic. Science 171:1113–1116.

Tilman, D., K. G. Cassman, P. A. Matson, and R. L. Naylor. 2002. Agricultural sustainability and intensive production practices. Nature 418:671–677.

Tivy, J. 1990. Agricultural ecology. Longman Scientific Technical, New York, New York, USA.

Tubiello, F. N., J.-F. Soussana, and S. M. Howden. 2007. Crop and pasture response to climate change. Proceedings of the National Academy of Sciences USA 104:19686–19690.

Twine, T. E., and C. J. Kucharik. 2009. Climate impacts on net primary productivity trends in natural and managed ecosystems of the central and eastern United States. Agricultural and Forest Meteorology 149:2143–2161.

Van Den Bosch, R. 1978. The pesticide conspiracy. University of California Press, Berkeley, California, USA.

Walsh, J., D. Wuebbles, K. Hayhoe, J. Kossin, K. Kunkel, G. Stephens, P. Thorne, R. Vose, M. Wehner, J. Willis, D. Anderson, S. Doney, R. Feely, P. Hennon, V. Kharin, T. Knutson, F. Landerer, T. Lenton, J. Kennedy, and R. Somerville. 2014. Chapter 2: Our changing climate. Pages 19–67 in J. M. Melillo, Terese (T.C.) Richmond, and G. W. Yohe, editors. Climate change impacts in the United States: the Third National Climate Assessment, U.S. Global Change Research Program. doi:10.7930/J0KW5CXT.

Weaver, H. L. 1946. A developmental study of maize with particular reference to hybrid vigor. American Journal of Botany 33:615–624.

5

Soil Organic Matter Dynamics
Controls and Management for Sustainable Ecosystem Functioning

Eldor A. Paul, Alexandra Kravchenko,
A. Stuart Grandy, and Sherri Morris

The composition of soil, particularly its organic matter content, reflects its role as a major controller of ecosystem functioning and soil fertility (Paul and Collins 1998, Basso et al. 2011, Bhardwaj et al. 2011). Soil organic matter (SOM), the largest global reservoir of terrestrial organic carbon (C), contains three to four times as much stored C as either the atmosphere or plant biomass. The soil biota, consisting of microorganisms as well as fauna, account for 1–3% of total soil C and complete the terrestrial C cycle by mineralizing SOM to carbon dioxide (CO_2) (Paul and Collins 1998, Robertson and Paul 2000). Soil organic matter is a complex, multi-structured, multicomponent pool of organic materials including decomposing plant residues, associated microorganisms and their products, and a biochemically transformed fraction, sometimes called humic material, that is complex in structure and often associated with soil minerals. Microbially derived C, because of its unique chemical structure and intimate association with minerals, is selectively protected and represents a particularly important component of total SOM. This pool serves as a dynamic source of labile nutrients and contributes to soil aggregate formation and erosion resistance (Robertson and Paul 1998, Grandy and Neff 2008).

The investigation of SOM dynamics provides a wealth of information on how organisms, including vegetation and soil biota, interact with climate, parent material, landscape, and management over time to influence ecosystem functioning (Collins et al. 1997). One of the most promising approaches for understanding SOM dynamics uses long-term incubations, together with tracers and density fractionation, to interpret SOM pools and their turnover.

The multiple components that make up SOM can be divided into three pools based on their turnover times (Paul et al. 2001a, b). The most labile, active pool is small at ~5% of total SOM. It includes some of the interaggregate fraction (i.e., not contained within soil aggregates) of plant residues and a portion of the soil biotic biomass, and has mean residence times (MRTs) of months to years. The slow pool, with MRTs ranging from months to decades, accounts for ~40% of total SOM and is a major source of soil nutrients that change with long-term management. Slow pool dynamics are controlled by aggregation and the association of microbial products with the calcium (Ca) and sesquioxide minerals as well as some silt and clay components. The SOM within soil aggregates (intraaggregate fraction) has both young and old constituents, with MRTs of decades. The oldest SOM is associated with silt and clay and is defined as the resistant pool; it is measured as the SOM that remains after acid hydrolysis. This third pool has the longest MRTs (century to millennia) and is best measured by carbon dating (Leavitt et al. 1996, Paul et al. 1997a).

Because SOM is a heterogeneous material including components with a wide range of turnover in response to abiotic and biotic controls, we currently cannot directly isolate and analyze SOM fractions based on their turnover dynamics. For example, there appears to be some old C (possibly charcoal) associated with the actively decomposing, recently added plant particulate material. On the other hand, old fractions such as the clays (Haile-Mariam et al. 2008) have some recently absorbed young microbial products. Old, resistant, non-hydrolyzable C is also known to contain recent plant-derived lignin that is not soluble in acid, even though on average the pool is very old (Paul et al. 2006).

Soils of the Kellogg Biological Station Long-term Ecological Research Site (KBS LTER) are ideal for studying how organisms, climate, parent material, landscape, and management influence SOM composition, dynamics, and ecosystem functioning. KBS soils developed in a moderately humid, temperate climate on glacial outwash over a period of ~18,000 years. They are moderately fertile Typic Hapludalfs (NRCS 1999) developed in an environment with a mean annual temperature of 9.0°C and annual precipitation of ~1,000 mm, under a northern, mixed hardwood forest with grassland openings attributed to fires promoted by native inhabitants ~700 C.E. (Robertson and Hamilton 2015, Chapter 1 in this volume). In the late nineteenth and early twentieth centuries, these soils were further disturbed by widespread deforestation and cultivation.

The soils underlying the KBS LTER consist of two main series: (1) Oshtemo: coarse-loamy, mixed, mesic, Typic Hapludalfs and (2) Kalamazoo: fine-loamy, mixed, mesic Typic Hapludalfs (Austin 1979, Mokma and Doolittle 1993, Crum and Collins 1995). The 0- to 20-cm surface layer has a texture of 39% sand, 43% silt, and 18% clay. This grades to 87% sand at a depth of 50–100 cm with 5% silt and 8% clay. Surface horizons can contain a small amount of inorganic C arising from the application of agricultural lime (mainly calcium and magnesium carbonates); in deeper horizons the presence of inorganic C reflects the calcareous origin of parent materials (Hamilton et al. 2007). Soil organic matter in surface horizons ranges from <10 to more than 30 g C kg^{-1} depending on landscape position, vegetation, and management history (Syswerda et al. 2011).

The organic matter content of soils reflects the balance between photosynthesis and decomposition, the two major driving factors in the global C cycle.

Decomposition transforms plant residues into microbial products that can become stabilized in the soil matrix, thereby maintaining 5–10% of the plant residue C as SOM (Follett et al. 1997). As the recipients of decomposition processes, soils represent a valuable storehouse of information on past vegetation, climate, and disturbance. They are also a major source of nutrients, especially nitrogen (N), and provide physical structure and moisture retention. Interactions of SOM within the soil matrix—particularly of silts and clays, but also aggregation with sand particles—determine the rooting environment for plants and both the storage and movement of soil moisture essential for plant growth.

The major factors controlling SOM dynamics are: (1) the quality of the incoming substrates, (2) the role of the soil biota and especially the microorganisms, (3) physical protection such as in aggregation, (4) interaction with the soil matrix such as the silts and clays as well as Ca and sesquioxides, and (5) the chemical nature of the SOM itself. These factors interact and are best studied together.

However, because soil takes so long to form and because SOM is such a small fraction of bulk soils, examination of these factors must involve long-term studies on well-characterized sites. Sites such as those in the LTER Network can supply the continuity and replication required to investigate SOM transformations and dynamics. In this chapter, we synthesize studies from the KBS LTER that focus on SOM dynamics in agricultural soils. We begin by examining soils of the Main Cropping System Experiment (MCSE, Table 5.1) and discuss the effect of landscape, vegetation, and agricultural management on SOM dynamics. We then discuss the biochemical controls on SOM dynamics, particularly as characterized by examining the size and turnover rates of important SOM constituents. And we end with an examination of the role of C inputs and microbial activities on the fate of C in agricultural soils.

Effect of Landscape, Vegetation, and Management on Soil Organic Matter Dynamics

Soils at the KBS LTER are prone to periodic drought, with some historical erosion, and thus have lower SOM content compared to some other nonglaciated cultivated soils in the Great Lakes region (Table 5.2). The Deciduous Forest system of the KBS LTER MCSE (Table 5.1) is representative of the native soil of the region (see Robertson and Hamilton 2015, Chapter 1 in this volume, for a description). The C in the SOM (hereafter referred to as soil organic carbon, SOC) of surface soils (0–20 cm; Table 5.2) has a MRT of 422 years, as determined by carbon dating (Paul et al. 2001a). This increases to 1712 years at 50- to 100-cm depth. Using acid hydrolysis to identify the old, resistant SOC (Paul et al. 2006, Plante et al. 2006), which is called non-hydrolyzable C (NHC), Paul et al. (2001a) found that 56% of the total SOC in the surface layer is NHC and is 977 years old (Table 5.2). Deeper in the forest soil profile (50- to 100-cm layer), the NHC accounts for only 23% of the SOC but is 4406 years old.

Such soil characterizations are most informative when examined in comparisons. Unifying concepts such as the content and structure of SOM and the role of the soil

Table 5.1. Description of the KBS LTER Main Cropping System Experiment (MCSE).[a]

Cropping System/Community	Dominant Growth Form	Management
Annual Cropping Systems		
Conventional (T1)	Herbaceous annual	Prevailing norm for tilled corn–soybean–winter wheat (c–s–w) rotation; standard chemical inputs, chisel-plowed, no cover crops, no manure or compost
No-till (T2)	Herbaceous annual	Prevailing norm for no-till c–s–w rotation; standard chemical inputs, permanent no-till, no cover crops, no manure or compost
Reduced Input (T3)	Herbaceous annual	Biologically based c–s–w rotation managed to reduce synthetic chemical inputs; chisel-plowed, winter cover crop of red clover or annual rye, no manure or compost
Biologically Based (T4)	Herbaceous annual	Biologically based c–s–w rotation managed without synthetic chemical inputs; chisel-plowed, mechanical weed control, winter cover crop of red clover or annual rye, no manure or compost; certified organic
Perennial Cropping Systems		
Alfalfa (T6)	Herbaceous perennial	5- to 6-year rotation with winter wheat as a 1-year break crop
Poplar (T5)	Woody perennial	Hybrid poplar trees on a ca. 10-year harvest cycle, either replanted or coppiced after harvest
Coniferous Forest (CF)	Woody perennial	Planted conifers periodically thinned
Successional and Reference Communities		
Early Successional (T7)	Herbaceous perennial	Historically tilled cropland abandoned in 1988; unmanaged but for annual spring burn to control woody species
Mown Grassland (never tilled) (T8)	Herbaceous perennial	Cleared woodlot (late 1950s) never tilled, unmanaged but for annual fall mowing to control woody species
Mid-successional (SF)	Herbaceous annual + woody perennial	Historically tilled cropland abandoned ca. 1955; unmanaged, with regrowth in transition to forest
Deciduous Forest (DF)	Woody perennial	Late successional native forest never cleared (two sites) or logged once ca. 1900 (one site); unmanaged

[a]Site codes that have been used throughout the project's history are given in parentheses. Systems T1–T7 are replicated within the LTER main site; others are replicated in the surrounding landscape. For further details, see Robertson and Hamilton (2015, Chapter 1 in this volume).

biota in ecosystem functioning are best studied by examining a range of soils under long-term management (Paul et al. 1997b). This is illustrated by comparing KBS native soil with two additional, related soils from the Great Lakes region of North America. The Hoytville, Ohio, soil also developed under Oak-Hickory vegetation in a similar manner to KBS, but on lower landscape elevations that had more moisture and higher clay contents. It has 17.8 g C kg^{-1} soil with an MRT of 920 years in

Table 5.2. Soil organic carbon (SOC) characterization of total and non-hydrolyzable fractions in forest and agricultural soils.[a]

Soil[b]	SOC (g kg⁻¹)	MRT (years)	NHC (%)	MRT NHC (years)
KBS LTER, MI (Forest)				
0–20 cm	10.7 (0.5)	422 (51)	56	977 (50)
25–50 cm	2.6 (0.4)	933 (67)	23	895 (54)
50–100 cm	1.3 (0.1)	1712 (50)	21	4406 (65)
Hoytville, OH (CT)				
0–20 cm	17.8 (0.7)	920 (53)	46	1770 (45)
25–50 cm	8.6 (0.40)	2627 (55)	45	5660 (870)
50–100 cm	4.3 (0.1)	6607 (79)	44	9875 (75)
Lamberton, MN (CT)				
0–20 cm	17.9 (1.0)	1100 (53)	49	1510 (45)
25–50 cm	8.7(0.7)	3100 (55)	45	3965 (65)
50–100 cm	4.3(0.2)	6107 (75)	48	7285 (90)

[a]Includes SOC content, mean residence time (MRT), non-hydrolyzable C fraction (NHC), and MRT of the NHC. Data are means (±SE, when applicable).
[b]The KBS LTER site is the Deciduous Forest system of the MCSE (Table 5.1) and the other two sites are agricultural row-crop production sites with conventional tillage (CT) and chemical use.
Source: Paul et al. (2001a, b).

the surface layer (Table 5.2). Soil at 50- to 100-cm depth contains 4.3 g C kg⁻¹, with an MRT of 6607 years, and thus has higher SOC content and is older than KBS soil. The long time required for SOM development is reflected in the MRT of 9875 years for the oldest NHC fraction, which accounts for 44% of total SOC at 50–100 cm. The grassland-derived soil from Lamberton, Minnesota, shows similar SOC levels and proportions of NHC as the forest-derived Hoytville soil. The Lamberton SOC has somewhat longer MRTs than Hoytville and substantially longer MRTs than those of KBS. In all cases, a strong relationship exists between the MRT of the SOC and the amount of SOC present, indicating the time it takes for SOM to accumulate.

Long-term incubations of these soils (Collins et al. 2000) showed that although SOC from deeper depths in the profile had very long MRTs, when brought into the laboratory and disturbed, the samples released nearly as much CO_2 per unit of SOC as did soil from the surface horizons. This implies that the long MRTs were the result of matrix interactions, such as those with silt and clay, and profile position and not necessarily intrinsic chemistry. Labile constituents are decomposed early in the incubation and thus largely control the size of the active SOC pool. This was confirmed by the observation that the stable C isotope ratios of CO_2 produced in incubations of soils obtained from cultivated sites moved toward the ratio produced by native-site soils late in their long-term incubation. At that point in the incubation, microbes are decomposing the older, native-derived materials. The stable C isotope ratio of the respired CO_2 of the cultivated sites' SOC did not become equal to that of CO_2 from the native sites' SOC because the total CO_2 evolved was less than 10% of the total SOC (Collins et al. 1999).

Effect of Management on Soil Organic Matter Levels, Temporal Trends, and Spatial Variability Patterns

Management strongly affects SOM accumulation, especially in the uppermost soil horizons (Robertson et al. 1993, 1997) and is primarily restricted to the top 7 cm of the soil profile (De Gryze et al. 2004, Syswerda et al. 2011). While using concentration (e.g., g C kg^{-1}) to describe SOC content is suitable in many situations, estimates of ecosystem C storage must be made on a surface-area basis (e.g., g C m^{-2}, kg C ha^{-1}), which incorporates soil bulk density and depth of soil sampling. Changes in both bulk density and distribution of SOC in the soil profile are known to occur in response to changes in land use, including agricultural management (Morris et al. 2007).

Twelve years after establishment of the seven MCSE systems on previously cropped soils at the LTER main site (Table 5.1), the Early Successional system had gained 380 g C m^{-2} in the upper 5 cm of surface soil, representing an increase of 61% compared to the Conventional system (Grandy and Robertson 2007). Other systems also gained SOC, in the order Alfalfa (341 g C m^{-2}, 55% greater) > No-till (264 g C m^{-2}, 43% greater) > Poplar (229 g m^{-2} C, 37% greater) > Biologically Based (148 g m^{-2} C, 24% greater) > Reduced Input (107 g m^{-2} C, 17% greater). In contrast, the soils of the Deciduous Forest and Mown Grassland (never tilled) systems had about 2.5 times more SOC than the Conventional system in the uppermost 5 cm.

Annual SOC accumulation rates of the MCSE main site systems relative to the Conventional system ranged from 32 g C m^{-2} yr^{-1} in the Early Successional community down to 8.9 g C m^{-2} yr^{-1} in the Reduced Input system. Whether these accumulation rates represent absolute gains of SOC or slower losses of SOC relative to the Conventional system depends on whether Conventional system soils were stable or were losing C over the 12-year study period. In 2006 and 2007, Senthilkumar et al. (2009a, b) remeasured SOC at geo-referenced locations in the MCSE, where SOC had previously been measured in 1988, and in the nearby Interactions Experiment, where SOC had previously been measured in 1986. The studied sites had a common history of conventional agricultural management for at least the prior 70–100 years. In both experiments, SOC content appeared to decline under conventional tillage management (Table 5.3). Carbon losses were less in the No-till and Biologically Based systems, consistent with other studies (Grandy and Robertson 2007; Syswerda et al. 2011) that show SOC gains in these systems relative to the Conventional system.

That C should be lost from these long-cultivated soils is surprising, and may be related to wintertime warming; Senthilkumar et al. (2009a) documented a significant local increase in the number of days per year with above-freezing air temperatures over the 20-year study period (see also Robertson and Hamilton 2015, Chapter 1 in this volume). These findings are consistent with other studies around the world; increasing temperatures are associated with observed soil C losses (Bellamy et al. 2005, Stevens and van Wesemael 2008). However, the contribution of accelerated erosion over this period cannot be ruled out (Wischmeir and Smith 1961), although the adoption of chisel in place of moldboard plowing in 1996 should have slowed rather than increased erosion in these well-drained soils. With continued erosion and the projected further increase in global temperature, the adoption of no-till management or the inclusion of cover crops in the crop rotation (Willson et al. 2001) may be necessary to sustain present soil C levels.

Results from KBS show that C losses in near surface soils can be slowed or even reversed with no-till and biologically based farming practices. Rates of SOC accumulation in the top 5–7 cm of no-till relative to conventional tillage are ~22–30 g C m^{-2} yr^{-1} (Robertson et al. 2000, Grandy and Robertson 2007) and are similar to those for the entire Ap horizon (33 g C m^{-2} yr^{-1}, Syswerda et al. 2011), indicating that most change occurs in the upper few centimeters. This is consistent with other estimates for U.S. Midwestern cropping systems (West and Marland 2002). There is no evidence for C change in deeper horizons of no-till soils at KBS (to 1 m; Syswerda et al. 2011), which means that whole-profile C change is dominated by changes in surface soils. That coefficients of variation for soil C increase from 7% in Ap to 29% in B/Bt to 92% in Bt2/C horizons (Syswerda et al. 2011) underscores the difficulty with which subsurface soil C change can be detected if it occurs at all (Kravchenko and Robertson 2011).

Table 5.3. Baseline and contemporary SOC content and changes in SOC in the MCSE and Interactions Experiment.[†]

System	SOC[‡] (g kg^{-1} soil)[*]		Change in SOC (g kg^{-1} soil)[††]
MCSE	1988	2006	From 1988 to 2006
Conventional	9.3 (0.7)[a]	8.2 (0.3)[a]	−1.15 (0.27)[a]
No-till	10.6 (0.5)[a]	9.9 (0.5)[b]	−0.71 (0.48)[ab] NS
Biologically Based	10.0 (0.5)[a]	9.8 (0.3)[b]	−0.16 (0.29)[b] NS
P value	0.13	0.002	0.06
Interactions Experiment	1986	2007	From 1986 to 2007
Conventional tillage			
N-fertilized	9.8 (1.4)[a]	9.5 (0.5)[a]	−0.22 (0.46)[ab] NS
Not N-fertilized	9.8 (1.4)[a]	8.2 (0.5)[b]	−1.54 (0.51)[a]
No-till			
N-fertilized	7.4 (1.4)[a]	8.3 (0.5)[ab]	0.31 (0.48)[b] NS
Not N-fertilized	8.7 (1.4)[a]	8.9 (0.5)[ab]	0.19 (0.51)[b] NS
P value	0.11	0.07	0.07
Mown Grassland (never tilled)	23.0 (0.9)	15.3 (0.6)	−7.7 (1.0)

[†]Mean SOC content (±SE) in 1988 and 2006 for the MCSE (see Table 5.1) and in 1986 and 2007 for the Interactions Experiment and Mown Grassland. Interactions Experiment systems are also corn–soybean–wheat rotations, but either tilled or no-till and N-fertilized or not N-fertilized in a full factorial design (see http://lter.kbs.msu.edu for a description). The Mown Grassland (never tilled) system is part of the MCSE (Table 5.1) and is adjacent to the Interactions Experiment. P values are from an ANCOVA for testing the significance of the factor effects.
[‡]Sampled to 15-cm depth in the MCSE and to 20 cm in the Interactions Experiment.
[*]Within the same column, values followed by the same letter for a given experiment are not significantly different at $\alpha = 0.1$.
[††]Within this column, values followed by the same letter for a given experiment are not significantly different at $\alpha = 0.05$.
Note: NS = the mean value is not significantly different from zero ($P < 0.05$).
Source: Senthilkumar et al. (2009a).

The soil C increases of 50 g C m^{-2} yr^{-1} in the Biologically Based system (Syswerda et al. 2011) are particularly intriguing. Syswerda et al. examined SOC in the full soil profile (to 1 m), whereas most studies including the aforementioned one by Senthilkumar et al. (2009a) sampled surface soils (to 15 or 20 cm depth: Table 5.3). The soils in the Biologically Based system received no manure or compost and soil disturbance in this system was more intensive than in the Conventional system because weed populations were controlled with cultivation. Although this system has a legume cover crop, total biomass inputs are not higher than those in the Conventional system, which produces more grain crop biomass. Thus, C accumulation in this system cannot be attributed to reduced soil disturbance or the quantity of biomass or other C inputs, but is instead due to some other factor that is changing the processing of SOM. One possible explanation is that the incorporation of legume cover crops in the rotation cycle (Table 5.1) is altering the decomposition and stabilization of SOM (Willson et al. 2001). Differences in soil communities, the biochemistry of plant inputs, or their interaction may have influenced soil aggregation or other factors that regulate SOM dynamics (Grandy and Robertson 2007). Although the mechanism is not yet clear, the most likely explanation for enhanced aggregation and soil C in the Biologically Based system is related to the red clover cover crop.

Additionally, SOM change in the Biologically Based system appears to vary with topography. On undulating glacial terrain, topography is one of the most influential factors governing spatial patterns of SOM at field scales, while strongly interacting with land use and management. Senthilkumar et al. (2009b) observed a tendency for greater gain in SOC in topographic depressions ("valleys") of the Biologically Based system than in higher areas of plots (Fig. 5.1). Muñoz et al. (2008) concluded that greater crop biomass inputs together with differences in residue quality contributed to greater SOC accumulation in depressions in spite of higher levels of soil moisture for decomposition. Considering that these results were obtained from small-scale (1-ha) plots of the MCSE, where terrain variations are substantially less than those typical of whole watersheds (e.g., average MCSE terrain slopes are around 1°, ranging from 0–5°), one can suspect that the interactions between management strategies and topographical effects are even more important elsewhere.

Erosion has not yet been explicitly measured in KBS soils, which is an important omission. Voroney et al. (1981), by using the van Veen and Paul (1981) model to predict long-term levels of SOC, found that the inclusion of the universal soil loss equation (Wischmeier and Smith 1961) greatly altered the predicted levels of SOC.

Irrespective of topography and historical soil erosion, there are other noticeable differences in the spatial patterns of SOC under different land management practices at KBS LTER (Kravchenko et al. 2006). Semivariance analysis shows that SOC in the MCSE No-till and Biologically Based systems is more spatially autocorrelated than in the Conventional system at scales of tens of meters (Fig. 5.2). Another study observed stronger spatial structure in the Poplar system than in the Conventional system even at 1–2 m scales (Stoyan et al. 2000).

That no differences in spatial patterns among MCSE plot locations were evident in the 1988 spatial analyses (prior to establishment of the experimental plots in 1989; Robertson et al. 1997) indicates that these system-specific patterns have

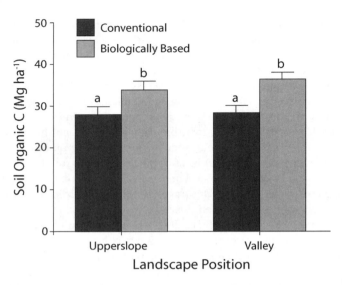

Figure 5.1. Differences in soil soil organic carbon (SOC) between the Conventional and Biologically Based systems at two contrasting landscape positions in the KBS Main Cropping System Experiment (MCSE, described in Table 5.1; data represent 0-30 cm depth, indicate mean ± SE. Similar letters indicate similar values ($\alpha = 0.05$.). Modified from Senthilkumar et al. (2009b).

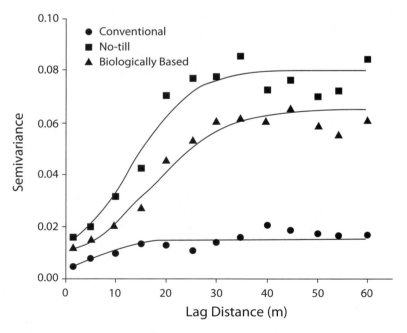

Figure 5.2. Variograms with model fits (solid lines) for SOC in the Conventional, No-till, and Biologically Based systems of the MCSE. Modified from Kravchenko et al. (2006).

developed since 1988 in response to different management practices. In particular, the reduction in soil disturbance in the No-till and Poplar systems apparently led to more pronounced site-specific variations in factors affecting soil C storage and C and N mineralization (Kravchenko and Hoa 2008). Likewise, replacement of uniform fertilizer applications under Conventional management with cover crops under Biologically Based management may also have led to differences in spatial variability, in this case because of a greater spatial variability of biomass inputs (Kravchenko et al. 2006, Muñoz et al. 2008).

Roots also play a major role in controlling SOM dynamics through their effects on aggregation and the movement of C to depth (Kavdir and Smucker 2005). Crop type and growth habits can potentially affect root-derived SOM, though probably over long time scales. Genetically engineered Bt corn, which is thought to produce litter that decomposes more slowly than traditional varieties, did not alter soil C and N pools relative to that of nonengineered corn over 7 years at the KBS LTER (Kravchenko et al. 2009). In another example, an herbivorous insect infestation of the Poplar system was found to affect the N use and mineralization of the defoliated poplar plants, but did not detectably alter soil N pools (Russell et al. 2004).

Research at KBS LTER has also documented the destabilizing effect of land-use intensification on soil aggregation and SOC mineralization. The conversion of long-term grasslands and other set-aside lands in the USDA Conservation Reserve Program (CRP) to row-crop agriculture (Feng and Babcock 2010) poses a potential threat to stored SOC (CAST 2011). Recent experiments at KBS demonstrated the rapid and destabilizing effect of tilling grasslands on aggregation and the potential implications for SOM dynamics. Grandy and Robertson (2006a, b) plowed a portion of the Mown Grassland (never tilled) community and followed changes in soil aggregation, CO_2 emissions, and microbial activity. They found that after a single spring tillage event, aggregates in the 2000–8000 μm size class declined substantially from 34% to 19% of total aggregates. Associated with these changes, the newly cultivated sites lost an average of 1.4 g C m^{-2} d^{-1}, derived from both plant litter inputs and native SOM, between May and October over 3 years. Such results raise concerns about the long-term environmental implications of expanding crop production into CRP and other grasslands to support the biofuel industry (Robertson et al. 2011, Gelfand and Robertson 2015, Chapter 12 in this volume). More work is needed to better understand and perhaps mitigate changes in SOM under different scenarios of land-use change.

Afforestation is another land-use change that impacts the terrestrial C cycle through changes in SOM (Morris and Paul 2003). Soil carbon accrual in afforested former agricultural lands at KBS, at the nearby Kellogg Forest, and at the Russ Forest in Cass County, Michigan—all on the same soil series and with stands of 50–60 years age—are shown in Fig. 5.3 (Morris et al. 2007). Changes were ascertained by comparison with adjacent agricultural fields. In contrast to C, the N content in afforested soils appeared to decrease in the MCSE and Kellogg Forest soils, but increase in the Russ Forest soils. The N accumulation at Russ Forest was in excess of that expected based on localized atmospheric N deposition (Morris et al. 2007). One explanation for the different responses is that afforested soils at KBS may have higher N concentrations because of manure applications when the

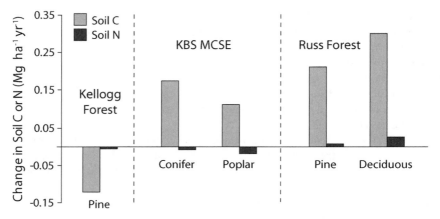

Figure 5.3. Change in soil organic carbon (C) and nitrogen (N) following afforestation of previously cropped Oshtemo/Kalamazoo loam soils at the Kellogg Forest, two tree plantations in the MCSE, and the Russ Forest in nearby Cass County, Michigan. From S. Morris (unpublished data).

Table 5.4. A comparison of SOC, total nitrogen (N), and C:N mass ratio of soils from the MCSE Conventional system and six privately owned sites in Cass County, Michigan.[a]

Site	Number of Samples	Profile SOC (Mg ha⁻¹)	Profile N (Mg ha⁻¹)	C:N Ratio
MCSE	6	59.1 (7.3)[bc]	8.7 (0.71)[c]	6.73 (0.24)[a]
Field 1	5	37.6 (3.6)[a]	4.1 (0.4)[a]	9.1 (0.09)[bc]
Field 2	5	63.4 (5.0)[ac]	6.0 (0.6)[a]	10.6 (0.26)[c]
Field 3	5	59.7 (9.9)[bc]	5.0 (0.7)[ab]	11.9 (0.66)[c]
Field 4	5	44.0 (1.7)[ab]	4.6 (0.2)[a]	9.7 (0.43)[b]
Field 5	5	46.6 (4.6)[ab]	5.0 (0.4)[ab]	9.4 (0.84)[b]
Field 6	5	73.7 (2.8)[c]	7.1 (0.8)[bc]	10.7 (0.78)[c]

[a]Samples to a soil profile depth of 1 m, corrected for equivalent weights. See Table 5.1 for a description of the MCSE Conventional system. Field sites 1–6 on the same soil type, under conventional till corn–soybean rotation. Data are means (±SE); means followed by similar letters do not differ significantly at $a = 0.05$.
Source: Morris et al. (2007).

land was first cleared, and thus less capacity to accumulate N. Total N content is somewhat higher in KBS soils than on local commercial farms with similar soil C contents (Table 5.4), resulting in a lower C:N ratio for the KBS soils in this comparison. These data further demonstrate the capacity for these soils to accrue C and N during afforestation in most cases, depending on tree species composition and SOM content. That C accrued in these soils indicates that they have not reached C saturation—the concept that a particular soil under a specific set of controls (such as the level of aggregation) can only sequester a specific amount of C. Soil C saturation been observed in both agricultural and forested soils (Six et al. 2002a, b).

When afforestation includes pines (*Pinus* spp.), soil calcium (Ca) content appears to support SOM accrual in a manner not found under deciduous species (Paul et al. 2003). Addition of Ca to soils following afforestation with pines, and with the addition of pine litter, increased the amount of stabilized SOM and decreased its decomposition rate, confirming the importance of Ca in regulating litter decomposition and SOM dynamics (Brewer 2004). Long-term stabilization of SOM through increased Ca on soils planted to pines could improve site fertility on this and other former agricultural sites on similar soil types. Calcium forms bridges between the SOM and the soil matrix such that its interaction with silt and clay-sized particles stabilizes and increases SOM levels (Baldock and Nelson 2000).

Pools and Fluxes That Control Soil Organic Matter Dynamics

A number of studies have provided insights into the mechanisms for SOM accrual in agricultural soils, including: (1) physical protection of partially decomposed plant residues found in inter- and intraaggregate fractions of SOM; (2) biochemical protection of SOM due to a molecular structure that is difficult for microbes to metabolize and is therefore recalcitrant; (3) chemical protection due to SOM associations with silt and clay; and (4) differences in soil biota that affect SOM sequestration and turnover.

Whether the plant residues and their associated biota are free or protected within soil aggregates controls their short-term breakdown (Fortuna et al. 2003). The incorporation of plant and microbial-derived C into aggregates protects plant residues (De Gryze et al. 2004), associated microbial C (Smith and Paul 1990), and microbial products (Paul and Clark 1996) for periods of weeks to decades depending on soil conditions (e.g., clay content and water availability) and management (e.g., tillage and crop rotation). Changes in land use that alter patterns of soil disturbance and the quantity and quality of plant residues will influence aggregate formation and stabilization, resulting in long-term effects on SOM dynamics (Willson et al. 2001).

KBS represents a unique environment in which to study soil aggregation because of the diversity of managed and unmanaged ecosystems that occur in close proximity on a common sand- and silt-dominated soil. There are few comparable places where relationships between aggregation and SOM dynamics can be studied in replicated systems that represent a range of management and chemical inputs. Most studies of soil structural processes and organic-mineral interactions have examined soils with high clay content. Differences in SOM across the seven MCSE systems at the LTER main site, developed since its inception in 1989, are closely associated with changes in soil aggregates, particularly the distribution of SOC in different aggregate size fractions, as indicated by an index of aggregation known as the mean weight-diameter (Fig. 5.4). The Biologically Based and No-till systems have more aggregates in the >2000-μm size class. Strong relationships between aggregation and SOM content indicate that physical protection is a key factor controlling short-term SOM dynamics in KBS soils (Grandy and Robertson 2007).

Density, particle size, and incubation-based measurements have been used at KBS to identify fractions enriched in SOM and sensitive to changes in management

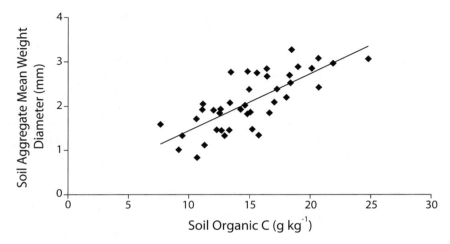

Figure 5.4. Relationship between SOC and soil aggregate size (mean weight-diameter; mm) in the MCSE. From A.S. Grandy (unpublished data).

(Sollins et al. 1999). These methods have shown that the greatest C sequestration potential lies in the mineral-associated fractions and in fine, intraaggregate particulate C. For example, Grandy and Robertson (2007) found that 82% of the C increase in the 2000- to 8000-μm size class of the No-till system was due to mineral-associated SOM. Incubation-based methods for separating fractions into active, slow, and resistant pools demonstrated the greatest potential C accumulation in the slow pool (6.5 to 21.4 g C m^{-2} yr^{-1}) followed by the resistant (0.66 to 16.05 g C m^{-2} yr^{-1}) and active pools (0 to 0.24 g C m^{-2} yr^{-1}).

Consistent with these results, De Gryze et al. (2004) found that in the Early Successional system, mineral-associated SOM accumulated C at 5- to 13-fold higher rates than did other fractions, while in the Poplar stands the fastest C accumulation rates occurred in the fine intraaggregate fraction. Soil organic matter pools with higher turnover times, such as the interaggregate pool, are extremely sensitive to disturbance and may show large, rapid changes in response to management. However, these pools do not have the long-term sequestration potential of mineral-associated pools with slower turnover times (Six et al. 2002a, b).

Sollins et al. (2009) used sequential density fractionation to separate KBS soil particles collected from 0- to 25-cm depth into "light" predominately mineral-free organic matter from the "heavy" particles associated with the soil minerals. Quartz, alkali feldspars, vermiculite, and kaolinite are major components of the fraction with density < 2.6 g cm^{-3}. Quartz dominates the 2.6–2.8 g cm^{-3} fraction, accounting for 73% of soil dry weight. The primary minerals hornblende, hematite, and epidote account for 2% of soil dry weight. The density of the particles was found to depend on the thickness of their SOM coating as well as the density of the mineral particles themselves.

The <1.65 g cm^{-3} density fraction represented 16% of total soil C with a total lignin-phenol content that accounted for 5.5% of total C. The 1.65–1.85 g cm^{-3}

density fraction accounted for 8% of soil C and the 1.85–2.00 g cm^{-3} fraction for 14%. Phenol and lignin concentrations were lower in the higher-density fractions and the C:N ratio decreased with density. The ^{14}C-based MRTs measured by Sollins et al. (2009) ranged from 108 years in the 2.00–2.30 g cm^{-3} density fraction to 165 years in the <1.65 g cm^{-3} fraction and 225 years in the 1.65–1.85 g cm^{-3} fraction. The heaviest densities (> 2.80 g cm^{-3}) had an average MRT of 1050 years. The average MRT for all densities was similar to the MRT of 420 years shown for KBS soil in Table 5.2.

Incubation-based measurements can also identify fractions enriched in SOM and sensitive to changes in management. As noted earlier, Paul et al. (1999a) used long-term incubations to show how only 5 years of MCSE management affected SOM fractions. From 10 to 15% of total SOC was oxidized during their 365-day incubations, with the highest CO_2 production occurring in soils with the higher SOC contents (Table 5.5). The active fraction, as defined by curve fitting of the CO_2 evolved early during incubation (Paul et al. 1995, 2001b; Horwath et al. 1996), represented 1.8–3.3% of the SOC. This is equivalent to half the C content of the interaggregate fraction of SOC isolated by density fractionation, agreeing with the analysis of Haile-Mariam et al. (2008) that the interaggregate fraction contains some nonlabile constituents. The slow pool represented 41–45% of the SOC and serves as the primary basis for soil fertility and ecosystem SOM dynamics. It is made up, in part, by materials in the intraaggregate fraction and by the more labile components of the silt and clay fractions (18–21%) shown to be corn-derived, based on stable carbon isotope ratios as explained in more detail below (Table 5.6). The interaggregate fraction, measured by density gradient fractionation before aggregate dispersion, is also referred to as the light fraction. The intraaggregate fraction, measured by sieving the >53-μm soil fraction after dispersion, is also often referred to as the particulate organic material.

The 6-*M* acid hydrolysis treatment used to measure the size of the resistant (non-hydrolyzable) fraction of SOC does not remove modern lignin plant

Table 5.5. Analytically defined pools and mean residence time (MRT) of SOC 5 years after the initiation of the MCSE.^{a}

System	SOC Pool (g kg^{-1})	CO_2–C Produced (% of SOC)	Active C Pool		Slow C Pool		Resistant C Pool	
			(% of SOC)	MRT (days)	(% of SOC)	MRT (years)	(% of SOC)	MRT (years)
Conventional	8.7	10	2.0	45	42	13.1	56	1435
Poplar	9.1	13	3.3	66	41	11.0	ND	ND
Early Successional	9.0	15	2.5	36	42	9.0	ND	ND
Mown Grassland (never tilled)	14.6	15	1.8	30	45	9.0	53	170

^{a}Long-term incubations (365 days) and acid hydrolysis used to determine pools and residence times. See Table 5.1 for a description of MCSE systems.

Note: CO_2–C = carbon dioxide–C; ND = not determined.

Source: Paul et al. (1999a).

constituents or some microbial cell walls (Paul et al. 2006, Plante et al. 2006), and thus, the non-hydrolyzable fraction contains some young C. However, it contains mostly old C and provides a reasonable estimate of the size of the resistant fraction required for modeling purposes. Carbon dating of this fraction in native KBS forest surface soils yields a MRT of 977 years (Table 5.2). The MRT of bulk soil C in the upper 20 cm of cultivated KBS soil was 546 years (Paul et al. 2001a) relative to 422 years for the native forest soil (Table 5.2), indicating that some labile C had been lost on cultivation and not replaced by more recent plant inputs.

Plants with the C_4 photosynthetic pathway have higher $^{13}C{:}^{12}C$ isotope ratios (expressed as $\delta^{13}C$ in parts per million or ‰) as compared with C_3 plants. Typically, C_4 plants such as corn, sorghum, and switchgrass have $\delta^{13}C$ values of -10 to -14‰ as compared to values of -22 to -30‰ in C_3 plants such as soybean, wheat, and alfalfa. This difference has been used to measure the dynamics of the inter- and intraaggregate particulate material and the SOM associated with silt and clay fractions. The growth of corn for 10 years on a KBS site adjacent to the MCSE that was formerly dominated by C_3 plants allowed Haile-Mariam et al. (2008) to examine SOM dynamics using δ ^{13}C analysis of soils from the field before and after 800 days of incubation. The interaggregate fraction, which represents partially decomposed plant residues (as shown by C:N ratios of 21:1), still had 45% noncorn C after 10 years of continuous corn, showing that even interaggregate fractions contain some old C (Table 5.6). This increased to 72% noncorn C by the end of the incubation, showing the difference in lability between the corn and noncorn C. Mid-infrared spectroscopic analysis of this fraction showed that a small

Table 5.6. Properties of SOC and $\delta^{13}C$ analysis of soils from continuous corn sites adjacent to the MCSE.[a]

Property	Interaggregate	Intraaggregate	Silt	Clay	Whole Soil
		Field soil, day zero			
% of soil C	4.5	20.4	17.5	48.5	100
C:N (by mass)	21	15.2	10.0	8.5	10.4
$\delta^{13}C$ ‰	−18.2	−20.7	−23.5	−23.2	−23.4
Corn-derived %C	55.7	38.1	18.4	21.2	21.3
		Incubated, day 800			
% of soil C	2.9	25	18.4	47.4	100
C:N by mass	20.4	14.0	8.8	7.7	8.8
$\delta^{13}C$ ‰	−22.2	−22.9	−24.4	−23.6	−23.2
Corn-derived %C	27.7	22.7	11.9	19.8	20.5
Corn C MRT (years)	3.9	11.4	10.9	16.5	
Non-corn C MRT (years)	19.7	33.4	47.1	40	

[a]SOC fractions determined by long-term incubation.
Note: MRT = mean residence time.
Source: Haile-Mariam et al. (2008).

proportion is clay-associated (Calderón et al. 2011) and some may be charcoal (Janik et al. 2007).

The intraaggregate SOC fraction obtained by breaking apart aggregates to release the SOC protected by aggregation showed that only 38% was corn-derived C (Table 5.6). Thus, 62% of the intraaggregate fraction was more than 10 years old. A C:N ratio of 15.2 indicates that accumulation of microbial biomass lowered the C:N ratio from that of the interaggregate fraction. The clay fraction was enriched in SOC relative to the silt fraction and 21% of its C was corn-derived in contrast to 18% in the silt fraction. This indicates that the SOM components of these fractions are primarily composed of materials greater than 10 years old, with a mixture of both young and old SOM components.

Data from this 800-day incubation confirm $\delta^{13}C$ field data in that the interaggregate fraction lost nearly half its C during the 800 days (Table 5.6). But there was an increase in the proportion of intraaggregate C, indicating the formation of new aggregates during the incubation. The drop in corn-derived C in the silt fraction during incubation showed turnover in this fraction even though most of it was old. Harris et al. (1997) used $\delta^{13}C$ to calculate the MRT of corn-derived vs. noncorn, older C. The MRT of the corn-derived interaggregate C was 3.9 years vs. 19.7 years for the noncorn C (Table 5.6), showing that although older, the noncorn residues were still decomposing. The corn-derived C in the intraaggregate fraction, with an MRT of 11.4 years, is part of the slow SOM pool as defined by incubation and curve fitting for kinetic analysis (Paul et al. 2001b). The 34.4-year MRT of the noncorn C in the interaggregate fraction was older than that of the slow pool determined by incubation (Table 5.6). The MRTs of the corn and noncorn C were determined using $\delta^{13}C$, as described in Harris et al. (1997). The MRTs of the corn-C in the silt and clay fractions (10.9 and 16.5, respectively) reflected the length of time that corn was grown (10 years), while the noncorn C MRT for both fractions was greater than 40 years.

A comparison of Tables 5.2 and 5.6 shows that much longer MRTs are obtained for the soil C from ^{14}C dating than from incubation and $\delta^{13}C$ analysis following a C_3–C_4 plant switch. The radiocarbon dates represent the accrual of SOM over a pedogenic (soil formation) time scale. The $\delta^{13}C$ values are a function of the length of time since native C_3 plants were replaced by C_4 corn. Both dates are correct, but the history of the switch and the turnover of the SOC pools must be taken into account when interpreting and modeling tracer data (Andrén et al. 2008).

The pool size and MRT data in Table 5.5 were used to mathematically model the emission of CO_2 from the Conventional system in 1994 (Fig. 5.5). The timing of the CO_2 flux was influenced by rainfall and temperature in the field. Predicted CO_2 emission based on the data in Table 5.5 and using the CENTURY model corresponded well to measured values during the summer (Paul et al. 1999a). This shows the reliability of the concepts used for modeling and the measurements of pool sizes and their MRTs (Basso et al. 2011) based on long-term incubation and tracers. However, the model did not accurately predict field fluxes for the fall season. This is likely because the model assumes that once harvest is complete, the residue is available for decomposition, whereas in fact aboveground residues are not immediately incorporated into soil; there is also a physical conditioning period before decomposition that is not adequately represented in the models.

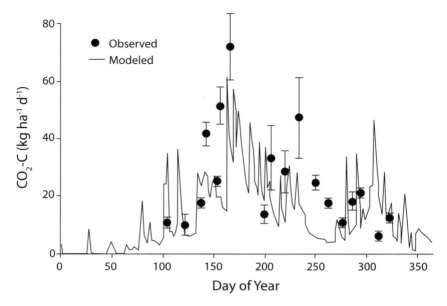

Figure 5.5. Simulated field carbon dioxide (CO_2) fluxes from the MCSE Conventional system in 1994, based on the soil organic matter pools and fluxes shown in Table 5.5. Bars on observed values indicate standard errors. Redrawn from Paul et al. (1999a) with permission from Elsevier.

Biochemical Controls on Soil Organic Matter Dynamics

A number of approaches are available to examine the molecular-biochemical structure of SOM components (Clapp et al. 2005). Two of the newest and most useful techniques are mid-infrared (mid-Ir) spectroscopy and pyrolysis followed by mass spectroscopy.

Mid-infrared (mid-IR) spectroscopy is a nondestructive measurement of the organic functional groups in both bulk soil and its fractions (Fig. 5.6). Mid-IR spectra show selective absorption by specific SOM functional groups, including OH, NH, aliphatic OH, carboxyl OH stretch, carbohydrates, aromatics, and N compounds, and indicate differences between soil fractions (Calderón et al. 2011). The mid-IR spectra show preferential absorptions of OH, NH, and aliphatic CH bonds in the interaggregate fraction. These are most often associated with partially decomposed plant residues. The intraaggregate and silt fractions show little absorption in these regions, but much higher absorption at 2000–1200 cm^{-1}, indicating a greater presence of carboxylic and aromatic groups. The high, mid-IR absorption in the 1700–1000 cm^{-1} range is attributable to polysaccharides, phenols, aromatics, and protein amides. Some of the minerals present in clays (particularly silica) also absorb in these regions (Janik et al. 2007, Calderón et al. 2011).

Principal components analysis of the mid-IR signal of soil fractions from KBS, Lamberton, Minnesota, Wooster, Ohio, and Hoytville, Ohio, shows similar

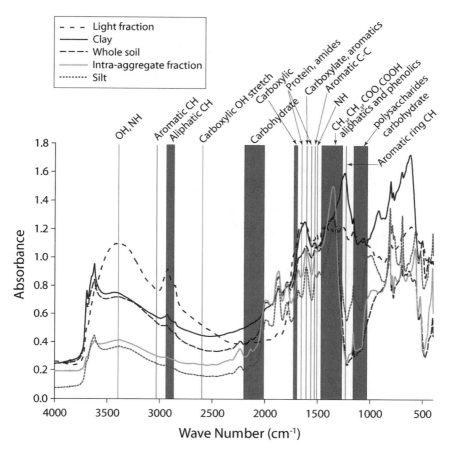

Figure 5.6. Fourier transformed mid-infrared spectra of KBS soil, including the whole soil and its various fractions. Absorbance bands of different organic functional groups are shown as vertical bars. From F. Calderón (unpublished data).

differentiation among SOM fractions for all four soils (Fig. 5.7), despite differ-ent MRTs across these sites (Table 5.2). The interaggregate fraction was well sep-arated from all others, while the intraaggregate and silt fractions were grouped together. The clays clearly separated from other fractions, and the clay fraction in the prairie-derived SOM from Lamberton separated from the forest-derived clays of KBS, Wooster, and Hoytville. Haile-Mariam et al. (2008) used $\delta^{13}C$ to determine the MRT of the SOM fractions of the Hoytville and Lamberton soils. Their analysis showed that the inter- and intraaggregate C_3-derived SOC (i.e., noncorn SOC) of the KBS soil was the youngest, which is consistent with the ^{14}C dating data (Table 5.2). The KBS soil also had higher SOM turnover rates than the other soils examined, as determined by incubation, carbon dating, and ^{13}C field analysis (Collins et al. 1999, 2000; Paul et al. 2001a). These regional comparisons offer a powerful means for identifying general controls on SOM dynamics in larger landscapes (Fierer et al. 2009, Morris et al. 2010).

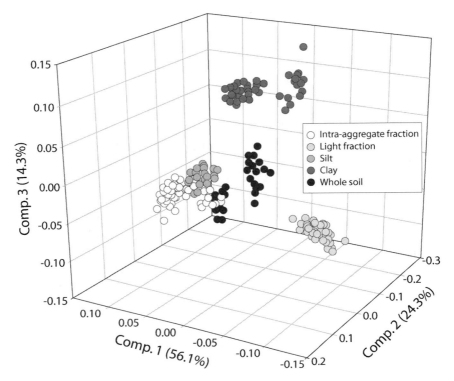

Figure 5.7. Principal Components Analysis of the infrared spectra of unincubated surface soil and particle size fractions from the KBS (Michigan), Wooster (Ohio), Hoytville (Ohio), and Lamberton (Minnesota) sites shown in Table 5.2. The percentages of spectral variance accounted for by each component are in parentheses. Modified from Calderón et al. (2011).

Pyrolysis followed by mass spectroscopy, sometimes with prior separation in a gas chromatographic column, identifies molecular breakdown products following heating in an inert environment. These products can be related back to SOM biochemical structures through the use of standards (Grandy et al. 2007, 2008; Plante et al. 2009). As determined by pyrolysis mass spectrometry, agriculture changes the chemistry of SOM, making it more heterogeneous than the SOM in native soil, and the chemical characteristics of the SOM can be related to both soil type and climate (Haddix et al. 2011). Pyrolysis-gas chromatography/mass spectroscopy has been used to determine changes in the chemistry of plant litter during decomposition under various MCSE systems. Wickings et al. (2011) observed changes in litter chemistry and in the composition and activity of the decomposer community during decomposition of corn and grass litter in the Conventional, No-till, and Early Successional systems. After one season in the field, grass litter in the Conventional and No-till systems was enriched in total polysaccharides, whereas in the Early Successional system, it was enriched in N-bearing compounds and lipids. Differences in the soil communities—in particular, microarthropods and fungal/bacterial ratios—were associated with differences in chemistry, indicating that the activity and structure of the decomposer community can influence chemical changes during decomposition (Wickings et al. 2011). Results demonstrate that efforts to predict long-term SOM turnover and stabilization dynamics may therefore need to consider the influence of different decomposer communities on changes in litter chemistry during decomposition.

Plante et al. (2009) used pyrolysis-molecular beam mass spectrometry to identify carbohydrates, N compounds, lignin derivatives, aliphatics, and sterols in SOM. They found similar overall constituents for forested and prairie soils. Grandy et al. (2007, 2008) also found similar chemical patterns in the silt and clay fractions from different ecosystems, supporting the hypothesis of Fierer et al. (2009) that all soils share similar overall characteristics that are a function of the biota and their physiological constraints. Much of the N resistant to acid hydrolysis was identified as amino N, indicating the protected proteinaceous nature of the resistant N. This supports the conclusions of Di Costy et al. (2003) who, by using ^{15}N-enriched clover additions and nuclear magnetic resonance analysis, found that the majority of N in soil is proteinaceous in nature.

The Role of Carbon Inputs and Microbial Activity in Soil Organic Matter Dynamics

The Fate of Plant Carbon Inputs

The amount and quality of SOM are dependent on the amount and type of plant inputs as well as their microbial turnover before stabilization. One of the best ways to examine these factors is the use of carbon isotope tracers. In previous sections, we noted the importance of the Poplar system for understanding both sequestration of SOM and aggregate formation. The amount of C that is provided as litter and from the roots is important in these transformations. Horwath (1993) and Horwath et al. (1994) used ^{14}C to observe the distribution of photosynthesis-derived C in the poplar plant, its movement belowground, and decomposition by the microbial biomass (Table 5.7). Trees were uniformly labeled in July or September by exposing the foliage to $^{14}CO_2$ for a single day and the movement of ^{14}C was traced through the plant–soil system by sampling trees 14 and 372 days after labeling.

Roots that accounted for 18.4% of total-tree C accounted for 9.8% of the ^{14}C label (Table 5.7). Some of the photosynthate was stored in coarse roots and moved to fine roots the following spring. During the 3-week labeling period, belowground respiration accounted for 7.7% of the label taken up by photosynthesis. Microbial biomass C, accounting for 1.5% of the soil C, received 0.4% of the ^{14}C, showing significant turnover during the labeling period. The 0.3% of the label present in the biomass after 328 days shows a slow turnover of root and microbial C during the period after labeling, with some transfer to the soil C component. Labeled litter, added to the soil in a separate experiment, lost 67% of its ^{14}C content during the first year, and 73% over a 2-year period (Horwath 1993; Horwath et al. 1994).

Table 5.7. Distribution of total carbon (C) in the MCSE Poplar system and a radiocarbon tracer (^{14}C) 14 and 328 days after a 1-day labeling event.[a]

Component	Total C		^{14}C Distribution (%)	
	(g)	(% of total)	After 14 days	After 371 days
Leaves + litter	448	37.0	24.0	0.2
Stems + branches	520	44.0	48.0	26.0
Roots <0.5-mm dia.	52	4.4	1.7	1.3
Roots >0.5-mm dia.	166	14.0	8.1	8.4
Root + soil respiration			7.7	
Microbial biomass	103	1.5	0.4	0.3
Soluble C	75			
Soil C	6788		1.2	1.4

[a]Distribution of ^{14}C expressed as % of recovered label. See Table 5.1 for a description of the MCSE Poplar system. *Source:* Horwath (1993) and Horwath et al. (1994).

The uptake of $^{14}CO_2$ by photosynthesis and its later distribution in the plant–soil system in a laboratory-grown sorghum crop showed results quite similar to those found for poplars in the field (Calderón 1997). Fifty percent of the ^{14}C remained in the aboveground biomass, 30% in the roots, and 6% was transferred to the soil in a 24-day period. Belowground respiration accounted for 12% and shoot respiration for 5% of the label in plants with symbiotic, mycorrhizal fungi. Non-mycorrhizal plants had more aboveground allocation and less allocation to the roots and soil respiration. The mycorrhizal plants did not quite compensate for the needs of their microbial partners by increased photosynthesis, indicating that the fungi can act as both symbionts and parasites (Calderón et al. 2011, 2012).

The Role of Microbial Biomass and Composition

Microbial biomass is an important component of SOM and, in particular, the active fraction that is so important to soil fertility (Paul et al. 1999a, b). Microbes also provide the enzymes required for decomposition and the microbial products that interact with the soil matrix to stabilize SOM (Paul and Clark 1996). Microbial biomass in KBS soils was measured from 1993 to 1996 (Horwath and Paul 1994, Horwath et al. 1996). The bacterial and fungal biomass of the Deciduous Forest soils constituted, on average, 107 mg C kg^{-1} soil and 179 mg C kg^{-1} soil, respectively. In MCSE agronomic systems, the biomass of the bacteria and fungi was nearly equal at 85–89 mg C kg^{-1} soil. The microbial biomass represented 1.2–1.8% of the total soil organic C, and MCSE systems with greater SOM accumulation (No-till and Biologically Based) had higher percentages of microbial biomass.

These values are similar to the 1.5% of soil C measured as microbial biomass in the Poplar system (Table 5.7). Fungal and bacterial biomass in this system were similar during the first few years of growth, but this changed as Poplar mycorrhizal associations switched from endomycorrhizal (AM) to a mixed endomycorrhizal-ectomycorrhizal (AM-EM) symbiosis. Fungal biomass increased with the addition of extra-radical mycelia associated with ectotrophic mycorrhiza. When the trees were sampled in 1997, both AM and EM fungi were present. Kosola et al. (2004) showed the differential response of these symbionts to environmental conditions: N fertilization reduced AM colonization of roots from 16% to 14% and increased EM colonization from 15% to 18%.

Bacterial communities are also affected by environmental conditions (Schmidt and Waldron 2015, Chapter 6 in this volume) and can be influenced by SOM differences associated with aggregates. Blackwood et al. (2006), for example, found greater inter- than intraaggregate variability in bacterial community structure in a comparison of soils from KBS and Wooster, Ohio. They found a higher number of active bacteria (as indicated by a cell volume greater than 0.18 μm^3) within aggregates than elsewhere, indicating potentially higher microbial activity within aggregates. The number of bacteria and the proportion of large, active cells were not affected by the cropping system; the larger bacteria accounted for 30–50% of the number of cells, but composed 85–90% of the biomass (Blackwood and Paul 2003).

The fungi showed more microspatial variability than bacteria (Horwath et al. 1994, 1996). They were larger and more often associated with plant residues than the bacteria. And whereas bacterial biomass varied by a factor of 50% during the growing season, fungal biomass varied by a factor of 100%. Fungi, however, tend to have cytoplasm-free cells, and thus the DNA content of soils may be 80% bacterial even though the two populations may have similar biomass. Harris and Paul (1994) used the incorporation of ^{3}H-thymidine to measure bacterial growth rates and found that the bacteria of the Conventional system soils doubled every 160 days, while those in soils of the Mown Grassland (never tilled) community, which have a higher SOM content, doubled every 107 days.

High seasonal variation in bacterial and fungal biomass during the growing season (Horwath and Paul 1994, Horwath et al 1996) indicates that the microbial biomass is a significant proportion of the active pool of SOM and changes in microbial biomass might be used to manage SOM dynamics and N fertility (Fortuna et al. 2003). This could be especially important when combined with the catalytic effect of rhizosphere microbiota on SOM decomposition. In the KBS Living Field Lab experiment (Snapp et al. 2015, Chapter 15 in this volume), Sánchez et al. (2002, 2004) found that cover crops such as clover increased both intraaggregate SOM and microbial biomass. A corn crop that followed soil N enrichment by a clover cover crop mineralized 168 kg N ha^{-1} from SOM pools, whereas wheat mineralized only 116 kg N ha^{-1}, and bare soils with roots excluded mineralized 108 kg N ha^{-1} from SOM pools (Table 5.8).

A corn crop after a clover cover crop contained 168 kg N ha^{-1} mineralized from SOM pools, as compared to 108 kg N ha^{-1} of mineralized N in microplots where roots were excluded and 116 kg N ha^{-1} of mineralized N in a wheat crop (Table 5.8). Soil planted with corn mineralized more C than bare soil with roots excluded This suggests that microbes stimulated by labile C from the corn rhizosphere specifically degraded N compounds. The role that legumes and cover crops (Harris et al. 1994),

Table 5.8. Effect of corn and wheat roots on nitrogen (N) uptake during crop growth and nitrogen and carbon (C) mineralization subsequent to cropping.[a]

	N Supplied by Soil (kg ha^{-1})	N Mineralized after Crop (mg N kg^{-1} soil)	C Mineralized after Crop (mg C kg^{-1} soil)
Bare soil	108	70	525
Corn	168	22	780
Wheat	116	45	680

[a]The field study was conducted at the KBS Living Field Lab, adjacent to the MCSE.
Note: Bare soil = microplots with roots excluded.
Source: Sánchez et al. (2002).

including clover, play in SOM accumulation was described earlier in this chapter. Their ability to provide N for subsequent crops indicates that such rotations should continue to be a dominant component of sustainable cropping systems.

Summary

Twenty years of KBS LTER research have produced a remarkable consensus about controls on SOM dynamics and has furthered our general understanding of long-term changes in SOM in agricultural ecosystems. This understanding includes the concept that although SOM takes many years to develop and includes a continuum of young and old materials, it is sensitive to management changes that can either improve or hinder its role as one of the major controllers of soil fertility and ecosystem functioning. Differences in SOM levels within the soil profile and across landscapes can foster greater biological diversity both above- and belowground, but the high spatial variability of SOM makes it especially difficult to interpret data collected at low frequencies, particularly from only one or two sampling points in time. Thus, long-term studies that allow resampling of the same locations across time offer one of the best opportunities to detect gradual changes in SOM levels, characteristics, and dynamics relative to vegetation influences, cultivation, fertilizers, and cover crops (Robertson and Paul 2000). These will all be affected by global environmental change and the need for increased production of food and biofuel crops in the future.

Results from KBS show that variation in SOM content and dynamics occurs across the landscape even on relatively flat terrain with only shallow depressions. Lower, wetter areas accumulate SOM most rapidly. Soils in these areas receive higher inputs of crop residue and also receive eroded materials from upper slopes, thus increasing their SOM content and water-holding capacities. Soils in so-called steady state after many years of cultivation can continue to slowly lose SOM. Both erosion and the change in SOM, with consequent alterations in aggregation, can significantly affect soil bulk density, an important soil physical property. Changes in bulk density can introduce uncertainty into estimates of soil C change when measurements made from soil cores are compared across time.

The complex, multistructured, multicomponent pool of organic materials constituting SOM includes: (1) decomposing plant residues, (2) associated microorganisms and their products, and (3) the biochemically transformed ("humic") fraction. KBS LTER studies have shown that all these components can be associated with Ca, minerals, and sesquioxides that greatly increase their MRTs. Long-term incubations to allow microorganisms to metabolize labile components, together with stable isotopic tracers to identify sources and acid hydrolysis to quantify the resistant fraction, have provided a useful framework to interpret SOM dynamics at KBS.

Interpretation of the long-term incubation results distinguishes three pools for modeling purposes—the active, slow, and resistant fractions—although SOM is known to represent an oxidation continuum (Paul et al. 2001b). The active pool representing the most labile components is small at ~5% of the SOM, and is composed of some of the interaggregate plant residues measured by physical fractionation. The active pool also contains a portion of the microbial biomass and has MRTs of months to a few years. This pool is important in the management of nutrients on a growing season basis. The slow pool, with MRTs ranging from months to decades, accounts for ~40% of total SOM. This is the major source of soil nutrients that change with management, often requiring 6 to 7 years to show measurable differences in yield and soil sustainability. The dynamics of the slow pool are controlled by aggregation (intraaggregate SOM) and the association of microbial products with Ca, as well as the silt and clay fractions. The intraaggregate fraction measured by physical separation also has young and old constituents, but on average has MRTs of decades. The silt and clay SOM fractions and the nonacid hydrolyzable residues have the longest turnover times. They also contain the largest concentrations of aluminosilicate minerals and sesquioxides. The actual mechanism of stabilization of the oldest SOM in soils (the resistant pool) has yet to be determined.

Soil organic matter is a dynamic system with some components that can change rapidly with changes in management and environmental and biotic controls, while other components respond very slowly. The fractions we have identified, while reflecting these controls, are not discrete. For example, in young fractions, such as those protected by aggregates, there is a small amount of old C, possibly charcoal. The older component, such as the SOM associated with clays, can also have a small amount of young C, such as that coming from absorbed microbial biomass constituents. The MRTs and pool sizes we have measured have been useful for interpreting the effects of management practices, as well as providing fundamental knowledge of the basic controls affecting SOM dynamics.

The recent use of molecular-structure analysis to measure the biochemical composition of SOM provides insights into SOM dynamics (Paul et al. 2008). The major components derived from plant and microbial products such as carbohydrates, proteins, and related N compounds, lignins, sterols, aromatics, and fatty acids are readily identified by pyrolysis-mass spectrometry. These constituents vary with ecosystem type and management. However, they also illustrate the effect of microbial processing under unifying controls in that the molecular structure of SOM also shows significant similarities among different soils. In addition, there are some unidentified organic components that have century-to-millennia turnover rates.

The old, resistant SOM fraction, identified by its resistance to acid hydrolysis, consists of some aromatics and long-chain aliphatic materials as well as protected carbohydrates and proteins closely associated with sesquioxides and clays. Mid-IR spectroscopy that identifies functional groups in both the organic and inorganic soil constituents clearly shows different functional groups in the inter- and intraaggregate fractions, with higher plant residue components in the interaggregate fraction. Clay and silt components are also clearly differentiated. Principal Components Analysis of KBS soil fractions, relative to other soils in our landscape comparison, shows more similarities within fractions than between soils, except that the prairie-derived clays were clearly separated from those in the forest-derived sites, such as those from KBS.

Microbial biomass, shown in our studies to represent approximately 1.5 to 3% of the soil C, plays a major role in decomposition and provides the intermediate products that are stabilized as organic matter. Fungi, which are equal in biomass to bacteria in many of our agricultural systems, are more variable over time and in their management responses. They provide a larger proportion of the biomass in the forested sites, where ectomycorrhizae are important. The soil biota, together with their breakdown products, are especially significant in the legume-based systems. Our studies have shown that the use of a clover cover crop in rotation results in a greater active SOM pool (Nannipieri and Paul 2009). This SOM pool can be accessed by the rhizosphere organisms of corn, and to some extent wheat, to provide much of the N required for plant growth. Clover cover crops should thus continue to be studied and used for their capacity to both fix N and release available N to succeeding crops (McSwiney et al. 2010).

The challenges of providing a sustainable food supply and a variety of resilient ecosystem services (Paustian et al. 1996, 1997; Robertson and Paul 1998; Bhardwaj et al. 2011) in the face of global climate change are closely related to the molecular composition of SOM and its interactions with physical, biotic, and chemical controls. The studies summarized in this chapter have laid an important baseline for many years to come as interacting processes and controls continue to be elucidated.

References

Andrén, O., H. Kirchmann, T. Katterer, J. Magid, E. A. Paul, and D. C. Coleman. 2008. Visions of a more precise soil biology. European Journal of Soil Science 59:380–390.

Austin, F. R. 1979. Soil survey of Kalamazoo County, Michigan. USDA Soil Conservation Service and Michigan Agricultural Experiment Station, East Lansing, Michigan, USA.

Baldock, J. A., and P. N. Nelson. 2000. Soil organic matter. Pages B, 25–84 in M. E. Sumner, editor. Handbook of soil science. CRC Press, Boca Raton, Florida, USA.

Basso, B., O. Gargiulo, K. Paustian, G. P. Robertson, C. Porter, P. R. Grace, and J. W. Jones. 2011. Procedures for initializing organic carbon pools in the DSSAT-CENTURY model for agricultural systems. Soil Science Society of America Journal 75:69–78.

Bellamy, P. H., P. J. Loveland, R. I. Bradley, R. M. Lark, and G. J. D. Kirk. 2005. Carbon losses from all soils across England and Wales 1978–2003. Nature 437:245–248.

Bhardwaj, A. K., P. Jasrotia, S. K. Hamilton, and G. P. Robertson. 2011. Ecological management of intensively cropped agro-ecosystems improves soil quality with sustained productivity. Agriculture, Ecosystems & Environment 140:419–429.

Blackwood, C. B., C. J. Dell, E. A. Paul, and A. J. M. Smucker. 2006. Eubacterial communities in different soil macroaggregate environments and cropping systems. Soil Biology & Biochemistry 38:720–728.

Blackwood, C. B., and E. A. Paul. 2003. Eubacterial community structure and population size within the soil light fraction, rhizosphere, and heavy fraction of several agricultural systems. Soil Biology & Biochemistry 35:1245–1255.

Brewer, E. A. 2004. Impacts of calcium and nitrogen on carbon stabilization in soils of an afforested red pine stand. Thesis, Bradley University, Peoria, Illinois, USA.

Calderón, F. J. 1997. Lipids: their value as molecular markers and their role in the carbon cycle of arbuscular mycorrhizae. Dissertation, Michigan State University, East Lansing, Michigan, USA.

Calderón, F. J., J. B. Reeves, H. P. Collins, and E. A. Paul. 2011. Chemical differences in soil organic matter fractions determined by diffuse-reflectance mid-infrared spectroscopy. Soil Science Society of America Journal 75:568–579.

Calderón, F. J., D. Schultz, and E. A. Paul. 2012. Carbon allocation, below ground transfers and lipid turnover in a plant-microbial association. Soil Science Society of America Journal 76:1614–1623.

CAST (Council for Agricultural Science and Technology). 2011. Carbon sequestration and greenhouse gas fluxes in agriculture: challenges and opportunities. Task Force Report 142, CAST, Ames, Iowa, USA.

Clapp, C. E., M. H. B. Hayes, A. J. Simpson, and W. L. Kingery. 2005. The chemistry of soil organic matter. Pages 1–150 in M. A. Tabatabai and D. L. Sparks, editors. Chemical processes in soils. American Society of Agronomy, Madison, Wisconsin, USA.

Collins, H. P., R. L. Blevins, L. G. Bundy, D. R. Christenson, W. A. Dick, D. R. Huggins, and E. A. Paul. 1999. Soil carbon dynamics in corn-based agroecosystems: results from carbon-13 natural abundance. Soil Science Society of America Journal 63:584–591.

Collins, H. P., E. T. Elliott, K. Paustian, L. G. Bundy, W. A. Dick, D. R. Huggins, A. J. M. Smucker, and E. A. Paul. 2000. Soil carbon pools and fluxes in long-term corn belt agroecosystems. Soil Biology & Biochemistry 32:157–168.

Collins, H. P., E. A. Paul, K. Paustian, and E. T. Elliott. 1997. Characterization of soil organic carbon relative to its stability and turnover. Pages 51–72 in E. A. Paul, K. Paustian, E. T. Elliott, and C. V. Cole, editors. Soil organic matter in temperate agroecosystems: long-term experiments in North America. CRC Press, Boca Raton, Florida, USA.

Crum, J. R., and H. P. Collins. 1995. KBS soils. Kellogg Biological Station Long-term Ecological Research, Michigan State University, Hickory Corners, Michigan, USA. <http://lter.kbs.msu.edu/about/site_description/soils.php>

De Gryze, S., J. Six, K. Paustian, S. J. Morris, E. A. Paul, and R. Merckx. 2004. Soil organic carbon pool changes following land use conversions. Global Change Biology 10:1120–1132.

DiCosty, R. J., D. P. Weliky, S. J. Anderson, and E. A. Paul. 2003. ^{15}N-CPMAS nuclear magnetic resonance spectroscopy and biological stability of soil organic matter in whole soil and particle size fractions. Organic Geochemistry 34:1635–1650.

Feng, H., and B. A. Babcock. 2010. Impacts of ethanol on planted acreage in market equilibrium. American Journal of Agricultural Economics 92:789–802.

Fierer, N., A. S. Grandy, J. Six, and E. A. Paul. 2009. Searching for unifying principles in soil ecology. Soil Biology & Biochemistry 41:2249–2256.

Follett, R. F., E. A. Paul, S. W. Leavitt, A. D. Halvorson, D. Lyon, and G. A. Peterson. 1997. Carbon isotope ratios of Great Plains soils and in wheat-fallow systems. Soil Science Society of America Journal 61:1068–1077.

Fortuna, A. M., E. A. Paul, and R. R. Harwood. 2003. The effects of compost and crop rotations on carbon turnover and the particulate organic matter fraction. Soil Science 168:434–444.

Gelfand, I., and G. P. Robertson. 2015. Mitigation of greenhouse gas emissions in agricultural ecosystems. Pages 310–339 in S. K. Hamilton, J. E. Doll, and G. P. Robertson, editors. The ecology of agricultural Landscapes: long-term research on the path to sustainability. Oxford University Press, New York, New York, USA.

Grandy, A. S., and J. C. Neff. 2008. Molecular soil C dynamics downstream: the biochemical decomposition sequence and its effects on soil organic matter structure and function. Science of the Total Environment 404:297–307.

Grandy, A. S., J. C. Neff, and M. N. Weintraub. 2007. Carbon structure and enzyme activities in alpine and forest ecosystems. Soil Biology & Biochemistry 39:2701–2711.

Grandy, A. S., and G. P. Robertson. 2006a. Aggregation and organic matter protection following tillage of a previously uncultivated soil. Soil Science Society of America Journal 70:1398–1406.

Grandy, A. S., and G. P. Robertson. 2006b. Initial cultivation of a temperate-region soil immediately accelerates aggregate turnover and CO_2 and N_2O fluxes. Global Change Biology 12:1507–1520.

Grandy, A. S., and G. P. Robertson. 2007. Land-use intensity effects on soil organic carbon accumulation rates and mechanisms. Ecosystems 10:58–73.

Grandy, A. S., R. L. Sinsabaugh, J. C. Neff, M. Stursova, and D. R. Zak. 2008. Nitrogen deposition effects on soil organic matter chemistry are linked to variation in enzymes, ecosystems and size fractions. Biogeochemistry 91:37–49.

Haddix, M. L., A. F. Plante, R. T. Conant, J. Six, J. M. Steinweg, K. Magrini-Bair, R. A. Drijber, S. J. Morris, and E. A. Paul. 2011. The role of soil characteristics on temperature sensitivity of soil organic matter. Soil Science Society of America Journal 75:56–86.

Haile-Mariam, S., H. P. Collins, S. Wright, and E. A. Paul. 2008. Fractionation and long-term laboratory incubation to measure soil organic matter dynamics. Soil Science Society of America Journal 72:370–378.

Hamilton, S. K., A. L. Kurzman, C. Arango, L. Jin, and G. P. Robertson. 2007. Evidence for carbon sequestration by agricultural liming. Global Biogeochemical Cycles 21:GB2021.

Harris, D., and E. A. Paul. 1994. Measurement of bacterial growth rates in soil. Applied Soil Ecology 1:277–290.

Harris, D., L. K. Porter, and E. A. Paul. 1997. Continuous flow isotope ratio mass spectrometry of carbon dioxide trapped as strontium carbonate. Communications in Soil Science and Plant Analysis 28:747–757.

Harris, G. H., O. B. Hesterman, E. A. Paul, S. E. Peters, and R. R. Janke. 1994. Fate of legume and fertilizer nitrogen-15 in a long-term cropping systems experiment. Agronomy Journal 86:910–915.

Horwath, W. R. 1993. The dynamics of carbon, nitrogen, and soil organic matter in *Populus* plantations. Dissertation, Michigan State University, East Lansing, Michigan, USA.

Horwath, W. R., and E. A. Paul. 1994. Microbial biomass. Pages 753–774 in R. W. Weaver, J. S. Angle, P. J. Bottomley, D. F. Bezdicek, M. S. Smith, M. A. Tabatabai, and A. G. Wollum, editors. Methods of soil analysis. Part 2, Microbiological and biochemical properties. Soil Science Society of America, Madison, Wisconsin, USA.

Horwath, W. R., E. A. Paul, D. Harris, J. Norton, L. Jagger, and K. A. Horton. 1996. Defining a realistic control for the chloroform fumigation incubation method using microscopic counting and ^{14}C substrates. Canadian Journal of Soil Science 76:459–467.

Horwath, W. R., K. S. Pregitzer, and E. A. Paul. 1994. ^{14}C allocation in tree soil-systems. Tree Physiology 14:1163–1176.

Janik, L. J., J. O. Skjemstad, K. D. Shepherd, and L. R. Spouncer. 2007. The prediction of soil carbon fractions using mid-infrared-partial least square analysis. Australian Journal of Soil Research 45:73–81.

Kavdir, Y., and A.J.M. Smucker. 2005. Soil aggregate sequestration of cover crop root and shoot-derived nitrogen. Plant and Soil 272:263–276.

Kosola, K. R., D. M. Durall, G. P. Robertson, D. I. Dickmann, D. Parry, C. A. Russell, and E. A. Paul. 2004. Resilience of mycorrhizal fungi on defoliated and fertilized hybrid poplars. Canadian Journal of Botany 82:671–680.

Kravchenko, A. N., and X. Hao. 2008. Management practice effects on spatial variability characteristics of surface mineralizable C. Geoderma 144:387–394.

Kravchenko, A. N., X. Hao, and G. P. Robertson. 2009. Seven years of continuously planted Bt corn did not affect mineralizable and total soil C and total N in surface soil. Plant and Soil 318:269–274.

Kravchenko, A. N., and G. P. Robertson. 2011. Whole-profile soil carbon stocks: the danger of assuming too much from analyses of too little. Soil Science Society America Journal 75:235–240.

Kravchenko, A. N., G. P. Robertson, X. Hao, and D. G. Bullock. 2006. Management practice effects on surface total carbon: difference in spatial variability patterns. Agronomy Journal 98:1559–1568.

Leavitt, S. W., R. F. Follett, and E. A. Paul. 1996. Estimation of slow- and fast-cycling soil organic carbon pools from 6N HCl hydrolysis. Radiocarbon 38:231–239.

McSwiney, C. P., S. S. Snapp, and L. E. Gentry. 2010. Use of N immobilization to tighten the N cycle in conventional agroecosystems. Ecological Applications 20:648–662.

Mokma, D. L., and J. A. Doolittle. 1993. Mapping some loamy alfisols in southwestern Michigan using ground-penetrating radar. Soil Survey Horizons 34:71–77.

Morris, S. J., S. Bohm, S. Haile-Mariam, and E. A. Paul. 2007. Evaluation of carbon accrual in afforested agricultural soils. Global Change Biology 13:1145–1156.

Morris, S. J., R. Conant, N. Mellor, E. A. Brewer, and E. A. Paul. 2010. Controls on soil carbon sequestration and dynamics: lessons from land-use change. Journal of Nematology 42:78–83.

Morris, S. J., and E. A. Paul. 2003. Forest soil ecology and soil organic carbon. Pages 109–125 in J. Kimble, J. Heath, R. A. Birdsey, and R. Lal, editors. The potential of U.S. forest soils to sequester carbon and mitigate the greenhouse effect. CRC Press, Boca Raton, Florida, USA.

Muñoz, J. D., R. Gehl, S. S. Snapp, and A. N. Kravchenko. 2008. Identifying the factors affecting cover crop performance in row crops. 9th International Conference on Precision Agriculture, Denver, Colorado, USA.

Nannipieri, P., and E. A. Paul. 2009. The chemical and functional characterization of soil N and its biotic components. Soil Biology & Biochemistry 41:2357–2369.

NRCS (Natural Resource Conservation Service). 1999. Soil taxonomy: a basic system of soil classification for making and interpreting soil surveys. U.S. Department of Agriculture, Washington, DC, USA.

Paul, E. A., and F. E. Clark. 1996. Soil microbiology and biochemistry. Second edition. Academic Press, San Diego, California, USA.

Paul, E. A., and H. P. Collins. 1998. The characteristics of soil organic matter relative to nutrient cycling. Pages 181–197 in R. Lal, W. H. Blum, and C. Valentine, editors. Methods for assessment of soil degradation. CRC Press, Boca Raton, Florida, USA.

Paul, E. A., H. P. Collins, and S. W. Leavitt. 2001a. Dynamics of resistant soil carbon of Midwestern agricultural soils measured by naturally-occurring ^{14}C abundance. Geoderma 104:239–256.

Paul, E. A., R. F. Follett, S. W. Leavitt, A. D. Halvorson, G. A. Peterson, and D. Lyon. 1997a. Radiocarbon dating for determination of soil organic matter pool sizes and dynamics. Soil Science Society of America Journal 61:1058–1067.

Paul, E. A., D. Harris, H. P. Collins, U. Schulthess, and G. P. Robertson. 1999a. Evolution of CO_2 and soil carbon dynamics in biologically managed, row-crop agroecosystems. Applied Soil Ecology 11:53–65.

Paul, E. A., D. Harris, M. Klug, and R. Ruess. 1999b. The determination of microbial biomass. Pages 291–317 in G. P. Robertson, C. S. Bledsoe, D. C. Coleman, and P. Sollins, editors. Standard soil methods for long-term ecological research. Oxford University Press, New York, New York, USA.

Paul, E. A., W. R. Horwath, D. Harris, R. Follett, S. W. Leavitt, B. A. Kimball, and K. Pregitzer. 1995. Establishing the pool sizes and fluxes in CO_2 emissions from soil organic matter turnover. Pages 297–305 in R. Lal, J. Kimble, E. Levine, and B. A. Stewarts, editors. Soils and global change. CRC Press, Boca Raton, Florida, USA.

Paul, E. A., K. Magrini-Baer, R. Conant, R. F. Follet, and S. J. Morris. 2008. Biological and molecular structure analysis of the controls on soil organic matter dynamics. Pages 167–170 in I. V. Permanova and N. A. Kulikova, editors. International Humic Substance Society Proceedings, Moscow, Russia.

Paul, E. A., S. J. Morris, and S. Bohm. 2001b. The determination of soil C pool sizes and turnover rates: biophysical fractionation and tracers. Pages 193–206 in R. Lal, J. M. Kimble, R. F. Follett, and B. A. Stewart, editors. Assessment methods for soil carbon. CRC Press, Boca Raton, Florida, USA.

Paul, E. A., S. J. Morris, R. T. Conant, and A. F. Plante. 2006. Does the acid hydrolysis incubation method measure meaningful soil organic carbon pools? Soil Science Society of America Journal 70:1023–1035.

Paul, E. A., S. J. Morris, J. Six, K. Paustian, and E. G. Gregorich. 2003. Interpretation of soil carbon and nitrogen dynamics in agricultural and afforested soils. Soil Science Society of America Journal 67:1620–1628.

Paul, E. A., K. Paustian, E. T. Elliott, and C. V. Cole, editors. 1997b. Soil organic matter in temperate agroecosystems: long-term experiments in North America. CRC Press, Boca Raton, Florida, USA.

Paustian, K., H. P. Collins, and E. A. Paul. 1997. Management controls on soil carbon. Pages 15–49 in E. A. Paul, K. Paustian, E. T. Elliott, and C. V. Cole, editors. Soil organic matter in temperate agroecosystems: long-term experiments in North America. CRC Press, Boca Raton, Florida, USA.

Paustian, K., E. T. Elliott, E. A. Paul, H. P. Collins, C. V. Cole, and S. D. Frey. 1996. The North American Site Network. Pages 37–54 in D. S. Powlson, P. Smith, and J. U. Smith, editors. Evaluation of soil organic matter models using existing, long-term data sets. Springer-Verlag, Berlin, Germany.

Plante, A. F., R. T. Conant, E. A. Paul, K. Paustian, and J. Six. 2006. Acid hydrolysis of easily dispersible and microaggregate derived-silt and clay sized fractions to isolate resistant soil organic matter. European Journal of Soil Science 57:456–467.

Plante, A. F., K. Magrini-Baer, M. F. Vigil, and E. A. Paul. 2009. Pyrolysis-molecular beam mass spectrometry to characterize soil organic matter composition in chemically isolated fractions from differing land uses. Biogeochemistry 92:145–161.

Robertson, G. P., J. R. Crum, and B. G. Ellis. 1993. The spatial variability of soil resources following long-term disturbance. Oecologia 96:451–456.

Robertson, G. P., and S. K. Hamilton. 2015. Long-term ecological research at the Kellogg Biological Station LTER Site: conceptual and experimental framework. Pages 1–32 in S. K. Hamilton, J. E. Doll, and G. P. Robertson, editors. The ecology of agricultural Landscapes: long-term research on the path to sustainability. Oxford University Press, New York, New York, USA.

Robertson, G. P., S. K. Hamilton, S. J. Del Grosso, and W. J. Parton. 2011. The biogeochemistry of bioenergy landscapes: carbon, nitrogen, and water considerations. Ecological Applications 21:1055–1067.

Robertson, G. P., K. M. Klingensmith, M. J. Klug, E. A. Paul, J. R. Crum, and B. G. Ellis. 1997. Soil resources, microbial activity, and primary production across an agricultural ecosystem. Ecological Applications 7:158–170.

Robertson, G. P., and E. A. Paul. 1998. Ecological research in agricultural ecosystems: contributions to ecosystem science and to the management of agronomic resources. Pages 142–164 in M. L. Pace and P. M. Groffman, editors. Successes, limitations and frontiers in ecosystem science. Springer-Verlag, New York, New York, USA.

Robertson, G. P., and E. A. Paul. 2000. Decomposition and soil organic matter dynamics. Pages 104–116 in E. S. Osvaldo, R. B. Jackson, H. A. Mooney, and R. W. Howarth, editors. Methods in ecosystem science. Springer-Verlag, New York, New York, USA.

Robertson, G. P., E. A. Paul, and R. R. Harwood. 2000. Greenhouse gases in intensive agriculture: contributions of individual gases to the radiative forcing of the atmosphere. Science 289:1922–1925.

Russell, C. A., K. R. Kosola, E. A. Paul, and G. P. Robertson. 2004. Nitrogen cycling in poplar stands defoliated by insects. Biogeochemistry 68:365–381.

Sánchez, J. E., R. R. Harwood, T. C. Willson, K. Kizilkaya, J. Smeenk, E. Parker, E. A. Paul, B. D. Knezek, and G. P. Robertson. 2004. Managing soil carbon and nitrogen for productivity and environmental quality. Agronomy Journal 96:769–775.

Sánchez, J. E., E. A. Paul, T. C. Willson, J. Smeenk, and R. R. Harwood. 2002. Corn root effects on the nitrogen-supplying capacity of a conditioned soil. Agronomy Journal 94:391–396.

Schmidt, T. M., and C. Waldron. 2015. Microbial diversity in agricultural soils and its relation to ecosystem function. Pages 135–157 in S. K. Hamilton, J. E. Doll, and G. P. Robertson, editors. The ecology of agricultural Landscapes: long-term research on the path to sustainability. Oxford University Press, New York, New York, USA.

Senthilkumar, S., B. Basso, A. N. Kravchenko, and G. P. Robertson. 2009a. Contemporary evidence for soil carbon loss in the U.S. corn belt. Soil Science Society of America Journal 73:2078–2086.

Senthilkumar, S., A. N. Kravchenko, and G. P. Robertson. 2009b. Topography influences management system effects on total soil carbon and nitrogen. Soil Science Society of America Journal 73:2059–2067.

Six, J., P. Callewaert, S. Landers, S. D. Gryze, S. J. Morris, R. G. Gregorich, E. A. Paul, and K. Paustian. 2002a. Measuring and understanding carbon storage in afforested soils by physical fractionation. Soil Science Society of America Journal 66:1981–1987.

Six, J., R. T. Conant, E. A. Paul, and K. Paustian. 2002b. Stabilization mechanisms of soil organic matter: implications for C-saturation of soils. Plant and Soil 241:155–176.

Smith, J. L., and E. A. Paul. 1990. The significance of soil microbial biomass estimations. Pages 357–396 in J. M. Bollag and G. Stotzky, editors. Soil biochemistry. Marcel Dekker, Inc., New York, New York, USA.

Snapp, S. S., R. G. Smith, and G. P. Robertson. 2015. Designing cropping systems for ecosystem services. Pages 378–408 in S. K. Hamilton, J. E. Doll, and G. P. Robertson, editors. The ecology of agricultural Landscapes: long-term research on the path to sustainability. Oxford University Press, New York, New York, USA.

Sollins, P., C. Glassman, E. A. Paul, C. Swanston, K. Lajtha, J. Heil, and T. E. Elliott. 1999. Soil carbon and nitrogen pools and fractions. Pages 89–105 in G. P. Robertson, C. S. Bledsoe, D. C. Coleman, and P. Sollins, editors. Standard soil methods for long-term ecological research. Oxford University Press, New York, New York, USA.

Sollins, P., M. G. Kramer, C. Swanston, K. Lajtha, T. R. Filley, A. K. Aufdenkampe, R. Wagai, and R. D. Bowden. 2009. Sequential density fractionation across soils of contrasting mineralogy: evidence for both microbial- and mineral-controlled soil organic matter stabilization. Biogeochemistry 96:209–231.

Stevens, A., and B. van Wesemael. 2008. Soil organic carbon stock in the Belgian Ardennes as affected by afforestation and deforestation from 1868 to 2005. Forest Ecology and Management 256:1527–1539.

Stoyan, H., H. De-Polli, S. Böhm, G. P. Robertson, and E. A. Paul. 2000. Spatial heterogeneity of soil respiration and related soil properties at the plant scale. Plant and Soil 222:203–214.

Syswerda, S. P., A. T. Corbin, D. L. Mokma, A. N. Kravchenko, and G. P. Robertson. 2011. Agricultural management and soil carbon storage in surface vs. deep layers. Soil Science Society of America Journal 75:92–101.

van Veen, J. A., and E. A. Paul. 1981. Organic carbon dynamics in grassland soils. I. Background information and computer simulation. Canadian Journal of Soil Science 61:185–201.

Voroney, R. P., J. A. van Veen, and E. A. Paul. 1981. Organic C dynamics in grassland soils. 2. Model validation and simulation of the long-term effects of cultivation and rainfall erosion. Canadian Journal of Soil Science 61:211–224.

West, T. O., and G. Marland. 2002. A synthesis of carbon sequestration, carbon emissions, and net carbon flux in agriculture: comparing tillage practices in the United States. Agriculture, Ecosystems, & Environment 91:217–232.

Wickings, K., A. S. Grandy, S. Reed, and C. Cleveland. 2011. Management intensity alters decomposition via biological pathways. Biogeochemistry 104:365–379.

Willson, T. C., E. A. Paul, and R. R. Harwood. 2001. Biologically active soil organic matter fractions in sustainable cropping systems. Applied Soil Ecology 16:63–76.

Wischmeier, W. H., and D. D. Smith. 1961. A universal soil loss equation to guide conservation farm planning. 7th Transaction, Congress International Soil Science Society, Madison, Wisconsin, USA.

6

Microbial Diversity in Soils of Agricultural Landscapes and Its Relation to Ecosystem Function

Thomas M. Schmidt and Clive Waldron

The taxonomic and functional diversity of microbes in soil is stunning. Although they are largely invisible to the naked eye, microbes are pervasive in nature and have a profound impact on Earth's habitability. Like many other terrestrial environments, the top layer of soil at the Kellogg Biological Station Long-Term Ecological Research site (KBS LTER) typically contains 1×10^9 microbes per gram of dry soil, representing between 10,000 and a million species (Gans et al. 2005). Extrapolating the estimated number of microbes in individual soil samples to a global scale yields an estimate of an incredible 26×10^{28} microbes in terrestrial habitats (Whitman et al. 1998). Combined with estimates of microbial abundance in aquatic and subsurface environments, Whitman and colleagues (1998) estimated that microbes contain at least half of the amount of carbon (C) stored in plants and 10 times more nitrogen (N) and phosphorus (P).

The cycling of C, N, P, and other less abundant elements that pass through soil microbial communities influences environments as small as soil aggregates and as far-reaching as global climate. Here, we focus on the composition and function of microbes that drive the cycling of C and N in soils and discuss the ecological significance of their diversity and physiologies within, across, and beyond agricultural landscapes. Compounds containing C and N are major determinants of both crop productivity and climate change, so understanding what controls their transformations will be important for developing agricultural strategies that balance crop yield and ecosystem services against environmental harm.

Our emphasis is on discoveries made in the past two decades at KBS LTER, where experiments have maintained different management practices for >20 years and aspects of C and N cycling are well documented (Robertson and Hamilton

2015, Paul et al. 2015, Millar and Robertson 2015; Chapters 1, 5, and 9 in this volume). Of particular interest is the partitioning of C and N between bioavailable compounds in the soil and the exchange of greenhouse gases between soils and the atmosphere. These processes are heavily influenced by microbial activity, so it is possible that altered partitioning of C and N could result from changes in the microbial community that are generated by different land uses and management practices. The first step to understanding potential relationships between soil microbes and the production and consumption of greenhouse gases is to determine the structure of microbial communities in different soils.

We begin this chapter with an overview of the composition of microbial communities in soils under different land management practices. We then discuss the relationships between subsets of these communities and the fluxes of methane (CH_4), nitrous oxide (N_2O), and carbon dioxide (CO_2)—the three greenhouse gases that contribute most to climate change (IPCC 2007). Third, we consider the life histories of bacteria in soil and discuss how an ecological perspective has informed new models relating microbial community structure to function. We end the chapter with some thoughts about long-term research to test these models and evaluate the potential for restoring ecological function by manipulating bacterial diversity.

Microbial Diversity in Soil

Molecular Surveys

As in the majority of complex microbial communities in nature, most microbes in soil have yet to be cultured in the laboratory: fewer than 1% of the microbes that can be visualized microscopically grow under traditional cultivation conditions (Staley and Konopka 1985). Cultivated strains provide valuable insights into the metabolism and behavior of microbes from soil (discussed below), but the composition of microbial communities can be determined more rapidly and comprehensively through the characterization of extracted nucleic acids without cultivation. Research at KBS LTER has used some of the most common strategies for these molecular surveys (Fig. 6.1).

The key to comparing these complex microbial communities is obtaining DNA sequences of conserved genes. The nucleotide sequence of a conserved gene is similar in all members because the encoded function of the gene is retained. However, certain nucleotides can be substituted for others without disrupting the gene function. Over time these changes accumulate, so organisms can be grouped according to the location and number of substitutions in their sequences, with the closest relatives having fewest differences. To provide enough material for sequencing, the target gene in an environmental sample must be amplified from each microbe, typically by the process of Polymerase Chain Reaction (PCR). Surveys are often based on comparative analyses of the gene encoding the small subunit ribosomal RNA (SSU rRNA), which has a sedimentation coefficient of 16S in bacteria and Archaea and 18S in eukaryotes.

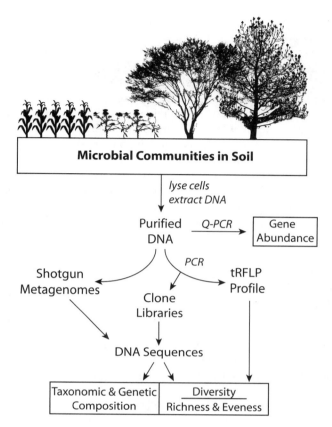

Figure 6.1. An overview of molecular approaches presented in this chapter that have been used to assess the composition and diversity of microbial communities in KBS LTER soils. The polymerase chain reaction (PCR) is used commonly to obtain sufficient quantities of a target gene for subsequent analysis. Quantitative PCR (Q-PCR) is used to estimate gene abundance, while terminal restriction fragment length polymorphism (tRFLP) is a finger-printing method that provides an overview of gene diversity.

The 16S and 18S rRNA-encoding genes contain regions of strict sequence conservation that are valuable for the design of universal primers to amplify the genes via PCR. These conserved regions are interspersed with regions of sequence variability that are useful for comparative analysis. A major advantage of targeting the SSU rRNA gene for surveys of diverse communities is the opportunity for comparison with the approximately 1.5 million other SSU rRNA gene sequences in curated public databases (Pruesse et al. 2007, Cole et al. 2009). Amplified 16S genes can either be cloned into plasmid vectors and sequenced individually, or sequenced directly with massively parallel "next-generation" sequencing. Because of their high capacity, next-generation technologies offer the potential for considerable in-depth exploration of diversity (Sogin et al. 2006). The 16S gene sequences can be compared at several thresholds of sequence similarity, providing insight into the composition of microbial communities at different taxonomic levels.

While 16S rRNA genes provide a taxonomic description of microbial communities, sequencing of DNA from microbial communities without initial amplification (also known as "shotgun" sequencing) offers insight into the metabolic potential of communities. The collection of sequences from shotgun libraries constitutes a metagenome. Advances have been made in the application of metagenomic methods in many environments (Dinsdale et al. 2008, Antonopoulos et al. 2009, Dethlefsen and Relman 2011), but microbial communities in soil have received considerably less attention. This is due in part to the remarkable diversity of microbes in soil and in part to the difficulty of defining and sampling microbial communities in such a heterogeneous structural matrix.

Heterogeneity in the physical structure of soil, including aggregates of recalcitrant organic matter, minerals, and microbes, complicates the task of collecting representative samples of the soil environment. Initial characterization of KBS LTER soils suggested that the microbial communities changed dramatically over relatively short distances; major variations in total microbial biomass, respiration, and the number of cultured bacteria were detected over distances as small as 20 cm (Robertson et al. 1997). The importance of spatial heterogeneity to microbial communities at KBS LTER was reinforced through analysis of 16S rRNA genes of *Burkholderia* isolates from the rhizospheres of nearby corn plants in the Biologically Based cropping system of the Main Cropping System Experiment (MCSE; Table 6.1; Robertson and Hamilton 2015, Chapter 1 in this volume). This study revealed dramatic differences in community composition and abundance between bacterial communities on individual plants (Ramette et al. 2005), highlighting the challenge presented by spatial variability in even a single field intensively managed for row-crop production. Variability in bacterial community structure was even detected among soil aggregates in KBS LTER soils. These studies were carried out before high-throughput sequencing became available. Sequence differences among 16S genes were identified indirectly by detecting specific sites for cleavage by restriction enzymes in a process known as Terminal Restriction Fragment Length Polymorphism (tRFLP; Fig. 6.1). As many as half of the differences between tRFLP profiles of 16S rDNA genes from soil could be explained by interaggregate variation, even when particles were less than 2 mm in diameter (Blackwood et al. 2006). This means microbial communities vary significantly from particle to particle—even in the same soil sample. This is a potential source of sampling bias when comparing microbial distributions across field sites or systems, but fortunately, spatial variability in microbial communities can be addressed with sufficient sample size and replication in sampling. For example, RFLP profiles of the genes encoding the final enzymatic step in denitrification (*nosZ*) were identical when triplicate 3-g samples were analyzed from either unmanaged successional communities or agricultural systems at KBS LTER (Stres et al. 2004).

In addition to spatial variability as a potential source of uncertainty, bias can also be introduced in certain steps of processing samples for molecular surveys. Several of these technical challenges have been addressed by research conducted with soils from KBS LTER. Differences in cell lysis and therefore DNA extraction efficiency can distort molecular surveys, revealing only a fraction of the tremendous diversity present in soil communities (Feinstein et al. 2009). One strategy for reducing the

Table 6.1. Description of the KBS LTER Main Cropping System Experiment (MCSE).[a]

Cropping System/Community	Dominant Growth Form	Management
Annual Cropping Systems		
Conventional (T1)	Herbaceous annual	Prevailing norm for tilled corn–soybean–winter wheat (c–s–w) rotation; standard chemical inputs, chisel-plowed, no cover crops, no manure or compost
No-till (T2)	Herbaceous annual	Prevailing norm for no-till c–s–w rotation; standard chemical inputs, permanent no-till, no cover crops, no manure or compost
Reduced Input (T3)	Herbaceous annual	Biologically based c–s–w rotation managed to reduce synthetic chemical inputs; chisel-plowed, winter cover crop of red clover or annual rye, no manure or compost
Biologically Based (T4)	Herbaceous annual	Biologically based c–s–w rotation managed without synthetic chemical inputs; chisel-plowed, mechanical weed control, winter cover crop of red clover or annual rye, no manure or compost; certified organic
Perennial Cropping Systems		
Alfalfa (T6)	Herbaceous perennial	5- to 6-year rotation with winter wheat as a 1-year break crop
Poplar (T5)	Woody perennial	Hybrid poplar trees on a ca. 10-year harvest cycle, either replanted or coppiced after harvest
Coniferous Forest (CF)	Woody perennial	Planted conifers periodically thinned
Successional and Reference Communities		
Early Successional (T7)	Herbaceous perennial	Historically tilled cropland abandoned in 1988; unmanaged but for annual spring burn to control woody species
Mown Grassland (never tilled) (T8)	Herbaceous perennial	Cleared woodlot (late 1950s) never tilled, unmanaged but for annual fall mowing to control woody species
Mid-successional (SF)	Herbaceous annual + woody perennial	Historically tilled cropland abandoned ca. 1955; unmanaged, with regrowth in transition to forest
Deciduous Forest (DF)	Woody perennial	Late successional native forest never cleared (two sites) or logged once ca. 1900 (one site); unmanaged

[a]Site codes that have been used throughout the project's history are given in parentheses. Systems T1–T7 are replicated within the LTER main site; others are replicated in the surrounding landscape. For further details, see Robertson and Hamilton (2015, Chapter 1 in this volume).

complexity to more manageable levels is through fractionation of extracted DNA according to its buoyant density, which is determined by the percentage of base pairs that are guanine:cytosine (G + C content). This strategy increases the capacity to detect bacterial species (defined as >97% identity of the 16S rRNA-encoding gene) in clone libraries (Morales et al. 2009). Using clone libraries from G + C-enrichment, Morales et al. (2009) subsequently designed and tested a collection of primers for PCR. They discovered significant variability in nontarget detection and demonstrated that rigorous empirical validation is necessary before new primers can be used to analyze complex communities using either regular (saturation) or quantitative PCR (Morales and Holben 2009).

While the construction of clone libraries has been a primary tool for assessing genetic landscapes in soil, the recent development of next-generation sequencing techniques offers the opportunity to explore the composition of complex microbial communities in much greater detail. These techniques can document not only the dominant members but also the newly revealed "rare biosphere" (Sogin et al. 2006)—those species in very low numbers that would otherwise go undetected. They can also identify other microbes such as Archaea and Fungi that are not detected in surveys of bacterial 16S genes. Application of next-generation sequencing to DNA extracted from KBS LTER soils has provided one of the first in-depth views of how terrestrial microbial communities differ functionally from complex communities in other ecosystems (Fig. 6.2).

Taxonomic and Functional Diversity

Greater taxonomic diversity within soil microbial communities has been revealed as the resolving power of analytical methods has improved—from culture collections to biochemical profiling (fatty acid methyl esters) to DNA sequences (clone libraries, then metagenomes). The remarkable diversity in soil results in large part from a wide variety of microbes present in low numbers. For example, analysis of a 5000-member clone library from the MCSE Conventional system (Table 6.1) led to an estimate of 3500 bacterial species (defined by 97% sequence identity of 16S rRNA genes) (Morales et al. 2009). Yet 80% of the 1700 species actually identified were encountered three times or less, so the projection that the soil contained about 3500 species is almost certainly an underestimate. Similarly, a study of 409 clones of 18S rDNA from basidiomycete fungi yielded a surprising variety of 241 species of basidiomycetes (Lynch and Thorn 2006). While there is not yet a complete description of the microbial diversity in any soil environment, statistically rigorous comparisons of the more abundant members of the community are now possible.

An in-depth assessment of the effect of row-crop agriculture on taxonomic diversity in microbial communities is currently under way across the KBS LTER landscape. Initial results, based on approximately 15,000 sequences of 16S rRNA encoding genes for replicated plots in the MCSE Deciduous Forest and Conventional systems of the MCSE (Table 6.1), revealed approximately 10,000 species in each treatment (Schmidt et al. unpublished). Surprisingly, despite the dramatic differences in these ecosystems, there is not an obvious difference in the phylum level composition of communities. As with other studies of bacterial diversity in soil (Janssen

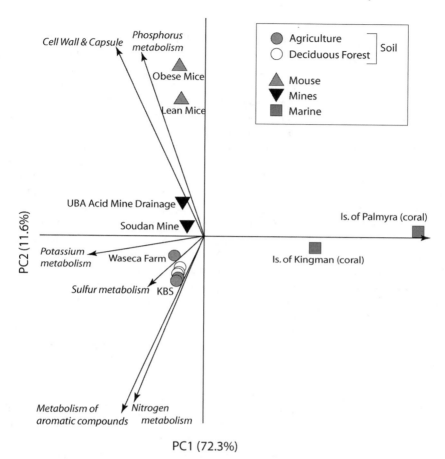

Figure 6.2. Principal components analysis of annotated shotgun metagenomes of microbial communities from soils (KBS LTER and Waseca Farm in Wisconsin), subterranean mines, marine coral reefs, and mouse guts. Axes depict the first two principal components (PCs) and indicate the % of variation explained by each. The functional diversity in soils is clearly distinguished from other biomes, driven in part by genes involved in the nitrogen cycle and the degradation of aromatic compounds. The length of each vector is proportional to the impact of the metabolic pathways in discriminating shotgun metagenomes.

2006), the phyla Proteobacteria and Acidobacteria are most abundant, constituting approximately 75% of the bacteria in soils at KBS LTER. The Proteobacteria are of particular interest because they include the majority of species that catalyze the consumption of CH_4 and production of N_2O—two important greenhouse gases whose emission to the atmosphere is influenced by land management (Gelfand and Robertson 2015, Chapter 12 in this volume). Given the reproducible differences in the fluxes of these gases among MCSE systems (Robertson et al. 2000), the composition of the Proteobacteria community may have an especially important influence on greenhouse gas production in managed landscapes, as discussed in the next section.

The most common bacterial phyla identified by 16S gene sequences (Proteobacteria, Actinobacteria, Cytophagales, Planctomycetes, Verrucomicrobia, and the Acidobacteria) were also investigated by the characterization of RNA extracted from KBS LTER soils. Because the concentration of rRNA in a cell is positively correlated with growth rate, direct probing of RNA provides an estimate of changes in the overall metabolic status of microbes. Changes in rRNA abundance reveal that soil microbial communities are dynamic and capable of responding to seasonal events. The relative abundance of microbial groups is also affected by local environments, so recognizable patterns of community structure can be related to land management (Buckley and Schmidt 2003). It is also worth noting that a low ratio of rRNA to rRNA-encoding genes suggests low overall metabolic activity, leading Jones and Lennon (2010) to propose that dormancy contributes to the maintenance of microbial diversity in lakes. Given the extensive diversity and low growth rates in soil microbes, it is worth considering dormancy as a major mechanism for the preservation of diversity in terrestrial habitats as well (Lennon and Jones 2011).

While detailed taxonomic characterization of communities can be derived by targeting the 16S rRNA encoding genes, analysis of shotgun metagenomes is currently the best approach for identifying the metabolic potential of a community. This provides a comprehensive catalog of DNA sequences in the soil and can indicate, through similarity to known genes, the relative abundance of metabolic functions and pathways that are encoded in that soil. Such data from KBS LTER revealed a previously unknown and systematic artifact in metagenomes (Gomez-Alvarez et al. 2009) that can be identified and removed with an online tool (Teal and Schmidt 2010)—a critical step in making quantitative metagenome comparisons. With this artifact removed, an initial assessment of the functional diversity in KBS LTER soils was made from replicate plots of the MCSE Deciduous Forest and Conventional corn–soybean–wheat systems. The metagenomes were annotated using the MG-RAST tool developed at Argonne National Laboratories (http://metagenomics. nmpdr.org) and compared to metagenomes from other biomes. Based on a principal components analysis of the annotated metagenomes, the functional diversity in soils was clearly distinguished from other biomes (Fig. 6.2). Nitrogen metabolism was one of the major features driving the distinction between the microbial communities in soils from those in other environments.

Environmental Drivers of Diversity

Chemical and physical factors that affect the distribution of microbes in soil are poorly understood. However, the application of molecular techniques is providing the capacity to identify environmental factors that influence microbial distributions in nature. Culture-independent approaches (Pace 2009) are particularly useful for exploring the biology of bacteria from phyla that are poorly represented in culture collections. These include one of the most abundant phyla in soil, the Acidobacteria. In two recent studies (Eichorst et al. 2007, 2011), the distributions of Acidobacteria in relation to physical and chemical characteristics of soil were determined across the MCSE using partial sequencing of cloned 16S rRNA genes. The percentage of subdivision 1 Acidobacteria was correlated with soil pH, being highest in the most acidic soils (Eichorst et al. 2007, 2011). To determine if this relationship was significant for other terrestrial environments, previously published sequences from a variety of soil environments were similarly analyzed. The combined datasets revealed a significant correlation ($p < 0.004$) between pH and the percentage of Acidobacteria in subdivision 1 (Fig. 6.3A). The potential for plant polymers to influence the distribution of Acidobacteria was assessed in a molecular survey using clone libraries

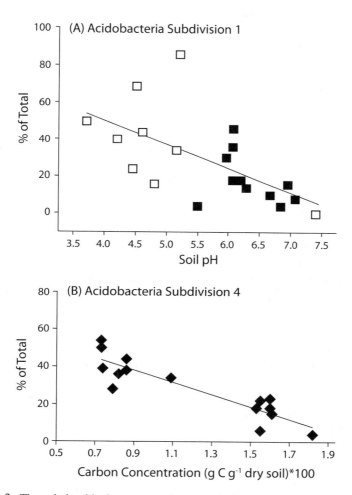

Figure 6.3. The relationship between environmental characteristics and Acidobacteria, one of the most abundant phyla in soils. (A) The proportion of Acidobacteria belonging to Subdivision 1 is related to the pH of the soil environment both at KBS LTER (solid symbols) and in soils worldwide (open symbols). Adapted from Eichorst et al. (2007). (B) The proportion of Acidobacteria in Subdivision 4 correlates with soil carbon in the agricultural and managed grassland systems at the KBS LTER. Adapted from Eichorst et al. (2011). Both panels demonstrate how data from molecular surveys are laying a foundation for understanding factors that influence the distribution of microbes in soil.

from native and agricultural soils at KBS LTER (Eichorst et al. 2011). The distribution of Acidobacteria varied with the C content of soil. In particular, subdivision 4 of Acidobacteria was most abundant in agricultural soils (Fig. 6.3B), which contain less organic C than forested soils (Paul et al. 2015, Chapter 5 in this volume).

Given that plants are the major source of organic C for soil microbes, it is reasonable to expect a coupling between the composition of plant and microbial communities in terrestrial ecosystems. Both physiological phospholipid fatty acid (PLFA) profiles and metabolic Biolog assays of rhizosphere soils obtained from different plant species indicate that plant species composition impacts microbial activity and functional diversity within soil microbial communities (Broughton and Gross 2000). Further, tRFLP fingerprints of DNA extracted from various soil fractions indicate that the highest diversity of microorganisms is associated with rapidly cycling soil C from freshly deposited plant residues (Paul et al. 2015, Chapter 5 in this volume). These studies also reveal that heterogeneity in soil microbial communities results from small-scale spatial heterogeneity in the availability and composition of plant residues, which is consistent with plant biomass as a driver of microbial community structure. Molecular surveys of microbial communities in soil provide a means to identify environmental factors that influence community structure and help set the stage for studies that relate the structure of microbial communities with their functions.

Relating Structure and Function of Microbial Communities

Relationships between Biological Diversity and Function

Ongoing studies at the Cedar Creek LTER (e.g., Tilman et al. 2001) and elsewhere (e.g., Suding et al. 2005) are exploring relationships between plant species diversity and ecosystem function. These studies have demonstrated, for instance, that plant diversity is positively correlated with net primary productivity (Tilman et al. 2001) and C sequestration in soil (Adair et al. 2009). Although the underlying explanation for such diversity–function relationships is vigorously debated, complementary contributions from different plant species appear to be of fundamental importance (Fargione et al. 2007). Given the positive relationship between the diversity of plant species and the magnitude of ecosystem processes, we asked if there might be similar relationships between bacterial communities and the processes they catalyze in KBS LTER ecosystems. Evidence from elsewhere suggests that such a relationship is unlikely for those microbially catalyzed processes, such as biomass decomposition, involving a wide variety of organisms (Schimel 1995, Groffman and Bohlen 1999). However, the number of species in a bacterial community (species richness) may be important for those processes catalyzed by fewer, more specialized species (Cavigelli and Robertson 2000). Given that microbes are responsible for the exchange of greenhouse gases with the atmosphere, including the consumption of atmospheric CH_4 and production of N_2O, the relationships among bacterial diversity and these processes seem especially important to explore.

Methane and Methanotrophs

The decrease in CH_4 consumption that accompanies conversion of forest or grassland to row-crop agriculture is well documented (Smith et al. 2000, Robertson et al. 2000), but had not been tied to the diversity of CH_4-oxidizing microbes (methanotrophs) in these soils. Levine et al. (2011) compared molecular surveys of bacteria in KBS LTER soils to *in-situ* measurements of CH_4 fluxes. These surveys were based on *pmoA*—a gene that codes for one of the subunits of CH_4 monooxygenase, the first enzyme in the pathway of CH_4 oxidation. Across MCSE systems, CH_4 consumption varied by ~7-fold and was greater in soils with a higher number of methanotroph species (Fig. 6.4A). Additionally, the temporal stability of CH_4 oxidation throughout the year increased with methanotroph richness: in different MCSE systems, CH_4 oxidation was less variable (there was less variance among system replicates) in treatments harboring the highest methanotroph richness (Fig. 6.4B). Levine et al. (2011) attributed increased stability to a greater capacity for diverse methanotroph communities to oxidize CH_4 under a broader set of environmental conditions.

The MCSE also provides an opportunity to examine the recovery of CH_4 oxidation and methanotroph diversity following abandonment from agriculture. The rate of CH_4 consumption and the number of methanotroph species both increase following the cessation of agricultural activities. Extrapolating from the current rate at which methanotroph richness and CH_4 consumption are being reestablished, Levine et al. (2011) estimate that approximately 80 years from the time of abandonment will be needed for CH_4 oxidation to return to the levels of native undisturbed soils. This relationship also suggests that managing lands to conserve or restore methanotroph richness (see Gelfand and Robertson 2015, Chapter 12 in this volume) could help mitigate increasing atmospheric concentrations of this potent greenhouse gas.

Bacteria and Nitrous Oxide Production

Nitrous oxide is another potent greenhouse gas of biological origin. Approximately half of contemporary anthropogenic N_2O emitted to the atmosphere is from agricultural soils (IPCC 2007) and its emission is accelerated by N fertilizer use (Millar and Robertson 2015, Chapter 9 in this volume). Nitrous oxide can be produced by both nitrifying and denitrifying bacteria (Robertson and Groffman 2015), but stable isotope tracing indicates that in agricultural soils at KBS LTER, it is made primarily by denitrifiers (Ostrom et al. 2010). During denitrification, microbes use nitrate (NO_3^-) in place of oxygen (O_2) as a terminal electron acceptor for respiratory metabolism. A key enzyme in denitrification is nitrite reductase (*nir*) encoded by either *nirK* or *nirS* genes (Fig. 6.5). Huizinga (2006) found that denitrifiers in KBS LTER soils primarily carry *nirK*, but her molecular surveys did not find any patterns in the distribution of denitrifiers with *nirK* that could be linked to N_2O flux.

However, there may be a pattern in the distribution of denitrifiers that carry a gene that codes for another enzyme involved in N_2O fluxes, N_2O reductase (*nos*). The net production of N_2O from denitrification is dependent not only on the activities of the enzymes nitrite reductase (nir) and nitric oxide reductase

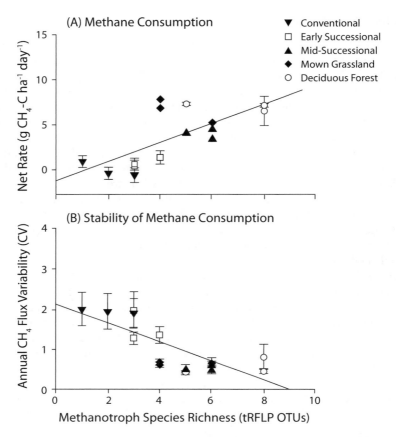

Figure 6.4. The number of bacterial species in soil that oxidize methane (methanotrophs) is directly related to both the rate (A) and stability of methane consumption (B) at KBS LTER. Methanotroph species richness is based on the number of Operational Taxonomic Units (OTUs) defined by tRFLP analysis of methane (CH_4) monooxygenase, the first enzyme in the CH_4 oxidation pathway. Methane fluxes were estimated *in situ* using chambers placed over the soil. The temporal stability of CH_4 consumption rates is expressed as the coefficient of variation (CV) over the annual period of measurements. Redrawn from Levine et al. (2011).

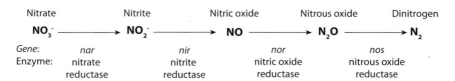

Figure 6.5. The bacterial denitrification pathway, including metabolic intermediates, essential genes and enzymes. A key enzyme in denitrification, producing the first gaseous metabolite, is nitrite reductase (*nir*) encoded by the *nirK* or *nirS* genes. Both genes have been used to survey the diversity of denitrifying bacteria.

(nor) that make N_2O, but also on the activity of N_2O reductase (nos) that reduces N_2O to the innocuous gas N_2 (Fig. 6.5). A tRFLP analysis of cloned *nosZ* genes (Stres et al. 2004) revealed greater diversity in the denitrifying community in the MCSE Conventional system than in the Mown Grassland (never tilled) treatment. Furthermore, even though nirS was less abundant than nirK, quantitative PCR revealed that this agricultural community had a higher ratio of *nirS* genes to *nosZ* genes and so had greater potential to produce N_2O (rather than N_2) from the same amount of NO_3^- (Morales et al. 2010). This represents the intriguing possibility that agricultural land use has selected for a subset of denitrifiers that may accelerate N_2O production.

Another possible functional difference between denitrifying bacteria from agricultural and successional soils at KBS LTER was suggested by a study of N_2O reductases from cultured representatives. Even though these cultured species represent only a small fraction of the denitrifying bacteria, the average O_2-sensitivity of their N_2O reductases varied significantly between the two soils (Cavigelli and Robertson 2001). Next-generation sequencing technologies will enable analysis of all the bacterial *nosZ* genes that code for N_2O reductase in agricultural and successional communities.

Microbial Respiration and Carbon Dioxide Flux

Increasing worldwide demand for agriculture to produce food, fuel, and fiber affects reservoirs of soil organic matter (SOM), which are highly responsive to both changing land use and shifts in climate (Paul et al. 2015, Chapter 5 in this volume; Robertson et al. 2015, Chapter 2 in this volume). Microorganisms play a crucial role in determining the turnover of SOM: they rapidly assimilate and respire labile fractions to CO_2 or transform organic matter to more recalcitrant compounds that are critical to C sequestration and long-term soil productivity. Approximately half of the C lost as CO_2 from soils is due to the metabolism of heterotrophic microbes, with the remainder ascribed primarily to plant root respiration (Hanson et al. 2000). Predicting the fate of microbially processed C (i.e., assimilation into biomass vs. respiratory loss) is thus critical to the development of robust models that accurately predict terrestrial C transformations.

Carbon dioxide emission from soils varies across MCSE systems (Paul et al. 1999), but unlike the specialized functions of CH_4 consumption or N_2O production, no direct relationship exists between CO_2 emission and bacterial richness in KBS LTER soils (Levine et al. 2011). Although the number of bacterial species does not vary dramatically with land management, the composition of the microbial heterotroph communities does. And because C cycling involves a broad diversity of microbes, it may be in the changing composition of heterotrophic microbes that we find explanations for the variation in respiratory CO_2 production. Most biogeochemical models of C cycling assume that microbial communities assimilate C at a fixed rate. For example, there are a number of transformations of C in the widely used CENTURY model (Fig. 6.6; Parton et al. 1987), and it is assumed that in each transformation, 55% of the C consumed by microbes is oxidized to CO_2, with the remainder incorporated into

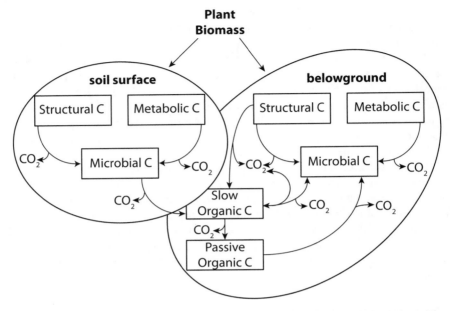

Figure 6.6. The soil subsystem of the Century Model (Parton et al. 1987) includes microbial transformations of plant biomass where assimilated carbon is apportioned between carbon dioxide (CO_2) and microbial biomass both at the soil surface and belowground. In both cases, the model assumes that 55% of the carbon assimilated by microbes is oxidized to CO_2 with the remainder incorporated into cell biomass, which is equivalent to a bacterial growth efficiency of 0.45.

cell biomass. This assumption has profound consequences for the predicted fate of C, including its retention in soil and the potential for global-scale feedbacks.

While the assumption that 55% of the C metabolized by microbes is released as CO_2 may be reasonable for pure cultures provided with substrates and nutrients in optimal proportions, the proportion of carbon respired to CO_2 changes under substrate or nutrient limitation (del Giorgio and Cole 1998). The efficiency by which microbial communities use resources to produce more biomass is not determined by a single enzyme or pathway, as the assumption implies, but rather by their ability to coordinate cellular activity with environmental signals. The net result of this integrated cellular system is a balance between growth and respiration that is characteristic of each microbe. To understand the rate and controls of microbial transformations of C in soil, we need to advance our knowledge of the processes that influence C metabolism in different members of the community. In other words, it is critical that we better understand the fundamental physiological and ecological mechanisms that control the assimilation and fate of C processed by soil microbes. Studies of microbial populations in KBS LTER soils are contributing to this understanding and are the subject of the next section.

Ecology of Soil Microbes

Cultivation and Characterization of Ecologically Relevant Microbes

While molecular surveys are helping to delineate the distribution and temporal dynamics of microbes in the environment, understanding the ecology of microbes requires studying the functions of the entire organism, not just its DNA sequences. Culture-based approaches—in which individual strains or consortia of microbes are isolated and grown—provide the most information about life history traits of microbes that thrive in the soil environment. Life history traits are characteristics that influence the growth, reproduction, or survival of an organism. Few studies of microbes embrace the notion of life histories, perhaps because they do not have obvious counterparts to plant or animal life history traits such as clutch size or parental care. Since natural selection favors individuals who are best able to survive and leave viable progeny regardless of whether they are single- or multicellular, the study of life history traits in microbes should be as useful as it has been in plants and animals and can provide a common currency to compare ecological strategies.

Microbiologists have traditionally used media that contain abundant organic compounds and other nutrients to grow and study microbes. These "rich" media formulations select for copiotrophic microbes—organisms that capitalize on the availability of abundant resources and grow quickly. Cultivating oligotrophs—organisms that thrive in resource-poor environments—requires a different strategy. Key culturing elements are growth media with limited nutrients, atmospheres with less than ambient O_2 concentrations, and sufficient time for slow-growing colonies to form (e.g., Janssen et al. 2002, Stevenson et al. 2004). This approach has enabled many never or rarely cultured bacteria to be grown in the lab, including representatives of the abundant but poorly characterized phyla Acidobacteria and Verrucomicrobia.

Acidobacteria are present in soils worldwide (Janssen 2006) and abundant in many KBS LTER soils, as noted earlier, yet few cultivated representatives of this cosmopolitan phylogenetic group exist. Presumably, they are unable to compete with the fast-growing bacteria that prosper on the rich media commonly used in microbiology laboratories. To learn more about the metabolic properties and potential ecological roles of members of this phylum, Eichorst et al. (2007) isolated Acidobacteria strains from KBS LTER soils using incubation conditions and media designed to mimic their natural environment. Cultivation conditions included low concentrations of nutrients, plant polymers as sole C and energy sources, and extended (3 to 4-week) incubation periods. Altered incubation atmospheres with decreased concentrations of O_2 and elevated levels of CO_2 resulted in a slightly acidified medium with a pH similar to *in-situ* measurements of soil pH at KBS LTER.

When plant polymers were used as a C and energy source, the diversity of Acidobacteria growing in culture increased relative to those cultured on simple sugars (Eichorst et al. 2011). All the cultivated strains of Acidobacteria contained either one or two copies of the 16S ribosomal RNA-encoding gene that, along with a relatively slow doubling time (10–15 hours at ~23°C), suggests an oligotrophic lifestyle. Several of these strains produce a carotenoid pigment that is thought to protect against oxidative stress when exposed to ambient O_2 concentrations. The need for its production in ambient atmospheres suggests a sensitivity to O_2 that would explain optimal growth rates under reduced O_2 atmospheres. The optimal growth of these Acidobacteria under slightly acidic pH and low O_2 concentrations, conditions commonly observed in many soils, is consistent with their widespread distribution and abundance in soils.

The strains in one collection from KBS LTER are sufficiently similar, but distinct enough from previously named Acidobacteria, to warrant creation of a new genus, *Terriglobus. T. roseus*—the pigment-producing strain—has been defined as the type species of the genus (Eichorst et al. 2007). Studies are under way to explore the role of the extensive extracellular polysaccharide produced by *Terriglobus* strains (Fig. 6.7) in the formation of soil aggregates and to document the capacity of the strains to degrade complex plant polymers.

The characterization of microbes in culture is a time-intensive endeavor that has faded in popularity as access to molecular approaches has widened. As a result, half of the 70+ known bacterial phyla have been identified solely from rRNA gene sequences (Pace 2009), making it difficult to draw inferences about their genetics, physiology, or ecology. The study of single cells (including deriving their complete genome sequences) offers an intriguing new possibility for advancing our knowledge of microbial physiology and the lifestyle trade-offs that underlie the distribution and activities of bacteria in nature.

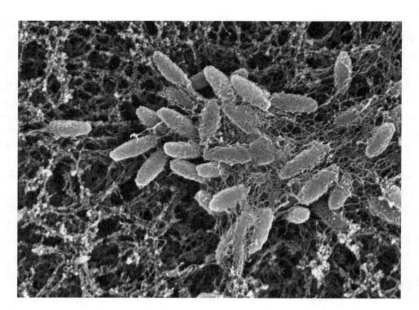

Figure 6.7. A scanning electron micrograph of an isolate of the phylum Acidobacterium revealing an extensive extracellular polysaccharide matrix that may be important in the formation of soil aggregates. From Eichorst et al. (2007).

Trade-Offs between Power and Efficiency

Little is known about the selective pressures that shape microbial communities in soil and, in particular, about fluctuations in environmental conditions that might trigger the growth of distinct microbes. Ecological strategies of microbes are commonly described on a spectrum of alternative responses to nutrient supply. At one extreme are oligotrophs that are most competitive when organic resources are scarce. At the other end are copiotrophs that thrive when nutrients are suddenly abundant. Klappenbach et al. (2000) isolated a collection of bacteria from KBS LTER soils to develop and test a model positing that these opposing strategies involve a trade-off between "power" in the rapid response to an influx of resources and "efficiency" in the use of scarce resources. A copiotroph's capacity for rapid growth in response to an influx of resources may be the selective pressure driving the maintenance of multiple copies of rRNA operons in its genome. The cost of maintaining this capacity for rapid response and growth would be detrimental to efficiency in nutrient-poor environments, so oligotrophs may be under selective pressure to minimize the number of their rRNA-encoding genes.

Considerable evidence has accumulated from KBS LTER isolates to support a conceptual model in which the number of rRNA operons encoded by a bacterium, which ranges from 1 to 15 (Lee et al. 2009), is indicative of where an organism lies on a spectrum of ecological strategies between oligotrophy (few rRNA operons) and copiotrophy (many rRNA operons; Fig. 6.8) (Klappenbach et al. 2000,

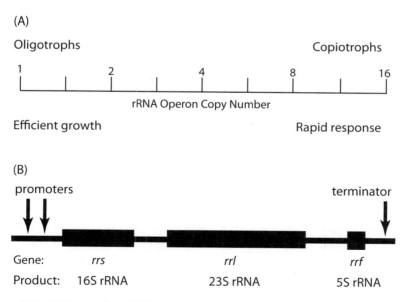

Figure 6.8. A) The number of rRNA operons encoded by a bacterium ranges from one to 15 and is proposed to be indicative of where an organism lies on a spectrum of ecological strategies between oligotrophy (few rRNA operons) and copiotrophy (many rRNA operons). B) The typical structure of an rRNA operon in bacteria.

Stevenson 2000, Dethlefsen and Schmidt 2007). For instance, bacteria that form visible colonies rapidly (<48 hours) upon exposure to nutritionally complex media contain an average of 5.5 copies of the SSU rRNA encoding gene, whereas bacteria that respond slowly contained an average of 1.4 copies (Klappenbach et al. 2000). These findings reveal phenotypic effects associated with the number of rRNA operons that underlie the distribution and abundance of bacterial populations in soil.

In addition to determining the number of rRNA operons, we have examined the efficiency with which the protein-synthesizing machinery operates in microbes. This machinery typically makes up more than half of a microbe's dry weight and consumes a majority of the cell's energy during rapid growth. The translation of mRNA into protein has been studied extensively in model organisms; nevertheless, while the translational apparatus is qualitatively similar in structure and function across all known life, little is known about variations between organisms in translational performance. The macromolecular composition of phylogenetically diverse oligotrophic and copiotrophic soil bacteria suggests that differences in translational power (normalized rate of protein synthesis) and associated maintenance costs help explain the fundamental trade-offs between the rapid growth in copiotrophs vs. the efficient use of resources in oligotrophs (Dethlefsen and Schmidt 2007). Analysis of bacterial genomes, in particular the patterns of codon usage in protein-coding genes, supports this model (Dethlefsen and Schmidt 2005).

When this model is applied to denitrifiers, the positive relationship between the number of rRNA-coding genes and maximum growth rate becomes apparent (Fig. 6.9). It suggests that the influence of land management on the composition of denitrifiers results from disturbance favoring fast-responding (copiotrophic) denitrifiers.

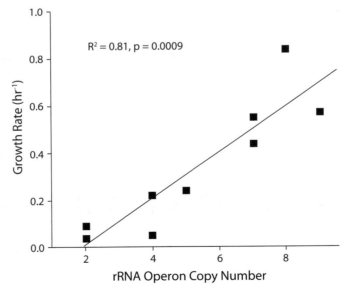

Figure 6.9. The relationship between growth rate and the number of rRNA operon copies in sequenced genomes of cultured denitrifying bacteria.

Based on this body of research, one might expect that bacteria that make efficient use of resources are favored in habitats where the concentration of resources is perpetually low. A minimum concentration of resources is required to provide maintenance energy, which is the sum of energy expenditures that are not directed toward growth, for example, maintaining a charged membrane or motility. At very low resource concentrations, the majority of assimilated C is therefore respired for maintenance purposes. Under these conditions, most C resources will be oxidized to CO_2 to provide energy rather than biomass. Consequently, bacterial growth efficiency should be low in resource-limited environments. When C and nutrient resources are more abundant, a greater proportion of assimilated C can be allocated to biomass production, making growth efficiency higher. Studies to measure the efficiency of C utilization in different MCSE systems are under way and will be a reasonable test of this model's ability to accurately predict the composition of bacterial communities in soil.

Summary

Microbial communities in soil are critical to the productivity and health of the biosphere and, in particular, to the cycling of C and N that underpin agricultural productivity and climate change. However, we are only just beginning to understand the composition and function of these complex soil microbial assemblages. The replicated array of managed and unmanaged ecosystems at KBS LTER has been an invaluable resource for studying the effects of land use on soil microbial communities and their impact on the biosphere. Molecular surveys reveal a tremendous taxonomic and functional diversity of microbes in KBS LTER soils—a diversity so large that even state-of-the-art, large-scale DNA sequencing methods have yet to reveal its full extent. Despite the challenges associated with measuring this enormous diversity, there are clear patterns in the distribution of microbes across the KBS LTER landscape, and the relationships between the structure of microbial communities and ecosystem-level processes such as C and N cycling are being revealed. One of the most striking relationships uncovered to date is the positive correlation between methanotroph diversity and CH_4 consumption.

As we continue gathering fundamental information about the structure of different bacterial communities, we must ask more questions about the connections between their varying compositions and the major ecological functions that are altered by agronomic management. It is particularly important that we begin to address cause and effect: Is a function modified directly by a change in the bacterial community, or are they independently influenced by land use?

The answers to these questions will affect future microbial research at KBS LTER in two key ways. The first will be to accelerate evolution of the current, largely observational approach to a more experimental strategy that manipulates variables *in-situ* to test hypotheses about the role of the bacterial community. These experiments will then lead to the second phase of research: developing and evaluating practices to maximize agricultural productivity while reducing environmental impact. If these practical applications involve direct manipulation of bacterial communities, we will have some major new challenges to overcome. Perhaps the most pressing, in light of discoveries that microbial communities may take decades to recover their diversity, will be to find ways to accelerate changes in the composition—and hence the function—of microbial communities in agricultural soils. The availability of large-scale replicated experimental systems reflecting different land management practices will ensure that the KBS LTER continues to play a pivotal role in both fundamental and applied aspects of microbial ecology research.

References

Adair, E. C., P. B. Reich, S. E. Hobbie, and J. Knops. 2009. Interactive effects of time, CO_2, N, and diversity on total belowground carbon allocation and ecosystem carbon storage in a grassland community. Ecosystems 12:1037–1052.

Atonopoulos, D. A., S. M. Huse, H. G. Morrison, T. M. Schmidt, M. L. Sogin, and V. B. Young. 2009. Reproducible community dynamics of the gastrointestinal microbiota following antibiotic perturbation. Infection and Immunity 77:2367–2375.

Blackwood, C. B., C. J. Dell, E. A. Paul, and A. J. M. Smucker. 2006. Eubacterial communities in different soil macroaggregate environments and cropping systems. Soil Biology & Biochemistry 38:720–728.

Broughton, L. C., and K. L. Gross. 2000. Patterns of diversity in plant and soil microbial communities along a productivity gradient in a Michigan old field. Oecologia 125:420–427.

Buckley, D. H., and T. M. Schmidt. 2003. Diversity and dynamics of microbial communities in soils from agroecosystems. Environmental Microbiology 5:441–452.

Cavigelli, M. A., and G. P. Robertson. 2000. The functional significance of denitrifier community composition in a terrestrial ecosystem. Ecology 81:1402–1414.

Cavigelli, M. A., and G. P. Robertson. 2001. Role of denitrifier diversity in rates of nitrous oxide consumption in a terrestrial ecosystem. Soil Biology & Biochemistry 33:297–310.

Cole, J. R., Q. Wang, E. Cardenas, J. Fish, B. Chai, R. J. Farris, A. S. Kulam-Syed-Mohideen, D. M. McGarrell, T. Marsh, G. M. Garrity, and J. M. Tiedje. 2009. The Ribosomal Database Project: improved alignments and new tools for rRNA analysis. Nucleic Acids Research 37:D141–D145.

del Giorgio, P. A., and J. J. Cole. 1998. Bacterial growth efficiency in natural aquatic systems. Annual Review of Ecology and Systematics 29:503–541.

Dethlefsen, L., and D. A. Relman. 2011. Incomplete recovery and individualized responses of the human distal gut microbiota to repeated antibiotic perturbation. Proceedings of the National Academy of Sciences USA 108:4554–4561.

Dethlefsen, L., and T. M. Schmidt. 2005. Differences in codon bias cannot explain differences in translational power among microbes. BMC Bioinformatics 6:3.

Dethlefsen, L., and T. M. Schmidt. 2007. Performance of the translational apparatus varies with the ecological strategies of bacteria. Journal of Bacteriology 189:3237–3245.

Dinsdale, E. A., R. A. Edwards, D. Hall, F. Angly, M. Breitbart, J. M. Brulc, M. Furlan, C. Desnues, M. Haynes, L. L. Li, L. McDaniel, M. A. Moran, K. E. Nelson, C. Nilsson, R. Olson, J. Paul, B. R. Brito, Y. J. Ruan, B. K. Swan, R. Stevens, D. L. Valentine, R. V. Thurber, L. Wegley, B. A. White, and F. Rohwer. 2008. Functional metagenomic profiling of nine biomes. Nature 455:830–830.

Eichorst, S. A., J. A. Breznak, and T. M. Schmidt. 2007. Isolation and characterization of soil bacteria that define *Terriglobus* gen. nov., in the phylum Acidobacteria. Applied and Environmental Microbiology 73:2708–2717.

Eichorst, S. A., C. R. Kuske, and T. M. Schmidt. 2011. Influence of plant polymers on the distribution and cultivation of bacteria in the phylum Acidobacteria. Applied and Environmental Microbiology 77:586–596.

Fargione, J., D. Tilman, R. Dybzinski, J. H. R. Lambers, C. Clark, W. S. Harpole, J. M. H. Knops, P. B. Reich, and M. Loreau. 2007. From selection to complementarity: shifts in the causes of biodiversity-productivity relationships in a long-term biodiversity experiment. Proceedings of the Royal Society B: Biological Sciences 274:871–876.

Feinstein, L. M., W. J. Sul, and C. B. Blackwood. 2009. Assessment of bias associated with incomplete extraction of microbial DNA from soil. Applied and Environmental Microbiology 75:5428–5433.

Gans, J., M. Wolinsky, and J. M. Dunbar. 2005. Computational improvements reveal great bacterial diversity and high metal toxicity in soil. Science 309:1387–1390.

Gelfand, I., and G. P. Robertson. 2015. Mitigation of greenhouse gas emissions in agricultural ecosystems. Pages 310–339 in S. K. Hamilton, J. E. Doll, and G. P. Robertson, editors. The ecology of agricultural Landscapes: long-term research on the path to sustainability. Oxford University Press, New York, New York, USA.

Gomez-Alvarez, V., T. K. Teal, and T. M. Schmidt. 2009. Systematic artifacts in metagenomes from complex microbial communities. The ISME Journal 3:1314–1317.

Groffman, P. M., and P. J. Bohlen. 1999. Soil and sediment biodiversity. BioScience 49:139–148.

Hanson, P. J., N. T. Edwards, C. T. Garten, and J. A. Andrews. 2000. Separating root and soil microbial contributions to soil respiration: a review of methods and observations. Biogeochemistry 48:115–146.

Huizinga, K. M. 2006. The diversity of dissimilatory nitrate reducers in an agroecosystem. Dissertation, Michigan State University, East Lansing, Michigan, USA.

IPCC (Intergovernmental Panel on Climate Change). 2007. Climate change 2007: synthesis report. Contribution of Working Groups I, II and III to the Fourth Assessment Report of the IPCC [Core Writing Team, R. K. Pachauri, and A. Resinger, editors]. IPCC, Geneva, Switzerland.

Janssen, P. H. 2006. Identifying the dominant soil bacterial taxa in libraries of 16S rRNA and 16S rRNA genes. Applied and Environmental Microbiology 72:1719–1728.

Janssen, P. H., P. S. Yates, B. E. Grinton, P. M. Taylor, and M. Sait. 2002. Improved culturability of soil bacteria and isolation in pure culture of novel members of the divisions Acidobacteria, Actinobacteria, Proteobacteria, and Verrucomicrobia. Applied and Environmental Microbiology 68:2391–2396.

Jones, S. E., and J. T. Lennon. 2010. Dormancy contributes to the maintenance of microbial diversity. Proceedings of the National Academy of Sciences USA 107:5881–5886.

Klappenbach, J., J. M. Dunbar, and T. M. Schmidt. 2000. rRNA gene copy number reflects ecological strategies in bacteria. Applied and Environmental Microbiology 66:1328–1333.

Lee, Z. M.-P., C. Bussema, III, and T. M. Schmidt. 2009. rrnDB: documenting the number of rRNA and tRNA genes in bacteria and archaea. Nucleic Acids Research 37:D489-D493.

Lennon, J. T., and S. E. Jones. 2011. Microbial seed banks: the ecological and evolutionary implications of dormancy. Nature Reviews: Microbiology 9:119–130.

Levine, U., T. K. Teal, G. P. Robertson, and T. M. Schmidt. 2011. Agriculture's impact on microbial diversity and associated fluxes of carbon dioxide and methane. The ISME Journal 5:1683–1691.

Lynch, M. D., and R. G. Thorn. 2006. Diversity of basidiomycetes in Michigan agricultural soils. Applied and Environmental Microbiology 72:7050–7056.

Millar, N., and G. P. Robertson. 2015. Nitrogen transfers and transformations in row-crop ecosystems. Pages 213–251 in S. K. Hamilton, J. E. Doll, and G. P. Robertson, editors. The ecology of agricultural Landscapes: long-term research on the path to sustainability. Oxford University Press, New York, New York, USA.

Morales, S. E., T. F. Cosart, and W. E. Holben. 2010. Bacterial gene abundances as indicators of greenhouse gas emission in soils. The ISME Journal 4:799–808.

Morales, S. E., T. F. Cosart, J. V. Johnson, and W. E. Holben. 2009. Extensive phylogenetic analysis of a soil bacterial community illustrates extreme taxon evenness and the effects of amplicon length, degree of coverage, and DNA fractionation on classification and ecological parameters. Applied and Environmental Microbiology 75:668–675.

Morales, S. E., and W. E. Holben. 2009. Empirical testing of 16S rRNA gene PCR primer pairs reveals variance in target specificity and efficacy not suggested by *in silico* analysis. Applied and Environmental Microbiology 75:2677–2683.

Ostrom, N. E., R. Sutka, P. H. Ostrom, A. S. Grandy, K. H. Huizinga, H. Gandhi, J. C. von Fisher, and G. P. Robertson. 2010. Isotopologue data reveal bacterial denitrification as the primary source of N_2O during a high flux event following cultivation of a native temperate grassland. Soil Biology & Biochemistry 42:499–506.

Pace, N. R. 2009. Mapping the tree of life: progress and prospects. Microbiology and Molecular Biology Reviews 73:565–576.

Parton, W. J., D. S. Schimel, C. V. Cole, and D. S. Ojima. 1987. Analysis of factors controlling soil organic matter levels in Great Plains Grasslands. Soil Science Society of America Journal 51:1173–1179.

Paul, E. A., D. Harris, H. P. Collins, U. Schulthess, and G. P. Robertson. 1999. Evolution of CO_2 and soil carbon dynamics in biologically managed, row-crop agroecosystems. Applied Soil Ecology 11:53–65.

Paul, E. A., A. Kravchenko, A. S. Grandy, and S. Morris. 2015. Soil organic matter dynamics: controls and management for sustainable ecosystem functioning. Pages 104–134 in S. K. Hamilton, J. E. Doll, and G. P. Robertson, editors. The ecology of agricultural Landscapes: long-term research on the path to sustainability. Oxford University Press, New York, New York, USA.

Pruesse, E., C. Quast, K. Knittel, B. M. Fuchs, W. Ludwig, J. Peplies, and F. O. Glöckner. 2007. SILVA: a comprehensive online resource for quality checked and aligned ribosomal RNA sequence data compatible with ARB. Nucleic Acids Research 35:7188–7196.

Ramette, A., J. J. LiPuma, and J. M. Tiedje. 2005. Species abundance and diversity of *Burkholderia cepacia* complex in the environment. Applied and Environmental Microbiology 71:1193–1201.

Robertson, G. P., and P. M. Groffman. 2015. Nitrogen transformations. Pages 421–426 in E. A. Paul, editor. Soil microbiology, ecology, and biochemistry. Fourth edition. Academic Press, Burlington, Massachusetts, USA.

Robertson, G. P., and S. K. Hamilton. 2015. Long-term ecological research at the Kellogg Biological Station LTER Site: conceptual and experimental framework. Pages 1–32 in S. K. Hamilton, J. E. Doll, and G. P. Robertson, editors. The ecology of agricultural Landscapes: long-term research on the path to sustainability. Oxford University Press, New York, New York, USA.

Robertson, G. P., K. M. Klingensmith, M. J. Klug, E. A. Paul, J. R. Crum, and B. G. Ellis. 1997. Soil resources, microbial activity, and primary production across an agricultural ecosystem. Ecological Applications 7:158–170.

Robertson, G. P., E. A. Paul, and R. R. Harwood. 2000. Greenhouse gases in intensive agriculture: contributions of individual gases to the radiative forcing of the atmosphere. Science 289:1922–1925.

Schimel, J. P. 1995. Ecosystem consequences of microbial diversity and community structure. Pages 239–269 in F. S. Chapin, III and C. Korner, editors. Arctic and alpine biodiversity: patterns, causes, and ecosystem consequences. Springer-Verlag, Berlin, Germany.

Smith, K. A., K. E. Dobbie, B. C. Ball, L. R. Bakken, B. K. Situala, S. Hansen, and R. Brumme. 2000. Oxidation of atmospheric methane in Northern European soils, comparison with other ecosystems, and uncertainties in the global terrestrial sink. Global Change Biology 6:791–803.

Sogin, M. L., H. G. Morrison, J. A. Huber, D. Mark Welch, S. M. Huse, P. R. Neal, J. M. Arrieta, and G. J. Herndl. 2006. Microbial diversity in the deep sea and the under-explored "rare biosphere." Proceedings of the National Academy of Sciences USA 103:12115–12120.

Staley, J. T., and A. E. Konopka. 1985. Measurement of in situ activities of nonphotosynthetic microorganisms in aquatic and terrestrial habitats. Annual Review Microbiology 39:321–346.

Stevenson, B. S. 2000. Microbiology and molecular genetics. Dissertation, Michigan State University, East Lansing, Michigan, USA.

Stevenson, B. S., S. A. Eichorst, J. T. Wertz, T. M. Schmidt, and J. A. Breznak. 2004. New strategies for cultivation and detection of previously uncultured microbes. Applied and Environmental Microbiology 70:4748–4755.

Stres, B., I. Mahne, G. Avguštin, and J. M. Tiedje. 2004. Nitrous oxide reductase (nosZ) gene fragments differ between native and cultivated Michigan soils. Applied and Environmental Microbiology 70:301–309.

Suding, K. N., S. L. Collins, L. Gough, C. Clark, E. E. Cleland, K. L. Gross, D. G. Milchunas, and S. Pennings. 2005. Functional- and abundance-based mechanisms explain diversity loss due to N fertilization. Proceedings of the National Academy of Sciences USA 102:4387–4392.

Teal, T. K., and T. M. Schmidt. 2010. Identifying and removing artificial replicates from 454 pyrosequencing data. Cold Spring Harbor Protocols. doi:10.1101/pdb.prot5409.

Tilman, D., P. B. Reich, J. Knops, D. A. Wedin, T. Mielke, and C. Lehman. 2001. Diversity and productivity in a long-term grassland experiment. Science 294:843–845.

Whitman, W. B., D. C. Coleman, and W. J. Wiebe. 1998. Prokaryotes: the unseen majority. Proceedings of the National Academy of Sciences USA 95:6578–6583.

7

Plant Community Dynamics in Agricultural and Successional Fields

Katherine L. Gross, Sarah Emery, Adam S. Davis,
Richard G. Smith, and Todd M. P. Robinson

Understanding the drivers and consequences of diversity and productivity in plant communities remains a central challenge in ecological research (Thompson et al. 2001, Mittelbach 2012). Interest in how the diversity and composition of plant communities regulate ecological processes and ecosystem services that different ecosystems provide has expanded over the past two decades (Loreau et al. 2001). This question has remained relatively unexplored in agricultural systems—particularly the annual row crops that supply much of the world's food (Power 2010). This is likely because factors known to influence weed and crop production—such as soil fertility, precipitation, and pests (weeds, pathogens, and insects)—are primarily managed with external inputs (fertilizer, irrigation, and pesticides) rather than by relying on ecological processes (Robertson and Swinton 2005). Growing concerns about the negative environmental impacts of using external inputs to sustain crop productivity have stimulated interest in the development of an ecological framework for agricultural management (Robertson and Swinton 2005, Swinton et al. 2006, 2007, Robertson and Hamilton 2015, Chapter 1 in this volume). In addition to crop yield, an ecological framework would consider other ecosystem services that can be managed and enhanced both in the field and in surrounding landscapes (Swinton et al. 2006, 2007, Power 2010). Plant diversity and composition are likely to play an important role in the actualization and sustainability of these services, particularly from landscapes that surround crop fields (Power 2010, Egan and Mortensen 2012).

Row-crop systems are designed and managed to maintain the dominance of a particular species (the crop), with the goal of maximizing productivity (crop yield). Although row-crop systems can provide an array of other ecosystem services (Swinton et al. 2006, 2007, Power 2010), promotion or enhancement of these services is rarely an explicit goal of intensive agriculture (Costanza et al. 1997, Daily et al. 2000). An exception may be high value crops, such as fruits and vegetables, where management practices such as planting or maintaining diverse plant communities along field edges have enhanced pollinators and fruit set (NRC 2007, Ricketts et al. 2008, Garibaldi et al. 2011). Communities in the landscape surrounding crops are important for biological control services, especially for beneficial insects that rely on the floral resources and habitat that plants provide (Landis et al. 2008, Meehan et al. 2011, Landis and Gage 2015, Chapter 8 in this volume). Thus, both economic and environmental incentives exist for ecological research on the functioning of row-crop systems and the contribution of plant diversity within and surrounding row-crop fields to the ecosystem services from agricultural landscapes.

What are the ecological factors that control diversity and productivity in plant communities, and how do they interact to affect the ecosystem services provided by row crops? For plant communities in general, much evidence exists that disturbance regimes (frequency, magnitude, and timing), soil fertility, and biotic interactions (competitors and consumers; see Mittelbach 2012) influence local diversity, and that these local factors interact with regional factors such as seed sources, landscape connectivity, and climate to determine species composition and diversity (Davis et al. 2000, Vellend 2010). But for weed communities, much less is known about how these factors—both local and regional—interact to determine their diversity, composition, and abundance in agricultural systems (Ryan et al. 2010, Egan and Mortensen 2012).

In this chapter, we examine how disturbance and nutrient additions influence plant species diversity, composition, and productivity of herbaceous plant communities typical of agricultural landscapes in the upper midwestern United States. We focus primarily on studies at the Kellogg Biological Station Long-Term Ecological Research site (KBS LTER) in successional fields and row crops. We compare results from the Main Cropping System Experiment (MCSE, Table 7.1; details in Robertson and Hamilton 2015, Chapter 1 in this volume) with smaller-scale experimental studies established within and adjacent to the MCSE. We also provide a broader context for our research on fertilizer manipulations by summarizing results from cross-site analyses of resource enrichments in herbaceous communities across a broad geographic gradient in North America, including a number of other LTER sites. We end the chapter by discussing how interacting processes might shape the future of agriculture, particularly in the context of global climate change and grassland restoration and management. Understanding how disturbance and nutrient availability interact and affect herbaceous plant communities is fundamental to the development of biologically based management of row crops and other agricultural systems.

Experimental Design and Research Approaches

The annual cropping systems of the MCSE provide us with the opportunity to compare the effects of disturbance (tillage) and nutrient input (cover crops vs. inorganic fertilizers), and their interaction, on weed communities and crop yield. Other KBS LTER researchers have evaluated how these management practices affect

Table 7.1. Description of the KBS LTER Main Cropping System Experiment (MCSE).[a]

Cropping System/Community	Dominant Growth Form	Management
Annual Cropping Systems		
Conventional (T1)	Herbaceous annual	Prevailing norm for tilled corn–soybean–winter wheat (c–s–w) rotation; standard chemical inputs, chisel-plowed, no cover crops, no manure or compost
No-till (T2)	Herbaceous annual	Prevailing norm for no-till c–s–w rotation; standard chemical inputs, permanent no-till, no cover crops, no manure or compost
Reduced Input (T3)	Herbaceous annual	Biologically based c–s–w rotation managed to reduce synthetic chemical inputs; chisel-plowed, winter cover crop of red clover or annual rye, no manure or compost
Biologically Based (T4)	Herbaceous annual	Biologically based c–s–w rotation managed without synthetic chemical inputs; chisel-plowed, mechanical weed control, winter cover crop of red clover or annual rye, no manure or compost; certified organic
Perennial Cropping Systems		
Alfalfa (T6)	Herbaceous perennial	5- to 6-year rotation with winter wheat as a 1-year break crop
Poplar (T5)	Woody perennial	Hybrid poplar trees on a ca. 10-year harvest cycle, either replanted or coppiced after harvest
Coniferous Forest (CF)	Woody perennial	Planted conifers periodically thinned
Successional and Reference Communities		
Early Successional (T7)	Herbaceous perennial	Historically tilled cropland abandoned in 1988; unmanaged but for annual spring burn to control woody species
Mown Grassland (never tilled) (T8)	Herbaceous perennial	Cleared woodlot (late 1950s) never tilled, unmanaged but for annual fall mowing to control woody species
Mid-successional (SF)	Herbaceous annual + woody perennial	Historically tilled cropland abandoned ca. 1955; unmanaged, with regrowth in transition to forest
Deciduous Forest (DF)	Woody perennial	Late successional native forest never cleared (two sites) or logged once ca. 1900 (one site); unmanaged

[a]Site codes that have been used throughout the project's history are given in parentheses. Systems T1–T7 are replicated within the LTER main site; others are replicated in the surrounding landscape. For further details, see Robertson and Hamilton (2015, Chapter 1 in this volume).

soil fertility and biogeochemical processes (Cavigelli et al. 1998, Harwood 2002, Snapp et al. 2010, Robertson et al. 2015, Chapter 2 in this volume). Using the Early Successional system as a reference community, we evaluate the long-term effects of disturbance and nutrient enrichment on species diversity, composition, and productivity in herbaceous plant communities.

Row Crops and Weed Communities

Research at KBS LTER has documented how management practices affect the composition and diversity of weeds in agricultural systems, both aboveground and in the soil seed bank (Smith and Gross 2006, Davis et al. 2005). While many studies have examined the effects of different cropping systems on weed communities, few have followed these changes over decades. This extended temporal focus allows us to detect whether weed community composition and productivity respond to longer-term drivers, including changes in climatic factors such as precipitation (Robinson 2011).

Although the MCSE allows us to make such comparisons, the lack of rotation "entry point" replication limits our ability to compare systems across years (i.e., each year has only one crop in the rotation). Also, the design of the MCSE (see Table 7.1) limits our ability to draw conclusions about the role of cropping system diversity in enhancing ecosystem services from agriculture. To address these constraints, in 2000 we established the Biodiversity Gradient Experiment to directly examine how variations in crop diversity (number of crops in a rotation) affect weeds, crop yields, and other agronomic and ecological factors.

Biodiversity Gradient Experiment

The Biodiversity Gradient Experiment includes a total of 21 treatments with monocultures and rotations of three grain crops (corn, soybean, and wheat), with and without cover crops, as well as spring and fall fallow treatments and a bare soil treatment (Table 7.2). All entry points of the rotations are included in the design, so we can quantify treatment effects on all crop yields in every year and directly determine how interannual variation in climatic factors affects crop yield and weed biomass. Crop treatments are classified into six systems that differ in the number of annual grain and cover crop species in the rotation (Table 7.2). Additional details on the management and design of this experiment are described in Smith et al. (2008) and at http://lter.kbs.msu.edu.

Early Successional Plant Communities

The MCSE Early Successional system allows us to quantify successional trajectories and dynamics in a midwestern U.S. landscape (Huberty et al. 1998, Gross and Emery 2007). Since 1997 these plots have been burned annually (or nearly so) in early spring to prevent colonization by trees and shrubs (see Foster and Gross 1999). Although historically the frequency and season of burning of midwestern grasslands likely varied depending on climate and other factors (Andersen and Bowles 1999), today annual spring fires are used to manage them (Packard and Mutel 1997).

When the MCSE was initiated (1989), an experiment was established within the Early Successional system with subplot manipulations of disturbance (tillage) and nitrogen (N) fertilization (the Disturbance by N-Fertilization Experiment). This experiment allows us to examine how disturbance and resource enrichment affect (1) productivity and species richness in successional communities (Gough et al. 2000, Dickson and Gross 2013), (2) the composition and stability of aboveground

Table 7.2. Species composition and rotational diversity treatments of the KBS biodiversity gradient experiment.[a]

Treatment Description[b]	Number of Treatments[c]	Crops per Year	Cover Crops per Year	Total Species per Year	Crops over Rotation	Cover Crops over Rotation	Total Species over Rotation
Spring or fall plowed fallow[d]	2	0	0	10–12	0	0	20+
C–S–W with 2 cover crops	3	1	1–2	2–3	3	3	6
C–S–W with 1 cover crop	3	1	1	2	3	2	5
C–S–W rotation	3	1	0	1	3	0	3
C–S, S–C, W–S rotations	3	1	0	1	2	0	2
C, S, or W with 1 cover crop	3	1	1	2	1	1	2
C, S, or W monoculture	3	1	0	1	1	0	1
Bare soil[e]	1	0	0	0	0	0	0

[a]All treatments replicated in each of four blocks; see Smith et al. (2008) for a detailed description of treatments and rotations.
[b]Crops planted in rotations are Corn = C, Soybean = S, and Wheat = W and, when included, either 1(legume) or 2 (legume and small grain) cover crops. Rotations are indicated by a hyphen; all entry points of rotations are planted each year.
[c]Number of treatments = number of entry points.
[d]Fallow treatments are tilled once a year (spring or fall), allowing weeds to establish.
[e]Bare soil treatment is repeatedly tilled to prevent weed establishment and to serve as a "no plant" reference for soil and microbial studies.

production (Grman et al. 2010), and (3) successional trajectories (Huberty et al. 1998). This experiment has shown the influence of landscape position or initial colonization events on successional trajectories (Foster and Gross 1999) and how these factors can constrain the restoration of native grasslands (Gross and Emery 2007, Suding and Gross 2006a, b; Suding et al. 2004). Participation in cross-site synthesis projects across the LTER Network has allowed us to compare results from the KBS LTER to those observed in other grasslands and has broadened our understanding of the response of herbaceous plant communities across North America to increases in N deposition and why these responses may differ across sites (e.g., Gough et al. 2000, 2012; Suding et al. 2005; Clark et al. 2007; Cleland et al. 2013).

Disturbance as a Driver of Plant Community Diversity

Disturbance, particularly fire and grazing, has been shown to be important in determining the composition and diversity of a variety of grasslands (Huston 1979, Miller 1982, Pickett and White 1985). Fire frequency (Collins 1992,

Howe 2000) and grazing intensity and the type of grazer (Collins et al. 1998, Knapp et al. 1999, Burns et al. 2009) have all been shown to affect the diversity, composition, and productivity in prairies. Reduction in fire frequency has been linked to the conversion of grasslands to woodlands and the loss of native species (Anderson and Bowles 1999, Packard and Mutel 1997, McPherson 1997). As a result, fire and the reintroduction of grazing are important management tools for restoring and maintaining native diversity in grasslands (Suding and Gross 2006a, b, Martin and Wilsey 2006).

In cropping systems, tillage and herbicide applications are disturbances that, like fire and grazing, affect not only the composition and diversity of existing weed communities but also those of the subsequent emergent weed community ("emergent" refers to a germinated and established weed in a crop field, as opposed to the potential weed community in the seed bank; Johnson et al. 2009, Hilgenfeld et al. 2004, also Mortensen et al. 2012). While such management changes can alter the composition and diversity of weed communities (see Smith and Gross 2006, 2007; Smith et al. 2010), growers are generally less interested in how management affects diversity and more interested in the effect on crop yield. Nevertheless, to manage for ecosystem services from agriculture, we need a better understanding of how the disturbance from agronomic practices affects the diversity and productivity of the overall plant communities—weeds as well as crops—in agricultural landscapes.

Effects of Disturbance on Weed Communities

The MCSE annual cropping systems provide the opportunity to compare the impact of tillage and herbicides on weed community structure under four different management regimes (Smith and Gross 2007). However, in these four systems, it is difficult to distinguish the effect of tillage alone because herbicides and fertilizer are also included in the management (see Table 7.1). The annually tilled plots in the Disturbance by N-Fertilization Experiment and the Biodiversity Gradient Experiment (Table 7.2) thus serve as reference communities to examine the effects of tillage alone (Smith and Gross 2007), or of tillage plus N fertilizer, on weed communities (Grman et al. 2010), and to relate long-term changes in species composition and dominance not only to annual disturbance (tillage), but also to longer-term drivers such as climate change (Robinson 2011, Cleland et al. 2013, Dickson and Gross 2013).

Despite major differences in management—including chemical inputs and tillage (Table 7.1)—differences in weed biomass and composition among the four annual cropping systems of the MCSE have been relatively small (Davis et al. 2005). This suggests that disturbance, whether created by tillage or herbicide, has similar effects on the emergent weed community. Ordinations of aboveground weed biomass and composition over the first 13 years of the study (1990–2002) did not show a strong association with management, although overall weed biomass was lower in the Conventional and No-till systems than in the Reduced Input and Biologically Based systems (Davis et al. 2005). There is, however, considerable interannual variation in weed biomass (Table 7.3) and weed species composition (Fig. 7.1) in the four annual cropping systems. This may reflect differences in what

Table 7.3. Temporal changes in seed bank density and aboveground weed biomass across systems of the MCSE.[a]

Variable/System	1990	1993	1996	1999	2002	2008
Seed bank(10^3 seeds m^{-2})[b]						
Conventional	5.9 (1.2)	2.2 (0.5)	1.4 (0.6)	19.7 (2.3)	23.2 (2.5)	34.2 (2.5)
No-till	13.9 (3.1)	6.0 (1.4)	3.1 (1.6)	38.9 (4.2)	22.8 (2.7)	13.1 (3.1)
Reduced Input	11.3 (1.5)	6.5 (1.3)	1.6 (0.3)	28.1 (3.2)	29.7 (1.9)	24.7 (4.2)
Biologically Based	6.2 (0.8)	10.7 (2.0)	0.4 (0.1)	19.1 (1.5)	19.7 (2.2)	16.9 (1.6)
Early Successional	15.1 (6.0)	26.0 (5.3)	110 (14.0)	47.7 (8.2)	21.6 (2.4)	29.6 (6.7)
Aboveground biomass (g m^{-2})[c]						
Conventional	46.6 (12.3)	34.7 (5.6)	3.3 (1.5)	4.9 (1.6)	23.5 (6.0)	0.0
No-till	5.2 (3.2)	156 (45.7)	59.8 (20.2)	227 (41.0)	16.2 (4.8)	0.0
Reduced Input	147.8 (55)	148 (36.7)	2.6 (0.9)	11.4 (4.5)	42.7 (8.4)	21.5 (5.8)
Biologically Based	184 (65.9)	161 (20.2)	83.8 (15.7)	20.6 (6.0)	154 (27.0)	56.9 (16.7)
Early Successional	416 (53.5)	450 (44.3)	340 (23.8)	642 (74.3)	701 (73.7)	772 (38.9)

[a]Corn planted in all of these years, except 1990 when soybean was planted in the Conventional and No-till systems. Values are mean (SE), n = 6 replicated plots. For annual crops, biomass is for weeds only and does not include crop or cover crop production.
[b]Seed bank density determined by elutriation (Gross and Renner 1989); sampling occurred in the spring (April).
[c]Aboveground biomass determined at peak weed biomass in each system; in August–September for annual cropping systems and early August for the Early Successional system.

crop was planted (i.e., stage of the rotation; see Fig. 7.1), interannual differences in total precipitation or its timing, and/or how those factors interact with the timing of management efforts to control weeds.

In contrast, seed banks in the Reduced Input and Biologically Based systems have diverged in species composition from those in the Conventional and No-till systems (Fig. 7.2), indicating that in the MCSE annual systems, herbicides can be a stronger determinant (or filter) of weed species composition than tillage. When examined alone, however, tillage has been shown to have either strong (Murphy et al. 2006, Sosnoskie et al. 2006) or weak (Thomas et al. 2004) effects on weed community composition and diversity. This makes it difficult to predict how the trend toward reduced tillage—and consequent increased herbicide use (e.g., shifts to no-till and planting crops genetically modified for herbicide resistance)—will impact weed communities in annual row crops. Further studies comparing management systems and their effects on emergent weed communities are needed to elucidate the long-term effects of herbicide use and tillage on weed communities in row crops (Davis et al. 2005).

The Biodiversity Gradient Experiment allows us to examine the effects of tillage timing on weed communities in row crops, and over the first 5 years of this experiment, weed community composition was strongly affected by the timing of primary tillage (Smith 2006, Smith and Gross 2007). Spring tillage (coinciding with corn and soybean planting) favored the establishment of spring-emerging annual

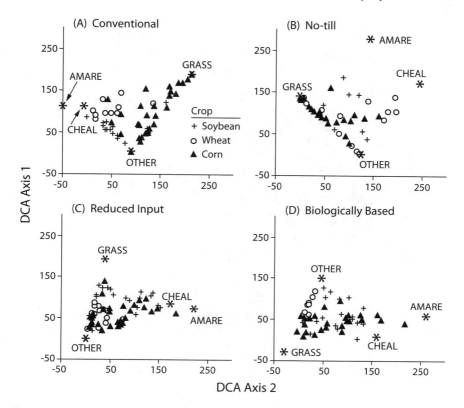

Figure 7.1. Variation in weed species composition in relation to crop grown in the four annual cropping systems of the Main Cropping System Experiment (MCSE). Plot scores are from detrended correspondence analysis (DCA) of weed species composition from 1990–2002 for each crop in A) Conventional, B) No-till, C) Reduced Input, and D) Biologically Based systems. Symbol legend for all panels appears in panel A. Asterisks indicate scores for the four dominant weeds used in the ordination. AMARE = *Amaranthus retroflexus;* CHEAL = *Chenopodium album;* Other = Other dicots; Grass = all grass species. Modified from Davis et al. (2005).

forbs and C_4 grasses, while fall tillage (coinciding with the winter wheat planting) favored winter-annual forbs and C_3 grass species (Smith 2006). The importance of tillage timing in determining weed species composition is indicated by strong similarities in species composition between corn, soybean, and the spring fallow treatment (spring tillage) as well as between wheat and the fall fallow treatment (fall tillage) (Fig. 7.3).

In the MCSE Early Successional system, a small area (20 × 30 m) at the northern border of each replicate plot has been annually tilled to maintain dominance by annual weeds as part of the Disturbance by N-Fertilization Experiment. Although the species composition of the annually tilled plots has varied over time (Grman et al. 2010), they are consistently dominated by giant foxtail (*Setaria faberi*; Table 7.4), a C_4 annual grass that is a common weed in corn and soybean in the

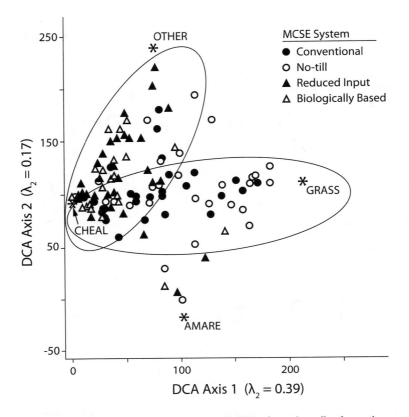

Figure 7.2. Detrended correspondence analysis (DCA) of weed seedbank species composition in four annual row-crop systems on the MCSE. Data are scores for replicate plots (n=6) of each system for the 5 years sampled (every three years, 1990–2002). Asterisks indicate scores for the four dominant weeds used in the ordination. AMARE = *Amaranthus retroflexus;* CHEAL = *Chenopodium album;* Other = Other dicots; Grass = all grass species. Ovals group data points (majority) of Conventional and No-till and of Reduced Input and Biologically Based systems to highlight divergence in weed seed banks. Modified from Davis et al. (2005).

upper U.S. Midwest (Nurse et al. 2009). This species dominates in the both the fertilized and unfertilized tilled plots (48% and 60% of total biomass, respectively; Table 7.4). And while precipitation does not predict plot biomass production (fertilized and not), the abundance of *S. faberi* across years is correlated with early spring rainfall and temperature (Robinson 2011). Interestingly, *S. faberi* was not a dominant weed in either the emergent or seed bank communities of the adjacent annual row-crop systems (Davis et al. 2005), even though the abundance of grass weed species, in general, differed among crops and cropping systems (Figs. 7.1 and 7.2), suggesting that the presence of a crop—or management of these systems—inhibits or reduces the abundance of this species.

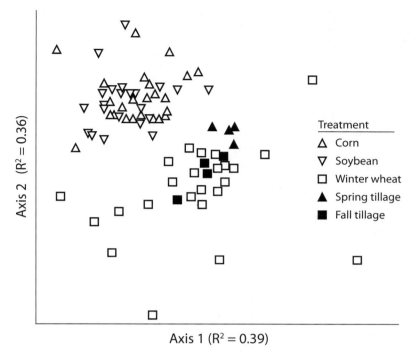

Figure 7.3. Weed species composition in response to agronomic and fallow treatments in the Biodiversity Gradient Experiment. Plot scores are from non-metric multidimensional scaling (NMDS) ordination of weed community composition and abundance in 2004 in relation to crop type (soybean, wheat, corn) and tillage time (spring, fall) for fallow treatments. Ordination based on Bray-Curtis dissimilarity in species composition.

Temporal Dynamics and Community Assembly

The plant community that assembles in response to a particular disturbance regime, whether in a row crop or an abandoned agricultural field, will depend on the nature of the disturbance; the local and regional species pool; and climatic (e.g., temperature and precipitation), abiotic (soil fertility), and biotic factors (e.g., competitors, mutualists, predators, and pathogens). Long-term KBS LTER studies allow us to compare, in replicated plots, how different management practices (Table 7.1) affect the colonization, establishment, persistence, and extinction of plant species in both successional and row-crop communities. When chronic or repeated disturbances cease, sites undergo a successional sequence of changes in both species composition and traits (Connell and Slatyer 1977). Long-term experiments within MCSE plots allow us to examine how nutrient enrichment (fertilization) and climatic factors interact to affect these trajectories. We can also determine how variation in chemical inputs to row crops affects the composition and diversity of weeds in agronomic systems.

Table 7.4. Indicator species in (A) untilled and (B) annually tilled treatments, both with (fert) and without (no fert [control]) N fertilization, in the disturbance by N-fertilization experiment in the MCSE Early Successional system.

Species[a]	Indicator Group[a]	Unfertilized Plots (% total biomass)[a]	Fertilized Plots (% total biomass)[a]	Life History[b]	Life Form[b]	Native or Introduced[b]
(A) Untilled (Perennial-dominated community)						
Solidago canadensis	Fert	25.90	36.87	P	F	N
Phleum pretense	No fert	10.06	2.70	P	G(C$_3$)	I
Trifolium pretense	No fert	7.80	0.66	P	L	I
Apocynum cannabinum	Fert	6.02	8.98	P	F	N
Elymus repens	Fert	4.57	6.43	P	G(C$_3$)	I
Hieracium spp.	No fert	1.31	0.01	P	F	I
Achilleamilli folium	No fert	0.95	0.21	P	F	N
Hypericum perforatum	No fert	0.88	0.24	P	F	I
Rumex crispus	Fert	0.85	2.72	P	F	I
Trifolium hybridum	No fert	0.84	0.06	P	L	I
Poa compressa	No fert	0.83	0.07	P	G(C$_3$)	I
Potentilla recta	No fert	0.72	0.21	P	F	I
Lotus corniculatus	No fert	0.60	0.00	P	L	I
Solidago juncea	No fert	0.54	0.00	P	F	N
Trifolium repens	No fert	0.19	0.03	P	L	I
Melandrium album	Fert	0.15	2.03	P	F	I
Asclepias syriaca	Fert	0.14	1.33	P	F	N
Ambrosia artemisifolia	Fert	0.07	0.41	A	F	I
Chenopodium album	Fert	0.00	0.13	A	F	I
Lactuca serriola	Fert	0.00	0.09	A/B	F	I
Rubus occidentalis	Fert	0.00	1.97	P	F	N
(B) Tilled (Annual-dominated community)						
Setaria faberi	Fert	47.80	59.46	A	G(C$_4$)	I
Chenopodium album	Fert	12.22	15.58	A	F	I
Ambrosia artemisifolia	Fert	6.96	8.69	A	F	I
Amaranthus retroflexus	Fert	1.38	3.27	A	F	N
Apocynum cannabinum	No fert	0.59	0.01	P	F	N
Panicum capillare	No fert	0.25	0.02	A	G(C$_3$)	N
Erigeron annuus	No fert	0.22	0.01	A	F	I
Echinochloa crus-galli	Fert	0.12	0.81	A	G(C$_3$)	I
Taraxacum officinale	No fert	0.12	0.01	P	F	N,I

[a]All species were significant indicator species (at $p = 0.01$) in a given treatment. For both treatments, species are listed in rank order (most to least % total biomass) in the control (No Fert) plots. Percentage total biomass determined from the average biomass over 18 years (1992–2009). Species names are accepted nomenclature (USDA PLANTS Profile: http://plants.usda.gov).
[b]Life history, life form, and native status determined from databases (e.g., USDA PLANTS) and field observations; A = annual, B = biennial, P = perennial; F = forb, G = grass (C$_3$ or C$_4$), L = legume; and Native = N; Introduced = I. See also Cleland et al. (2008).

Temporal Dynamics in Weed Communities

The seed bank is an important source for weed infestations in agricultural fields (Buhler et al. 1997). Although the linkage between composition of the weed seed bank and weed pressure can be difficult to gauge (Davis 2006), understanding the factors affecting the persistence and species composition of weed seed banks in arable soils is important for weed management (Buhler 2002, Davis 2006).

When the MCSE was established, weed seed banks in all four annual row-crop systems were dominated by *Chenopodium album* (lambsquarters), likely reflecting its ability to persist with the previous 20 + years of herbicide use typical of weed management in conventional row-crop agriculture (Davis et al 2005). Over time, the composition of the weed seed banks in the four annual row-crop systems diverged in response to management (Menalled et al. 2001, Davis et al. 2005). In the Conventional and No-till systems, seed banks shifted to dominance by annual, C_4 grasses (mainly *Panicum dichotomiflorum* (fall panicgrass) and *Digitaria sanguinalis* (large or hairy crabgrass), and some *S. faberi*, whereas the Reduced Input and Biologically Based systems became dominated by small-seeded dicot species (Fig. 7.2) such as *Stellaria media* (common chickweed), *Veronica perigrina* (purslane speedwell), and *Arabidopsis thaliana* (mouse-ear cress). Despite the important role that seed banks play in determining weed communities and the observed divergence in weed seed bank composition (Fig. 7.2) and differences in emergent weed biomass (Table 7.3), there is little divergence in the weed species composition among annual systems (Fig. 7.1; see also Davis et al. 2005). This suggests that interannual variation in these communities is strongly controlled by cropping system management (Davis et al. 2005) and climatic variation (Robinson 2011).

The timing of tillage and the use of cover crops can have dramatic effects on the composition of the emergent weed community in row crops (Smith 2006, Smith and Gross 2007), which may result from interactions between the disturbance regime (e.g., whether and how tillage or herbicides are used to control weeds) and the source of N (legume cover crop vs. inorganic fertilizer). Both the type (herbicides vs. interrow cultivation) and timing of weed management disturbances can alter the composition of the weed community, which in turn affects its response to differences in N availability.

Temporal Dynamics in Successional Communities

Within 4–5 years after abandonment from agriculture, the MCSE Early Successional system underwent a typical shift in species composition from initial dominance by annual weeds to dominance by herbaceous perennials (Huberty et al. 1998, Gross and Emery 2007). Although species composition initially differed among the six replicate plots, within 5 years all plots converged to a similar composition (Fig. 7.4) and were dominated by perennial forbs and relatively few grasses (see also untilled treatment, Table 7.4).

Although native species are relatively rare in these communities (Table 7.4), they produce about 50% of the aboveground biomass (Gross and Emery 2007). The low number of native species in these fields probably results from the lack

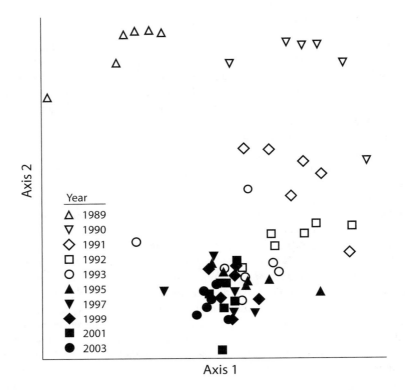

Figure 7.4. Changes in species composition in the first decade following abandonment of agricultural practices in the Early Successional system of the MCSE. Data are for each of 6 replicate plots with year indicated by different symbols. Species composition compared using non-metric multidimensional scaling (NMDS) analysis of annual biomass harvest averaged across 4–5 sampling stations in each replicate (see Gross and Emery (2007) for details). From *Old Fields*, edited by Viki A. Kramer and Richard J. Hobbs. Copyright © 2007 Island Press. Reproduced with permission of Island Press, Washington, D.C.

of native-dominated communities in the surrounding landscape (Burbank et al. 1992, Foster 1999, Gross and Emery 2007) as dispersal limitation can be an important controller of diversity in abandoned fields in this region (Suding and Gross 2006a, b; Houseman and Gross 2011). The introduction of spring burns to control colonization by woody species in this system (1997) has had no effect on the establishment of native species (Gross and Emery 2007) or on the composition of these communities (Dickson and Gross 2013). This is consistent with results from Suding and Gross (2006b) who found that only when seeds of native species are added to burned areas is there an increase in recruitment of native species. Fire may have promoted the convergence (greater similarity) in species composition among replicates (Fig. 7.4) by selecting for species that were favored by annual burning, but did not promote native species recruitment (Gross and Emery 2007). Native C_4 grasses that are consistently favored by annual burning in other midwestern grasslands (Symstad et al. 2003, Collins et al. 1998) remain rare in KBS MCSE Early Successional communities (Table 7.4), likely because of their absence from the surrounding landscapes.

Controls on Productivity

Voluminous evidence exists that the productivity of terrestrial ecosystems is limited by nutrients (Chapin et al. 1986, Elser et al. 2007) and N has repeatedly been shown to be a critical limiting nutrient in both natural and agricultural temperate ecosystems (Drinkwater and Snapp 2007, LeBauer and Treseder 2008). Across North American grasslands and other "low-stature" herbaceous plant communities, the response to N-fertilization can depend on species composition, soil nutrient status (Clark et al. 2007), and interannual variation in precipitation (Cleland et al. 2013). While the magnitude of a productivity response to N-fertilization can vary across communities, generally there is an increase in aboveground biomass production and a decrease in species richness (Gough et al. 2000, Suding et al. 2005). Thus, N fertilizing agricultural systems to enhance productivity may come at the expense of diversity, which may reduce or limit the ecosystem services they provide (Robertson et al. 2015, Chapter 2 in this volume). Few studies have examined how enhancing plant species diversity can increase crop productivity or yield. In fact, increasing the diversity of weed species is generally assumed to have a negative effect on crop yield, the primary ecosystem service expected from row-crop agriculture.

Results from the MCSE cropping systems and Early Successional communities have been included in several meta-analyses and cross-site syntheses of fertilization experiments, allowing our results to be interpreted in a broader regional context (Gough et al. 2000, Davis 2005, Suding et al. 2005, Smith 2006, Clark et al. 2007, Smith and Gross 2007, Gough et al. 2012). We summarize studies from both the cropping and Early Successional systems here to address our overall goal of applying lessons learned and insights gained from research in noncrop plant communities to the management of cropping systems, and vice versa.

Productivity in Successional Grasslands

The Disturbance by N-Fertilization Experiment, established within the MCSE Early Successional plots, provides clear evidence that productivity in these systems is limited by N (Fig. 7.5). Although the magnitude of this effect varied across years— likely driven by variation in seasonal precipitation (Robinson 2011, see Cleland et al. 2013)—on average, the addition of fertilizer increased aboveground production in both the untilled and annually tilled plots by approximately 50% (Dickson and Gross 2013). There was a significant correlation between aboveground production in the fertilized and unfertilized plots across years in the untilled treatment (Fig. 7.5A; $r = 0.60$), but not in the annually tilled treatment. Annual precipitation is a significant predictor of productivity in both fertilized and unfertilized plots in the untilled treatment ($r = 0.49$ and 0.37, $p < 0.025$ and 0.05, respectively), but not in

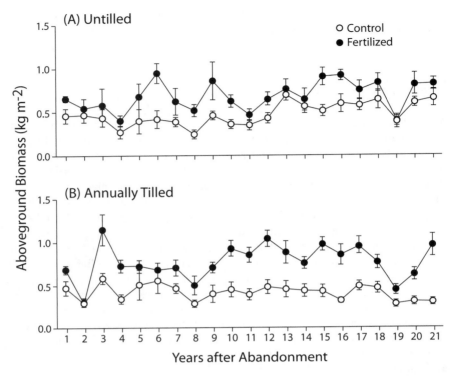

Figure 7.5. Interannual variability in aboveground biomass production in the Disturbance by N-Fertilization Experiment in the MCSE Early Successional system following abandonment (Year 1=1989) from control (unfertilized) and fertilized (nitrogen added) in A) untilled, successional and B) annually tilled treatments. Samples were harvested at peak biomass, typically late-July (untilled) and mid-August (tilled). Values are means ± SE, n = 6.

the annually tilled treatment, suggesting that different external drivers controlled productivity in these communities. This difference may be due to species-specific differences in recruitment of annual weeds in response to precipitation, as exemplified by the dominant grass species *S. faberi,* which as discussed earlier, accounts for much of the weed productivity in these systems (Robinson 2011).

Nitrogen fertilization reduced species richness approximately 20% in both the untilled and annually tilled treatments (Fig. 7.6). While this response was relatively rapid in the annually tilled community (Fig 7.6B), it took 14 years before fertilization had a detectable effect on species richness in the untilled community (Fig. 7.6A). A recent meta-analysis of fertilization experiments in grasslands suggests that community composition, specifically the presence of "tall runners" (i.e., clonal populations of plants of tall stature interconnected underground by horizontal roots), can influence the magnitude of fertilization-driven changes in species diversity (Gough et al. 2012). Tall-runner species appeared in MCSE fields a few years after abandonment from agriculture, but as a functional group, they

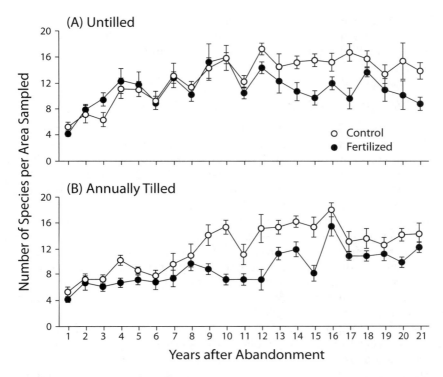

Figure 7.6. Temporal changes in species richness (number of species per harvested sample) in the Disturbance by N-Fertilization Experiment in the MCSE Early Successional system in control (unfertilized) and fertilized (nitrogen added) treatments in the A) untilled, successional plots and B) annually tilled plots. The area sampled varied for the first 3 years (Yr. 1 = 0.2 m², Yrs. 2 and 3 = 0.3 m²); from Yr. 4 onward sampling area was 1.0 m²; Year 1 = 1989. Values are means ± SE, n=6.

varied in abundance over time (Dickson and Gross 2013). *Solidago* (goldenrod) species, which initially made up over 80% of the tall-runner biomass in these fields, declined in abundance after 5 years. The reemergence of *S. canadensis* and other tall runners in these fields after 14 years coincided with a decline in species richness in fertilized treatments (Dickson and Gross 2013).

The delayed effect of fertilization on species richness in untilled plots of the MCSE Early Successional system may be a consequence of the low abundance of C_4 grasses and greater abundance of herbaceous perennial dicots and C_3 grasses (Table 7.4), as compared to other successional grasslands in the area (see Gross and Emery 2007; Clark et al. 2007). Cross-site synthesis work has shown that there is an environmental context to species responses to N addition (Pennings et al. 2005). For example, *Elymus (*formerly *Agropyron) repens* (quackgrass), a nonnative C_3 grass that dominates following fertilization of successional fields at Cedar Creek LTER in Minnesota (Tilman 1984, 1987), occurs in—but does not dominate—the fertilized MCSE Early Successional plots at the KBS LTER

(Table 7.4). Native C_4 grasses that dominate tallgrass prairies of Konza LTER in Kansas and show a strong positive response to fertilization (Clark et al. 2007) are rare at KBS LTER, likely reflecting their absence in the surrounding landscape (Foster 1999).

Cross-Site Analyses of Fertilization Effects on Grasslands

Many sites in the LTER Network have established and maintained long-term N addition experiments in grasslands and similar herbaceous communities, providing opportunity for cross-site analysis of the relationship between productivity and diversity across wide geographic and climatic gradients (Gross et al. 2000, Gough et al. 2000, Suding et al. 2005). An initial synthesis of these data showed a unimodal relationship between productivity and plant species diversity across sites (Gross et al. 2000) and that N addition had similar effects on herbaceous communities ranging from Arctic heathlands to tallgrass prairie and coastal marshes, although the magnitude of their responses differed (Gough et al. 2000, Suding et al. 2005). Although these experiments differed in sampling area, similar amounts of N were added (10–12 g m^{-2}), so it was possible to identify mechanisms that drive the magnitude of the response to fertilization across communities (Suding et al. 2005, Clark et al. 2007, Gough et al. 2012).

On average, N addition resulted in a 50% increase in aboveground production and a consistent decline in species richness across sites (except for coastal marshes) despite a broad range in initial aboveground productivity (Suding et al. 2005, Clark et al. 2007). The magnitude of the productivity increase was strongly correlated with the magnitude of the decrease in species richness, except in several of the coastal marsh systems (Suding et al. 2005). Although functional groups differed in their probability of being lost from a fertilized plot, overall species abundance in unfertilized control plots was the strongest predictor of species loss in response to fertilization. Species that were rare in the unfertilized community were more likely to be excluded in fertilized plots, regardless of their functional group (Suding et al. 2005). Subsequent analyses of this dataset showed that the loss of species following N addition was greatest in communities with lower soil cation exchange capacity, colder regional temperature, and a larger production increase following N addition (Clark et al. 2007).

Species composition also was an important determinant of the productivity response, specifically the abundance of C_4 grasses (Clark et al. 2007); however, the photosynthetic pathway (C_3 vs. C_4) did not appear to be the causal factor (Suding et al. 2005). In a recent meta-analysis, Gough et al. (2012) found that the form of clonal growth (having a spreading or clumping growth form vs. nonclonal) combined with height (relative position in the canopy) were strong predictors of both species and community responses to N addition. However, neither clonality nor height alone predicted the probability of species loss following N addition (Suding et al. 2005).

A shift from soil resource limitation to light limitation is often assumed to be important in determining plant species composition following nutrient enrichment. However, that plant communities become less diverse with N addition, regardless of their initial productivity (Gough et al. 2000, Suding et al. 2005) or life history composition (Suding et al. 2005), suggests that other processes besides light limitation may be mediating species and community responses to increase N, such as species associations with soil biota (e.g., Johnson et al. 2008, Johnson 2010) and plant–soil feedbacks (Bever et al. 2010). Plant–soil interactions are likely also important determinants for agricultural weed communities (Kremer 1993, Kremer and Li 2003, Jordan and Vatovec 2004), both because agronomic management typically (but not always) keeps weed abundance below thresholds where strong competitive interactions can occur and because low diversity of cropping systems can promote pathogens specific to particular species (Bever et al. 2010, Johnson 2010).

Effects of Weed Abundance and Diversity on Crop Yield

Agricultural management systems are designed to increase crop yield by reducing soil resource limitation and competition from weeds, and often achieve this by combining control (herbicides and tillage) with fertilization. This combination, however, confounds our ability to distinguish the effects of disturbance (herbicides and tillage) from fertilization on crop yield, and limits our ability to determine how weed production and composition may interact to influence crop yields. For example, the higher weed biomass that is usually found in the lower chemical input MCSE systems (the Reduced Input and Biologically Based systems) compared to the Conventional and No-till systems (Table 7.3) may reflect the efficacy of herbicides for weed control compared to tillage. However, in some years, weed biomass in the Reduced Input system was equal to that in the Conventional system (tillage and herbicide), and much higher in the No-till (herbicide only) system (e.g., 1996, Table 7.3). Nonetheless, herbicide use clearly is important in the overall control of weed abundance in row crops of the MCSE, although other factors also play a role in determining the production and composition of weed communities and their effects on yield.

Of agronomic importance in lower chemical input and organic systems (particularly those using manure) is knowing how crop yield is affected by competition with weeds under relatively nutrient-limited conditions (Smith and Gross 2006, Posner et al. 2008, Smith et al. 2010). While high weed biomass generally has a negative effect on crop yield (Zimdahl 2004), some evidence exists that management-induced changes in weed species composition can also prevent potentially dominant weed species from reaching abundances where they reduce crop yield (Davis et al. 2005, Pollnac et al. 2009). There is limited evidence that more diverse weed communities may have less of an effect on crop yields than low diversity weed communities with a few dominant (and abundant) species (Smith and Gross 2006; see also Smith et al. 2010). However, experimental studies in grasslands have shown that more diverse communities can result in increased total productivity (Fargione and Tilman 2005), and every additional weed species occurring in the community increases the possibility of introducing a species that is highly competitive with the crop.

Cropping System Diversity and Yield

Species diversity is important is determining the productivity of unmanaged herbaceous communities, but what about managed systems? How does cropping system diversity, via crop rotation and cover crops, affect productivity and agronomic yields? An important component of an ecological framework for understanding crop production and other ecosystem services from agriculture involves understanding how cropping system diversity affects these services. The annual row-crop systems in the MCSE provide insights into the potential for biological processes (e.g., N-fixation or weed suppression by cover crops) to replace reliance on chemical inputs (e.g., pesticides and fertilizers) in row crops. However, they cannot be used to evaluate the role of cropping system diversity (via crop rotation or cover crops), because they differ in a variety of inputs and all follow the same crop rotation (Table 7.1). The Biodiversity Gradient Experiment was established to explicitly test the effects of crop species and diversity (Table 7.2), not only on yield but also on a suite of ecosystem variables. Because no fertilizers or pesticides are used in these treatments, variation in crop yield and other system responses is directly attributable to the number of different crops planted in the rotation (Smith et al. 2008), providing insight into the potential for biological processes (e.g., N-fixation or weed suppression by cover crops) to replace or reduce reliance on chemical inputs.

For the first 3 years of the Biodiversity Gradient Experiment, cropping system diversity showed no effect on the yield of any of the crops (Smith et al. 2008). But by the fourth year (2003), the number of species in the rotation had a significant effect on grain yield in corn (Fig. 7.7). Although the magnitude of this effect has varied annually, typically corn grain yields have been highest in the two highest diversity treatments (five and six species over a 3-year rotation; Fig. 7.7). In

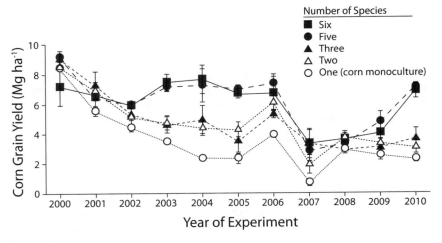

Figure 7.7. Effects of rotational diversity treatments on average annual corn grain yield from 2000–2010 in the Biodiversity Gradient Experiment. Treatments are coded based on the number of species in the rotation. Values are means ± SE, n = 4. See Table 7.2 for description of treatments.

contrast, corn grain yield has steadily declined in the monoculture treatment over time (R^2 = 0.56). A severe drought in 2007 reduced yields in all treatments, but yields in the two highest diversity systems rebounded to predrought levels the next year (Fig. 7.7). This suggests that more diverse cropping systems (four to six species) may be more resilient (*sensu* Scheffer et al. 2001) to drought (and presumably other environmental perturbations) than continuous monocultures.

What causes these differences in overall yield and in capacity to recover from stress? That remains to be determined. Smith et al. (2008) proposed that the higher grain yield in corn, but not in other crops, may result from greater reliance on spring soil N levels, which tend to be higher in the more diverse cropping systems. Spring soil N levels and cropping system diversity are positively correlated, and the strength of the relationship depends on the number of legumes (grain crops and cover crops) in the rotation (Smith et al. 2008). Parker (2011) confirmed that soils from the more diverse corn treatments had higher N-mineralization rates, but in a greenhouse experiment detected no effect of N fertilizer on corn grown in soils from these different treatments, suggesting that some factor other than N must be responsible for reduced yields in less diverse cropping systems. Although disease and/or pest buildup can be a major concern in continuous monocultures, to date we have seen no evidence that pathogens and/or pests are higher in the less diverse systems. Instead, it may be that changes in the diversity and/or composition of the soil microbial community—and its ability to process carbon and nitrogen—are important determinants of corn grain yield across these treatments.

Other Factors Affecting Diversity and Productivity of Agricultural Landscapes

Landscape Structure and Community Composition

Landscape structure, past land use, and management history are increasingly recognized as important drivers of local species diversity that affect successional trajectories (e.g., Myster and Pickett 1993, Foster and Gross 1999), the restoration of native ecosystems (Suding et al. 2004, Gross and Emery 2007), and weed composition in crop fields (Poggio et al. 2010). Overcoming seed limitation may be as or more important than reestablishing natural disturbance regimes for the successful restoration of a native plant community (Suding et al. 2004, Suding and Gross 2006b, Houseman and Gross 2006, 2011). Intentionally seeding restoration areas with native species may be necessary to overcome their dispersal limitations and to increase the ratio of native to nonnative plants in these communities (Suding and Gross 2006b).

Past land use, the absence of fire, and changes in surrounding landscape diversity have all been shown to influence the composition and diversity of restored and successional grasslands in the U.S. Midwest. Although seed addition and fire are often used in grassland restoration (Leach and Givnish 1996), experimental studies in degraded grasslands near KBS found that neither fire nor seed addition alone increased native species richness. In some sites, fire increased the number of nonnative species, but only when fire and seed addition were combined was there an increase in the number of native species relative to nonnatives (Suding and Gross 2006b).

Our understanding of how landscape factors regulate weed community dynamics and composition in agricultural systems is still in its infancy (Gabriel et al. 2005). Much of the research in this area has been conducted outside of the United States. In these studies, local plant species and genetic richness in agricultural fields have been shown to be strongly affected by processes operating at landscape scales, even across distances as short as 2 km (Gabriel et al. 2005, Poggio et al. 2010). Recent studies in the midwestern USA have found evidence that weedy species in the landscape surrounding an agricultural field may provide ecosystem services, such as biocontrol and pollinator services (Isaacs et al. 2009, Gardiner et al. 2009, Landis and Gage 2015, Chapter 8 in this volume). This has sparked interest in understanding how an agricultural landscape that supports multiple functions and ecosystem services can be established. Understanding the economic, social, and ecological processes to promote this type of landscape is an important focus of agroecological research in the United States (Jordan and Warner 2010).

Climate Change and Precipitation

At the global scale, there is a strong correlation between primary productivity and mean annual precipitation (MAP) in terrestrial plant communities in general (Melillo et al. 1993), and in grasslands in particular (Knapp and Smith 2001, Cleland et al. 2013, Robinson et al. 2013). How plant communities respond to altered precipitation patterns—particularly, increases in precipitation variability, as predicted by global change models—has heightened interest in this relationship (Knapp and Smith 2001, Huxman et al. 2004). Although in a cross-site analysis, Knapp and Smith (2001) found a positive correlation between aboveground net primary production (ANPP) and MAP across temperate biomes at a continental scale, they found no relationship between interannual variation in productivity and annual precipitation at the local scale. Their analysis revealed that some biomes—specifically, temperate grasslands—were more responsive to pulses (maxima) in precipitation than others and that this was driven by abundant, highly responsive species in ecosystems where precipitation and evapotranspiration were approximately balanced. A more recent cross-site synthesis (Cleland et al. 2013) across a broad range of grasslands showed that while species richness was strongly correlated with MAP, only the most xeric sites were responsive to interannual variation in MAP. Much of this response was driven by annual species whose emergence was sensitive to precipitation variation, suggesting that annual and perennial communities may respond differently to changes in precipitation variability.

Although the relationship between MAP and productivity is well studied in both grasslands and agricultural systems (e.g., Laurenroth and Sala 1992, Knapp and Smith 2001, Motha and Baier 2005), considerably less is known about how predicted changes in precipitation variability, particularly seasonal distribution, will affect not only productivity but other ecosystem processes as well (Cleland et al. 2013, Robinson et al. 2013). Only a few studies at KBS have manipulated precipitation patterns directly (see Aanderud et al. 2011, Robinson 2011), but long-term data on precipitation and productivity across the broad range of MCSE plant communities provide insight into how the changes in precipitation patterns predicted for this region may influence their productivity and diversity. For example, in the untilled successional treatment of the Diversity by N-Fertilization Experiment, which is dominated by perennial species, MAP is positively related to ANPP in both unfertilized ($R^2 = 0.25$, $p < 0.025$) and fertilized ($R^2 = 0.14$, $p < 0.05$) plots. However, in the annually tilled plots where annual species dominate, there is no relationship, regardless of fertilizer addition. Instead of MAP, one might expect growing season precipitation to be a better predictor of ANPP in tilled communities because their growth is strongly controlled by tillage, which is a seasonal event. But it is not—there is no significant relationship between growing season precipitation (i.e., April–September) and ANPP. Precipitation totals during specific periods of the growing season prove to be better predictors of aboveground productivity than either annual and growing season totals (Robinson et al. 2013). It is not surprising that the amount of precipitation during specific life stages (e.g., germination) is a key driver for annual communities. Analyses of long-term data and of short-term manipulation experiments show that precipitation during the first weeks of the growing season has long-lasting effects on annual community development (Robinson 2011).

Only in the No-till system, which includes both annual and perennial weed species, is weed biomass related to precipitation variation (growing season: $R^2 = 0.44$; annual: $R^2 = 0.14$). The lack of a correlation between precipitation and weed biomass in the Conventional, Reduced Input, and Biologically Based cropping systems may result from differences in weed management (Table 7.1). All three include tillage as part of their management, although the timing and frequency of tillage events differ among them, whereas the No-till system relies only on herbicides for weed control. This difference between systems suggests that management, particularly the timing and implementation of weed control practices, affects the response of weed communities to external drivers such as variability in the amount and seasonal distribution of precipitation. The response, however, may be due more to changes in the composition of the weed community than in its total biomass.

That no relationship exists between annual or growing season precipitation and weed biomass in the row-crop systems or plant biomass in annually disturbed successional plots stands in direct contrast to the strong relationship observed in more water-limited systems (deserts and grasslands) (Noy-Meir 1973). In more mesic systems such as KBS, it is likely that the timing and intensity of precipitation events, as well as the intervals between them, impact productivity (Robinson et al. 2013). Climate shifts that affect the timing of snowmelt and the frequency and intensity of storms (Easterling et al. 2000, Weltzin et al. 2003, IPCC 2007) will likely affect productivity, as well as composition, of annual weed communities in agricultural systems. Annual communities may be particularly responsive to the frequency and intensity of precipitation events, as this can affect the timing and percentage of seed germination in annual species, which can differ in their response to variability in precipitation (see Pake and Venable 1996, Robinson and Gross 2010, Robinson 2011). Because the germination of many annual weed species varies with temperature and moisture (Baskin and Baskin 1999), understanding how early season precipitation and temperature interact with tillage (disturbance timing and frequency) is important for determining how climate change may affect the composition and abundance of weed species in row crops.

Summary

Understanding the processes that determine the diversity and productivity of plant communities remains an important challenge in plant community ecology, and is fundamental to the sustainable management of agroecosystems. In this chapter, we have focused primarily on comparisons of ecological processes in annual row crops (corn, soybean, and wheat) and successional fields, which are important components of the agricultural landscape of the upper U.S. Midwest.

Research at the KBS LTER has shown that agroecosystems and successional grasslands generally conform to our understanding of how disturbance and nutrient availability interact to determine productivity and species diversity in terrestrial plant communities. Disturbance, whether caused by tillage or herbicide use, has a very strong effect on plant community composition in both the Early Successional community and the weed communities of annual row crop systems. Fertilization generally increases production and decreases species diversity in grasslands (Gough et al. 2000, Clark et al. 2007), and while nutrient inputs certainly increase crop yield, the nutrient source (inorganic or legume-based) confounds our interpretation of the fertilizer effect of the abundance and composition of weed communities. This constrains our ability to use results from unmanaged successional grasslands to predict how crop grain yield or weed biomass will respond to particular changes in agricultural management. However, research on the ecology of weeds in agricultural ecosystems may provide insights into how to manage invasive species in remnant, degraded, or restored ecosystems (Smith et al. 2006).

Our experiments on row-crop and successional systems at the KBS LTER have provided important insights into the mechanisms by which diversity may influence crop yield. For example, studies of seed bank dynamics in the MCSE annual row-crop systems have shown that disturbance and fertilization interact with soil biota to influence seed mortality (Davis et al. 2005). Although this mechanism has not been widely explored in natural plant communities, it may be among the plant–soil feedbacks (Bever et al. 2010) that can be managed in reduced input or organic cropping systems. Findings such as these can lead to the development of management practices that rely less on chemical inputs and more on manipulation of ecosystem processes. These insights inform future research on factors influencing plant communities in sustainable agricultural systems, as well as natural systems. However, while there is growing evidence of the importance of regional species pools and other landscape factors to the composition and diversity of grassland communities, considerably less is known about how these regional processes influence the composition of weed communities.

Predicting how species and communities will respond to changes in global climate (particularly, temperature and precipitation patterns) remains one of the grand challenges in ecology. Improving our ability to make these predictions has important consequences for agriculture because climate changes are likely to affect crop production not just directly, but also indirectly by affecting the type and abundance of pests. As crops become either more intensively managed or more widely planted across the landscape to meet increasing demand for food and fuel, we will be challenged to better understand how landscape factors influence the dynamics of plant communities in agricultural landscapes. Increasing temporal variability in precipitation and other environmental factors may make it more difficult to manage these systems and to predict how they will respond to changes in both biotic and abiotic drivers, including crop management practices. The work to date at the KBS LTER—and cross-site syntheses to place it in a continental scope—provides a context for further investigation on how plant communities in agricultural landscapes can be managed to provide a wide range of ecosystem services.

References

Aanderud, Z. T., D. M. Schoolmaster Jr., and J. T. Lennon. 2011. Plants mediate the sensitivity of soil respiration to rainfall variability. Ecosystems *14*:156–167.

Anderson, R. C., and M. L. Bowles. 1999. Deep-soil savannas and barrens of the Midwestern United States. Pages 155–170 in R. C. Anderson, J. S. Fralish, and J. M. Baskin, editors. Savannas, barrens and rock outcrop plant communities of North America. Cambridge University Press, Cambridge, UK.

Baskin, C. C., and J. M. Baskin. 1999. Seeds: ecology, biogeography, and evolution of dormancy and germination. Second edition. Academic Press, San Diego, California, USA.

Bever, J. D., I. A. Dickie, E. Facelli, J. M. Facelli, J. N. Klironomos, M. Moora, M. C. Rilling, W. D. Stock, M. Tibbett, and M. Zobel. 2010. Rooting theories of plant community ecology in microbial interactions. Trends in Ecology and Evolution *25*:468–478.

Buhler, D. D. 2002. Challenges and opportunities for integrated weed management. Weed Science *50*:273–280.

Buhler, D. D., R. G. Hartzler, and F. Forcella. 1997. Implications of weed seedbank dynamics to weed management. Weed Science *45*:329–336.

Burbank, D. H., K. S. Pregitzer, and K. L. Gross. 1992. Vegetation of the W.K. Kellogg Biological Station. Research Report 510, Michigan State University Agricultural Experiment Station, East Lansing, Michigan, USA.

Burns, C. E., S. L. Collins, and M. D. Smith. 2009. Plant community response to loss of large herbivores: comparing consequences in a South African and a North American grassland. Biodiversity and Conservation *18*:2327–2342.

Cavigelli, M. A., S. R. Deming, L. K. Probyn, and R. R. Harwood, editors. 1998. Michigan field crop ecology: managing biological processes for productivity and environmental quality. Michigan State University Extension Bulletin E-2646, East Lansing, Michigan, USA.

Chapin, F. S., III, P. M. Vitousek, and K. VanCleve. 1986. The nature of nutrient limitation in plant communities. American Naturalist *127*:48–58.

Clark, C. M., E. E. Cleland, S. L. Collins, J. E. Fargione, L. Gough, K. L. Gross, S. C. Pennings, K. N. Suding, and J. B. Grace. 2007. Environmental and plant community determinants of species loss following nitrogen enrichment. Ecology Letters *10*:596–607.

Cleland, E. E., C. M. Clark, S. L. Collins, J. E. Fargione, L. Gough, K. L. Gross, D. G. Milchunas, S. C. Pennings, W. D. Bowman, I. C. Burke, W. K. Lauenroth, G. P. Robertson, J. C. Simpson, D. Tilman, and K. N. Suding. 2008. Species responses to nitrogen fertilization in herbaceous plant communities, and associated species traits. Ecology *89*:1175.

Cleland, E. E., S. L. Collins, T. L. Dickson, E. C. Farrer, K. L. Gross, E. A. Gherardi, L. M. Hallett, R. J. Hobbs, J. S. Hsu, L. Turnball, and K. N. Suding. 2013. Sensitivity of grassland plant community composition to spatial vs. temporal variation in precipitation. Ecology *94*:1687–1696.

Collins, S. L. 1992. Fire frequency and community heterogeneity in tallgrass prairie vegetation. Ecology *73*:2001–2006.

Collins, S. L., A.K., Knapp, J. M. Briggs, J. M. Blair, and E. A. Steinauer. 1998. Modulation of diversity by grazing and mowing in native tallgrass prairie. Science *280*:745–747.

Connell, J. H., and R. O. Slatyer. 1977. Mechanisms of succession in natural communities and their role in community stability and organization. American Naturalist *111*:1119–1144.

Costanza, R., R. d'Arge, R. de Groot, S. Farber, M. Grasso, B. Hannon, K. Limburg, S. Naeem, R. V. O'Neill, J. Paruelo, R. G. Raskin, P. Sutton, and M. van den Belt. 1997. The value of the world's ecosystem services and natural capital. Nature *387*:253–260.

Daily, G. C., T. Söderqvist, S. Aniyar, K. Arrow, P. Dasgupta, P. R. Ehrlich, C. Folke, A. M. Jansson, B.-O. Jansson, N. Kautsky, S. Levin, J. Lubchenco, K.-G. Mäler, D. Simpson, D. Starrett, D. Tilman, and B. Walker. 2000. The value of nature and the nature of value. Science *289*:395–396.

Davis, A. S. 2006. When does it make sense to target the weed seed bank. Weed Science *54*:558–565.

Davis, A. S., K. Renner, and K. L. Gross. 2005. Weed seedbank and community shifts in a long-term cropping systems experiment. Weed Science *53*:296–306.

Davis, M. A., J. P. Grime, and K. Thompson. 2000. Fluctuating resources in plant communities: a general theory of invasibility. Journal of Ecology *88*:528–534.

Dickson, T. L., and K. L. Gross. 2013. Plant community responses to long-term fertilization: changes in functional group abundance drive changes in species richness. Oecologia *173*:1513–1520.

Drinkwater, L. E., and S. S. Snapp. 2007. Nutrients in agroecosystems: rethinking the management paradigm. Advances in Agronomy *92*:163–186.

Easterling, D. R., G. A. Meehl, C. Parmesan, S. A. Changnon, T. R. Karl, and L. O. Mearns. 2000. Climate extremes: observations, modeling, and impacts. Science *289*:2068–2074.

Egan, J., and D. A. Mortensen. 2012. A comparison of land-sharing and land-sparing strategies for plant richness conservation in agricultural landscapes. Ecological Applications *22*:459–471.

Elser, J. J., M. E. S. Bracken, E. E. Cleland, D. S. Gruner, W. S. Harpole, H. Hillebrand, J. T. Ngai, E. W. Seabloom, J. B. Shurin, and J. E. Smith. 2007. Global analysis of nitrogen and phosphorus limitation of primary production in freshwater, marine, and terrestrial ecosystems. Ecology Letters *10*:1135–1142.

Fargione, J. E., and D. Tilman. 2005. Diversity decreases invasion via both sampling and complementarity effects. Ecology Letters *8*:604–611.

Foster, B. L. 1999. Establishment, competition, and the distribution of native grasses among Michigan old-fields. Journal of Ecology *87*:476–489.

Foster, B. L., and K. L. Gross. 1999. Temporal and spatial patterns of woody plant establishments in Michigan old fields. American Midland Naturalist *142*:229–243.

Gabriel, D., C. Thies, and T. Tscharntke. 2005. Local diversity of arable weeds increases with landscape complexity. Perspectives in Plant Ecology, Evolution and Systematics *7*:85–93.

Gardiner, M. M., D. A. Landis, C. Gratton, C. D. DiFonzo, M. O'Neal, J. M. Chacon, M. T. Wayo, N. P. Schmidt, E. E. Mueller, and G. E. Heimpel. 2009. Landscape diversity enhances the biological control of an introduced crop pest in the north-central USA. Ecological Applications *19*:143–154.

Garibaldi, L. A., I. Steffan-Dewenter, C. Kremen, J. M. Morales, R. Bommarco, S. A. Cunningham, L. G. Carvalheiro, N. P. Chacoff, J. H. Dedenhöffer, S. S. Greenleaf, A. Holzschuh, R. Isaacs, K. Krewenka, Y. Mandelik, L. A. Morandin, S. G. Potts, T. H. Ricketts, H. Szentgyörgyi, B. F. Viana, C. Westphal, R. Winfree, and A. M. Klein. 2011. Stability of pollination services decreases with isolation from natural areas despite honey bee visits. Ecology Letters *14*:1062–1072.

Gough, L., K. L. Gross, E. E. Cleland, C. M. Clark, S. L. Collins, J. E. Fargione, S. C. Pennings, and K. N. Suding. 2012. Incorporating clonal growth form clarifies the role of plant height in response to nitrogen addition. Oecologia *169*:1053–1062.

Gough, L., C. W. Osenberg, K. L. Gross, and S. L. Collins. 2000. Fertilization effects on species density and primary productivity in herbaceous plant communities. Oikos *89*:428–439.

Grman, E., J. A. Lau, D. R. Schoolmaster, and K. L. Gross. 2010. Mechanisms contributing to stability in ecosystem function depend on the environmental context. Ecology Letters *13*:1400–1410.

Gross, K. L., and S. A. Emery. 2007. Succession and restoration in Michigan old-field communities. Pages 162–179 in V. Cramer and R. J. Hobbs, editors. Old fields: dynamics and restoration of abandoned farmland. Island Press, Washington, DC, USA.

Gross, K. L., M. R. Willig, L. Gough, R. Inouye, and S. B. Cox. 2000. Species density and productivity at different spatial scales in herbaceous plant communities. Oikos *89*:417–427.

Harwood, R. R. 2002. Sustainable agriculture on a populous, industrialized landscape: building ecosystem vitality and productivity. Pages 305–315 in R. Lal, D. Hansen, N. Uphoff, and S. Slack, editors. Food security and environmental quality in the developing world. CRC Press, Boca Raton, Florida, USA.

Hilgenfeld, K. L., A. R. Martin, D. A. Mortensen, and S. C. Mason. 2004. Weed management in a glyphosate resistant soybean system: weed species shifts. Weed Technology *18*:284–291.

Houseman, G. R., and K. L. Gross. 2006. Does ecological filtering across a productivity gradient explain variation in species pool-richness relationships? Oikos *115*:148–154.

Houseman, G. R., and K. L. Gross. 2011. Linking grassland plant diversity to species pools, sorting, and plant traits. Journal of Ecology *99*:464–472.

Howe, H. F. 2000. Grass response to seasonal burns in experimental plantings. Journal of Range Management *53*:437–441.

Huberty, L. E., K. L. Gross, and C. J. Miller. 1998. Effects of nitrogen addition on successional dynamics and species diversity in Michigan old-fields. Journal of Ecology *86*:794–803.

Huston, M. 1979. A general hypothesis of species diversity. American Naturalist *113*:81–101.

Huxman, T. E., M. D. Smith, P. A. Fay, A. K. Knapp, M. R. Shaw, M. E. Loik, S. D. Smith, D. T. Tissue, J. C. Zak, J. F. Weltzin, W. T. Pockman, O. E. Sala, B. M. Haddad, J. Harte, G. W. Koch, S. Schwinning, E. E. Small, and D. G. Williams. 2004. Convergence across biomes to a common rain-use efficiency. Nature *429*:651–654.

IPCC (Intergovernmental Panel on Climate Change). 2007. Climate change 2007: the physical science basis. Contribution of Working Group I to the Fourth Assessment Report of the Intergovernmental Panel on Climate Change. Cambridge University Press, New York, New York, USA.

Isaacs, R., J. Tuell, A. Fiedler, M. Gardiner, and D. Landis. 2009. Maximizing arthropod-mediated ecosystem services in agricultural landscapes: the role of native plants. Frontiers in Ecology and the Environment 7:196–203.

Johnson, N. C. 2010. Resource stoichiometry elucidates the structure and function of arbuscular mycorrhizas across scales. New Phytologist 185:631–647.

Johnson, N. C., D. L. Rowland, L. Corkidi, and E. B. Allen. 2008. Winners and losers during grassland N-eutrophication differ in biomass allocation and mychorrizas. Ecology 89:2868–2878.

Johnson, W. G., V. M. Davis, G. R. Kruger, and S. C. Weller. 2009. Influence of glyphosate-resistant cropping systems on weed species shifts and glyphosate-resistant weed populations. European Journal of Agronomy 31:162–172.

Jordan, N., and C. Vatovec. 2004. Agroecological benefits from weeds. Pages 137–158 in Inderjit, editor. Weed biology and management. Kluwer Academic Publishers, Dordrecht, The Netherlands.

Jordan, N., and K. D. Warner. 2010. Enhancing the multifunctionality of US agriculture. BioScience 60:60–66.

Knapp, A., J. M. Blair, J. M. Briggs, S. L. Collins, L. C. Johnson, and E. G. Towne. 1999. The keystone role of bison in North American tallgrass prairie. Bioscience 49:39–50.

Knapp, A. K., and M. D. Smith. 2001. Variation among biomes in temporal dynamics of aboveground primary production. Science 291:481–484.

Kremer, R. J. 1993. Management of weed seed banks with microorganisms. Ecological Applications 3:42–52.

Kremer, R. J., and J. M. Li. 2003. Developing weed-suppressive soils through improved soil quality management. Soil & Tillage Research 72:193–202.

Landis, D. A., and S. H. Gage. 2015. Arthropod diversity and pest suppression in agricultural landscapes. Pages 188–212 in S. K. Hamilton, J. E. Doll, and G. P. Robertson, editors. The ecology of agricultural ecosystems: long-term research on the path to sustainability. Oxford University Press, New York, New York, USA.

Landis, D. A., M. M. Gardiner, W. van der Werf, and S. M. Swinton. 2008. Increasing corn for biofuel production reduces biocontrol services in agricultural landscapes. Proceedings of the National Academy of Sciences USA 105:20552–20557.

Laurenroth, W. K., and O. E. Sala. 1992. Long-term forage production of North American shortgrass steppe. Ecological Applications 2:397–403.

Leach, M. K., and T. J. Givnish. 1996. Ecological determinants of species loss in remnant prairies. Science 273:1555–1558.

LeBauer, D. S., and K. K. Treseder. 2008. Nitrogen limitation of net primary production in terrestrial ecosystems is globally distributed. Ecology 89:371–379.

Loreau, M., S. Naeem, P. Inchausti, J. Bengtsson, J. P. Grime, A. Hector, D. U. Hooper, M. A. Huston, D. Raffaelli, B. Schmid, D. Tilman, and D. A. Wardle. 2001. Biodiversity and ecosystem functioning: current knowledge and future challenges. Science 294:804–808.

Martin, L. M., and B. J. Wilsey. 2006. Assessing grassland restoration success: relative roles of seed additions and native ungulate activities. Journal of Applied Ecology 43:1098–1109.

McPherson, G. R. 1997. Ecology and management of North American savannas. University of Arizona Press, Tucson, Arizona, USA.

Meehan, T. D., B. P. Werling, D. A. Landis, and C. Gratton. 2011. Agricultural landscape simplification and insecticide use in the Midwestern United States. Proceedings of the National Academy of Sciences USA *108*:11500–11505.

Melillo, J. M., A. D. McGuire, D. W. Kicklighter, B. I. Moore, C. J. Varosmarty, and A. L. Schloss. 1993. Global climate change and terrestrial net primary production. Nature *363*:234–336.

Menalled, F. D., K. L. Gross, and M. Hammond. 2001. Weed aboveground and seedbank community responses to agricultural management systems. Ecological Applications *11*:1586–1601.

Miller, T. E. 1982. Community diversity and interactions between size and frequency of disturbance. American Naturalist *120*:533–536.

Mittelbach, G. G. 2012. Community ecology. Sinauer Associates, Inc., Sunderland, Massachusetts, USA.

Mortensen, D. A., J. F. Egan, B. D. Maxwell, M. R. Ryan, and R. G. Smith. 2012. Navigating a critical juncture for sustainable weed management. BioScience *62*:75–84.

Motha, R. P., and W. Baier. 2005. Impacts of present and future climate change and climate variability on agriculture in the temperate regions: North America. Climatic Change *70*:137–164.

Murphy, S. D., D. R. Clements, S. Belaoussoff, P. G. Kevan, and C. J. Swanton. 2006. Promotion of weed species diversity and reduction of weed seedbanks with conservation tillage and crop rotation. Weed Science *54*:69–77.

Myster, R. W., and S. T. A. Pickett. 1993. Effects of litter, distance, density and vegetation patch type on postdispersal tree seed predation in old fields. Oikos *66*:381–388.

Noy-Meir, I. 1973. Desert ecosystems: environment and producers. Annual Review of Ecology and Systematics *4*:25–51.

NRC (National Research Council). 2007. Status of pollinators in North America. National Academy Press, Washington, DC, USA.

Nurse, R. E., S. J. Darbyshire, C. Bertin, and A. DiTommaso. 2009. The biology of Canadian weeds. 141. *Setaria faberi* Herm. Canadian Journal of Plant Science *89*:379–404.

Packard, S., and C. F. Mutel. 1997. The tallgrass restoration handbook: for prairies, savannas, and woodlands. Island Press, Washington, DC, USA.

Pake, C. E., and D. L. Venable. 1996. Seed banks in desert annuals: implications for persistence and coexistence in variable environments. Ecology *77*:1427–1435.

Parker, T. C. 2011. Investigating the mechanisms behind corn yield patterns under different rotation diversity treatments. Thesis, University of York, York, UK.

Pennings, S. C., C. M. Clark, E. E. Cleland, S. L. Collins, L. Gough, K. L. Gross, D. G. Milchunas, and K. N. Suding. 2005. Do individual plant species show predictable responses to nitrogen addition across multiple experiments? Oikos 110:547–555.

Pickett, S. T. A., and P. S. White, editors. 1985. The ecology of natural disturbance and patch dynamics. Academic Press, San Diego, California, USA.

Poggio, S. L., E. J. Chaneton, and C. M. Ghersa. 2010. Landscape complexity differentially affects alpha, beta, and gamma diversities of plants occurring in fencerows and crop fields. Biological Conservation *143*:2477–2486.

Pollnac, F. W., B. D. Maxwell, and F. D. Menalled. 2009. Weed community characteristics and crop performance: a neighbourhood approach. Weed Research *49*:242–250.

Posner, J. L., J. O. Baldock, and J. L. Hedtcke. 2008. Organic and conventional production systems in the Wisconsin Integrated Systems Trials: I. Productivity 1990–2002. Agronomy Journal *100*:253–260.

Power, A. G. 2010. Ecosystem services and agriculture: tradeoffs and synergies. Philosophical Transactions of the Royal Society B: Biological Sciences *365*:2959–2971.

Ricketts, T. H., J. Regetz, I. Steffan-Dewenter, S. A. Cunningham, C. Kremen, A. Bobdanski, B. Gemmill-Herren, S. S. Greenleaf, A. M. Klein, M. M. Mayfield, L. A. Morandin, S. G. Potts, and B. F. Viana. 2008. Landscape effects on crop pollination services: are there general patterns? Ecology Letters *11*:499–515.

Robertson, G. P., K. L. Gross, S. K. Hamilton, D. A. Landis, T. M. Schmidt, S. S. Snapp, and S. M. Swinton. 2015. Farming for ecosystem services: an ecological approach to row-crop agriculture. Pages 33–53 in S. K. Hamilton, J. E. Doll, and G. P. Robertson, editors. The ecology of agricultural ecosystems: long-term research on the path to sustainability. Oxford University Press, New York, New York, USA.

Robertson, G. P., and S. K. Hamilton. 2015. Long-term ecological research at the Kellogg Biological Station LTER Site: conceptual and experimental framework. Pages 1–32 in S. K. Hamilton, J. E. Doll, and G. P. Robertson, editors. The ecology of agricultural ecosystems: long-term research on the path to sustainability. Oxford University Press, New York, New York, USA.

Robertson, G. P., and S. M. Swinton. 2005. Reconciling agricultural productivity and environmental integrity: a grand challenge for agriculture. Frontiers in Ecology and the Environment *3*:38–46.

Robinson, T. M. P. 2011. Impacts of precipitation variability on plant communities. Dissertation, Michigan State University, East Lansing, Michigan, USA.

Robinson, T. M. P., and K. L. Gross. 2010. The impact of altered precipitation variability on annual weed species. American Journal of Botany *97*:1625–1629.

Robinson, T. M. P., K. J. La Pierre, M. A. Vadeboncoeur, K. M. Byrne, M. L. Thomey, and S. Colby. 2013. Seasonal, not annual precipitation, drives community productivity across ecosystems. Oikos *122*:727–738.

Ryan, M. R., R. G. Smith, S. B. Mirsky, D. A. Mortensen, and R. Seidel. 2010. Management filters and species traits: weed community assembly in long-term organic rotational trials. Weed Science *58*: 265–277.

Scheffer, M., S. Carpenter, J. A. Foley, C. Folke, and B. Walker. 2001. Catastrophic shifts in ecosystems. Nature *413*:591–596.

Smith, R. G. 2006. Timing of tillage is an important filter on the assembly of weed communities. Weed Science *54*:705–712.

Smith, R. G., and K. L. Gross. 2006. Weed community and corn yield variability in diverse management systems. Weed Science *54*:106–113.

Smith, R. G., and K. L. Gross. 2007. Assembly of weed communities along a crop diversity gradient. Journal of Applied Ecology *44*:1046–1056.

Smith, R. G., K. L. Gross, and G. P. Robertson. 2008. Effects of crop diversity on agroecosystem function: crop yield responses. Ecosystems *11*: 355–366.

Smith, R. G., B. D. Maxwell, F. D. Menalled, and L. J. Rew. 2006. Lessons from agriculture may improve the management of invasive plants in wildland systems. Frontiers in Ecology and the Environment *4*:428–434.

Smith, R. G., D. A. Mortensen, and M. R. Ryan. 2010. A new hypothesis for the functional role of diversity in mediating resource pools and weed-crop competition in agroecosystems. Weed Research *50*:37–48.

Snapp, S. S., L. E. Gentry, and R. R. Harwood. 2010. Management intensity—not biodiversity—the driver of ecosystem services in a long-term row crop experiment. Agriculture, Ecosystems and Environment *138*:242–248.

Sosnoskie, L. M., C. P. Herms, and J. Cardina. 2006. Weed seedbank community composition in a 35-yr-old tillage and rotation experiment. Weed Science *54*:263–273.

Suding, K. N., S. L. Collins, L. Gough, C. Clark, E. E. Cleland, K. L. Gross, D. G. Milchunas, and S. Pennings. 2005. Functional- and abundance-based mechanisms explain diversity loss due to N fertilization. Proceedings of the National Academy of Sciences USA *102*:4387–4392.

Suding, K. N., and K. L. Gross. 2006a. How systems change: succession, multiple states and restoration trajectories. Pages 190–209 in D. A. Falk, M. A. Palmer, and J. B. Zedler, editors. Foundations of restoration ecology. Island Press, Washington, DC, USA.

Suding, K. N., and K. L. Gross. 2006b. Modifying native and exotic species richness correlations: the influence of fire and seed addition. Ecological Applications *16*:1319–1326.

Suding, K. N., K. L. Gross, and G. Houseman. 2004. Alternative states and positive feedbacks in restoration ecology. Trends in Ecology & Evolution *19*:46–53.

Swinton, S. M., F. Lupi, G. P. Robertson, and S. K. Hamilton. 2007. Ecosystem services and agriculture: cultivating agricultural ecosystems for diverse benefits. Ecological Economics *64*:245–252.

Swinton, S. M., F. Lupi, G. P. Robertson, and D. A. Landis. 2006. Ecosystem services from agriculture: Looking beyond the usual suspects. American Journal of Agricultural Economics *88*: 1160–1166.

Symstad, A. J., F. S. Chapin III, D. H. Wall, K. L. Gross, L. F. Huenneke, G. G. Mittelbach, D. P. C. Peters, and D. Tilman. 2003. Long-term and large-scale perspectives on the relationship between biodiversity and ecosystem functioning. BioScience *53*: 89–98.

Thomas, A. G., D. A. Derksen, R. E. Blackshaw, R. C. Van Acker, A. Légère, P. R. Watson, and G. C. Turnbull. 2004. A multistudy approach to understanding weed population shifts in medium- to long-term tillage systems. Weed Science *52*:874–880.

Thompson, J. N., O. J. Reichman, P. J. Morin, G. A. Polis, M. E. Power, R. W. Sterner, C. A. Couch, L. Gough, R. Holt, D. U. Hooper, F. Keesing, C. R. Lovell, B. T. Milne, M. C. Molles, D. W. Roberts, and S. Y. Strauss. 2001. Frontiers of ecology. BioScience *51*:15–24.

Tilman, D. 1984. Plant dominance along an experimental nutrient gradient. Ecology *65*:1445–1453.

Tilman, D. 1987. Secondary succession and the pattern of plant dominance along experimental nitrogen gradients. Ecological Monographs *57*:189–214.

Vellend, M. 2010. Conceptual synthesis in community ecology. Quarterly Review of Biology *85*:183–206.

Weltzin, J. F., M. E. Loik, S. Schwinning, D. G. Williams, P. A. Fay, B. M. Haddad, J. Harte, T. E. Huxman, A. K. Knapp, G. Lin, W. T. Pockman, M. R. Shaw, E. E. Small, M. D. Smith, S. D. Smith, D. T. Tissue, and J. C. Zak. 2003. Assessing the response of terrestrial ecosystems to potential changes in precipitation. BioScience *53*:941–952.

Zimdahl, R. L. 2004. Weed-crop competition: a review. Blackwell Publishing, Ames, Iowa, USA.

8

Arthropod Diversity and Pest Suppression in Agricultural Landscapes

Douglas A. Landis and Stuart H. Gage

Research at the Kellogg Biological Station Long-Term Ecological Research site (KBS LTER) is focused on understanding the ecological interactions underlying the productivity of row-crop ecosystems. Within these systems, insect pests and weeds represent two major groups of organisms that farmers must consistently and effectively manage. Since its inception in 1989, entomologists associated with KBS LTER have sought to develop a better understanding of the ecology of beneficial insects and the crop pests they control within agricultural landscapes. As a group, we have specifically focused on key taxa involved in pest suppression, namely, predators and parasitoids of insect herbivores and predators of weed seeds. The long-term goal of this work has been to inform agricultural practices that might enhance natural pest suppression and thus reduce the need for chemical pest controls. Working toward this goal has involved long-term observations coupled with shorter-term, hypothesis-driven experiments. This combination has proven a fruitful model for advancing science at KBS and the LTER Network in general (Knapp et al. 2012).

Shifting Systems of Pest Management

For millennia, farmers have battled with weeds and insects to avoid crop losses. During the first half of the twentieth century, U.S. row-crop farmers primarily relied on natural enemies (predators and parasitoids of herbivores), cultural practices (e.g., tillage, rotation, variety selection), and a limited number of inorganic insecticides to help control insect pests. As a result, literature from that time is full of careful observations on the biology and ecology of both crop pests and their natural enemies. However, following the discovery of organochlorine insecticides during

World War II, research on insect management rapidly shifted to a narrow focus on chemical control. These long-lasting and highly effective insecticides seemed a panacea, providing nearly complete control of even the most troublesome pests. Unfortunately, near sole reliance on chemical controls resulted in the development of insecticide resistance and decimation of natural enemy communities, and pest outbreaks followed, as did growing concerns over environmental impacts.

As early as 1959, Vernon Stern and colleagues began to call for the integration of chemical insecticides into a more holistic set of practices they termed "integrated control" (Stern et al. 1959). Their concept, now known as Integrated Pest Management (IPM), sought to combine cultural, biological, and chemical pest control in a systems approach. In response, researchers developed IPM systems that combined cultural tools like rotation and resistant varieties with biological controls including importation, conservation, and augmentation of natural enemies. Additionally, crops were regularly scouted and chemicals applied only after a pest population exceeded an economic threshold, that is, the population level at which action is needed to prevent an economic loss (Radcliffe et al. 2009).

More recently, the advent of genetically modified (GM) crops has once again shifted the focus in pest management. The development of field crops with built-in resistance to broad-spectrum herbicides—for example, glyphosate-tolerant soybean and corn—allowed a very different approach to weed management. Rather than scouting fields for weed species composition and growth stage and using selective herbicides, growers can now spray a single broad-spectrum herbicide whenever weeds reach critical levels. Moreover, farmers can also purchase "stacked" GM crop seeds that not only contain genes for herbicide resistance but also genes for producing bacterial toxins that confer resistance to multiple insect pests. Although it has simplified pest management for farmers, reliance on such a small set of tools has again yielded instances of resistance and concerns about environmental degradation (Ferry and Gatehouse 2009).

Ecologically Based Pest Management

At the same time that agricultural scientists were developing new methods of pest management, ecologists were beginning to study agriculture from an ecological perspective (Lowrance et al. 1984, Carroll et al. 1990, Gliessman 1998, Robertson et al. 2004). One aspect of agroecology has focused on the question of how the biodiversity of cropping systems might be managed to achieve improved pest management (Altieri 1994). The relationship between biotic diversity and ecological performance has been a key question for ecologists for more than a half-century. Rooted in the diversity–stability arguments of the late twentieth century (MacArthur 1955, Elton 1958, Odum 1959, May 1973) and more recently in the study of biodiversity and ecosystem function (Schulze and Mooney 1993, Loreau et al. 2002), our understanding of the ways in which biodiversity influences ecosystem services continues to evolve. The study of predator–prey interactions has produced a rich body of theoretical and empirical work elucidating the influence of biotic diversity on herbivore population regulation (Ives et al. 2005, Bruno and Cardinale 2008, Letourneau et al. 2009).

Within this broader context, KBS LTER scientists have sought a deeper understanding of the interactions of beneficial insects and crop pests in agricultural landscapes.

Determining Which Insects and Processes to Quantify

The KBS LTER Main Cropping System Experiment (MCSE) design was established in 1989 (Robertson and Hamilton 2015, Chapter 1 in this volume) and includes annual systems made up of corn (*Zea mays* L.)—soybean (*Glycine max* L.)—winter wheat (*Triticum aestivum* [L.]) rotations, alfalfa (*Medicago sativa* L.), hybrid poplars (*Populus* sp.), and unmanaged successional ecosystems (Table 8.1). Due to the long-term nature of the research, large 1-ha plots were established with five permanently located monitoring sites within each. From an entomological perspective, this sampling design presented both advantages and constraints. Fixed sampling points provide the opportunity to follow spatial patterns over time, but limit other types of investigations such as those involving dispersal and predator–prey interactions. From the outset, a nontrivial question has been which insects and ecological processes could best be studied within this framework.

A direct focus on insect herbivores was initially considered but ultimately set aside, primarily because each crop can support multiple species of insect pests and differences in pest life cycles and behaviors would require different, labor-intensive sampling strategies. In addition, many insect sampling approaches require destructive methods that would be at odds with other study objectives and themselves represent a disturbance to the ecosystems. Instead, we focused on insect predators and parasitoids that engage in biological regulation of insect herbivores, and in particular on predatory ladybird beetles (Coleoptera: Coccinellidae). Since 1989 coccinellids have been monitored in all MCSE systems at the main site and in selected portions of the surrounding landscape.

Logistical requirements suggested adoption of a simple sampling method that would minimally disturb the plant community but would capture the dynamics of coccinellid predators and foster understanding of their ecological function, both in time and space. A previous investigation of apple maggot (*Rhagoletis pomonella* Walsh) dispersal at KBS had successfully used transects of yellow sticky traps to determine flight paths of adult flies dispersing to an isolated orchard adjacent to the LTER main site (Ryan 1990). These sticky traps also captured many species of dispersing Coccinellidae. Maredia et al. (1992a) determined the optimal trap color to attract coccinellids and other key predators to be yellow, which was most attractive to *Coccinella septempunctata* (L.), the most abundant coccinellid, and equally as attractive as other colors to *Hippodamia parenthesis* (Say) and *Chrysoperla carnea* (Stephens) (Neuroptera: Chrysopidae). As a result, yellow sticky traps (PHEROCON AM, Great Lakes IPM, Vestaburg, Michigan) have been deployed since 1988. A pole supports the traps 1 m above the soil surface (Maredia et al. 1992b) at each permanent sampling location. Traps are deployed for a minimum of 8 weeks each year from May to September in each replicate plot of the MCSE systems, for a total of 255 sample sites. Each is visited weekly to record the abundance of 17 species of Coccinellidae.

Table 8.1. Description of the KBS LTER Main Cropping System Experiment (MCSE).[a]

Cropping System/Community	Dominant Growth Form	Management
Annual Cropping Systems		
Conventional (T1)	Herbaceous annual	Prevailing norm for tilled corn–soybean–winter wheat (c–s–w) rotation; standard chemical inputs, chisel-plowed, no cover crops, no manure or compost
No-till (T2)	Herbaceous annual	Prevailing norm for no-till c–s–w rotation; standard chemical inputs, permanent no-till, no cover crops, no manure or compost
Reduced Input (T3)	Herbaceous annual	Biologically based c–s–w rotation managed to reduce synthetic chemical inputs; chisel-plowed, winter cover crop of red clover or annual rye, no manure or compost
Biologically Based (T4)	Herbaceous annual	Biologically based c–s–w rotation managed without synthetic chemical inputs; chisel-plowed, mechanical weed control, winter cover crop of red clover or annual rye, no manure or compost; certified organic
Perennial Cropping Systems		
Alfalfa (T6)	Herbaceous perennial	5- to 6-year rotation with winter wheat as a 1-year break crop
Poplar (T5)	Woody perennial	Hybrid poplar trees on a ca. 10-year harvest cycle, either replanted or coppiced after harvest
Coniferous Forest (CF)	Woody perennial	Planted conifers periodically thinned
Successional and Reference Communities		
Early Successional (T7)	Herbaceous perennial	Historically tilled cropland abandoned in 1988; unmanaged but for annual spring burn to control woody species
Mown Grassland (never tilled) (T8)	Herbaceous perennial	Cleared woodlot (late 1950s) never tilled, unmanaged but for annual fall mowing to control woody species
Mid-successional (SF)	Herbaceous annual + woody perennial	Historically tilled cropland abandoned ca. 1955; unmanaged, with regrowth in transition to forest
Deciduous Forest (DF)	Woody perennial	Late successional native forest never cleared (two sites) or logged once ca. 1900 (one site); unmanaged

[a]Site codes that have been used throughout the project's history are given in parentheses. Systems T1–T7 are replicated within the LTER main site; others are replicated in the surrounding landscape. For further details, see Robertson and Hamilton (2015, Chapter 1 in this volume).

Additional sampling regimes were overlaid on this basic design when other interesting pest–natural enemy associations were observed or uncovered by patterns in the predator trapping data. Over the years, pitfall traps have also been used to determine carabid (Coleoptera: Carabidae) beetle community structure (Clark et al. 1997), and the results coupled with long-term studies of the weed seedbank (Gross et al. 2015, Chapter 7 in this volume) to better understand the role of insects in shaping weed community dynamics. Also, new invasive insects have been studied as they entered KBS, including herbivores such as the gypsy moth (*Lymnatria dispar* L.; Parry 2000, Kosola et al. 2001, Agrawal et al. 2002, Kosola et al. 2006), the soybean aphid (*Aphis glycines* Matsumura; Noma and Brewer 2007, 2008), and the exotic predaceous coccinellids *C. septempunctata* (Maredia et al. 1992b) and *Harmonia axyridis* Pallas, the multicolored Asian ladybird beetle (Colunga-Garcia and Gage 1998). In the remainder of this chapter, we present three case studies focusing on carabids, coccinellids, and the soybean aphid. Together, these examples encompass the breadth of insect studies conducted on site and illustrate some of the key lessons to be drawn from this long-term effort.

Carabids in the KBS Landscape

Ground-dwelling beetles in the family Carabidae are a diverse and frequently studied taxon. With over 2500 species in North America, they inhabit nearly all terrestrial ecosystems and perform a variety of ecological functions as herbivores, carnivores, and omnivores. In agricultural ecosystems, carabids are best known as predators of insects, gastropods, and other invertebrates. However, many carabid species are omnivorous and some species feed mainly on seeds (granivores). In row-crop systems, carabids can thus provide significant pest suppression by consuming insect pests and weed seeds in the soil seed bank.

Carabid Response to Habitat

Clark et al. (1997) first characterized the carabid communities of KBS LTER in the 5th and 6th years following MCSE establishment in 1989. They recorded 18 species, but 4 predatory species dominated and comprised 87% of the total catch. Pronounced differences in carabid communities occurred between the annual and perennial plant systems and between the Conventional and No-till systems. Management practices influenced habitat characteristics and served to structure the carabid communities in particular ways. For example, annual crop habitats contained significantly more *Poecilus lucoblandus* (Say) and *Agonum placidum* (Say), while the No-till and perennial crop systems favored *Cyclotrachelus sodalis* (Leconte).

Overall, Clark et al. (1997) concluded that no single system or habitat could be characterized as favoring carabid communities as a whole; rather, some systems and practices (e.g., tillage) favor particular species and disfavor others. Because carabids were frequently associated with feeding on crop insect pests, Clark et al. suggested there is potential for managing for selected carabid communities to enhance pest suppression.

Carabids and Weed Seed Predation

Weed seedbanks can build over time and present a significant challenge for agronomic management of annual row crops. Menalled et al. (2001) observed that the total abundance and number of weed species in the soil seedbank were increasing over a 6-year period (1993–1998) in the Conventional and No-till systems but declining in the Reduced Input and Biologically Based systems. Seedbank increases in the Conventional and No-till systems were dominated by annual grasses, and a later study by Menalled et al. (2007) found that carabid abundance and community structures responded to these changes. They found more total carabids in the Conventional than in the No-till and Biologically Based systems. However, granivores made up 32% of the total individuals captured in the No-till system but only 4 and 10% of total carabids in the Biologically Based and Conventional systems, respectively—implying that more resources were present for weed seed predators in the No-till system.

Menalled et al. (2007) tested this hypothesis by conducting seed removal experiments in these systems and found that predation on seeds of fall panicum (*Panicum dichotomiflorum* Michx.) and common lambsquarters (*Chenopodium album* L.) was (1) often more than twice as high in No-till compared to the Conventional and Biologically Based systems, particularly for fall panicum (Fig. 8.1), and (2) was closely correlated with seed predator captures ($r > 0.94$). Overall, these studies

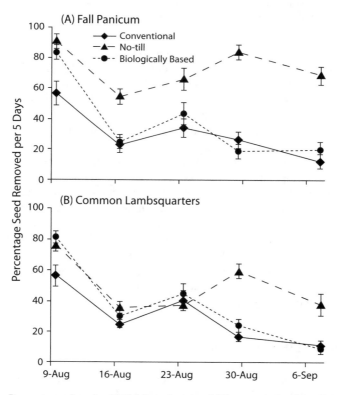

Figure 8.1. Percentage of seeds of (A) fall panicum and (B) common lambsquarters removed by invertebrate seed predators in three KBS LTER Main Cropping System Experiment (MCSE) systems during late summer of 2000 (mean ± SE, n = 6). Each data point represents a five-day period. Redrawn from Menalled et al. (2007) with permission from Elsevier.

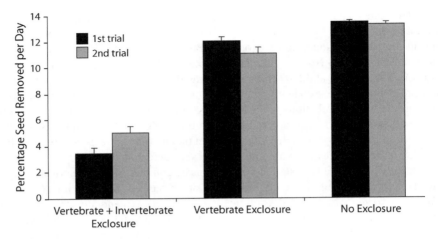

Figure 8.2. Percentage of weed seed removed per day in invertebrate + vertebrate exclosures, vertebrate exclosures, and without exclosure of seed predators (i.e., the control) in two trials averaged across field, species, and landscape type (mean ± SE). Redrawn from Menalled et al. (2000) with permission from Elsevier.

show that crop management affects carabid communities, which can in turn affect the weed seed bank through weed seed predation. Menalled et al. (2000) also studied weed seed predation by vertebrates vs. invertebrates in agricultural landscapes with increasing structural complexity. They found seed removal rates of between 7–12% per day, with invertebrates responsible for 50–66% of total predation (Fig. 8.2). They also identified a trend toward higher removal rates in more complex landscapes.

Overall, these studies suggest that carabid communities readily respond to changes in crop type and management (perennial vs. annual crops and tilled vs. no-till management). In turn, changes in community structure influence the ecosystem services that carabids provide—pest suppression and weed seed predation. However, probably due to the limited ability of carabids to disperse, changes in community structure at local scales do not always translate to similar effects at the landscape scale, as inconsistent impacts of landscape structure on weed seed predation have been observed. In the case of seed predation, this suggests that management efforts at the field and field-margin scale may more reliably influence carabid communities and services.

Coccinellids in the KBS Landscape

Ladybird beetles, in the family Coccinellidae, are a major group of arthropod predators in agricultural landscapes (Obrycki et al. 2009). In addition to feeding on insect prey, many also consume nonprey foods including plant pollen and nectar (Lundgren 2009). Most overwinter as adults in noncrop habitats and disperse into crops in the spring in search of resources. After consuming sufficient prey, females lay eggs; one or more generations occur per year based on the biology of individual species.

More than 20 years of sampling the coccinellid community of KBS has revealed a number of novel insights on the spatial and temporal patterns of insect predator responses to crop type and management practices. Our array of permanent geo-located sites has allowed scaling of analyses from microhabitat to landscape, and from individual weeks to seasonal to interannual change over two decades. Moreover, the spatial-temporal design of our insect observation program has provided the ability to quantify several unanticipated events, including the arrival of new herbivores and predators that, through long-range and local dispersal, entered the KBS landscape.

Characterizing the Coccinellid Community

Maredia et al. (1992b) used a combination of sweep net, sticky trap, and visual observations to characterize the occurrence and relative abundance of predatory Coccinellidae at KBS LTER in 1989 and 1990. During that time period, they recorded 12 native and 1 exotic species (Table 8.2). Subsequent sticky trap sampling at the site has revealed the occurrence of one additional native species (*Hippodamia glacialis* [Fab.]) and documented the arrival of three additional exotic species (*Harmonia axyridis, Hippodamia variegata* [Goeze], and *Propylaea quatuordecimpunctata* [L.]), bringing the total to 13 native and 4 exotic species by 2009. Several native species have apparently declined in abundance since 1989–1990. For example, *Adalia bipunctata* (L.), *Chilocorus stigma* (Say), *H. convergens* Guérin-Meneville, and *Hippodamia parenthesis* (Say), all reported as common in 1989–1990, became rare by 2009. In addition, several species that were listed as occasionally observed in 1989–1990 fell below detectable levels by 2009, including *Anatis labiculata* (Say), *Coccinella novemnotata* Herbst, *Hippodamia tredecimpunctata tibialis* (Say), and *Hyperaspis undulata* (Say). Because Maredia et al. (1992b) used multiple collection methods and subsequent sampling only used sticky traps, it is uncertain if these represent true declines or sampling biases.

The exotic species *C. septempunctata*, which was intentionally released in Michigan in 1985 for control of aphids (Maredia et al. 1992b) and rapidly became a dominant species (Sirota 1990), was of particular interest to early LTER researchers. Maredia et al. (1992c) confirmed Sirota's (1990) observations that *C. septempunctata* was a univoltine (one generation per year) species in Michigan with peak adult populations occurring in mid- to late June. Wheat and alfalfa were found to be important early season habitats for *C. septempunctata*, likely because they contained aphid prey prior to spring-planted annual crops like corn and soybean. Later in the season, *C. septempunctata* dispersed throughout the landscape and was found in all MCSE systems but particularly in the Early Successional and Poplar systems that tended to have late season aphid infestations (Maredia et al. 1992b).

The LTER database also allows for coccinellid habitat preferences to be studied over longer periods of time, and examination of habitat use by nine species in MCSE systems from 1989 to 2007 reveals distinct patterns (Fig. 8.3). With the exception of *H. axyridis*, most species are found in greater abundance in one or

Table 8.2. Coccinellid species observed in the KBS landscape from 1989 to 2009.

Species	Common Name, Lady Beetle	First Reported at LTER	1989–1990 Frequency of Observation[a]	2008–2009 Frequency of Observation[b]
Native				
Adalia bipunctata (L.)	2-spotted	1989	common	rare
Anatis labiculata (Say)	15-spotted	1989	occasional	not detected
Brachiacantha ursina (Fab.)	orange-spotted	1989	common	uncommon
Coleomegilla maculata (De Geer)	pink spotted	1989	common	common
Chilocorus stigma (Say)	twice-stabbed	1989	common	rare
Coccinella novemnotata Herbst	9-spotted	1989	occasional	not detected
C. trifaciata perplexa Mulsant	three-banded	1989	occasional	rare
Cycloneda munda (Say)	polished	1989	common	uncommon
Hippodamia glacialis (Fab.)	—	1994	na	uncommon
H. convergens Guérin-Meneville	convergent	1989	common	rare
H. parenthesis (Say)	parenthesis	1989	common	rare
H. tredecimpunctata tibialis (Say)	13-spotted	1989	occasional	not detected
Hyperaspis undulata (Say)	—	1989	occasional	not detected
Exotic				
Coccinella septempunctata (L.)	7-spotted	1989	common	common
Harmonia axyridis (Pallas)	multicolored Asian	1994	na	common
Hippodamia variegata (Goeze)	variegated	2000	na	uncommon
Propylaea quatuordecimpunctata (L.)	14-spotted	2006	na	uncommon

[a]From Maredia et al. 1992b. na = presumably not arrived yet.
[b]From LTER database. Common >5%, uncommon 1–5%, rare <1% of total captures.

two of the MCSE systems and found rarely in others. This is most striking for *Coleomegilla maculata* that is primarily found in corn, and for *H. convergens* that most commonly occurs in soybean. Other species like *C. septempunctata* and *Coccinella trifasciata perplexa* (Mulsant) are commonly found in multiple MCSE systems but only rarely in others.

Role of the Surrounding Landscape

Coccinellids use various habitats in the landscape as they move from overwintering sites to spring and summer feeding habitats. By sampling native plant communities at the interface of woodlots and crop fields, Colunga-Garcia (1996) documented the role of early flowering plants such as spring beauty (*Claytonia virginica* L.) and common dandelion (*Taraxacum officinale* F.H. Wigg) in providing spring pollen sources to adult coccinellids emerging from overwintering sites. He also developed a model to estimate the location of overwintering sites based on the position of woodlots and early spring pollen sources.

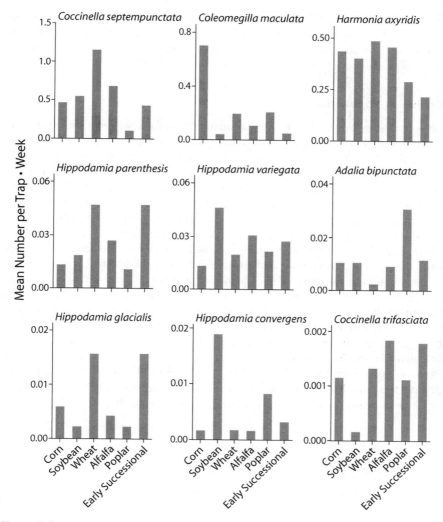

Figure 8.3. Summary of mean number of adults captured per weekly trapping interval for nine ladybird beetle species within different systems of the MCSE over 18 years (1989–2007). MCSE systems are described in Table 8.1.

Ostrom et al. (1997) used stable isotope techniques to show that the $\delta^{13}C$ and $\delta^{15}N$ ratios of coccinellids track those of their food sources and from this inferred patterns of coccinellid movement in the KBS landscape. In particular, they found that 32 and 68% of the diet of *C. maculata* were derived from alfalfa and corn pollen, respectively, which was consistent with the distribution of this species during their study. Subsequently, Colunga-Garcia et al. (1997) showed that the coccinellid community responded to overall landscape structure as measured by habitat diversity and patchiness. In concordance with prior studies, *C. maculata* was more abundant in a landscape that included corn, while *C. stigma* and

Brachiacantha ursina (Fab.) were more abundant in those that included deciduous forest habitats. Overall, coccinellid species richness increased in sites containing uncultivated habitats, demonstrating the importance of these habitats in
shaping predator communities (Woltz and Landis 2014).

Documenting Invasive Species

Long-term sampling at KBS has also been important for documenting the arrival of
exotic coccinellids and their impacts on the predator community. When the MCSE
was initiated in 1989, *C. septempunctata* was concluding its initial outbreak phase
(Sirota 1990) and was the dominant coccinellid species. Subsequent observations
show that this species exhibits roughly a 5-year population cycle (Fig. 8.4). In 1994
the KBS LTER trap network was the first to detect the occurrence of the exotic
species *H. axyridis* in Michigan (Colunga-Garcia and Gage 1998) (Fig. 8.5). In
contrast to *C. septempunctata*, which primarily inhabits field crops and herbaceous
plants in old-field habitats, *H. axyridis* is considered a semi-arboreal species (Koch
and Galvan 2008), inhabiting both trees and herbaceous habitats. These flexible
habitat requirements allowed *H. axyridis* to become a dominant species in all
MCSE habitats. Its occurrence in forested habitats was associated with a decline
in the abundance of *B. ursina, Cycloneda munda* (Say), and *C. stigma*—all species
that prefer wooded habitats—suggesting that competitive displacement may have
been occurring. In 2005 another exotic coccinellid, *H. variegata*, was reported in
Michigan for the first time at KBS and in three additional counties (Gardiner and
Parsons 2005), although a subsequent search of KBS LTER records showed it was
first detected in 2000. Finally, in 2006, the exotic 14-spotted lady beetle (*P. quatuordecimpunctata*) was discovered in Michigan (Gardiner et al. 2009a). This species

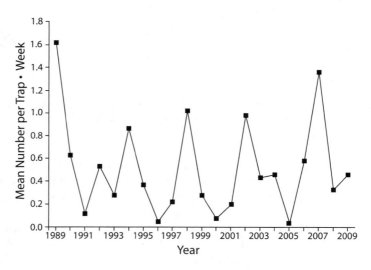

Figure 8.4. Mean number of *Coccinella septempunctata* ladybird beetle adults caught per
trap over week-long deployments at the MCSE between 1989–2009.

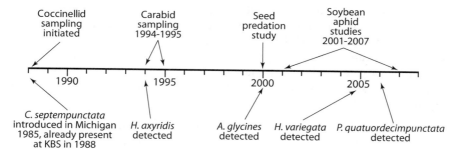

Figure 8.5. Time line showing insect sampling efforts (above line) and the arrival of key exotic coccinellid species (below line) to KBS LTER study sites. See Table 8.2 for full names of coccinellids; *A. glycines* is the soybean aphid *Aphis glycines*.

continued to increase and as early as 2008 was found to be the second most abundant coccinellid in corn after *C. maculata* (Gardiner et al. 2010).

Key lessons of these long-term coccinellid observations include a clearer understanding of the innate habitat preferences of different species (Fig. 8.3) and the seasonal movement of coccinellids from noncrop to crop habitats. As these predators move through the landscape (Isard and Gage 2001), they are influenced by the availability of prey and, as discussed below, can be important regulators of prey density. The addition of new exotic coccinellid species into the KBS landscape has shaped—and continues to shape—the structure and diversity of these communities (Bahlai et al. 2013, 2014).

Soybean Aphid: A New Herbivore Changes Everything

The arrival of the soybean aphid *A. glycines*, an exotic invasive herbivore, into the KBS landscape created an opportunity to evaluate how a new link in the existing food web alters system dynamics. *A. glycines* is an invasive insect pest from Asia that was first discovered in the United States in 2000 and rapidly became the nation's most significant threat to soybean production (Ragsdale et al. 2004). Prior to its arrival, soybean experienced relatively low insect herbivore pressure and was seldom treated with insecticides. The arrival of the soybean aphid fundamentally changed soybean production, with the aphid becoming a key pest, frequently requiring insecticide applications to control (Ragsdale et al. 2011).

The soybean aphid overwinters as an egg on several species of shrubs/small trees in the genus *Rhamnus*, principally common buckthorn (*R. cathartica* L.), which is itself an exotic invasive pest. Several generations of *A. glycines* occur on buckthorn in the spring before alates (winged, sexually mature individuals) are produced and migrate to soybean. On soybean plants, females reproduce asexually (parthenogenesis) and give birth to live young, with multiple generations occurring on a single soybean plant. Soybean aphid populations can reach 30,000 aphids per plant (DiFonzo 2006, as cited in Walter and DiFonzo 2007) and result in yield losses of up to 40% if left unchecked (Ragsdale et al. 2007). In September, alates are produced that return to *Rhamnus* spp., where mating occurs and eggs are laid.

Initial studies at KBS LTER and elsewhere found that *A. glycines* was attacked by a wide diversity of native and previously established predators (Fox et al. 2004, 2005; Rutledge et al. 2004) and parasitoids (Kaiser et al. 2007, Pike et al. 2007) with the potential to suppress *A. glycines* population growth. This provided the opportunity to ask several important questions:

(1) Is *A. glycines* primarily limited by top-down or bottom-up forces? Top-down forces represent the influence of higher trophic levels such as predation, whereas bottom-up forces represent the influence of lower trophic levels such as plant vigor or defense mechanisms.

(2) How do predators and parasitoids interact in the *A. glycines*–soybean system, and does intraguild predation (predation of potentially competing predators and parasitoids) alter the outcomes of these enemy interactions?

(3) Is predation/parasitism sufficient to cause a trophic cascade, whereby predators suppress herbivore prey, leading to increased crop yield?

(4) How does the occurrence of this new food source affect established coccinellid communities?

(5) How does landscape structure interact with enemy communities to alter *A. glycines* population dynamics?

Studies addressing these key questions were conducted at KBS as well as in commercial soybean fields in Michigan and throughout the U.S. North Central Region and are discussed below.

Top-Down vs. Bottom-Up Effects

In a series of studies conducted in the MCSE, Costamagna and colleagues explored the impact of crop management and natural enemies on soybean aphid population dynamics (Costamagna and Landis 2006; Costamagna et al. 2007a, b). By contrasting soybean aphid population growth in the Conventional, No-till, and Biologically Based systems (Table 8.1), they were able to examine a full range of potential bottom-up influences (fertility, soil moisture, induced host defences, etc.) that could be generated under realistic soybean growing conditions. In addition, by excluding natural enemies from selected plots, Costamagna et al. contrasted the relative importance of top-down and bottom-up forces for keeping aphid populations in check. They found that predation reduced initial aphid establishment by ~30% in 24 hours and that, overall, top-down influences provided a 4- to 7-fold suppression of aphid populations (Fig. 8.6).

In contrast, these investigators found no evidence for significant bottom-up forces across the range of agricultural practices, that is, there were no agricultural practices that differed in their abilities to check aphid populations in the absence of predators. The natural enemy community at KBS is dominated by generalist predators (lady beetles, anthocorid bugs, syrphid fly larvae) and generalist aphid parasitoids (Braconidae). Coccinellids appear particularly important for controlling aphids, and intraguild predation—where predators attack other predators who are

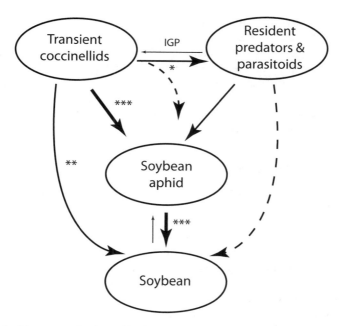

Figure 8.6. Summary of tri-trophic level interactions in the soybean aphid system. Thickness and direction of lines indicate the magnitude and direction of impacts. Dashed lines represent potential interactions that have not yet been shown to occur and asterisks represent increasing levels of statistical significance (P = 0.05 – 0.001). IGP = intraguild predation. Based on Costamagna and Landis (2006), Costamagna et al. (2007a, b, c; 2008), and Gardiner and Landis (2007).

their competitors—appears to be limiting the impacts of the parasitoid community (Costamagna and Landis 2006).

Intraguild Predation

In subsequent studies conducted at the KBS LTER Biodiversity Gradient Experiment (Robertson and Hamilton 2015, Chapter 1 in this volume) and other locales, Costamagna et al. (2008) explored the role of intraguild predation between generalist predators and parasitoids. In addition, they examined the potential for the community of natural enemies to cause a trophic cascade (Costamagna et al. 2007a). They used selective exclusion cages that allowed the exploration of how the soybean aphid was impacted by parasitoids (in the absence of most predators) and by the presence of both predators and parasitoids. Results demonstrated the potential for season-long suppression of soybean aphid by the community of generalist natural enemies and a resulting trophic cascade, leading to increased soybean yield. In both studies, parasitoids alone provided statistically significant but biologically modest suppression of soybean aphid populations; they delayed peak aphid populations but not for long enough to suppress populations below their threshold for economic harm.

Predators attacked parasitoids (thereby demonstrating intraguild predation), but even when protected from predation, parasitoids were unable to provide economically significant levels of aphid control. In contrast, predators alone or in combination with parasitoids were capable of suppressing aphids below economic thresholds. Coccinellids were again identified as the key predators. These results support theoretical predictions that key predators can provide strong herbivore suppression even when they prey on species from other guilds within the natural enemy community (Costamagna et al. 2008).

Modeling Population Growth

Modeling population dynamics can be a powerful tool for exploring scenarios that may be difficult to investigate empirically. While simple models frequently suffice, more complex species-specific models may be necessary to understand certain phenomena. Costamagna et al. (2007b) used the results of predator exclusion cage experiments at the KBS LTER and other sites to develop a series of models exploring soybean aphid population growth. Using a simple model, Costamagna and Landis (2006) estimated that in the absence of natural enemies the intrinsic rate of increase for *A. glycines* was very high ($r = 0.30–0.33$), consistent with previous studies in other portions of the aphid's exotic range (Indonesia). Subsequently, Costamagna et al. (2007b) showed that *A. glycines* population growth could be more accurately simulated by incorporating an intrinsic rate of increase that declines linearly with time following soybean planting. They interpreted the decline in intrinsic growth rate as a response to declining host quality (i.e., older soybean plants may become less nutritious: a bottom-up control) that could interact with other mortality factors to play an important role in our understanding of overall aphid dynamics. For example, generalist natural enemies that continually suppress colonies of aphids may delay the growth of aphid colonies to a time when soybean growth is less suitable for their reproduction (Rutledge and O'Neil 2006). In this way, the early season impact of generalist predators becomes magnified by the later season impact of declining host quality. Finally, Matis et al. (2009) extended the specific model to more generally address the population dynamics of any organism specializing in the exploitation of ephemeral resources.

Field-Level Response to Soybean Aphid

McKeown (2003) investigated the numerical response of the four dominant coccinellids in the KBS landscape (*H. axyridis, C. septempunctata, C. maculata, and C. munda*) to the presence of the soybean aphid and alternative prey at crop interfaces. Coccinellid predators were monitored for 18 weeks during the 2001 growing season in a field near the MCSE where corn and soybean were planted in alternated blocks (Fig. 8.7). Using an array of traps deployed within the two crops and at their interfaces, they showed that each of the four species displayed a marked preference for a particular habitat. Two species, *C. septempunctata* and *H. axyridis*, were significantly more likely to be found in soybean than in corn. In contrast, *C. maculata*, which is known to feed on corn pollen, exhibited an overwhelming preference

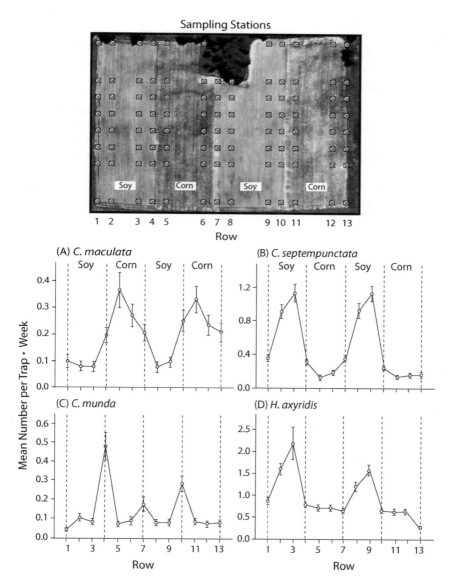

Figure 8.7. Average number of four coccinellid species captured per weekly trapping interval in soybean, corn, and interfacing areas within a large KBS field during 2001. A) *Coleomegilla maculata*, B) *Coccinella septempunctata*, C) *Cycloneda munda*, D) *Harmonia axyridis*. Aerial photo at top shows configuration of sampling stations. Figures indicate coccinellid abundance (mean ±SE) per crop habitat with rows 1, 4, 7, 10, and 13 representing edge habitats between crops; rows 2, 3, 8, and 9 habitats within soybean; and rows 5, 6, 11, and 12 habitats within corn.

for corn. Finally, *C. munda* was the only species that appeared to prefer the edge habitats. Within soybean where *A. glycines* was particularly abundant in 2001, only *C. septempunctata* and *H. axyridis* responded numerically to the presence of the aphid (McKeown 2003).

A Working Model of Pest Suppression

Through many of the above observations, KBS LTER researchers have developed a working model of how soybean aphid suppression occurs. At the time soybean aphids first arrive in soybean fields, resident predators such as anthocorids and carabids—coupled with the feeding of more transient predators such as coccinellid adults—result in the elimination of some incipient aphid colonies, and more commonly the repeated suppression of those colonies that do establish. Sustained predation pressure, in conjunction with declining host suitability later in the season, can suppress aphid population growth. This effectively reduces food resources for the subsequent generation of natural enemies and may also reduce the numbers or fitness of aphids as they overwinter. Alternatively, if there are insufficient predators, or if aphid immigration overwhelms the predators' capacity to suppress their growth, aphid colonies will grow to the point that they themselves begin to produce alates and aphid abundance in the crop field may reach outbreak levels. Such aphid outbreaks provide a nearly unlimited food source for subsequent natural enemy generations and may increase natural enemy numbers (Fig. 8.8) and their overwintering fitness (Heimpel et al. 2010).

A useful analogy is to consider the incipient aphid colonies in a field as "spot fires" and generalist natural enemies as somewhat inefficient "firefighters." The firefighters continually find these spot fires and attempt to extinguish them. Sometimes they succeed in completely eliminating a colony, but more frequently, a few aphids are left behind. Under the right conditions, these "embers" may rekindle and allow the colony to persist and grow. If colonies reach sufficient size that they themselves begin to shed "sparks" (alate aphids), the field may soon become a "wildfire" (aphid outbreak) that the predators are unable to control. Alternatively, with sufficient numbers of predators, even if individually inefficient, a predator community may be able to keep aphid numbers low for an extended period of time. This holding action delays aphid population growth into the later season when conditions become less favorable for population outbreaks to occur.

Landscape Effects on Soybean Aphid Suppression

The preceding analogy allows us to ask: What types of landscapes support a sufficient community of "firefighters" to result in effective soybean aphid suppression? Gardiner et al. (2009b) studied the impact of landscape structure on aphid-suppression services in soybean. In particular, they examined the community of mobile coccinellids that have repeatedly been shown vital to aphid suppression. Their studies demonstrate that these predators are responsive to landscape structure and that landscape diversity within 1.5 km of a soybean field is strongly related to the level of soybean aphid suppression. Landscapes with high proportions of land

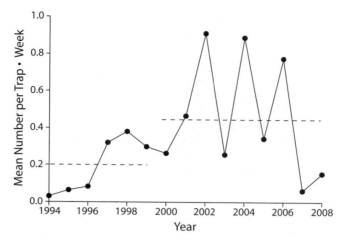

Figure 8.8. Mean number of *Harmonia axyridis* captured on yellow sticky card traps placed in multiple crop and non-crop habitats at the KBS LTER site, 1994–2008. Dotted lines show the mean numbers of *H. axyridis* from 1994 to 1999 (before soybean aphid arrival) and from 2000 to 2008 (after soybean aphid arrival). Note the response of *H. axyridis* following years of local *A. glycines* outbreaks (2001, 2003, 2005). Redrawn from Heimpel et al. (2010) with permission from Springer Science and Business Media.

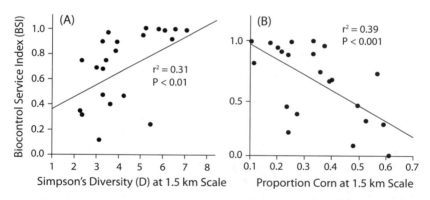

Figure 8.9. Biocontrol services from coccinellids as a function of landscape diversity (A) and the dominance of corn within 1.5 km of soybean fields (B). Panel (A) is redrawn from Gardiner et al. (2009b) with permission of the Ecological Society of America; permission conveyed through Copyright Clearance Center, Inc. Panel (B) is redrawn from Landis et al. (2008).

in corn production had low landscape diversity and significantly reduced biocontrol services in soybean fields (Fig. 8.9).

Landscape structure also altered coccinellid community structure, with the exotic *H. axyridis* more abundant in landscapes with patches of woody vegetation and native coccinellids more abundant in landscapes with abundant grasslands

(Gardiner et al. 2009a). An analysis of the value of biodiversity for aphid protection in these landscapes showed that for soybean producers using an integrated pest management strategy, natural suppression of aphids was worth ~$33 ha^{-1} in increased yield and decreased pesticide use in 2007, summing to >$239 million yr^{-1} for the four midwestern states studied (Landis et al. 2008).

KBS LTER research on soybean aphid–natural enemy interactions has yielded a number of key insights. First, results of observations and experiments support the hypothesis that communities of generalist natural enemies can provide effective herbivore suppression. At present, parasitoids are minor contributors to suppression in the soybean–aphid system, but could become more important with the importation of more effective parasitoid species (Wyckhuys et al. 2009), although intraguild predation could limit their effectiveness (Chacon et al. 2008). Second, soybean aphids serve as a food source and when abundant can support high coccinellid populations, particularly *H. axyridis*. This species, in turn, can act as an intraguild competitor, increasing its potential for negative impacts on native coccinellids (Colunga-Garcia and Gage 1998, Gardiner et al. 2011). Finally, based in part on data from KBS LTER, the soybean aphid system has been proposed as an example of "invasional meltdown" (*sensu* Ricciardi and MacIsaac 2000, Simberloff 2006), where the prior establishment of one exotic paves the way for others. Indeed, researchers investigating the soybean–soybean aphid system have documented such a cascade involving interactions among 11 Eurasian species (Heimpel et al. 2010).

Summary

Twenty years of arthropod studies at KBS LTER have yielded insights that both confirm and extend ideas about basic ecology and the ecosystem services and disservices that arthropods contribute to agroecosystems. In particular, studies of arthropods at KBS have yielded insights relevant to basic population biology, food web ecology, and invasion biology theory. Studies of the soybean aphid have contributed to our understanding of top-down vs. bottom-up forces (Costamagna and Landis 2006); intraguild predation (Costamagna and Landis 2007, Gardiner and Landis 2007, Costamagna et al. 2008); trophic cascades in food webs (Costamagna et al. 2007a); and landscape control on herbivore–natural enemy interactions (Landis et al. 2008; Gardiner et al. 2009a, b). Population modelers have also used the soybean aphid system to elucidate a novel formulation of exponential growth based on cumulative density-dependent feedback (Costamagna et al. 2007, Matis et al. 2009). Moreover, they suggest key ways in which such systems may be designed to enhance desirable ecosystem services in the future.

One of the earliest and perhaps most fundamental lessons learned is that cropping systems form the proximate template on which pest and natural enemy interactions play out. The distribution of carabid beetles is sensitive to soil disturbance (e.g., till vs. no-till) and crop persistence (e.g., annual vs. perennial), and this affects their potential importance as both insect and weed seed predators (Clark et al. 1997, Menalled et al. 2007). Similarly, some natural enemies appear to have innate preferences for specific crop types (e.g., *C. maculata* in corn), which influence their ability to provide predation services to other parts of the agricultural landscape. Spatial configuration of different crops within a local area also influences the distribution of insect predators, with coccinellids showing predictable patterns of movement from one crop to another throughout the season.

Taken as a whole, these findings suggest that at the farm scale growers have significant capabilities to manage their local landscape to promote more effective pest control services (Bianchi et al. 2006, Landis et al. 2000). However, landscape context also matters. Though landscape management is beyond the control of most farmers, except those with exceptionally large land holdings, it determines the regional pool of natural enemies that are present to move through individual fields (Gardiner et al. 2009a) and has critical implications for pest suppression (Gardiner et al. 2009b) and even crop profitability (Landis et al. 2008). Finally, the arrival of new exotic organisms has been a regular occurrence at KBS (Fig. 8.5) and has resulted in major shifts in plant productivity (Kosola et al. 2001), plant defense (Kosola et al. 2006), native insect communities (Colunga-Garcia and Gage 1998), and pest management (Costamagna and Landis 2006, 2007, Costamagna et al. 2007a, Costamagna et al. 2008).

Much remains to be learned about arthropod biodiversity and pest suppression in agricultural landscapes. Continuing studies at KBS LTER focus on understanding the role of transient generalist predators in regulating population levels of key herbivores (Woltz and Landis 2013) and the impact of landscape structure and keystone invaders such as *R. cathartica* in shaping these interactions. One of the most important forces poised to affect future agricultural landscapes is the creation of cellulosic biofuel cropping systems (Robertson et al. 2008). Future research at KBS LTER and elsewhere is needed to reveal how habitat type affects pest suppression services, and could provide a strong rationale for increasing landscape diversity via biofuel crop choice (Meehan et al. 2011).

In conclusion, arthropod predators and parasitoids play critical roles in regulating herbivore abundance and damage in agricultural systems. Likewise, there is evidence that arthropod seed predators may also influence weed population dynamics in row-crop systems. Work at KBS LTER has elucidated the relative influence of crop management and farm- and landscape-scale spatial heterogeneity on the ability of arthropod natural enemies to provide pest suppression services and thus reduce grower reliance on chemical pesticides. Case studies of carabid beetles, coccinellids, and soybean aphids illustrate a dynamic agricultural landscape where the arrival of exotic organisms—both herbivores and natural enemies—has had a major impact on insect dynamics and ecosystem performance. Moreover, they suggest that pest suppression services are influenced by features of both the crop production system and the broader landscape in which the crop is grown. Collectively, KBS LTER studies suggest that there is significant potential to understand and even design future agroecosystems to take better advantage of pest suppression services.

References

Agrawal, A. A., K. R. Kosola, and D. Parry. 2002. Gypsy moth defoliation and N fertilization affect hybrid poplar regeneration following coppicing. Canadian Journal of Forest Research 32:1491–1495.

Altieri, M. A. 1994. Biodiversity and pest management in agroecosystems. Hayworth Press, Binghamton, New York, USA.

Bahlai, C. A., M. Colunga-Garcia, S. H. Gage, and D. A. Landis. 2013. Long term community dynamics of aphidophagous coccinellids in response to repeated invasion in a diverse agricultural landscape. PLoS One. DOI: 10.1371/journal.pone.0083407.

Bahlai, C. A., M. Colunga-Garcia, S. H. Gage, and D. A. Landis. 2014. The role of exotic species in the decline of native ladybeetle populations: evidence from long-term monitoring. Biological Invasions. DOI: 10.1007/s10530-014-0772-4

Bianchi, F., C. J. H. Booij, and T. Tscharntke. 2006. Sustainable pest regulation in agricultural landscapes: a review on landscape composition, biodiversity, and natural pest control. Proceedings of the Royal Society B 273:1715–1727.

Bruno, J. F., and B. J. Cardinale. 2008. Cascading effects of predator richness. Frontiers in Ecology and the Environment 6:539–546.

Carroll, C. R., J. H. Vandermeer, and P. Rosset, editors. 1990. Agroecology. McGraw-Hill Publishing, New York, New York, USA.

Chacon, J. M., D. A. Landis, and G. E. Heimpel. 2008. Potential for biotic interference of a classical biological control agent of the soybean aphid. Biological Control 46:216–225.

Clark, M. S., S. H. Gage, and J. R. Spence. 1997. Habitats and management associated with common ground beetles (Coleoptera: Carabidae) in a Michigan agricultural landscape. Environmental Entomology 26:519–527.

Colunga-Garcia, M. 1996. Interactions between landscape structure and ladybird beetles (Coleoptera: Coccinellidae) in field crop agroecosystems. Dissertation, Michigan State University, East Lansing, Michigan, USA.

Colunga-Garcia, M., and S. H. Gage. 1998. Arrival, establishment, and habitat use of the multicolored Asian lady beetle (Coleoptera: Coccinellidae) in a Michigan landscape. Environmental Entomology 27:1574–1580.

Colunga-Garcia, M., S. H. Gage, and D. A. Landis. 1997. The response of an assemblage of Coccinellidae (Coleoptera) to a diverse agricultural landscape. Environmental Entomology 26:797–804.

Costamagna, A. C., and D. A. Landis. 2006. Predators exert top-down control of soybean aphid across a gradient of agricultural management systems. Ecological Applications 16:1619–1628.

Costamagna, A. C., and D. A. Landis. 2007. Quantifying predation on soybean aphid through direct field observations. Biological Control 42:16–24.

Costamagna, A. C., D. A. Landis, and M. J. Brewer. 2008. The role of natural enemy guilds in *Aphis glycines* suppression. Biological Control 45:368–379.

Costamagna, A. C., D. A. Landis, and C. D. DiFonzo. 2007a. Suppression of soybean aphid by generalist predators results in a trophic cascade in soybeans. Ecological Applications 17:441–451.

Costamagna, A. C., W. van der Werf, F. J. J. A. Bianchi, and D. A. Landis. 2007b. An exponential growth model with decreasing *r* captures bottom-up effects on the population growth *Aphis glycines* Matsumura (Hemiptera: Aphididae). Agriculture and Forest Entomology 9:297–305.

DiFonzo, C. D. 2006. Multi-state soybean aphid RAMP project: soybean aphid IPM on a landscape scale (SAILS). <http://www.soybeans.umn.edu/pdfs/2006aphid/RAMPNewsJune06.pdf>

Elton, C. S. 1958. Ecology of invasions by animals and plants. Methuen, London, UK.

Ferry, N., and A. M. R. Gatehouse, editors. 2009. Environmental impact of genetically modified crops. CABI, Wallingford, UK.

Fox, T. B., D. A. Landis, F. F. Cardoso, and C. D. DiFonzo. 2004. Predators suppress *Aphis glycines* Matsumura population growth in soybean. Environmental Entomology 33:608–618.

Fox, T. B., D. A. Landis, F. F. Cardoso, and C. D. DiFonzo. 2005. Impact of predation on establishment of the soybean aphid *Aphis glycines* Matsumura in soybean, *Glycine max* L. BioControl 50:545–563.

Gardiner, M. M., and D. A. Landis. 2007. Impact of intraguild predation by *Harmonia axyridis* (Coleoptera: Coccinellidae) on *Aphis glycines* (Hemiptera: Aphididae). Agricultural and Forest Entomology 9:297–305.

Gardiner, M. M., D. A. Landis, C. Gratton, C. D. DiFonzo, M. O'Neal, J. M. Chacon, M. T. Wayo, N. P. Schmidt, E. E. Mueller, and G. E. Heimpel. 2009b. Landscape diversity enhances the biological control of an introduced crop pest in the north-central USA. Ecological Applications 19:143–154.

Gardiner, M. M., D. A. Landis, C. Gratton, N. Schmidt, M. O'Neal, E. Mueller, J. Chacon, G. E. Heimpel, and C. D. DiFonzo. 2009a. Landscape composition influences patterns of native and exotic lady beetle abundance. Diversity and Distributions 15:554–564.

Gardiner, M. M., M. E. O'Neal, and D. A. Landis. 2011. Intraguild predation and native lady beetle decline PLoS ONE 6:e23576.

Gardiner, M. M., and G. L. Parsons. 2005. *Hippodamia variegata* (Goeze) (Coleoptera: Coccinellidae) detected in Michigan soybean fields. Great Lakes Entomologist 38:164–169.

Gardiner, M. M., J. K. Tuell, R. Isaacs, J. Gibbs, J. S. Ascher, and D. A. Landis. 2010. Implications of three biofuel crops for beneficial arthropods in agricultural landscapes. BioEnergy Research 3:6–19.

Gliessman, S. R., editor. 1998. Agroecology: ecological processes in sustainable agriculture. Ann Arbor Press, Chelsea, Michigan, USA.

Gross, K. L., S. Emery, A. S. Davis, R. G. Smith, and T. M. P. Robinson. 2015. Plant community dynamics in agricultural and successional fields. Pages 158–187 in S. K. Hamilton, J. E. Doll, and G. P. Robertson, editors. The ecology of agricultural ecosystems: long-term research on the path to sustainability. Oxford University Press, New York, New York, USA.

Heimpel, G. E., L. E. Frelich, D. A. Landis, K. R. Hopper, K. A. Hoelmer, Z. Sezen, M. K. Asplen, and K. Wu. 2010. European buckthorn and Asian soybean aphid as components of an extensive invasional meltdown in North America. Biological Invasions 12:2913–2931.

Isard, S. A., and S. H. Gage. 2001. Flow of life in the atmosphere: an airscape approach to understanding invasive organisms. Michigan State University Press, East Lansing, Michigan, USA.

Ives, A. R., B. J. Cardinale, and W. E. Snyder. 2005. A synthesis of subdisciplines: predator-prey interactions, and biodiversity and ecosystem functioning. Ecology Letters 8:102–116.

Kaiser, M. E., T. Noma, M. J. Brewer, K. S. Pike, J. R. Vockeroth, and S. D. Gaimari. 2007. Hymenopteran parasitoids and dipteran predators found utilizing soybean aphid after its Midwestern United States invasion. Annals of the Entomological Society of America 100:196–205.

Knapp, A. K., M. D. Smith, S. E. Hobbie, S. L. Collins, T. J. Fahey, G. Hansen, D. A. Landis, J. M. Mellilo, T. R. Seastadt, G. R. Shaver, and J. R. Webster. 2012. Past, present, and future roles of long-term experiments in the LTER network. BioScience 62:377–389.

Koch, R. L., and T. L. Galvan. 2008. Bad side of a good beetle: the North American experience with *Harmonia axyridis*. BioControl 53:23–35.

Kosola, K. R., D. I. Dickmann, E. A. Paul, and D. Parry. 2001. Repeated insect defoliation effects on growth, nitrogen acquisition, carbohydrates and root demography of poplars. Oecologia 129:65–74.

Kosola, K. R., D. Parry, and B. A. Workmaster. 2006. Response of condensed tannins in poplar roots to fertilization and gypsy moth defoliation. Tree Physiology 26:1607–1611.

Landis, D. A., M. M. Gardiner, W. van der Werf, and S. M. Swinton. 2008. Increasing corn for biofuel production reduces biocontrol services in agricultural landscapes. Proceedings of the National Academy of Sciences USA 105:20552–20557.

Landis, D. A., S. D. Wratten, and G. M. Gurr. 2000. Habitat management to conserve natural enemies of arthropod pests in agriculture. Annual Review of Entomology 45:175–201.

Letourneau, D. K., J. A. Jedlicka, S. G. Bothwell, and C. R. Moreno. 2009. Effects of natural enemy biodiversity on the suppression of arthropod herbivores in terrestrial ecosystems. Annual Review of Ecology, Evolution, and Systematics 40:573–592.

Loreau, M., S. Naeem, and P. Inchausti, editors. 2002. Biodiversity and ecosystem functioning: synthesis and perspectives. Oxford University Press, New York, New York, USA.

Lowrance, R., B. R. Stinner, and G. J. House, editors. 1984. Agricultural ecosystems: unifying concepts. John Wiley & Sons, New York, New York, USA.

Lundgren, J. G. 2009. Relationships of natural enemies and non-prey foods. Springer, Dordrecht, The Netherlands.

Macarthur, R. 1955. Fluctuations of animal populations, and a measure of community stability. Ecology 36:533–536.

Maredia, K. M., S. H. Gage, D. A. Landis, and J. M. Scriber. 1992c. Habitat use patterns by the seven-spotted lady beetle (Coleoptera: Coccinellidae) in a diverse agricultural landscape. Biological Control 2:159–165.

Maredia, K. M., S. H. Gage, D. A. Landis, and T. M. Wirth. 1992a. Visual response of *Coccinella septempunctata* (L.), *Hippodamia parenthesis* (Say), (Coleoptera: Coccinellidae), and *Chrysoperla carnea* (Stephens), (Neuroptera: Chrysopidae) to colors. Biological Control 2:253–256.

Maredia, K. M., S. H. Gage, D. L. Landis, and T. M. Wirth. 1992b. Ecological observations on predatory Coccinellidae (Coleoptera) in southwestern Michigan. Great Lakes Entomologist 25:265–270.

Matis, J. H., T. R. Kiffe, W. van der Werf, A. C. Costamagna, T. J. Matis, and W. E. Grant. 2009. Population dynamics models based on cumulative density dependent feedback: a link to the logistic growth curve and a test for symmetry using aphid data. Ecological Modelling 220:1745–1751.

May, R. M. 1973. Stability and complexity in model ecosystems. Princeton University Press, Princeton, New Jersey, USA.

McKeown, C. H. 2003. Quantifying the roles of competition and niche separation in native and exotic cocconellids, and the changes in the community in response to an exotic prey species. Thesis, Michigan State University, East Lansing, Michigan, USA.

Meehan, T. D., B. P. Werling, D. A. Landis, and C. Gratton. 2011. Agricultural landscape simplification and insecticide use in the Midwestern United States. Proceedings of the National Academy of Sciences USA 108:11500–11505.

Menalled, F. D., K. L. Gross, and M. Hammond. 2001. Weed aboveground and seedbank community responses to agricultural management systems. Ecological Applications 11:1586–1601.

Menalled, F. D., P. C. Marino, K. A. Renner, and D. A. Landis. 2000. Post-dispersal weed seed predation in Michigan crop fields as a function of agricultural landscape structure. Agriculture, Ecosystems & Environment 77:193–202.

Menalled, F. D., R. G. Smith, J. T. Dauer, and T. B. Fox. 2007. Impact of agricultural management systems on carabid beetle communities and weed seed predation. Agriculture, Ecosystems & Environment 118:49–54.

Noma, T., and M. J. Brewer. 2007. Fungal pathogens infecting soybean aphid and aphids on other crops grown in soybean production areas of Michigan. Great Lakes Entomologist 40:41–49.

Noma, T., and M. J. Brewer. 2008. Seasonal abundance of resident parasitoids and predatory flies and corresponding soybean aphid density, with comments on classical biological control of soybean aphid in the Midwest. Journal of Economic Entomology 101:278–287.

Obrycki, J. J., J. D. Harwood, T. J. Kring, and R. J. O'Neil. 2009. Aphidophagy by Coccinellidae: application of biological control in agroecosystems. Biological Control 51:244–254.

Odum, E. P. 1959. Fundamentals of ecology. Saunders, Philadelphia, Pennsylvania, USA.

Ostrom, P. H., M. Colunga, and S. H. Gage. 1997. Establishing pathways of energy flow for insect predators using stable isotope ratios: field and laboratory evidence. Oecologia 109:108–113.

Parry, D. 2000. Induced responses of poplars to defoliation and their effects on leaf-feeding Lepidoptera. Dissertation, Michigan State University, East Lansing, Michigan, USA.

Pike, K. S., P. Stary, M. J. Brewer, T. Noma, S. Langley, and M. E. Kaiser. 2007. A new species of *Binodoxys* (Hymenoptera: Braconidae, Aphidiinae), parasitoid of the soybean aphid, *Aphis glycines* Matsumura, with comments on biocontrol. Proceedings of the Entomological Society of Washington 109:359–365.

Radcliffe, E. B., W. D. Hutchinson, and R. E. Cancelado, editors. 2009. Integrated pest management: concepts, tactics, strategies and case studies. Cambridge University Press, Cambridge, UK.

Ragsdale, D. W., D. A. Landis, J. Brodeur, G. E. Heimpel, and N. Desneux. 2011. Ecology and management of the soybean aphid in North America. Annual Review of Entomology 56:375–399.

Ragsdale D. W., B. P. McCornack, R. C. Venette, B. D. Potter, I. V. Macrae, E. W. Hodgson, M. E. O'Neal, K. D. Johnson, R. J. O'Neil, C. D. DiFonzo, T. E. Hunt, P. A. Glogoza, and E. M. Cullen 2007. Economic threshold for soybean aphid (Hemiptera: Aphididae). Journal of Economic Entomology 100:1258–1267.

Ragsdale, D. W., D. J. Voegtlin, and R. J. O'Neil. 2004. Soybean aphid biology in North America. Annals of the Entomological Society of America 97:204–208.

Ricciardi, A., and H. J. MacIsaac. 2000. Recent mass invasion of the North American Great Lakes by Ponto-Caspian species. Trends in Ecology & Evolution 15:62–65.

Robertson, G. P., J. C. Broome, E. A. Chornesky, J. R. Frankenberger, P. Johnson, M. Lipson, J. A. Miranowski, E. D. Owens, D. Pimentel, and L. A. Thrupp. 2004. Rethinking the vision for environmental research in US agriculture. BioScience 54:61–65.

Robertson, G. P., V. H. Dale, O. C. Doering, S. P. Hamburg, J. M. Melillo, M. M. Wander, W. J. Parton, P. R. Adler, J. N. Barney, R. M. Cruse, C. S. Duke, P. M. Fearnside, R. F. Follett, H. K. Gibbs, J. Goldemberg, D. J. Miadenoff, D. Ojima, M. W. Palmer,

A. Sharpley, L. Wallace, K. C. Weathers, J. A. Wiens, and W. W. Wilhelm. 2008. Sustainable biofuels redux. Science 322:49.

Robertson, G. P., and S. K. Hamilton. 2015. Long-term ecological research at the Kellogg Biological Station LTER Site: conceptual and experimental framework. Pages 1–32 in S. K. Hamilton, J. E. Doll, and G. P. Robertson, editors. The ecology of agricultural ecosystems: long-term research on the path to sustainability. Oxford University Press, New York, New York, USA.

Rutledge, C. E., and R. J. O'Neil. 2006. Soybean plant stage and population growth of soybean aphid. Journal of Economic Entomology 99:60–66.

Rutledge, C. E., R. J. O'Neil, T. B. Fox, and D. A. Landis. 2004. Soybean aphid predators and their use in Integrated Pest Management. Annals of the Entomological Society of America 97:240–248.

Ryan, J. M. 1990. Facultative dispersal of the apple maggot, *Rhagoletis pomonella* (Walsh) (Diptera: Tephritidae). Thesis, Michigan State University, East Lansing, Michigan, USA.

Schulze, E.-D., and H. A. Mooney, editors. 1993. Biodiversity and ecosystem function. Springer-Verlag, Berlin, Germany.

Simberloff, D. 2006. Invasional meltdown 6 years later: important phenomenon, unfortunate metaphor, or both? Ecology Letters 9:912–919.

Sirota, J. M. 1990. The temporal and spatial distribution and sex ratio of summer adults of the seven spotted ladybird beetle (*Coccinella septempunctata* L. Coleoptera: Coccinellidae) in alfalfa. Thesis, Michigan State University, East Lansing, Michigan, USA.

Stern, V. M., R. F. Smith, R. van den Bosch, and G. E. Heimpel. 1959. The integrated control concept. Hilgardia 29:81–101.

Walter, A. J., and C. D. DiFonzo. 2007. Soil potassium deficiency affects soybean phloem nitrogen and soybean aphid populations. Environmental Entomology 36:26–33.

Woltz, J.M. and D.A. Landis. 2013. Coccinellid immigration to infested host patches influences suppression of *Aphis glycines* in soybean. Biological Control. 64:330-337.

Woltz, J.M., D.A. Landis. 2014. Coccinellid response to landscape composition and configuration. Agricultural and Forest Entomology. 16:341-349.

Wyckhuys, K. A. G., R. L. Koch, R. R. Kula, and G. E. Heimpel. 2009. Potential exposure of a classical biological control agent of the soybean aphid, *Aphis glycines*, on non-target aphids in North America. Biological Invasions 11:857–871.

9

Nitrogen Transfers and Transformations in Row-Crop Ecosystems

Neville Millar and G. Philip Robertson

Nitrogen (N) is an essential constituent of both proteins and nucleic acids, crucial components of all living things. Humans depend on agricultural systems to provide most of their daily protein, prompting Liebig (1840) to note that agriculture's principal objective is the production of digestible N. Today's intensive agriculture is built on a foundation of N augmentation via the use of synthetic fertilizers and cultivation of N-fixing crops on a massive scale. Globally, the addition of this fixed N to cropping systems is now greater than natural terrestrial N fixation (Galloway et al. 2004, Vitousek et al. 2013) and is a pervasive and fundamental feature of modern crop management (Robertson and Vitousek 2009). The net benefit to humans of this additional N to agriculture is immense—it has enabled greater food production and unprecedented increases in human population (Smil 2002). At the same time, however, this anthropogenic acceleration of the global N cycle has caused serious environmental problems including contributing to climate change and stratospheric ozone depletion, eutrophication and harmful algal blooms, poor air quality, biodiversity loss, and degradation of drinking water supplies (Galloway et al. 2008).

Here, we summarize findings on the cycling of N in agricultural ecosystems of the Kellogg Biological Station Long-term Ecological Research site (KBS LTER). KBS LTER results illustrate a number of opportunities to manage agricultural N in ways that improve N conservation and reduce the escape of fixed N to the environment; they also reveal needs for future research to fill knowledge gaps and more fully document the complex cycling of N in cropping systems that range from conventional to organic and from annual to perennial.

The Agricultural Nitrogen Cycle

A comprehensive understanding of the agricultural N cycle requires knowledge of the transfers (movement) of N into and out of the soil–plant system (e.g., precipitation, gas exchange, leaching) as well as the exchange of N between compartments within the system (e.g., biological assimilation and release). Transformations in the chemical form of N, often mediated by microbes, determine the availability and mobility of N in soils and water.

Nitrogen Inputs and Outputs

New N is added to cropping systems through biological N fixation, N deposition, and the application of compost, manure, and synthetic fertilizers (Fig. 9.1). In any cropping system, to remain sustainable, the amount of N added must replace the N removed in crop yield and in environmental losses. The amount of N removed with crop harvest varies widely, largely as a function of crop species and growing conditions. In high-yielding annual grain systems, 100–270 kg N ha⁻¹ are typically removed during harvest (Robertson 1997).

In annual cropping systems of the KBS LTER Main Cropping System Experiment (MCSE; Table 9.1), average harvest N removals range from 34 kg N ha⁻¹ yr⁻¹ by winter wheat (*Triticum aestivum* L.) in the Biologically Based system to 163 kg N ha⁻¹ yr⁻¹ by soybean in the No-till system (Table 9.2). Among perennial crops, Poplar (*Populus* spp.) harvest removes only 107 kg N ha⁻¹ after 10 years of growth, or an annualized average of 10.7 kg N ha⁻¹ yr⁻¹, whereas alfalfa harvest removes 215 kg N ha⁻¹ yr⁻¹. The Early Successional system is not harvested but with each annual burn loses

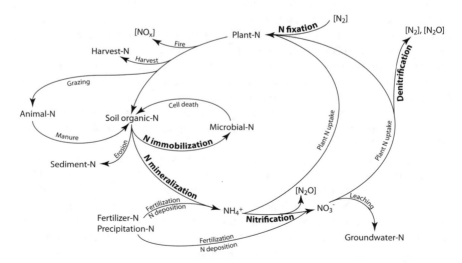

Figure 9.1. Schematic representation of the major elements of the terrestrial nitrogen (N) cycle. Processes mediated by soil microbes appear in bold, and gases appear in brackets. Redrawn from Robertson and Groffman (2015) with permission from Elsevier Limited.

Table 9.1. Description of the KBS LTER Main Cropping System Experiment (MCSE).[a]

Cropping System/Community	Dominant Growth Form	Management
Annual Cropping Systems		
Conventional (T1)	Herbaceous annual	Prevailing norm for tilled corn–soybean–winter wheat (c–s–w) rotation; standard chemical inputs, chisel-plowed, no cover crops, no manure or compost
No-till (T2)	Herbaceous annual	Prevailing norm for no-till c–s–w rotation; standard chemical inputs, permanent no-till, no cover crops, no manure or compost
Reduced Input (T3)	Herbaceous annual	Biologically based c–s–w rotation managed to reduce synthetic chemical inputs; chisel-plowed, winter cover crop of red clover or annual rye, no manure or compost
Biologically Based (T4)	Herbaceous annual	Biologically based c–s–w rotation managed without synthetic chemical inputs; chisel-plowed, mechanical weed control, winter cover crop of red clover or annual rye, no manure or compost; certified organic
Perennial Cropping Systems		
Alfalfa (T6)	Herbaceous perennial	5- to 6-year rotation with winter wheat as a 1-year break crop
Poplar (T5)	Woody perennial	Hybrid poplar trees on a ca. 10-year harvest cycle, either replanted or coppiced after harvest
Coniferous Forest (CF)	Woody perennial	Planted conifers periodically thinned
Successional and Reference Communities		
Early Successional (T7)	Herbaceous perennial	Historically tilled cropland abandoned in 1988; unmanaged but for annual spring burn to control woody species
Mown Grassland (never tilled) (T8)	Herbaceous perennial	Cleared woodlot (late 1950s) never tilled, unmanaged but for annual fall mowing to control woody species
Mid-successional (SF)	Herbaceous annual + woody perennial	Historically tilled cropland abandoned ca. 1955; unmanaged, with regrowth in transition to forest
Deciduous Forest (DF)	Woody perennial	Late successional native forest never cleared (two sites) or logged once ca. 1900 (one site); unmanaged

[a]Site codes that have been used throughout the project's history are given in parentheses. Systems T1–T7 are replicated within the LTER main site; others are replicated in the surrounding landscape. For further details, see Robertson and Hamilton (2015, Chapter 1 in this volume).

~41 kg N ha^{-1} to the atmosphere. These N removals represent the minimum amount of N that must be replaced annually to avoid N depletion—the minimum because not included are additional N losses by other pathways such as leaching and denitrification (Fig. 9.1). Although a small amount of N is added as deposition (see Nitrogen Deposition below), most N is replaced by either biological or industrial fixation of atmospheric N$_2$ or by application of organic N fertilizers such as compost or manure.

Table 9.2. Nitrogen (kg N ha^{-1} yr^{-1}) contained in crop components and removed in harvest of MCSE systems.[a]

Component	MCSE System						
	Annual Crops[b]				Perennial Crops		Successional
	Conventional	No-Till	Reduced Input	Biologically Based	Alfalfa[c] Poplar[d]		Early Successional[e]
Corn							
Grain	84	88	80	61			
Stover	22[f]	26[f]	25[f]	18[f]			
Soybean							
Grain	138	163	151	148			
Stover	21[f]	21[f]	19[f]	18[f]			
Wheat							
Grain	61	66	57	34			
Straw	9	10	10[f]	6[f]			
Harvested N[g]	97	109	96	81	215	11	41

[a]Mean, n = 6 replicated plots.
[b]Crop N in different components calculated as biomass × N content. Values are the average annual N amount in each crop component over 18 years (1993–2010), covering six 3-year rotations of the annual cropping systems.
[c]Annual N in alfalfa calculated as the sum of N harvested annually and averaged over 1993–2010 (15 years with crop present).
[d]N removed in aboveground woody biomass at clear-cut harvest of 1999 (107 kg N ha-1) divided by the number of growth years (10; 1989–1998).
[e]N in aboveground biomass volatilized by burning and calculated from aggregated postfrost aboveground biomass and surface litter.
[f]Not harvested (left on soil surface).
[g]N removed at harvest of annual crops (corn–soybean–wheat) determined as the annual sums of N removed in grain (corn, soybean, and wheat) plus harvested straw for wheat as noted, averaged over 18 years (1993–2010) covering six 3-year rotations. N removed in wheat straw is based on the straw N content and the ratio of grain to aboveground biomass at peak biomass.

Nitrogen Fertilizers

In 1900 agriculture used very little inorganic N fertilizer, with less than 0.5 Tg of N applied to crops worldwide, mainly as nitrate from Chile and as ammonium sulfate derived from coke-oven gas. The Haber–Bosch process—the production of ammonia from its constituent elements N_2 and H_2—was industrialized by 1913 in Germany (Leigh 2004). Global ammonia production was ~2.4 Tg yr^{-1} N in 1946 and today is ~133 Tg yr^{-1}, most of which is used to make N fertilizers (Kramer 2004).

Synthetic or industrially fixed N differs from organic N sources in that most synthetic N is immediately available for plant uptake, whereas most organic N must first be mineralized to NH_4^+ and then nitrified to NO_3^- (see Internal Nitrogen Transformations below) before it is available to plants (Robertson and Vitousek 2009). In the United States, synthetic fertilizer accounts for ~60% of the total N added to agricultural land; legume N fixation and manure make up most of the rest (IPNI 2012). In North America, most synthetic fertilizer N is applied as anhydrous ammonia (27% of inputs), urea ammonium nitrate (24%), or urea (23%) (IFA 2011).

Synthetic fertilizers used in the MCSE include different combinations of ammonium nitrate (applied in N–P–K formulations) and urea ammonium nitrate (UAN) (Table 9.3). Rates of N application in the Conventional and No-till systems have ranged from 112–163 kg N ha^{-1} yr^{-1} for corn (*Zea mays* L.) and from 56–90 kg N ha^{-1} yr^{-1} for winter wheat. The Reduced Input system receives ~one-third of the synthetic inputs of the Conventional system (Table 9.3). No synthetic N fertilizer is applied to the Biologically Based system or to soybean and Alfalfa (*Medicago sativa* L.), and no system receives manure or other organic N forms. Poplars receive between ~120 to 160 kg N ha^{-1} at or shortly after planting.

Decisions about N fertilizer rates in MCSE annual crops are guided by MSU Extension recommendations as well as past practice and best judgment. Prior to 2008, Extension recommendations for the Conventional and No-till corn and wheat rotations were based on the yield goal approach (Warncke et al. 2004), which calculates base rates from past yields and projected yield increases. Since 2008 recommendations for corn have been based on the Maximum Return to Nitrogen approach (MRTN; Warncke et al. 2009), now used by seven U.S. Corn Belt states including Michigan (ISU 2004). For MRTN, N is applied on the basis of statewide N response trials weighted by the price of fertilizer and corn to provide an economically optimized N rate. In 2011 we applied 156 kg N ha^{-1} to the Conventional and No-till corn in the MCSE. This is very close to the Economic Optimum Nitrogen Rate (EONR) of 155 kg N ha^{-1} based on 5-year average corn yields for different N-fertilizer levels in the adjacent Resource Gradient Experiment (Fig. 9.2; Robertson and Hamilton 2015, Chapter 1 in this volume). For wheat, a yield goal approach still guides N rate recommendations; in 2010 we applied 89 kg N ha^{-1} to wheat, which exceeded the EONR rate of 68 kg N ha^{-1} based on the average yields (2007 and 2010) for wheat in the Resource Gradient Experiment (Fig. 9.2). The EONR is always less than the agronomic maximum nitrogen rate, the rate at which agronomic yields are maximized (Fig 9.2), because at some point the cost of additional N fertilizer is greater than the income provided by more yield.

Over the 1993 to 2010 period, corn in the Conventional and No-till systems received 141 kg N ha^{-1} yr^{-1}, on average. This value is close to the statewide and national averages of 134 and 149 kg N ha^{-1} yr^{-1} for the same period (NASS 2014). For wheat over this period, we added 76 kg N ha^{-1} yr^{-1}, on average, ~29% lower than the average N rate applied to wheat in Michigan (107 kg N ha^{-1} yr^{-1}) between 2004 and 2009, but very close to the national average (75 kg N ha^{-1} yr^{-1}) for the same period (NASS 2014).

Nitrogen Fixation

Approximately 78% of the atmosphere is composed of dinitrogen gas (N$_2$), which is unusable by most organisms because of the strong triple bond between two

Table 9.3. Nitrogen fertilizer application rates (kg N ha^{-1} yr^{-1}) and cover crops in the MCSE annual cropping systems (1989–2010).

Year	Crop(s)[a]	Conventional	No-till	Reduced Input	Biologically Based
		Rate (Formulation; Timing)[b]		Rate (Formulation; Timing)[b] + Cover Crop[c]	Cover Crop[c]
1989	Corn[1] or wheat[2]	123 (34-0-0; Sd)	123 (34-0-0; Sd)	28 (34-0-0) + RC	RC
1990	Soybean[1] or corn[2]	0	0	28 (34-0-0) + CC	CC
1991	Corn[1] or soybean[2]	123 (34-0-0; Sd)	123 (34-0-0; Sd)	0	0
1992	Soybean[1] or wheat[2]	0	0	28 (34-0-0) + RC	RC
1993	Corn	28 (UAN; Pl) 84 (34-0-0; Sd)	28 (UAN; Pl) 84 (34-0-0; Sd)	28 (UAN; Pl) + CC	CC
1994	Soybean	0	0	0	0
1995	Wheat	56 (34-0-0)	56 (34-0-0)	28 (34-0-0) + RC	RC
1996	Corn	28 (UAN; Pl) 135 (34-0-0; Sd)	28 (UAN; Pl) 135 (34-0-0; Sd)	28 (UAN; Pl) + CC	CC
1997	Soybean	0	0	0	0
1998	Wheat	56/(34-0-0)	56 (34-0-0)	28 (34-0-0) + RC	RC
1999	Corn	28 (UAN; Pl) 87 (34-0-0; Sd)	28 (UAN; Pl) 87 (34-0-0; Sd)	28 (UAN; Pl) + CC	CC
2000	Soybean	0	0	0	0
2001	Wheat	78 (UAN; split)	78 (UAN; split)	47 (UAN) + RC	RC
2002	Corn	30 (19-17-0; Pl) 125 (UAN; Sd)	30 (19-17-0; Pl) 125 (UAN; Sd)	30 (19-17-0; Pl) + RC	RC
2003	Soybean	0	0	0	0
2004	Wheat	36 (19-19-19; Pl) 54 (UAN; Sd)	36 (19-19-19; Pl) 54 (UAN; Sd)	54 (UAN; Sd) + RC	RC
2005	Corn	34 (19-17-0)[Pl] 123 (UAN)[Sd]	34 (19-17-0)[Pl] 123 (UAN)[Sd]	34 (19-17-0; Pl) + RC	RC
2006	Soybean	0	0	0	0
2007	Wheat	35 (19-19-19; Pl) 54 (UAN; Sd)	35 (19-17-0; Pl) 54 (UAN; Sd)	54 (UAN; Sd) + RC	RC
2008	Corn	33 (19-17-0; Pl) 113 (UAN; Sd)	33 (19-17-0; Pl) 113 (UAN; Sd)	33 (19-17-0; Pl) + RC	RC
2009	Soybean	0	0	0	0
2010	Wheat	89 (UAN; split)	89 (UAN; split)	50 (UAN) + RC	RC

[a]From 1989 to 1992, different crops were grown in the Conventional and No-till systems (denoted as superscript 1) than in the Reduced Input and Biologically Based systems (denoted as superscript 2).

[b]Nitrogen was applied as ammonium nitrate (N–P–K content: 34-0-0, or 19-17-0, or 19-19-19) and urea ammonium nitrate (UAN; 28% N) at planting (Pl) and/or as a side-dressing (Sd). For wheat, a split application of UAN was applied in some years.

[c]Cover crops were red clover (RC, *Trifolium pratense* L.) and crimson clover (CC, *Trifolium incarnatum* L.) which are grown prior to corn and plowed under before planting in spring. Wilke (2010; Table 14) estimated aboveground N in red clover at spring harvest as 67 and 59 kg N ha^{-1} in the Reduced Input and Biologically Based systems, respectively. No data available for crimson clover.

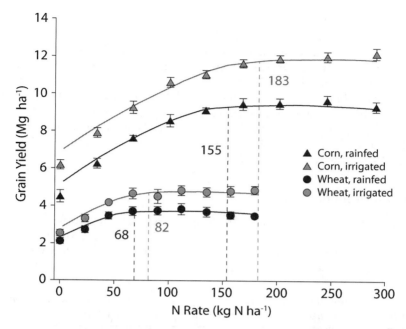

Figure 9.2. Yield responses to incremental increases in synthetic N fertilization rate in corn and winter wheat at the KBS LTER Resource Gradient Experiment. Economic Optimum Nitrogen Rates (EONR; dashed lines) were calculated as the weighted mean of four yield response curves (quadratic, quadratic-plateau, linear-plateau, and spherical) using the International Plant Nutrition Institute Crop Nutrient Response Tool (CNRT) v4.5; http://nane.ipni.net/article/NANE-3068). The EONRs for rainfed and irrigated corn are based on average yields for the 2003, 2004, 2005, 2008, and 2011 crop years, and EONRs for rainfed and irrigated wheat are based on average yields for the 2007 and 2010 crop years. The agronomic optimum nitrogen rates (AONR) were 191 and 198 kg N ha^{-1} for rainfed and irrigated corn, respectively, and for rainfed and irrigated wheat, 94 and 108 kg ha^{-1}, respectively (not shown for clarity).

N atoms. Biological N fixation (BNF) is the dominant process by which this N becomes reactive:

$$N_2 + 8\ H^+ + 6\ e^- \rightarrow 2\ NH_3 + H_2$$

Most BNF is carried out by bacteria possessing the nitrogenase enzyme, and in terrestrial ecosystems most BNF is plant-associated (Vitousek et al. 2013).

Plant-mediated N fixation is largely facultative: where soil N is readily available, plants allocate less fixed carbon (C) to BNF bacteria and less N is fixed. Thus, fixation rates can vary widely even within the same crop species. Soybeans when N-fertilized, for example, fix very little N, but when grown in infertile soil without added N, they can fix 98% of a 200 kg ha^{-1} N requirement (Ruschel et al. 1979). In the MCSE Conventional system, soybeans fix 56–58% of their aboveground N needs based on ^{15}N natural abundance experiments (Gelfand and Robertson, in press), or 92 kg N ha^{-1} yr^{-1} (Table 9.4). This percentage agrees closely with recent reviews (Herridge et al. 2008, Salvagiotti et al. 2008).

Cover crops are a second major source of BNF in the MCSE Reduced Input and Biologically Based systems (Table 9.4). Wilke (2010), based on ^{15}N natural abundance, estimated that BNF in red clover accounted for 55 and 72% of its total aboveground N in these respective systems. BNF in the successional communities of the MCSE has not been measured but based on legume biomass, and N content is likely ~10 kg N ha^{-1} yr^{-1} in the Early Successional community and negligible in the Mid-successional system. No more than 7 kg N ha^{-1} yr^{-1} is likely fixed in the mature Deciduous Forest system (Cleveland et al. 1999).

Nitrogen Deposition

Nitrogen oxides (NO and NO$_2$, referred to as NO$_x$) are emitted from industrial combustion and other sources, including agricultural soils. Oxidation of NO$_x$ results in increased atmospheric N deposition to ecosystems, which in turn causes higher regional watershed N fluxes (e.g., Jaworski et al. 1997), increased soil N$_2$O

Table 9.4. Biological N fixation (BNF) by legumes, aboveground N, and percentage of aboveground N derived from BNF in annually harvested MCSE systems.[a]

Variable	Annual Crops				Perennial Crops[d]
	Conventional	No-Till	Reduced Input	Biologically Based	Alfalfa
Legume aboveground N (kg N ha^{-1} yr^{-1})[b]					
Soybean	159	184	170	166	
Red clover			56	63	
Alfalfa					215
Total aboveground N (kg N ha^{-1})[c]	342	383	398	341	215
BNF (kg N ha^{-1} yr^{-1})[b]					
Soybean[e]	92	107	99	96	
Red clover[f]			31	45	
Alfalfa[g]					157
Aboveground N met by BNF (%)	27	28	33	41	73

[a]Mean, n = 6 replicated plots (SE not shown).
[b]N in legume aboveground biomass and N from BNF per year (de facto per 3-year rotation for soybean and red clover) for each N-fixing crop (as described below) averaged over six rotations (1993–2010).
[c]N in grain and stover/straw of corn, soybean, and wheat plus N in harvests of the cover crops red clover (1995–2008) and cereal rye (2006–2009) per 3-year rotation averaged over six rotations (1993–2010).
[d]For alfalfa, legume aboveground N, total aboveground N, and N from BNF determined per year averaged over 1993–2010 (15 years with crop present).
[e]BNF of soybean estimated as 58% of the total aboveground N uptake (Gelfand and Robertson, in press; Salvagiotti et al. 2008).
[f]BNF of red clover estimated from the percentage of BNF-derived N (55 and 72% for Reduced Input and Biologically Based systems, respectively; average of fall 2007 and spring 2008 harvest data from Wilke 2010; Figure 9).
[g]BNF of alfalfa estimated as 73% of the total aboveground N uptake (Yang et al. 2010).

emissions (Butterbach-Bahl et al. 2002, Ambus and Robertson 2006), and changes in species composition (Bobbink et al. 1998, 2010). Chronic increases in atmospheric N deposition have been shown to decrease plant species diversity even at moderate levels (e.g., 10 kg N ha^{-1} yr^{-1}; Stevens et al. 2004, Clark and Tilman 2008). In agricultural systems, N deposition could be considered a free fertilizer input and, in locations where deposition is high, may need to be considered when making fertilizer N recommendations.

Between 1989 and 2010, KBS received, on average, 6.3 ± 1.3 kg N ha^{-1} yr^{-1} via wet precipitation (NH$_4$ + NO$_3$; NADP/NTN 2011). At sites in Michigan, dry deposition of N (HNO$_3$ + NO$_3$ + NH$_4$) between 1989 and 2010 was ~one-third that of wet N deposition (CASTNET 2011). Using this ratio, an annual average total N deposition (wet and dry) at KBS during this period was ~8.4 kg N ha^{-1} yr^{-1}. This estimate is lower than the earlier estimate of 14.2 kg N ha^{-1} yr^{-1} by Rheaume (1990) for Kalamazoo County between 1986–1987. The declining trend for wet deposition of NO$_3^-$ at KBS (Fig. 9.3) is likely due to the drop in NO$_x$ emissions following adoption of air-quality regulations and emission-control technologies (Greaver et al. 2012; Hamilton 2015, Chapter 11 in this volume).

Nitrogen fertilizer rates for wheat and corn at the MCSE are ~10 to 15 times higher than atmospheric N deposition levels, so N deposition has little impact in these systems. In the successional and forested systems, however, the present rate of atmospheric N deposition represents a 2- to 3-fold increase over preindustrial levels and likely has profound ecological effects in these N-limited ecosystems (Galloway et al. 2004, Dentener et al. 2006).

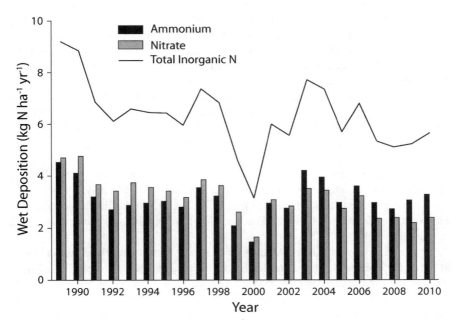

Figure 9.3. Wet nitrogen (NH$_4$-N, NO$_3$-N) and total wet nitrogen ([NH$_4$ + NO$_3$]-N) deposition (kg N ha^{-1} yr^{-1}) at KBS for 1989–2010. Data from NADP/NTN (2011).

Internal Nitrogen Transformations

Microbial activity mediates the major N transformations within the ecosystem, including mineralization, immobilization, nitrification, and denitrification (Fig. 9.1).

Mineralization and Immobilization

Nitrogen mineralization is the conversion of organic N to soluble inorganic forms that can be taken up by plants and other microbes (Robertson and Groffman 2015). If plant residues are rich in N, microbes release inorganic N in excess of their needs to the soil solution. Nitrogen immobilization is the uptake of N by soil microbes—the reverse of mineralization. If plant residues are low in N, microbes scavenge additional soluble N from their surroundings, immobilizing it in microbial biomass. In soil, both processes occur simultaneously, and the balance is known as net mineralization. When net mineralization is positive, inorganic N is added to the soil solution; when net mineralization is negative, inorganic N is removed from the soil solution.

At the MCSE, net N mineralization over the growing season (April to October) is greater in the Reduced Input and Biologically Based systems (178 and 163 kg N ha^{-1}) than in the Conventional and No-till systems (99 and 113 kg N ha^{-1}), respectively (Table 9.5; Robertson et al. 2000). Higher rates reflect the build-up of mineralizable N and soil organic matter (Syswerda et al. 2011) in these cover-cropped systems and, in particular, the importance of leguminous cover crops in providing inorganic N to the subsequent primary crop (Sánchez et al. 2001). The cover crop effect is also seen in the N content of soils—inorganic N concentrations are about as high in the Reduced Input and Biologically Based systems as they are in the fully fertilized Conventional and No-till systems (Fig. 9.4). Moreover, N appears to be more available, especially as nitrate, from June through August (Fig. 9.5), the portion of the growing season when plant uptake is greatest.

Among the perennial crops, net N mineralization between April and October (Table 9.5) is slightly higher in Alfalfa (192 kg N ha^{-1}) than in the annual cropping systems with legume cover crops, but considerably lower in Poplar (62 kg N ha^{-1}). Low N mineralization in Poplar may reflect this system's lower quality (higher C:N ratio) leaf litter, and is associated with low soil inorganic N pools (Figs. 9.4 and 9.5). Among the successional systems, net N mineralization generally followed the pattern Deciduous Forest > Mid-successional > Early Successional > Mown Grassland (never tilled) (Table 9.5). Net N immobilization appears to only occur in the Mown Grassland (never tilled) system and only during July and August (Fig. 9.6).

Total soil N contents of the A/Ap horizon of the Mown Grassland (never tilled) system (5.95 Mg N ha^{-1}) and the late successional Deciduous Forest (5.33 Mg N ha^{-1}) represent indigenous (precultivation) soil N, and are about 50–65% higher than total soil N in the annual cropping systems (~3.6 Mg N ha^{-1}; Table 9.5). This difference reflects the loss of soil organic N due to a century or more of cultivation, which accelerated the mineralization of soil organic matter (Paul et al. 2015, Chapter 5 in this volume) and subsequent N loss by various pathways—including

Table 9.5. Mean total soil N content, relative nitrification rate, and net N mineralization rate during the growing season in the A/Ap horizon of MCSE soils.[a]

System	Total N[b]		Relative Nitrification[c] (%)	Net N Mineralization Rate[d] (kg ha^{-1} season^{-1})
	(g kg soil^{-1})	(Mg ha^{-1})		
Annual Cropping Systems				
Conventional	1.12	3.58	79	99
No-till	1.21	3.63	76	113
Reduced Input	1.24	3.72	80	178
Biologically Based	1.17	3.51	79	163
Perennial Cropping Systems				
Poplar	1.17	3.28	32	62
Alfalfa	1.35	4.05	75	192
Coniferous Forest	Na	na	60	na
Successional and Reference Communities				
Early	1.30	3.90	42	90
Mown Grassland	2.48	5.95	28	27
Mid-successional	1.36	4.08	31	113
Deciduous Forest	2.05	5.33	72	137

[a]$n = 6$ plots for all systems, except Mown Grassland ($n = 4$) and the Mid-successional, Coniferous, and Deciduous Forest systems ($n = 3$). na = not available.
[b]Calculated from Syswerda et al. (2011).
[c]Percentage of net mineralizable N as nitrate at the end of *in situ* incubations averaged over 18 years (1993–2010), covering six 3-year rotations of the annual cropping systems (Robertson et al. 1999).
[d]Determined by extrapolation of daily net N mineralization rates (modified from Robertson et al. 2000, Table 1) over the growing season (April to October, 1989–1995).

harvest, which removes ~0.1 Mg ha^{-1} yr^{-1} from cropped systems (Table 9.2). The proportion of total N in the A/Ap horizon that is mineralized each year ranges from 0.5% (Mown Grassland never tilled system) to 4.8% (Reduced Input system; calculated from Table 9.5), which is consistent with values reported for cropping systems elsewhere in the U.S. North Central Region (Cassman et al. 2002).

Temporal and spatial patterns in rates of net N mineralization with respect to crop N requirements largely determine the degree of N synchrony within the system (Robertson 1997). Where N mineralization is asynchronous with plant growth—as might happen, for example, when N mineralization occurs largely in the spring prior to crop growth or in the fall after plant senescence—the mineralized N will be susceptible to loss. Such asynchrony is unfortunately a normal situation for annual row-crop production in temperate climates. Managing the system to maximize synchrony is an important strategy for conserving N in cropping systems (e.g., McSwiney et al. 2010). Synthetic fertilizers owe their effectiveness to adding a large pulse of available N to soil just as crops enter their prime growth phase. Reproducing this pulse with biological management is an extraordinary challenge, although evidence from the Reduced Input and

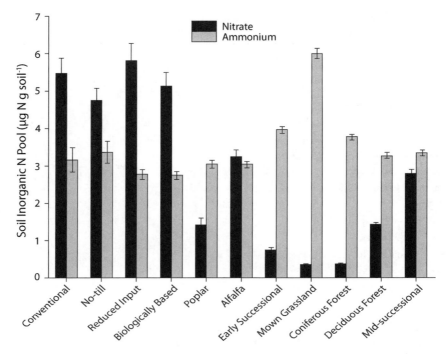

Figure 9.4. Soil inorganic N pools of nitrate and ammonium (μg N g soil^{-1}) in MCSE systems for 1993–2010. Mean ± SE for all measurements during the period.

Biologically Based systems suggests that the leguminous cover crop may provide much of this timely input (Fig. 9.5). Residue placement can also better ensure the timely mineralization of N from soil organic matter (Loecke and Robertson 2009).

Manipulating N mineralization offers a primary opportunity to improve N synchrony. Particulate organic matter (POM) measurements could be used to estimate N mineralization if combined with information about the previous year's crop and cover crop production (Willson et al. 2001). Permanganate Oxidizable Carbon (POXC) has also been shown to be a quick and inexpensive assay for assessing management changes associated with changes in the labile C pool, which can reflect key processes such as N mineralization (Culman et al. 2012).

Nitrification

Autotrophic nitrification is the microbial oxidation of NH_4^+ to nitrite (NO_2^-) by ammonia oxidizers and then to nitrate (NO_3^-) by nitrite oxidizers (Robertson and Groffman 2015). In most soils, the NO_2^- produced rarely accumulates as it is quickly oxidized to NO_3^-. Nitrification can also produce the gas nitric oxide (NO) as a by-product of the chemical breakdown of hydroxylamine (NH_2OH) during NH_3 oxidation, and, when oxygen (O_2) is limiting, produce both NO and N_2O

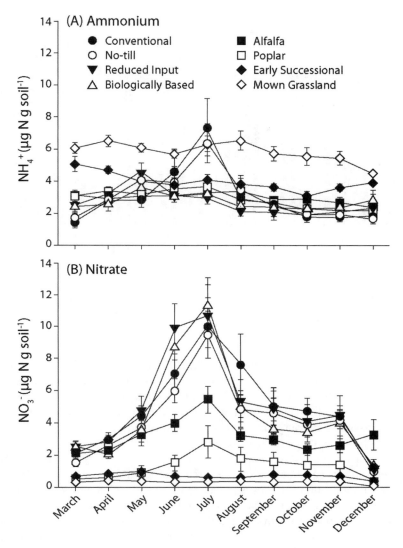

Figure 9.5. Monthly A) ammonium and B) nitrate concentrations (µg N g soil⁻¹) in MCSE systems for 1989–1995. Mean ± SE for all measurements within each month.

via NO_2^- reduction—effectively becoming denitrifying nitrifiers (Zhu et al. 2013, Robertson and Groffman 2015).

Phillips et al. (2000a) found larger populations of nitrifiers in the MCSE Conventional and No-till systems than in the Poplar and Early Successional systems, likely due to N fertilizer addition. However, nitrifier abundance, even when high, still constitutes a relatively small proportion of the total microbial population. Phillips et al. (2000b) found no detectable differences in the species composition of ammonia oxidizers in these systems—all were dominated by members of

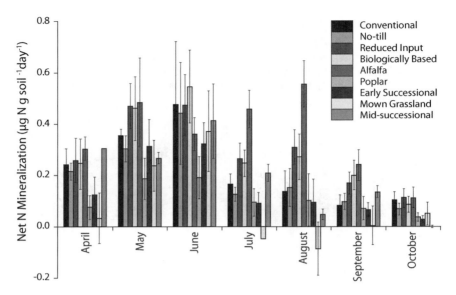

Figure 9.6. Net N mineralization rates (μg N g soil^{-1} day^{-1}) during the growing season (April to October) in the MCSE systems for 1989–1995 (except no data 1991). Mean ± SE for all measurements within each month.

Nitrosospira cluster 3. On the other hand, Bruns et al. (1999) did find more, but less diverse, culturable ammonia oxidizers (also *Nitrosospira*) in Conventional system soils than in Early Successional or Mown Grassland (never tilled) systems. The recent discovery of Archaeal nitrifiers and their likely prevalence in many soils promises to transform our community-level understanding of nitrifiers at the KBS LTER as elsewhere (Robertson and Groffman 2015).

In most aerated soils, nitrification is controlled primarily by factors that affect NH_4^+ availability and rates can be inferred by measuring changes in NH_4^+ and NO_3^- in soil incubations. In most soil incubations, where plant uptake and leaching losses are excluded, much of the mineralized N typically ends up as NO_3^- rather than NH_4^+ (Robertson and Groffman 2015). In all annual cropping systems as well as the Alfalfa and Deciduous Forest systems, the proportion of net mineralized N that exists as NO_3^- at the end of the incubation—here called relative nitrification (Robertson et al. 1999)—is high (70–80%, Table 9.5), indicating rapid nitrification of NH_4^+ as it is formed during mineralization. In contrast, relative nitrification is low (28–42%) in the Poplar, Early Successional, Mid-successional, and Mown Grassland (never tilled) systems. Low relative nitrification may reflect edaphic, plant, or other conditions that affect nitrifier communities or otherwise delay nitrification.

Nitrification is typically rapid in cultivated soils and may be regarded as a gateway to N loss (Robertson 1982); only after NH_4^+ is transformed into NO_3^- is N readily lost from most ecosystems. Nitrate can be quickly leached from soils and also serves as the substrate for denitrification, which produces the gases N_2 and N_2O.

Nitrogen Loss from Cropping Systems

Hydrologic Losses and Fate

A significant amount of the N fertilizer applied to cultivated crops is lost in agricultural drainage waters, primarily as highly mobile NO_3^-. Other forms of reactive N in the soil solution (e.g., NH_4^+, dissolved organic nitrogen [DON]) are typically present in such small quantities that they are unimportant sources of N loss, even in fertilized soils (Hamilton 2015, Chapter 11 in this volume; cf. van Kessel et al. 2009).

Syswerda et al. (2012) estimated NO_3^- leaching losses from MCSE systems by combining measured NO_3^- concentrations in water draining the root zone (sampled at 1.2-m depth) with modeled rates of water loss. Nitrate losses varied with tillage and the intensity of management inputs. Among the annual cropping systems, average annual losses followed the order Conventional (62.3 ± 9.5 kg N ha^{-1} yr^{-1}) > No-till (41.3 ± 3.0) > Reduced Input (24.3 ± 0.7) > Biologically Based (19.0 ± 0.8) management. Among the perennial and unmanaged ecosystems, NO_3^- losses followed the order Alfalfa (12.8 ± 1.8 kg N ha^{-1} yr^{-1}) = Deciduous Forest (11.0 ± 4.2) >> Early Successional (1.1 ± 0.4) = Mid-successional (0.9 ± 0.4) > Poplar ($<0.01 \pm 0.007$ kg N ha^{-1} yr^{-1}) systems (Fig. 9.7).

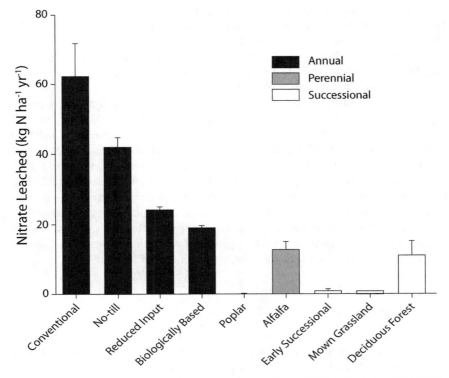

Figure 9.7. Nitrate leaching losses (kg NO_3^-–N ha^{-1} yr^{-1}) in MCSE systems for 1995–2006. Mean ± SE (n = 3 replicate locations). Modified from Syswerda et al. (2012).

These patterns suggest the potential for management intervention to significantly reduce NO_3^- leaching from cropping systems: (1) compared to conventional management, no-till cultivation reduced NO_3^- losses by 33%; (2) the incorporation of cover crops as a substitute for N fertilizer reduced losses by 60–70%; and (3) substituting perennial crops (alfalfa and hybrid poplars) for annual crops reduced losses by 80–100%. In annual crops, most NO_3^- loss takes place during periods when plants are absent—the fall and spring for those systems without cover crops (e.g., Syswerda et al. 2012). Management to conserve NO_3^- might thus be best focused on reducing losses during these periods.

Nitrate lost to groundwater is later discharged from seeps, springs, and drains into streams and rivers (Mueller and Helsel 1996, Crumpton et al. 2008), where it can be further transformed or transported to lakes, estuaries, and marine systems (Howarth et al. 1996, Alexander et al. 2000). In the U.S. Corn Belt, NO_3^- concentrations in ground and surface waters often exceed the 10 mg N L^{-1} maximum contaminant level for drinking water set by the U.S. Environmental Protection Agency (Jaynes et al. 1999, Mitchell et al. 2000).

Much of the NO_3^- entering wetlands and small headwater streams is likely to be transformed to the inert N_2 form (Hamilton 2015, Chapter 11 in this volume), albeit with some production and emission of N_2O (Paludan and Blicher-Mathiesen 1996, Stadmark and Leonardson 2005, Beaulieu et al. 2011). Nitrate concentrations in headwater streams often limit denitrification (Inwood et al. 2005) and are the best predictor of N_2O emissions rates from streams around KBS (Beaulieu et al. 2008). On an areal basis, these streams have higher N_2O emission rates (on average, 35.2 µg N_2O-N m^{-2} hr^{-1}; Beaulieu et al. 2008) than annual crops of the MCSE (on average, 14.5 µg N_2O-N m^{-2} hr^{-1}; Robertson et al. 2000), though their regional contribution is small because of a much smaller areal extent.

Losses via denitrification can be extremely high at the interface where emerging groundwater enters surface water bodies (e.g., Cooper 1990, Whitmire and Hamilton 2005) as well as along subsurface flow paths (e.g., Pinay et al. 1995). In a headwater stream at KBS, Hedin et al. (1998) and Ostrom et al. (2002) showed that substantial amounts of NO_3^- are removed from flow paths prior to stream entry when sufficient dissolved organic carbon (DOC) is available to support denitrification. Subsurface N chemistry and $\delta^{15}N$ natural abundance analyses suggest that a narrow near-stream region is functionally the most important location for denitrifier NO_3^- consumption. These studies suggest that managing these areas to provide sufficient DOC—for example, by planting perennial vegetation streamside—could be an effective mitigation strategy for reducing the impact of leached NO_3^- on aquatic systems.

Denitrification

Denitrification is the stepwise reduction of soil NO_3^- to the N gases NO, N_2O, and N_2. Four denitrification enzymes—nitrate reductase (Nar), nitrite reductase (Nir), nitric oxide reductase (Nor), and nitrous oxide reductase (Nos)—are usually induced sequentially under anaerobic conditions (Tiedje 1994, Robertson 2000). A wide variety of mostly heterotrophic bacteria can denitrify (Schmidt and Waldron

2015, Chapter 6 in this volume), using NO_3^- rather than O_2 as a terminal electron acceptor during respiration when O_2 is in short supply.

In well-aerated soils, denitrification mainly occurs within soil aggregates and particles of organic matter, where O_2 is depleted because its consumption by microbial activity is faster than its replacement by diffusion (Sexstone et al. 1985). Other factors can also be important, including denitrifier community composition (Schmidt and Waldron 2015, Chapter 6 in this volume). Cavigelli and Robertson (2000, 2001), for example, found numerically dominant denitrifying taxa in MCSE Conventional system soils that were absent in Mown Grassland (never tilled) soils, and vice versa. They also found that taxa differed in the sensitivities of their Nos enzymes to O_2, and thus in their abilities to reduce N_2O to N_2 under identical C, NO_3^-, and O_2 conditions (Fig. 9.8). This study provided an early example of the importance of microbial diversity for ecosystem function, which has subsequently become an important research topic (Schmidt and Waldron 2015, Chapter 6 in this volume).

Denitrification in normally unsaturated soils is highly episodic, occurring primarily after wetting events that create anoxic microsites in soil layers where NO_3^- and labile C are abundant. Because denitrifying enzymes are induced sequentially, there can be a lag period just after wetting when N_2O is a dominant end product (Robertson 2000). Bergsma et al. (2002), for example, used ^{15}N tracers at KBS to show that in soils from the Conventional system, the N_2O mole fraction $((N_2O/[N_2O + N_2])$ after a wetting event depended on whether or not the soils had been wetted

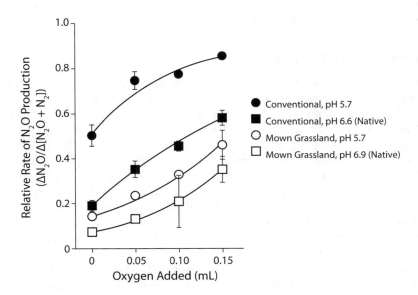

Figure 9.8. N_2O as a proportion of total N gas production $(N_2O / [N_2O + N_2])$ at four oxygen and two pH levels (native and adjusted) for soil from the Conventional and Mown Grassland (never tilled) MCSE systems. Mean ± SE ($n = 3$ for Conventional; $n = 2$ for Mown Grassland systems). Redrawn from Cavigelli et al. (2000) with permission of the Ecological Society of America; permission conveyed through Copyright Clearance Center, Inc.

2 days prior to the measurement period, which considerably reduced the proportion of N_2O emitted, vs. immediately prior to measurement. In contrast, prior wetting had no effect on the mole fraction in the Early Successional system—further illustrating presumed differences in denitrifier communities.

Measuring denitrification rates in the field is difficult (Groffman et al. 2006). Fluxes of N_2 cannot be assessed directly because the comparatively small amount of N_2 produced by denitrification cannot be readily differentiated from the very large atmospheric background. Consequently, we must rely on inference from laboratory incubations (e.g., Robertson and Tiedje 1985, Weier et al. 1993) or mass balance techniques where denitrification rates are assumed to be the difference between total N inputs (e.g., N fixation, fertilization, and deposition) and measurable outputs (e.g., leaching, harvest, and erosion).

In a novel approach to directly measure denitrification *in situ*, Bergsma et al. (2001) used an ^{15}N-gas nonequilibrium technique to measure simultaneous fluxes of N_2O and N_2 from an MCSE soil planted to winter wheat. The N_2O mole fraction ranged from <0.004 to 0.14, with an average of 0.008 ± 0.004 when both gases were above the detection limit, showing that N_2O production is only a small fraction of total N gas production ($N_2O + N_2$). If generalizable, this suggests that N_2 emissions from U.S. Midwest row crops may, on average, be substantially greater than N_2O emissions. If so, then extrapolation suggests that total losses of N from denitrification in these systems are of similar importance to the hydrologic losses discussed earlier (Fig. 9.7). Although similar findings have been estimated from mass balance approaches elsewhere (e.g., Gentry et al. 2009), N losses from denitrification in KBS LTER soils remain highly uncertain.

Nitrous Oxide (N_2O) Emissions

Nitrous oxide is produced primarily by denitrifying and nitrifying bacteria; other sources appear unimportant in agricultural soils (e.g., Robertson and Tiedje 1987, Crenshaw et al. 2008). The extent to which N_2O is produced by denitrifiers vs. nitrifiers in agronomic systems is a source of controversy; knowing this could be valuable for designing N_2O mitigation strategies. Nitrous oxide isotopomer analysis (the intramolecular distribution of the ^{14}N and ^{15}N isotopes; Ostrom and Ostrom 2011) is the only current field-based technique that can unambiguously differentiate between these two processes without significantly altering microbial activity. Ostrom et al. (2010a) measured isotopomer site preference to show that denitrification is the dominant source of N_2O in the Mown Grassland (never tilled) community following tillage of a subplot; in that experiment, denitrification accounted for 53–100% of the N_2O produced over a diurnal cycle. Using the same approach in no-till wheat of the Resource Gradient Experiment, Ostrom et al. (2010b) showed that denitrification dominated regardless of N fertilizer rate (0, 134, and 246 kg N ha^{-1}).

Nitrogen availability, on the other hand, is the single best predictor of overall N_2O emissions across both unmanaged and cropped ecosystems at KBS LTER (Gelfand and Robertson 2015, Chapter 12 in this volume) as elsewhere (Matson and Vitousek 1987; Bouwman et al. 2002a, b). In cropped systems, paired comparisons of N_2O production in fertilized vs. unfertilized soils suggest that ~1% of added N is converted to N_2O, and this is the factor currently used by most national greenhouse gas (GHG) inventories to estimate N_2O production (IPCC 2006; Gelfand and Robertson 2015, Chapter 12 in this volume). However, the response curves of N_2O emission vs. N fertilizer rate are beginning to illustrate a more nonlinear relationship that suggests N_2O fluxes increase disproportionately as fertilizer rates exceed the crop's capacity to utilize added N (Shcherbak et al. 2014).

McSwiney and Robertson (2005), for example, reported a nonlinear, exponentially increasing N_2O fertilizer response along the nine N fertilizer rates in corn at the KBS LTER Resource Gradient Experiment. They found that N_2O emissions more than doubled at N fertilizer rates greater than the level at which yield was maximized. Likewise, across six N fertilizer rates in winter wheat, Millar et al. (2014) also found an exponential increase in N_2O emissions with an increasing N rate. Hoben et al. (2011) confirmed this relationship in commercial corn fields across Michigan (Fig. 9.9), and Grace et al. (2011) used this relationship to revise estimates for N_2O emissions from corn in the U.S. North Central Region for 1964–2005 from 35% to 59% of all GHG emissions associated with corn, and the revised N_2O emission was equivalent to 1.75% of N fertilizer

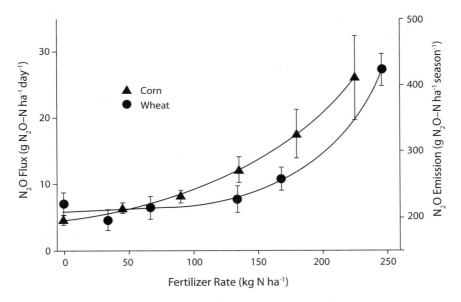

Figure 9.9. Nitrous oxide (N_2O) response to increasing N fertilizer rates for wheat (g N_2O-N ha^{-1} season^{-1}) at the KBS LTER Resource Gradient Experiment (adapted from Millar et al. 2014), and for corn (g N_2O-N ha^{-1} day^{-1}) in five commercial corn fields in Michigan (redrawn from data presented in Hoben et al. 2011). For wheat, emissions were measured using automated sampling chambers; values are means determined from sub-daily fluxes over 47 days (right axis) ± SE (n = 159 – 230, repeated measures). For corn, emissions were measured using manual sampling chambers; values are means (left axis) ± SE (n = 32, 8 site years × 4 replicate blocks).

inputs. This agrees well with the estimate by Griffis et al. (2013) of 1.8% based on tall tower measurements of atmospheric N_2O at an agricultural landscape in central Minnesota (USA). Millar et al. (2010, 2012) used this relationship to develop a methodology for C markets to better incentivize N_2O mitigation via improved N fertilizer management.

Temporal variability severely challenges accurate estimates of annual N_2O fluxes from agricultural systems. As for denitrification, N_2O emissions are often stimulated by episodic agronomic and environmental events, including fertilization, tillage, rainfall, and freeze-thaw cycles (e.g., Wagner-Riddle et al. 2007, Halvorson et al. 2008). Episodic N_2O fluxes can constitute a substantial portion of long-term total N_2O emissions (Parkin and Kaspar 2006), questioning the validity of annual emission estimates based on infrequent sampling (Smith and Dobbie 2001). Automated chamber methods where fluxes are measured several times a day address the issue of the temporal variability, and at KBS as elsewhere, such measurements reveal a substantial amount of diel and day-to-day variation (Ambus and Robertson 1998).

Agronomic Nitrogen Balances in Annual MCSE Ecosystems

Although we do not have sufficient knowledge of all N fluxes in the MCSE to construct complete N budgets for each system, we can construct simple agronomic budgets (e.g., Vitousek et al. 2009) for the annual cropping systems that are informative. Estimates of N in agronomic inputs (from fertilizer additions and BNF) less outputs (in harvest) provide a first-order measure of surplus N. The balance (Table 9.6) is instructive: the Conventional system has an overall balance of +7 kg N ha⁻¹ yr⁻¹, followed by No-till (0 kg N ha⁻¹ yr⁻¹), Reduced Input (–28 kg N ha⁻¹ yr⁻¹), and Biologically Based (–34 kg N ha⁻¹ yr⁻¹). The Conventional system closely compares to the balance of +10 kg N ha⁻¹ yr⁻¹ (Table 9.6) for a generalized U.S. Midwest (Illinois) corn–soybean rotation determined by Vitousek et al. (2009).

That the Conventional System is in near balance (only ~7% of estimated inputs are not removed by harvest) and the No-till is in exact balance suggests conservative N management in these systems (Table 9.6). As noted earlier, N fertilizer for corn is applied at rates recommended by university extension based on an EONR approach. The negative surpluses in the Reduced Input and Biologically Based systems are striking, and likely indicate cover crop scavenging of N that would otherwise be lost to the environment by leaching and denitrification. Alternative explanations are that soil organic matter could be providing additional N or that BNF by the cover crops could be underestimated. However, soil organic matter is accreting in these systems rather than declining (Syswerda et al. 2011), and thus is a sink not a source of N. And while rates of BNF in the Reduced Input and Biologically Based systems are only 12–16 kg N ha⁻¹ yr⁻¹ greater than in the Conventional system, which seems low, recall that leguminous cover crops are grown during only one of the MCSE's three rotation phases (preceding wheat). For this phase, rates of BNF for red clover are 31 and 45 kg N ha⁻¹ yr⁻¹ for the Reduced Input and Biologically Based systems, respectively (Table 9.4), which is a reasonable range for red clover (Schipanski and Drinkwater 2011) and for winter cover crops in general (Parr et al. 2011).

Table 9.6. Agronomic N inputs, N outputs and surplus (difference) for corn–soybean–wheat rotations in the U.S. Midwest[a] and the annual MCSE systems.[b]

Agronomic N Inputs, Outputs, and Balance	U.S. Midwest	MCSE System			
		Conventional	No-Till	Reduced Input	Biologically Based
Agronomic Inputs (kg N ha^{-1} yr^{-1})					
Fertilizer	93	73	73	25	0
BNF	62	31	36	43	47
Total	*155*	*104*	*109*	*68*	*47*
Agronomic Outputs (kg N ha^{-1} yr^{-1})					
Harvest	145	97	109	96	81
Total	*145*	*97*	*109*	*96*	*81*
Surplus	**+10**	**+7**	**0**	**−28**	**−34**

[a]Values for the U.S. Midwest from Vitousek et al. (2009) for a tile-drained corn–soybean rotation in Illinois.
[b]Values for MCSE systems from Tables 9.2, 9.3, and 9.4, annualized to crop rotation cycle.

Worth noting is that the overall balance for each system is the sum of balances for individual crops within each 3-year rotation, and that year-to-year balances differ by crop. In the Conventional and No-till systems, for example, only wheat has an agronomic N budget in approximate balance; corn has a significant N surplus, and soybean a significant N deficit. On average for 1993–2010, wheat in these systems had a 3 kg N excess (76 kg N fertilizer less 73 kg N harvest), corn had a 55 kg N ha^{-1} yr^{-1} excess (141 kg N from fertilizer inputs less 86 kg N harvested), and soybean had a 51 kg N deficit (100 kg N from BNF less 151 kg N harvested). N balance over the entire rotation, then, is the result of soybean's deficit being made up by corn's excess.

That leaching and presumably denitrification losses are significant in all systems, but especially in the annual cropping systems (see the prior section above), suggests that overall N budgets are substantially out of balance: in the annual cropping systems, more N appears to be lost via harvest, leaching, and denitrification than is being gained via N deposition, BNF, and N fertilizer. This suggests either that BNF has been underestimated, and that (1) more N is being mineralized from soil organic matter than is being immobilized annually; (2) N leaching losses have been overestimated; or (3) there is an unrecognized N source such as atmospheric NH$_3$ adsorption. This apparent imbalance is a major knowledge gap that deserves future attention.

Mitigation of Excess Nitrogen

The main management challenge for mitigating reactive N in the environment is to maximize the efficiency with which N is used in agricultural systems (CAST 2011)—for field crops, this means implementing practices that minimize fertilizer use and maximize N conservation. Robertson and Vitousek (2009) organized practices to improve the N Use Efficiency (NUE) of high-productivity cropping systems into four strategies: (1) provide farmers with decision support to better match fertilizer N rates to crop N requirements; (2) adjust the rotation to add crops and cover crops that will improve uptake of added N; (3) better manage the timing, placement, and formulation of N fertilizers—including organic—to better synchronize N additions with plant N needs; and (4) manage hydrologic flow paths to capture and process the N leached from farm fields. Tillage management provides a fifth option for some soils. KBS LTER results inform all strategies and as well the degree to which farmers are likely to adopt various strategies (Swinton et al. 2015, Chapter 13 in this volume).

Nitrogen Fertilizer Rate

As noted earlier, a direct relationship exists between crop yield and N fertilizer rate up to the point at which crop N needs are met. In recent years, hundreds of N-response trials conducted throughout the U.S. Midwest (e.g., Vanotti and Bundy 1994) have contributed to a database sufficient to support the new MRTN approach based on the EONR (ISU 2004, Sawyer et al. 2006).

At KBS LTER, the difference between the traditional Agronomic Optimum N Rate (AONR) and the EONR approach illustrates potential fertilizer savings. Results from the Resource Gradient Experiment (Fig. 9.2) suggest that applying N fertilizer at economic optimum rates could result in a potential savings of 23% on N fertilizer costs for corn and 36% for wheat, assuming a typical fertilizer to crop price ratio of 0.10. That NO_3^- leaching also increases substantially at N fertilizer rates in excess of crop N demand (e.g., Andraski et al. 2000, Gehl et al. 2006) suggests that the EONR approach could mitigate other forms of N in the environment in addition to N_2O.

Cover Crops

Winter cover crops can capture N that would otherwise be available for loss following annual crop harvest. During active growth in the fall and spring, winter cover crops take up mineralized N and/or residual N fertilizer, which can then become available to the next crop upon remineralization (Rasse et al. 2000, Strock et al. 2004). Much of the N immobilized by the cover crop presumably would be otherwise lost as leached NO_3^- (Feyereisen et al. 2006) or as N_2O and N_2 via denitrification (Baggs et al. 2000a).

In the KBS LTER Living Field Lab Experiment (Snapp et al. 2015, Chapter 15 in this volume), McSwiney et al. (2010) found that the inclusion of cereal rye as a winter cover crop in a conventional corn system maintained corn yield, while significantly reducing N_2O and NO_3^- losses, particularly at N fertilizer rates in excess of corn N requirements. They calculated apparent N recoveries of >80% for N fertilizer rates up to 101 kg N ha^{-1} when cover crops were included; typical estimates for row crops without cover crops range from 30% to 60% (e.g., Cassman et al. 2002). Cover crops also significantly reduced NO_3^- leaching compared to systems without cover crops (Syswerda et al. 2012). This reduction could be linked to increased evapotranspiration and soil nitrogen scavenging in cover crop systems.

When legumes are included as cover crops, as in the MCSE Reduced Input and Biologically Based systems, there is the potential benefit of N input by BNF. Although BNF will be low when adequate soil N is available, winter legumes can provide the same degree of soil inorganic N scavenging as their nonleguminous counterparts and have the additional advantage of a low C:N biomass that decomposes rapidly after spring killing, making more N available earlier for the growth of the summer crop (Fig. 9.5; Corak et al. 1991, Crandall et al. 2005). Crop residue quality (e.g., C:N ratio, lignin content) has also been shown to affect denitrification rates and N_2O emissions (Baggs et al. 2000b; Millar and Baggs 2004, 2005). Cost savings from avoided N fertilizer use could be put toward cover crop seed and planting expenses.

Fertilizer Formulation, Placement, and Timing

Applying an appropriate form of N when and where the crop can best use it can readily improve ecosystem NUE. Fertilizer N should ideally be applied in several doses to match the timing of crop N demand—this ensures the greatest synchrony between fertilizer addition and crop need. However, except where N might be applied continuously in irrigation water, weather and the availability of equipment and labor typically limit N applications to no more than two per season—in corn, a starter rate at planting and the remainder just before the rapid growth stage. Worse, for about one-third of U.S. cropland, fertilizer N is applied once in the fall—months before active crop growth (Randall and Sawyer 2008, Ribaudo et al. 2011). Long-term paired comparisons show lower corn yields and 15–40% greater NO_3^- losses with fall vs. spring fertilizer applications (e.g., Randall and Mulla 2001, Randall and Vetsch 2005). Bundy (1986) concluded that fall-applied N is usually 10–15% less effective for crop utilization than spring-applied N.

Fertilizer placement also affects NUE. The spatial arrangement of available N vis-à-vis the distribution of plants and their roots (e.g., Van Noordwijk et al. 1993) affects the likelihood of N uptake vs. N loss. Placing synthetic fertilizers in a concentrated band within or very close to crop rows, rather than between them as is more common, can increase NUE and reduce surface N loss (Malhi and Nyborg 1985, CAST 2011). Injecting anhydrous ammonia into soil near rows rather than broadcasting over the soil surface can decrease N leaching and volatilization by as much as 35% (Achorn and Broder 1984). And broadcasting it has been shown to double N_2O emissions (Venterea et al. 2010) when compared to broadcast urea.

The location and particle size of plant residue also have an impact on its N release and uptake. Loecke and Robertson (2009) found that N in red clover residue is more likely to be taken up by corn if the residue is sufficiently aggregated to concentrate mineralized N in small patches, but not so large as to inhibit decomposition by creating anoxic microsites.

Fertilizer placement at the field scale can also greatly influence NUE. That soil nitrogen availability is spatially variable in predictable patterns (Fig. 9.10) is well known from studies at the KBS LTER (Robertson et al. 1993, 1997; Senthilkumar

(A) Soil Nitrate   (B) Net N Mineralization

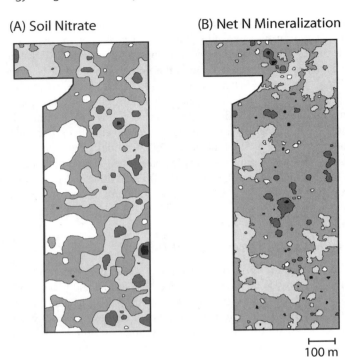

100 m

Figure 9.10. Isopleths for A) soil nitrate (5.7-15.1 mg N g soil⁻¹) and B) net N mineralization (0.30-0.70 mg N g⁻¹ d⁻¹) across the 48-ha MCSE prior to plot establishment. Shading
patterns denote five equal increments in levels for each variable across its range. Distance
unit in meters. Redrawn from Robertson et al. (1997) with permission of the Ecological
Society of America; permission conveyed through Copyright Clearance Center, Inc.

et al. 2009) and elsewhere, and yield patterns are concomitantly related to nutrient availability (Robertson 1997, Kravchenko and Bullock 2000, Kravchenko
et al. 2006, Kravchenko and Robertson 2007). Applying fertilizer in a spatially
targeted manner that recognizes this variability can concentrate N on those areas
of a field with the highest yield potential and avoid adding N to those areas not
likely to be N responsive. For most fields, less fertilizer N will be added using this
approach than applying a generally recommended rate to the entire field. (e.g.,
Mamo et al. 2003, Scharf et al. 2005). On-the-go fertilizer placement, which uses
spectral reflectance of the canopy to spatially judge real-time crop N needs (Raun
et al. 2002, Li et al. 2009, Scharf and Lory 2009), is a promising technology for
improving NUE.

Better management of crop residues can undergird, supplement, and even
replace synthetic fertilizers. The quality of organic residues plays an especially
important role and can influence N availability by (1) adding N to soils, (2) affecting N mineralization–immobilization patterns, (3) serving as an energy source for
microbial activities, and (4) acting as precursors to N sequestration in soil organic
matter (Palm and Rowland 1997; Paul et al. 2015, Chapter 5 in this volume).

Decomposition of crop residues and resultant N release are governed by climatic, edaphic, and resource quality factors (Swift et al. 1979).

Of these factors, resource quality is likely the easiest for farmers to manage, but its impact is often difficult to assess. Farmers use a variety of organic inputs, ranging from crop residues to manures, which vary widely in N content. And only a minority of the N from a winter cover crop may be available for the following summer crop (e.g., 4–35%; Ranells and Wagger 1997). Knowing how the quality of applied organic materials (e.g., C:N ratio, mineralizable C and N content, and lignin content) affects N mineralization and immobilization rates (Aulakh et al. 1991, Wagger et al. 1998) is critical for predicting the effect of organic residues on N availability for annual crops.

Due to its effect on available N, residue quality has also been shown to affect denitrification rates and N_2O emissions (Aulakh et al. 2001, Baggs et al. 2000a, Millar et al. 2004). In laboratory incubations using KBS soils, Ambus et al. (2001) found that high-quality (1.88% N) pea residues resulted in greater N_2O emissions than low-quality (0.63% N) barley residues, when both were separately incorporated into soil, either as ground or coarsely cut residues. That study also showed how residue particle size and placement affected both N_2O emissions and NO_3^- leaching potentials.

Managing Hydrologic Flow Paths to Retain or Remove Reactive N

Hydrologic export of reactive N from agricultural systems to ground and surface waters causes well-documented problems, including the degradation of drinking water by excessively high concentrations of NO_3^-, eutrophication of downstream surface waters including marine coastal zones, and additional emission of N_2O to the atmosphere (Galloway et al. 2008; Hamilton 2015, Chapter 11 in this volume). These problems have motivated research to understand how we can manage landscapes to retain or remove reactive N from hydrologic flow paths. Three strategies that specifically apply to agricultural watersheds are discussed in this section (Robertson et al. 2007, Robertson and Vitousek 2009).

First, riparian and other downslope conservation plantings can be managed to keep NO_3^- leached from cropped fields from entering local waterways (Liebman et al. 2013). Native or planted perennial vegetation in stream riparian (buffer) zones can immobilize N in growing biomass and soil organic matter (Lowrance 1998). It is well established that waterway grass (filter) strips can also trap soil particles that would otherwise erode organic N into surface waters. Such measures offer the additional benefits of mitigating sediment and phosphorus losses to surface waters.

Second, restoring stream channels and small wetlands in agricultural watersheds can promote denitrification and other microbial processes that convert NO_3^- to inert or less mobile forms of N (Mitsch et al. 2001). Denitrification is the main process through which streams can permanently remove N (Mulholland et al. 2009). The effectiveness of wetlands in reducing N export from agricultural fields is largely dependent on the magnitude and timing of NO_3^- inputs and the capacity of the system to denitrify or accumulate N in plant biomass and organic detritus. Channelization effectively turns headwater streams and wetlands into pipes that are less conducive to N retention (Opdyke et al. 2006) because water moves out faster and its N has less contact with stream edges and sediments, particularly during periods of high flow (Peterson et al. 2001, Royer et al. 2006, Alexander et al. 2009). Restoring stream channels and small wetlands to intercept leached N has the potential to significantly reduce downstream NO_3^- loadings.

Additionally, the third strategy involves targeting landscape positions that contribute disproportionately to watershed N fluxes (Robertson et al. 2007; Robertson and Vitousek 2009; Hamilton 2015, Chapter 11 in this volume). It is increasingly clear that much nonpoint source pollution from agriculture arises from relatively small fractions of the landscape (Giburek et al. 2002). Planting forage or other perennial crops such as cellulosic biofuels in these areas (Robertson et al. 2011) could reduce landscape-level N outputs. Restoring or expanding wetlands in low-lying areas could even convert these areas from N sources to N sinks.

Tillage Management to Mitigate N Loss

Tillage affects a number of factors that influence N conservation: physical factors such as soil bulk density, water-holding capacity, drainage, aeration, and aggregate stability; chemical factors such as C and N stores and availability; and biological factors such as microbial activity, rates of decomposition, the presence of earthworms and other invertebrates, and plant root distributions. Few other management practices have such far-reaching effects on cropping system N cycling.

Historically, tillage is responsible for most soil organic matter loss in cultivated ecosystems (Paul et al. 2015, Chapter 5 in this volume) and thereby the loss of most soil organic N stores. On conversion from native vegetation or long-term fallow, most of the N initially harvested in subsequent crops is derived from decomposition (Robertson 1997). Once soil organic N pools are depleted—typically, a few decades in temperate regions, more quickly in the tropics (Robertson and Grandy 2006)—legumes and fertilizers are required to replace the soil's lost capacity to supply N. Because herbicides can now provide weed control as effectively as tillage, conservation tillage (including no-till) can be practiced to conserve organic C and N in soil and thereby restore many of the fertility benefits of less disturbed soils (e.g., Franzluebbers and Arshad 1997, Lal 2004).

Although the many benefits of no-till are well known (e.g., Blevins et al. 1977, Phillips et al. 1980), benefits related to NO_3^- and N_2O conservation are less clear. Comparisons of no-till vs. conventional tillage systems have shown no significant difference in NO_3^- leaching (e.g., Cabrera et al. 1999, Mitsch et al. 1999, Smith et al. 1990), or have demonstrated that no-till leaches either more (Tyler and Thomas 1977, Chichester 1977) or less (Rasse and Smucker 1999, Ogden et al. 1999) NO_3^-. Syswerda et al. (2012) argue that much of this ambiguity is due to experiment duration. They note that most studies last only 2–3 years and begin shortly after no-till establishment. Short-term studies may mask long-term effects that emerge only over periods with both higher and lower rainfall levels (e.g., Cabrera et al. 1999). And studies conducted too soon after a change in management can misrepresent the long-term effects that emerge after equilibration (Rasmussen et al. 1998). In addition to short duration, many studies are performed in small plots, which cannot readily account for the effects of spatial variation at realistic field scales (Robertson et al. 2007).

Syswerda et al. (2012) made similar comparisons of no-till vs. conventional tillage, but examined NO_3^- losses over an 11-year period that began 6 years after no-till establishment. On average, they found NO_3^- leaching losses from the MCSE No-till and Conventional systems represented 50 and 76%, respectively, of the total N applied to these systems. NUE was higher in the No-till system despite 16% higher drainage losses (388 vs. 334 mm H_2O yr^{-1}, respectively), suggesting that channelized flow in the better-structured no-till soils allows water to leave the profile before it has equilibrated with NO_3^- in small pores (Rasse and Smucker 1999). Higher plant demand for NO_3^- may also have contributed to lower no-till fluxes as the No-till system had somewhat higher average yields (Grandy et al. 2006; Smith et al. 2007; Robertson et al. 2015, Chapter 2 in this volume) and therefore more N uptake.

No-till management can also affect N_2O emissions from soil, although such effects are not consistent. Comparisons of N_2O emissions in the MCSE Conventional and No-till systems have shown no consistently significant differences (e.g., Robertson et al. 2000, Grandy et al. 2006, Gelfand et al. 2013). This is in agreement with other (Parkin and Kaspar 2006, Dusenbury et al. 2008, Sey et al. 2008, Gregorich et al. 2008) but not all (e.g., Liu et al. 2006, Omonode et al. 2011) similar studies. Van Kessel et al. (2013) concluded through meta-analysis that it typically takes at least 10 years before no-till soils exhibit lower N_2O fluxes, in which case we can expect the MCSE No-till system to emit less N_2O in coming years.

Summary

Nitrogen is an essential nutrient for crop growth and the N demands of today's intensive cropping systems are met primarily by synthetic N fertilizer application. Direct consequences of over-applying N fertilizer are substantial losses of reactive N to the environment in the form of NO_3^- leached to ground and surface waters and N_2O emitted to the atmosphere. The environmental costs of excess N loading include coastal zone eutrophication, compromised drinking water and air quality, climate warming, stratospheric ozone depletion, and biodiversity loss. However, cropping systems can acquire N through legume N fixation, manure addition, and crop residue return—offering many options for N management at the farm scale.

Results from KBS LTER research underscore the value of practical agronomic practices that improve N retention in row crops. Potential interventions include increasing rotational complexity with different primary and cover crops; using no-till management; and improving N synchrony with better rate, timing, placement, and formulation of N fertilizers, so crop N needs are met more precisely. Residue management can also contribute to N conservation. Improved landscape management can partially mitigate N leaching losses from the farm field through measures such as maintaining or planting riparian vegetation, restoring stream channels and small wetlands, and the targeted planting of forage or other perennial crops such as cellulosic biofuels. The most poorly known N cycle fluxes at the KBS LTER are emissions of N_2 and NO_x, both from denitrification. Providing farmers with strategies or incentives that reduce N fertilizer use while maintaining high agronomic yield is a logical first step in mitigating agriculture's impact on the environment.

References

Achorn, F. P., and M. F. Broder. 1984. Mechanics of applying nitrogen fertilizer. Pages 493–506 in R. D. Hauck, editor. Nitrogen in crop production. Soil Science Society of America, Madison, Wisconsin, USA.

Alexander, R., R. Smith, and G. Schwarz. 2000. Effect of stream channel size on the delivery of nitrogen to the Gulf of Mexico. Nature 403:758–761.

Alexander, R. B., J. K. Bohlke, E. W. Boyer, M. B. David, J. W. Harvey, P. J. Mulholland, S. P. Seitzinger, C. R. Tobias, C. Tonitto, and W. M. Wollheim. 2009. Dynamic modeling of nitrogen losses in river networks unravels the coupled effects of hydrological and biogeochemical processes. Biogeochemistry 93:91–116.

Ambus, P., E. S. Jensen, and G. P. Robertson. 2001. Nitrous oxide and N-leaching losses from agricultural soil: influence of crop residue particle size, quality and placement. Phyton (Austria) 41:7–15.

Ambus, P., and G. P. Robertson. 1998. Automated near-continuous measurement of carbon dioxide and nitrous oxide fluxes from soil. Soil Science Society of America Journal 62:394–400.

Ambus, P., and G. P. Robertson. 2006. The effect of increased N deposition on nitrous oxide, methane, and carbon dioxide fluxes from unmanaged forest and grassland communities in Michigan. Biogeochemistry 79:315–337.

Andraski, T. W., L. G. Bundy, and K. R. Brye. 2000. Crop management and corn nitrogen rate effects on nitrate leaching. Journal of Environmental Quality 29:1095–1103.

Aulakh, M. S., J. W. Doran, D. T. Walters, A. R. Mosier, and D. D. Francis. 1991. Crop residue type and placement effects on denitrification and mineralization. Soil Science Society of America Journal 55:1020–1025.

Aulakh, M. S., T. Khera, J. Doran, and K. Bronson. 2001. Denitrification, N_2O and CO_2 fluxes in rice-wheat cropping system as affected by crop residues, fertilizer N and legume green manure. Biology and Fertility of Soils 34:375–389.

Baggs, E. M., R. M. Rees, K. A. Smith, and A. J. A. Vinten. 2000a. Nitrous oxide emission from soils after incorporating crop residues. Soil Use and Management 16:82–87.

Baggs, E. M., C. A. Watson, and R. M. Rees. 2000b. The fate of nitrogen from incorporated cover crop and green manure residues. Nutrient Cycling in Agroecosystems 56:153–163.

Beaulieu, J. J., C. P. Arango, S. K. Hamilton, and J. L. Tank. 2008. The production and emission of nitrous oxide from headwater streams in the Midwestern United States. Global Change Biology 14:878–894.

Beaulieu, J. J., J. L. Tank, S. K. Hamilton, W. M. Wollheim, R. O. Hall, Jr., P. J. Mulholland, B. J. Peterson, L. R. Ashkenas, L. W. Cooper, C. N. Dahm, W. K. Dodds, N. B. Grimm, S. L. Johnson, W. H. McDowell, G. C. Poole, H. M. Valett, C. P. Arango, M. J. Bernot, A. J. Burgin, C. L. Crenshaw, A. M. Helton, L. T. Johnson, J. M. O'Brien, J. D. Potter, R. W. Sheibley, D. J. Sobota, and S. M. Thomas. 2011. Nitrous oxide emission from denitrification in stream and river networks. Proceedings of the National Academy of Sciences USA 108:214–219.

Bergsma, T. T., N. E. Ostrom, M. Emmons, and G. P. Robertson. 2001. Measuring simultaneous fluxes from soil of N_2O and N_2 in the field using the ^{15}N-gas "nonequilibrium" technique. Environmental Science and Technology 35:4307–4312.

Bergsma, T. T., G. P. Robertson, and N. E. Ostrom. 2002. Influence of soil moisture and land use history on denitrification end-products. Journal of Environmental Quality 31:711–717.

Blevins, R. L., G. W. Thomas, and P. L. Cornelius. 1977. Influence of no-till and N-fertilization on certain soil properties after five years of continuous corn. Agronomy Journal 69:383–386.

Bobbink, R., K. Hicks, J. Galloway, T. Spranger, R. Alkemade, M. Ashmore, M. Bustamante, S. Cinderby, E. Davidson, F. Dentener, B. Emmett, J. W. Erisman, M. Fenn, F. Gilliam, A. Nordin, L. Pardo, and W. De Vries. 2010. Global assessment of nitrogen deposition effects on terrestrial plant diversity: a synthesis. Ecological Applications 20:30–59.

Bobbink, R., M. Hornung, and J. G. M. Roelofs. 1998. The effects of air-borne nitrogen pollutants on species diversity in natural and semi-natural European vegetation. Journal of Ecology 86:717–738.

Bouwman, A. F., L. J. M. Boumans, and N. H. Batjes. 2002a. Emissions of N_2O and NO from fertilized fields: summary of available measurement data. Global Biogeochemical Cycles 16:1058–1070.

Bouwman, A. F., L. J. M. Boumans, and N. H. Batjes. 2002b. Modeling global annual N_2O and NO emissions from fertilized fields. Global Biogeochemical Cycles 16:1080–1090.

Bruns, M. A., J. R. Stephen, J. I. Prosser, and E. A. Paul. 1999. Comparative diversity of ammonia-oxidizer 16S rRNA gene sequences in native, tilled, and successional soils. Applied and Environmental Microbiology 65:2994–3000.

Bundy, L. G. 1986. Timing nitrogen applications to maximize fertilizer efficiency and crop response in conventional crop production. Journal of Fertilizer Issues 3:99–106.

Butterbach-Bahl, K., L. Breuer, R. Gasche, G. Willibald, and H. Papen. 2002. Exchange of trace gases between soils and the atmosphere in Scots pine forest ecosystems of the northeastern German lowlands 1. Fluxes of N_2O, NO/NO_2 and CH_4 at forest sites with different N-deposition. Forest Ecology and Management 167:123–134.

Cabrera, M. L., D. M. Endale, D. E. Radcliffe, J. L. Steiner, W. K. Vencill, L. Lohr, and H. H. Schomberg. 1999. Tillage and fertilizer source effects on nitrate leaching in cotton production in Southern Piedmont. Pages 49–50 in J. E. Hook, editor. Proceedings of the 22nd Annual Southern Conservation Tillage Conference for Sustainable Agriculture. Special Publication 95, Georgia Agricultural Experiment Station, Tifton, Georgia, USA.

Cassman, K. G., A. Dobermann, and D. T. Walters. 2002. Agroecosystems, nitrogen-use efficiency, and nitrogen management. Ambio 31:132–140.

CAST (Council for Agricultural Science and Technology). 2011. Carbon sequestration and greenhouse gas fluxes in agriculture: challenges and opportunities. Task Force Report No.142, CAST, Ames, Iowa, USA.

CASTNET (Clean Air Status and Trends Network). 2011. Download data. U.S. Environmental Protection Agency, Washington, DC, USA. <http://java.epa.gov/castnet/reportPage.do> Accessed December 19, 2011.

Cavigelli, M. A., and G. P. Robertson. 2000. The functional significance of denitrifier community composition in a terrestrial ecosystem. Ecology 81:1402–1414.

Cavigelli, M. A., and G. P. Robertson. 2001. Role of denitrifier diversity in rates of nitrous oxide consumption in a terrestrial ecosystem. Soil Biology & Biochemistry 33:297–310.

Chichester, F. W. 1977. Effects of increased fertilization rates on nitrogen content of runoff and percolate from monolith lysimeters. Journal of Environmental Quality 6:211–217.

Clark, C. M., and D. Tilman. 2008. Loss of plant species after chronic low-level nitrogen deposition to prairie grasslands. Nature 451:712–715.

Cleveland, C. C., A. R. Townsend, D. S. Schimel, H. Fisher, R. W. Howarth, L. O. Hedin, S. S. Perakis, E. F. Latty, J. C. Von Fischer, A. Elseroad, and M. F. Wasson. 1999. Global patterns of terrestrial biological nitrogen (N_2) fixation in natural ecosystems. Global Biogeochemical Cycles 13:623–645.

Cooper, A. B. 1990. Nitrate depletion in the riparian zone and stream channel of a small headwater catchment. Hydrobiologia 202:13–26.

Corak, S. J., W. W. Frye, and M. S. Smith. 1991. Legume mulch and nitrogen fertilizer effects on soil water and corn production. Soil Science Society America Journal 55:1395–1400.

Crandall, S. M., M. L. Ruffo, and G. A. Bollero. 2005. Cropping systems and nitrogen dynamics under a cereal winter cover crop preceding corn. Plant and Soil 268:209–219.

Crenshaw, C. L., C. Lauber, R. L. Sinsabaugh, and L. K. Stavely. 2008. Fungal control of nitrous oxide production in semiarid grassland. Biogeochemistry 87:17–27.

Crumpton, W. G., D. A. Kovacic, D. L. Hey, and J. A. Kostel. 2008. Potential of restored and constructed wetlands to reduce nutrient export from agricultural watersheds in the Corn Belt. Pages 29–42 in Upper Mississippi River Sub-basin Hypoxia Nutrient Committee (UMRSHNC). Final report: Gulf Hypoxia and Local Water Quality Concerns Workshop, American Society of Agricultural and Biological Engineers (ASABE), St. Joseph, Michigan, USA.

Culman, S. W., S. S. Snapp, M. A. Freeman, M. Schipanski, J. Beniston, R. Lal, L. E. Drinkwater, A. J. Franzluebbers, J. D. Glover, A. S. Grandy, J. Lee, J. Six, J. E. Maul, S. B. Mirksy, J. T. Spargo, and M. M. Wander. 2012. Permanganate oxidizable carbon reflects a processed soil fraction that is sensitive to management. Soil Science Society America Journal 76:494–506.

Dentener, F., J. Drevet, J. F. Lamarque, I. Bey, B. Eickhout, A. M. Fiore, D. Hauglustaine, L. W. Horowitz, M. Krol, U. C. Kulshrestha, M. Lawrence, C. Galy-Lacaux, S. Rast, D. Shindell, D. Stevenson, T. Van Noije, C. Atherton, N. Bell, D. Bergman, T. Butler, J. Cofala, B. Collins, R. Doherty, K. Ellingsen, J. Galloway, M. Gauss, V. Montanaro, J. F. Müller, G. Pitari, J. Rodriguez, M. Sanderson, F. Solmon, S. Strahan, M. Schultz, K. Sudo, S. Szopa, and O. Wild. 2006. Nitrogen and sulfur deposition on regional and global scales: a multi-model evaluation. Global Biogeochemical Cycles 20:GB4003.

Dusenbury, M. P., R. E. Engel, P. R. Miller, R. L. Lemke, and R. Wallander. 2008. Nitrous oxide emissions from a northern great plains soil as influenced by nitrogen management and cropping systems. Journal of Environmental Quality 37:542–550.

Feyereisen, G. W., B. N. Wilson, G. R. Sands, J. S. Strock, and P. M. Porter. 2006. Potential for a rye cover crop to reduce nitrate loss in southwestern Minnesota. Agronomy Journal 98:1416–1426.

Franzluebbers, A. J., and M. A. Arshad. 1997. Particulate organic carbon content and potential mineralization as affected by tillage and texture. Soil Science Society of America Journal 61:1382–1386.

Galloway, J. N., F. J. Dentener, D. G. Capone, E. W. Boyer, R. W. Howarth, S. P. Seitzinger, G. P. Asner, C. C. Cleveland, P. A. Green, E. A. Holland, D. M. Karl, A. F. Michaels, J. H. Porter, A. R. Townsend, and C. J. Vöosmarty. 2004. Nitrogen cycles: past, present, and future. Biogeochemistry 70:153–226.

Galloway, J. N., A. R. Townsend, J. W. Erisman, M. Bekunda, Z. C. Cai, J. R. Freney, L. A. Martinelli, S. P. Seitzinger, and M. A. Sutton. 2008. Transformation of the nitrogen cycle: Recent trends, questions, and potential solutions. Science 320:889–892.

Gehl, R. J., J. P. Schmidt, C. B. Godsey, L. D. Maddux, and W. B. Gordon. 2006. Post-harvest soil nitrate in irrigated corn: variability among eight field sites and multiple nitrogen rates. Soil Science Society of America Journal 70:1922–1931.

Gelfand, I., and G. P. Robertson. 2015. Mitigation of greenhouse gas emissions in agricultural ecosystems. Pages 310–339 in S. K. Hamilton, J. E. Doll, and G. P. Robertson, editors. The ecology of agricultural ecosystems: long-term research on the path to sustainability. Oxford University Press, New York, New York, USA.

Gelfand, I., and G. P. Robertson. 2015. Nitrogen uptake, fixation, and response to N fertilizer by soybean in US Midwest Biogeochemistry (in press).

Gelfand, I., R. Sahajpal, X. Zhang, C. R. Izaurralde, K. L. Gross, and G. P. Robertson. 2013. Sustainable bioenergy production from marginal lands in the US Midwest. Nature 493:514–517.

Gentry, L. E., M. B. David, F. E. Below, T. V. Royer, and G. F. McLsaac. 2009. Nitrogen mass balance of a tile-drained agricultural watershed in east-central Illinois. Journal of Environmental Quality 38:1841–1847.

Giburek, W. J., C. C. Drungil, M. S. Srinivasan, B. A. Needleman, and D. E. Woodward. 2002. Variable source area controls on phosphorus transport: bridging the gap between research and design. Journal of Soil and Water Conservation 57:534–543.

Grace, P., G. P. Robertson, N. Millar, M. Colunga-Garcia, B. Basso, S. H. Gage, and J. Hoben. 2011. The contribution of maize cropping in the Midwest USA to global warming: a regional estimate. Agricultural Systems 104:292–296.

Grandy, A. S., T. D. Loecke, S. Parr, and G. P. Robertson. 2006. Long-term trends in nitrous oxide emissions, soil nitrogen, and crop yields of till and no-till cropping systems. Journal of Environmental Quality 35:1487–1495.

Greaver, T. L., T. J. Sullivan, J. D. Herrick, M. C. Barber, J. S. Baron, B. J. Cosby, M. E. Deerhake, R. L. Dennis, J.-J. B. Dubois, C. L. Goodale, A. T. Herlihy, G. B. Lawrence, L. Liu, J. A. Lynch, and K. J. Novak. 2012. Ecological effects of nitrogen and sulfur air pollution in the US: what do we know? Frontiers in Ecology and the Environment 10:365–372.

Gregorich, E. G., P. Rochette, P. St.-Georges, U. F. McKim, and C. Chan. 2008. Tillage effects on N_2O emission from soils under corn and soybeans in Eastern Canada. Canadian Journal of Soil Science 88:153–161.

Griffis, T. J., X. Lee, J. M. Baker, M. P. Russelle, X. Zhang, R. Venterea, and D. B. Millet. 2013. Reconciling the differences between top-down and bottom-up estimates of nitrous oxide emissions for the U.S. Corn Belt. Global Biogeochemical Cycles 27:746–754.

Groffman, P. M., M. A. Altabet, J. K. Böhlke, K. Butterbach-Bahl, M. B. David, M. K. Firestone, A. E. Giblin, T. M. Kana, L. P. Nielsen, and M. A. Voytek. 2006. Methods for measuring denitrification: diverse approaches to a difficult problem. Ecological Applications 16:2091–2122.

Halvorson, A. D., S. J. Del Grosso, and C. A. Reule. 2008. Nitrogen, tillage, and crop rotation effects on nitrous oxide emissions from irrigated cropping systems. Journal of Environmental Quality 37:1337–1344.

Hamilton, S. K. 2015. Water quality and movement in agricultural landscapes. Pages 275–309 in S. K. Hamilton, J. E. Doll, and G. P. Robertson, editors. The ecology of agricultural ecosystems: long-term research on the path to sustainability. Oxford University Press, New York, New York, USA.

Hedin, L. O., J. C. von Fischer, N. E. Ostrom, B. P. Kennedy, M. G. Brown, and G. P. Robertson. 1998. Thermodynamic constraints on nitrogen transformation and other biogeochemical processes at soil-stream interfaces. Ecology 79:684–703.

Herridge, D. F., M. B. Peoples, and R. M. Boddey. 2008. Global inputs of biological nitrogen fixation in agricultural systems. Plant and Soil 311:1–18.

Hoben, J. P., R. J. Gehl, N. Millar, P. R. Grace, and G. P. Robertson. 2011. Nonlinear nitrous oxide (N_2O) response to nitrogen fertilizer in on-farm corn crops of the US Midwest. Global Change Biology 17:1140–1152.

Howarth, R. W., G. Billen, D. Swaney, A. Townsend, N. Jaworski, K. Lajtha, J. A. Downing, R. Elmgren, N. Caraco, T. Jordan, F. Berendse, J. Freney, V. Kudeyarov, P. Murdoch, and Z. L. Zhu. 1996. Regional nitrogen budgets and riverine N&P fluxes for the drainages to the North Atlantic Ocean: natural and human influences. Biogeochemistry 35:75–139.

IFA (International Fertilizer Industry Association). 2011. IFADATA statistics. Paris, France. <http://www.fertilizer.org/ifa/ifadata/search>

Inwood, S. E., J. L. Tank, and M. J. Bernot. 2005. Patterns of denitrification associated with land use in 9 midwestern headwater streams. Journal of the North American Benthological Society 24:227–245.

IPCC (Intergovernmental Panel on Climate Change). 2006. 2006 IPCC guidelines for national greenhouse gas inventories. National Greenhouse Gas Inventories Programme, Institute for Global Environmental Strategies (IGES), Hayama, Japan.

IPNI (International Plant Nutrition Institute). 2012. 4R plant nutrition: a manual for improving the management of plant nutrition. T.W. Bruulsema, P. E. Fixen, and G. D. Sulewski, editors. IPNI, Norcross, Georgia, USA.

ISU (Iowa State University). 2004. Corn nitrogen rate calculator: a regional (Corn Belt) approach to nitrogen rate guidelines. Iowa State University Agronomy Extension, Ames, Iowa, USA. <http://extension.agron.iastate.edu/soilfertility/nrate.aspx>

Jaworski, N. A., R. W. Howarth, and L. J. Hetling. 1997. Atmospheric deposition of nitrogen oxides onto the landscape contributes to coastal eutrophication in the northeast United States. Environmental Science & Technology 31:1995–2004.

Jaynes, D. B., J. L. Hatfield, and D. W. Meek. 1999. Water quality in Walnut Creek Watershed: herbicides and nitrate in surface waters. Journal of Environmental Quality 28:45–59.

Kramer, D. A. 2004. Nitrogen. Pages 53.1–53.18 in Minerals yearbook, Metals and minerals 2004, Vol. 1, U.S. Geological Survey, Washington, DC, USA.

Kravchenko, A. N., and D. G. Bullock. 2000. Correlation of corn and soybean grain yield with topography and soil properties. Agronomy Journal 92:75–83.

Kravchenko, A. N., and G. P. Robertson. 2007. Can topographical and yield data substantially improve total soil carbon mapping by regression kriging? Agronomy Journal 99:12–17.

Kravchenko, A. N., G. P. Robertson, X. Hao, and D. G. Bullock. 2006. Management practice effects on surface total carbon: difference in spatial variability patterns. Agronomy Journal 98:1559–1568.

Lal, R. 2004. Soil carbon sequestration to mitigate climate change. Geoderma 123:1–22.

Leigh, G. J. 2004. The world's greatest fix: a history of nitrogen and agriculture. Oxford University Press, New York, New York, USA.

Li, F., Y. Miao, F. Zhang, Z. Cui, R. Li, X. Chen, H. Zhang, J. Schroder, W. R. Raun, and L. Jia. 2009. In-season optical sensing improves nitrogen-use efficiency for winter wheat. Soil Science Society of America Journal 73:1566–1574.

Liebig, J. 1840. Chemistry in its application to agriculture and physiology. Taylor & Walton, London, UK.

Liebman, M., M. J. Helmers, L. A. Schulte, and C. A. Chase. 2013. Using biodiversity to link agricultural productivity with environmental quality: results from three field experiments in Iowa. Renewable Agriculture and Food Systems 28:115–128.

Liu, X. J., A. R. Mosier, A. D. Halvorson, and F. S. Zhang. 2006. The impact of nitrogen placement and tillage on NO, N_2O, CH_4 and CO_2 fluxes from a clay loam soil. Plant and Soil 280:177–188.

Loecke, T. D., and G. P. Robertson. 2009. Soil resource heterogeneity in the form of aggregated litter alters maize productivity. Plant and Soil 325:231–241.

Lowrance, R. 1998. Riparian forest ecosystems as filters for nonpoint-source pollution. Pages 113–141 in M. L. Pace and P. M. Groffman, editors. Successes, limitations, and frontiers in ecosystem science. Springer-Verlag, New York, New York, USA.

Malhi, S. S., and M. Nyborg. 1985. Methods of placement for increasing the efficiency of nitrogen fertilizers applied in the fall. Agronomy Journal 77:27–32.

Mamo, M., G. L. Malzer, D. J. Mulla, D. R. Huggins, and J. Strock. 2003. Spatial and temporal variation in economically optimum nitrogen rate for corn. Agronomy Journal 95:958–964.

Matson, P. A., and P. M. Vitousek. 1987. Cross-system comparisons of soil nitrogen transformations and nitrous oxide flux in tropical forest ecosystem. Global Biogeochemical Cycles 1:163–170.

McSwiney, C. P., and G. P. Robertson. 2005. Nonlinear response of N_2O flux to incremental fertilizer addition in a continuous maize (*Zea mays* sp.) cropping system. Global Change Biology 11:1712–1719.

McSwiney, C. P., S. S. Snapp, and L. E. Gentry. 2010. Use of N immobilization to tighten the N cycle in conventional agroecosystems. Ecological Applications 20:648–662.

Millar, N., and E. M. Baggs. 2004. Chemical composition, or quality, of agroforestry residues influences N_2O emissions after their addition to soil. Soil Biology & Biochemistry 36:935–943.

Millar, N., and E. M. Baggs. 2005. Relationships between N_2O emissions and water-soluble C and N contents of agroforestry residues after their addition to soil. Soil Biology & Biochemistry 37:605–608.

Millar, N., J. K. Ndufa, G. Cadisch, and E. M. Baggs. 2004. Nitrous oxide emissions following incorporation of improved-fallow residues in the humid tropics. Global Biogeochemical Cycles 18:GB1032. doi:1010.1029/2003GB002114

Millar, N., G. P. Robertson, A. Diamant, R. J. Gehl, P. R. Grace, and J. P. Hoben. 2012. Methodology for quantifying nitrous oxide (N_2O) emissions reductions by reducing nitrogen fertilizer use on agricultural crops. American Carbon Registry, Winrock International, Little Rock, Arkansas, USA.

Millar, N., G. P. Robertson, P. R. Grace, R. J. Gehl, and J. P. Hoben. 2010. Nitrogen fertilizer management for nitrous oxide (N_2O) mitigation in intensive corn (Maize) production: an emissions reduction protocol for US Midwest agriculture. Mitigation and Adaptation Strategies for Global Change 15:185–204.

Millar, N., D. W. Rowlings, P. R. Grace, R. J. Gehl, and G. P. Robertson. 2014. Non-linear N_2O response to nitrogen fertilizer in winter wheat using automated chambers (in review).

Mitchell, J. K., G. F. McIsaac, S. E. Walker, and M. C. Hirshi. 2000. Nitrate in river and subsurface drainage flows from an east central Illinois watershed. Transactions of the ASAE 43:337–342.

Mitsch, W. J., J. W. Day, Jr., J. W. Gilliam, P. M. Groffman, D. L. Hey, G. W. Randall, and N. M. Wang. 2001. Reducing nitrogen loading to the Gulf of Mexico from the Mississippi River Basin: strategies to counter a persistent ecological problem. BioScience 51:373–388.

Mitsch, W. J., J. W. Day, J. W. Gilliam, P. M. Groffman, D. L. Hey, G. W. Randall, and N. Wang. 1999. Reducing nitrogen loads, especially nitrate-nitrogen, to surface water, groundwater, and the Gulf of Mexico: Topic 5 report for the integrated assessment on hypoxia in the Gulf of Mexico. NOAA Coastal Ocean Program Decision Analysis Series No. 19, Silver Spring, Maryland, USA.

Mueller, D. K., and D. R. Helsel. 1996. Nutrients in the nation's waters—too much of a good thing? National Water-Quality Assessment Program, U.S. Geological Survey, Washington, DC, USA.

Mulholland, P. J., R. O. Hall, D. J. Sobota, W. K. Dodds, S. E. G. Findlay, N. B. Grimm, S. K. Hamilton, W. H. McDowell, J. M. O'Brien, J. L. Tank, L. R. Ashkenas, L. W. Cooper, C. N. Dahm, S. V. Gregory, S. L. Johnson, J. L. Meyer, B. J. Peterson, G. C. Poole, H. M. Valett, J. R. Webster, C. P. Arango, J. J. Beaulieu, M. J. Bernot, A. J. Burgin, C. L. Crenshaw, A. M. Helton, L. T. Johnson, B. R. Niederlehner, J. D. Potter, R. W. Sheibley, and S. M. Thomas. 2009. Nitrate removal in stream ecosystems measured by N-15 addition experiments: denitrification. Limnology and Oceanography 54:666–680.

NADP/NTN (National Atmospheric Deposition Program/National Trends Network). 2011. NADP/NTN monitoring location MI26, Kellogg Biological Station. NADP/NTN Coordination Office, Illinois State Water Survey, Champaign, Illinois, USA. <http://nadp.sws.uiuc.edu/sites/siteinfo.asp?net=NTN&id=MI26>

NASS (National Agricultural Statistics Service). 2014. Homepage. U.S. Department of Agriculture, Washington, DC, USA. <http://www.nass.usda.gov/>

Ogden, C. B., H. M. Van Es, R. J. Wagenet, and T. S. Steenhuis. 1999. Spatial-temporal variability of preferential flow in a clay soil under no-till and plow-till. Journal of Environmental Quality 28:1264–1273.

Omonode, R. A., D. R. Smith, A. Gal, and T. J. Vyn. 2011. Soil nitrous oxide emissions in corn following three decades of tillage and rotation treatments. Soil Science Society America Journal 75:152–163.

Opdyke, M. R., M. B. David, and B. L. Rhoads. 2006. Influence of geomorphological variability in channel characteristics on sediment denitrification in agricultural streams. Journal of Environmental Quality 35:2103–2112.

Ostrom, N. E., L. O. Hedin, J. C. von Fischer, and G. P. Robertson. 2002. Nitrogen transformations and NO_3^- removal at a soil-stream interface: a stable isotope approach. Ecological Applications 12:1027–1043.

Ostrom, N. E., and P. H. Ostrom. 2011. The isotopomers of nitrous oxide: analytical considerations and application to resolution of microbial production pathways. Pages 453–476 in M. Baskaran, editor. Handbook of environmental isotope geochemistry. Springer Berlin, Heidelberg, Germany.

Ostrom, N. E., P. H. Ostrom, H. Gandhi, N. Millar, and G. P. Robertson. 2010b. Isotopologue data reveal denitrification as the primary source of nitrous oxide at nitrogen fertilization gradient in a temperate agricultural field. Geochimica et Cosmochimica Acta 74:A780.

Ostrom, N. E., R. Sutka, P. H. Ostrom, A. S. Grandy, K. H. Huizinga, H. Gandhi, J. C. von Fisher, and G. P. Robertson. 2010a. Isotopologue data reveal bacterial denitrification as the primary source of N_2O during a high flux event following cultivation of a native temperate grassland. Soil Biology & Biochemistry 42:499–506.

Palm, C. A., and A. P. Rowland. 1997. A minimum dataset for characterization of plant quality for decomposition. Pages 379–392 in G. Cadisch and K. E. Giller, editors. Driven by nature. CAB International, Wallingford, UK.

Paludan, C., and G. Blicher-Mathiesen. 1996. Losses of inorganic carbon and nitrous oxide from a temperate freshwater wetland in relation to nitrate loading. Biogeochemistry 35:305–326.

Parkin, T. B., and T. C. Kaspar. 2006. Nitrous oxide emissions from corn-soybean systems in the Midwest. Journal of Environmental Quality 35:1496–1506.

Parr, M., J. M. Grossman, S. C. Reberg-Horton, C. Brinton, and C. Crozier. 2011. Nitrogen delivery from legume cover crops in no-till organic corn production. Agronomy Journal 103:1578–1590.

Paul, E. A., A. Kravchenko, A. S. Grandy, and S. Morris. 2015. Soil organic matter dynamics: controls and management for sustainable ecosystem functioning. Pages 104–134 *in* S. K. Hamilton, J. E. Doll, and G. P. Robertson, editors. The ecology of agricultural ecosystems: long-term research on the path to sustainability. Oxford University Press, New York, New York, USA.

Peterson, B. J., W. M. Wollheim, P. J. Mulholland, J. R. Webster, J. L. Meyer, J. L. Tank, E. Marti, W. B. Bowden, H. M. Valett, A. E. Hershey, W. H. McDowell, W. K. Dodds, S. K. Hamilton, S. Gregory, and D. D. Morrall. 2001. Control of nitrogen export from watersheds by headwater streams. Science 292:86–90.

Phillips, C. J., D. Harris, S. L. Dollhopf, K. L. Gross, J. I. Prosser, and E. A. Paul. 2000b. Effects of agronomic treatments on the structure and function of ammonia oxidizing communities. Applied and Environmental Microbiology 66:5410–5418.

Phillips, C. J., E. A. Paul, and J. I. Prosser. 2000a. Quantitative analysis of ammonia oxidising bacteria using competitive PCR. FEMS Microbial Ecology 32:167–175.

Phillips, R. E., R. L. Blevins, G. W. Thomas, W. F. Frye, and S. H. Phillips. 1980. No-tillage agriculture. Science 208:1108–1114.

Pinay, G., C. Ruffinoni, and A. Fabre. 1995. Nitrogen cycling in two riparian forest soils under different geomorphic conditions. Biogeochemistry 30:9–29.

Randall, G. W., and D. J. Mulla. 2001. Nitrate-N in surface waters as influenced by climatic conditions and agricultural practices. Journal of Environmental Quality 30:337–344.

Randall, G. W., and J. E. Sawyer. 2008. Nitrogen application timing, forms, and additives. Pages 73–85 in Final report: Gulf Hypoxia and Local Water Quality Concerns Workshop. Upper Mississippi River Sub-basin Hypoxia Nutrient Committee (UMRSHNC), American Society of Agricultural and Biological Engineers (ASABE), St. Joseph, Michigan, USA.

Randall, G. W., and J. A. Vetsch. 2005. Nitrate losses in subsurface drainage from a corn-soybean rotation as affected by fall vs. spring application of nitrogen and nitrapyrin. Journal of Environmental Quality 34:590–597.

Ranells, N. N., and M. G. Wagger. 1997. Nitrogen-15 recovery and release by rye and crimson clover cover crops. Soil Science Society of America Journal 61:943–948.

Rasmussen, P. E., K. W. T. Goulding, J. R. Brown, P. R. Grace, H. H. Janzen, and M. Korschens. 1998. Long-term agroecosystem experiments: assessing agricultural sustainability and global change. Science 282:893–896.

Rasse, D. P., J. T. Ritchie, W. R. Peterson, J. Wei, and A. J. M. Smucker. 2000. Rye cover crop and nitrogen fertilization effects on nitrate leaching in inbred maize fields. Journal of Environmental Quality 29:298–304.

Rasse, D. P., and A. J. M. Smucker. 1999. Tillage effects on soil nitrogen and plant biomass in a corn alfalfa rotation. Journal of Environmental Quality 28:873–880.

Raun, W. R., J. B. Solie, G. V. Johnson, M. L. Stone, R. W. Mullen, K. W. Freeman, W. E. Thomason, and E. V. Lukina. 2002. Improving nitrogen use efficiency in cereal grain production with optical sensing and variable rate application. Agronomy Journal 94:815–820.

Rheaume, S. J. 1990. Geohydrology and water quality of Kalamazoo County, Michigan, 1986–88. Water-Resources Investigations Report 90–4028, U.S. Geological Survey, Books and Open-File Reports Section, Lansing Michigan, Denver, Colorado.

Ribaudo, M., J. A. Delgado, L. Hansen, M. Livingston, R. Mosheim, and J. Williamson. 2011. Nitrogen in agricultural systems: implications for conservation policy. U.S. Department of Agriculture, Economic Research Service., Washington, DC, USA.

Robertson, G. P. 1982. Nitrification in forested ecosystems. Philosophical Transactions of the Royal Society of London 296:445–457.

Robertson, G. P. 1997. Nitrogen use efficiency in row crop agriculture: crop nitrogen use and soil nitrogen loss. Pages 347–365 in L. Jackson, editor. Ecology in agriculture. Academic Press, New York, New York, USA.

Robertson, G. P. 2000. Denitrification. Pages C181–C190 in M. E. Sumner, editor. Handbook of soil science. CRC Press, Boca Raton, Florida, USA.

Robertson, G. P., L. W. Burger, C. L. Kling, R. Lowrance, and D. J. Mulla. 2007. New approaches to environmental management research at landscape and watershed scales. Pages 27–50 in M. Schnepf and C. Cox, editors. Managing agricultural landscapes for environmental quality. Soil and Water Conservation Society, Ankeny, Iowa, USA.

Robertson, G. P., J. R. Crum, and B. G. Ellis. 1993. The spatial variability of soil resources following long-term disturbance. Oecologia 96:451–456.

Robertson, G. P., and A. S. Grandy. 2006. Soil system management in temperate regions. Pages 27–39 in N. Uphoff, A. S. Ball, E. Fernandes, H. Herren, O. Husson, M. Laing, C. Palm, J. Pretty, P. Sanchez, N. Snaninga, and J. Thies, editors. Biological approaches to sustainable soil systems. CRC Press, Boca Raton, Florida, USA.

Robertson, G. P., and P. M. Groffman. 2015. Nitrogen transformations. Pages 421–426 in E. A. Paul, editor. Soil microbiology, ecology and biochemistry. Fourth edition. Academic Press, Burlington, Massachusetts, USA.

Robertson, G. P., K. L. Gross, S. K. Hamilton, D. A. Landis, T. M. Schmidt, S. S. Snapp, and S. M. Swinton. 2015. Farming for ecosystem services: an ecological approach to row-crop agriculture. Pages 33–53 in S. K. Hamilton, J. E. Doll, and G. P. Robertson, editors. The ecology of agricultural ecosystems: long-term research on the path to sustainability. Oxford University Press, New York, New York, USA.

Robertson, G. P., and S. K. Hamilton. 2015. Long-term ecological research at the Kellogg Biological Station LTER Site: conceptual and experimental framework. Pages 1–32 in S. K. Hamilton, J. E. Doll, and G. P. Robertson, editors. The ecology of agricultural ecosystems: long-term research on the path to sustainability. Oxford University Press, New York, New York, USA.

Robertson, G. P., S. K. Hamilton, S. J. Del Grosso, and W. J. Parton. 2011. The biogeochemistry of bioenergy landscapes: carbon, nitrogen, and water considerations. Ecological Applications 21:1055–1067.

Robertson, G. P., K. M. Klingensmith, M. J. Klug, E. A. Paul, J. R. Crum, and B. G. Ellis. 1997. Soil resources, microbial activity, and primary production across an agricultural ecosystem. Ecological Applications 7:158–170.

Robertson, G. P., E. A. Paul, and R. R. Harwood. 2000. Greenhouse gases in intensive agriculture: contributions of individual gases to the radiative forcing of the atmosphere. Science 289:1922–1925.

Robertson, G. P., and J. M. Tiedje. 1985. An automated technique for sampling the contents of stoppered gas-collection vials. Plant and Soil 83:453–457.

Robertson, G. P., and J. M. Tiedje. 1987. Nitrous oxide sources in aerobic soils: nitrification, denitrification, and other biological processes. Soil Biology and Biochemistry 19:187–193.

Robertson, G. P., and P. M. Vitousek. 2009. Nitrogen in agriculture: balancing the cost of an essential resource. Annual Review of Environment and Resources 34:97–125.

Robertson, G. P., D. A. Wedin, P. M. Groffman, J. M. Blair, E. Holland, D. Harris, and K. Nadelhoffer. 1999. Soil carbon and nitrogen availability: nitrogen mineralization, nitrification, and soil respiration potentials. Pages 258–271 in G. P. Robertson, C. S. Bledsoe, D. C. Coleman, and P. Sollins, editors. Standard soil methods for long-term ecological research. Oxford University Press, New York, USA.

Royer, T. V., M. B. David, and L. E. Gentry. 2006. Timing of riverine export of nitrate and phosphorus from agricultural watersheds in Illinois: implications for reducing nutrient loading to the Mississippi River. Environmental Science & Technology 40:4126–4131.

Ruschel, A. P., P. B. Vose, R. L. Victoria, and E. Salati. 1979. Comparison of isotope techniques and non-nodulating isolines to study the effect of ammonium fertilization on nitrogen fixation in soybean, *Glycine max*. Plant and Soil 53:513–525.

Salvagiotti, F., K. G. Cassman, J. E. Specht, D. T. Walters, and A. Dobermann. 2008. Nitrogen uptake, fixation and response to fertilizer N in soybeans: a review. Field Crops Research 108:1–13.

Sánchez, J. E., T. C. Willson, K. Kizilkaya, E. Parker, and R. R. Harwood. 2001. Enhancing the mineralizable nitrogen pool through substrate diversity in long term cropping systems. Soil Science Society of America Journal. 65:1442–1447.

Sawyer, J. E., E. D. Nafziger, G. W. Randall, L. G. Bundy, G. W. Rehm, and B. C. Joern. 2006. Concepts and rationale for regional nitrogen rate guidelines for corn. Iowa State University Extension Publication PM2015, Ames, Iowa, USA.

Scharf, P. C., N. R. Kitchen, K. A. Sudduth, J. G. Davis, V. C. Hubbard, and J. A. Lory. 2005. Field-scale variability in optimal nitrogen fertilizer rate for corn. Agronomy Journal 97:452–461.

Scharf, P. C., and J. A. Lory. 2009. Calibrating reflectance measurements to predict optimal sidedress nitrogen rate for corn. Agronomy Journal 101:615.

Schipanski, M. E., and L. E. Drinkwater. 2011. Nitrogen fixation of red clover interseeded with winter cereals across a management-induced fertility gradient. Nutrient Cycling in Agroecosystems 90:105–119.

Schmidt, T. M., and C. Waldron. 2015. Microbial diversity in agricultural soils and its relation to ecosystem function. Pages 135–157 in S. K. Hamilton, J. E. Doll, and G. P. Robertson, editors. The ecology of agricultural ecosystems: long-term research on the path to sustainability. Oxford University Press, New York, New York, USA.

Senthilkumar, S., A. N. Kravchenko, and G. P. Robertson. 2009. Topography influences management system effects on total soil carbon and nitrogen. Soil Science Society of America Journal 73:2059–2067.

Sexstone, A. J., N. P. Revsbech, T. P. Parkin, and J. M. Tiedje. 1985. Direct measurement of oxygen profiles and denitrification rates in soil aggregates. Soil Science Society of America Journal 49:645–651.

Sey, B. K., J. K. Whalen, E. G. Gregorich, P. Rochette, and R. I. Cue. 2008. Carbon dioxide and nitrous oxide content in soils under corn and soybean. Soil Science Society of America Journal 72:931–938.

Shcherbak, I., N. Millar, and G. P. Robertson. 2014. A global meta-analysis of the nonlinear response of soil nitrous oxide (N_2O) to fertilizer nitrogen. Proceedings of the National Academy of Sciences USA 111:9199–9204.

Smil, V. 2002. Nitrogen and food production: proteins for human diets. Ambio 31:126–131.

Smith, K. A., and K. E. Dobbie. 2001. The impact of sampling frequency and sampling times on chamber-based measurements of N_2O emissions from fertilized soils. Global Change Biology 7:933–945.

Smith, R. G., F. D. Menalled, and G. P. Robertson. 2007. Temporal yield variability under conventional and alternative management systems. Agronomy Journal 99:1629–1634.

Smith, S. J., J. S. Schepers, and L. K. Porter. 1990. Assessing and managing agricultural nitrogen losses to the environment. Advances in Soil Science 14:1–43.

Snapp, S. S., R. G. Smith, and G. P. Robertson. 2015. Designing cropping systems for ecosystem services. Pages 378–408 in S. K. Hamilton, J. E. Doll, and G. P. Robertson, editors. The ecology of agricultural ecosystems: long-term research on the path to sustainability. Oxford University Press, New York, New York, USA.

Stadmark, J., and L. Leonardson. 2005. Emissions of greenhouse gases from ponds constructed for nitrogen removal. Ecological Engineering 25:542–551.

Stevens, C. J., N. B. Dise, J. O. Mountford, and D. J. Gowing. 2004. Impact of nitrogen deposition on the species richness of grasslands. Science 303:1876–1879.

Strock, J. S., P. M. Porter, and M. P. Russelle. 2004. Cover cropping to reduce nitrate loss through subsurface drainage in the northern US Corn Belt. Journal of Environmental Quality 33:1010–1016.

Swift, M. J., O. W. Heal, and J. M. Anderson. 1979. Decomposition in terrestrial ecosystems. University of California Press, Berkeley, California, USA.

Swinton, S. M., N. Rector, G. P. Robertson, C. B. Jolejole-Foreman, and F. Lupi. 2015. Farmer decisions about adopting environmentally beneficial practices. Pages 340–359 in S. K. Hamilton, J. E. Doll, and G. P. Robertson, editors. The ecology of agricultural ecosystems: long-term research on the path to sustainability. Oxford University Press, New York, New York, USA.

Syswerda, S. P., B. Basso, S. K. Hamilton, J. B. Tausig, and G. P. Robertson. 2012. Long-term nitrate loss along an agricultural intensity gradient in the Upper Midwest USA. Agriculture, Ecosystems and Environment 149:10–19.

Syswerda, S. P., A. T. Corbin, D. L. Mokma, A. N. Kravchenko, and G. P. Robertson. 2011. Agricultural management and soil carbon storage in surface vs. deep layers. Soil Science Society of America Journal 75:92–101.

Tiedje, J. M. 1994. Denitrifiers. Pages 245–267 in R. W. Weaver, J. S. Angle, P. S. Bottomley, D. Bezdecik, S. Smith, A. Tabatabai, and A. Wollum, editors. Methods of soil analysis. Part 2, Microbiological and biochemical properties. Soil Science Society of America, Madison, Wisconsin, USA.

Tyler, D. D., and G. W. Thomas. 1977. Lysimeter measurements of nitrate and chloride losses from soil under conventional and no-tillage corn. Journal of Environmental Quality 6:63–66.

van Kessel, C., T. Clough, and J. W. van Groenigen. 2009. Dissolved organic nitrogen: an overlooked pathway of nitrogen loss from agricultural systems? Journal of Environmental Quality 38:393–401.

van Kessel, C., R. Venterea, J. Six, M. A. Adviento-Borbe, B. Linquist, and K. J. van Groenigen. 2013. Climate, duration, and N placement determine N_2O emissions in reduced tillage systems: a meta-analysis. Global Change Biology 19:33–44.

van Noordwijk, M., P. C. de Ruiter, K. B. Zwart, J. Bloem, J. C. Moore, H. G. van Faassen, and S. L. G. E. Burgers. 1993. Synlocation of biological activity, roots, cracks and recent organic inputs in a sugar beet field. Geoderma 56:265–276.

Vanotti, M. B., and L. G. Bundy. 1994. An alternative rationale for corn nitrogen-fertilizer recommendations. Journal of Production Agriculture 7:243–249.

Venterea, R. T., M. S. Dolan, and T. E. Ochsner. 2010. Urea decreases nitrous oxide emissions compared with anhydrous ammonia in a Minnesota corn cropping system. Soil Science Society America Journal 74:407–418.

Vitousek, P. M., D. N. L. Menge, S. C. Reed, and C. C. Cleveland. 2013. Biological nitrogen fixation: rates, patterns and ecological controls in terrestrial ecosystems. Philosophical

Transactions of the Royal Society of London Biological Sciences. <http://dx.doi.org/10.1098/rstb.2013.0119>

Vitousek, P. M., R. Naylor, T. Crews, M. B. David, L. E. Drinkwater, E. Holland, P. J. Johnes, J. Katzenberger, L. A. Martinelli, P. A. Matson, G. Nziguheba, D. Ojima, C. A. Palm, G. P. Robertson, P. A. Sanchez, A. R. Townsend, and F. S. Zhang. 2009. Nutrient imbalances in agricultural development. Science 324:1519–1520.

Wagger, M. G., M. L. Cabrera, and N. N. Ranells. 1998. Nitrogen and carbon cycling in relation to cover crop residue quality. Journal of Soil and Water Conservation 53:214–218.

Wagner-Riddle, C., A. Furon, N. L. McLaughlin, I. Lee, J. Barbeau, S. Jayasundara, G. Parkin, P. Von Bertoldi, and J. Warland. 2007. Intensive measurement of nitrous oxide emissions from a corn-soybean-wheat rotation under two contrasting management systems over 5 years. Global Change Biology 13:1722–1736.

Warncke, D., J. Dahl, and L. Jacobs. 2009. Nutrient recommendations for field crops in Michigan. Michigan State University Extension Bulletin E2904 (revised August 2009), Michigan State University, East Lansing, Michigan, USA. See also the Soil Fertility and Nutrient Management Program website at http://www.soil.msu.edu/

Warncke, D., J. Dahl, L. Jacobs, and C. Laboski. 2004. Nutrient recommendations for field crops in Michigan. Michigan State University Extension Bulletin E2904, Michigan State University, East Lansing, Michigan, USA. See also the Soil Fertility and Nutrient Management Program website at http://www.soil.msu.edu/

Weier, K. L., J. W. Doran, J. F. Power, and D. T. Walters. 1993. Denitrification and the dinitrogen/nitrous oxide ratio as affected by soil water, available carbon, and nitrate. Soil Science Society of America Journal 57:66–72.

Whitmire, S. L., and S. K. Hamilton. 2005. Rapid removal of nitrate and sulfate by freshwater wetland sediments. Journal of Environmental Quality 34:2062–2071.

Wilke, B. 2010. Challenges for developing sustainable nitrogen sources in agriculture: cover crops, nitrogen fixation and ecological principals. Dissertation, Michigan State University, East Lansing, Michigan, USA.

Willson, T. C., E. A. Paul, and R. R. Harwood. 2001. Biologically active soil organic matter fractions in sustainable cropping systems. Applied Soil Ecology 16:63–76.

Yang, J. Y., C. F. Drury, X. M. Yang, R. De Jong, E. C. Huffman, C. A. Campbell, and V. Kirkwood. 2010. Estimating biological N_2 fixation in Canadian agricultural land using legume yields. Agriculture, Ecosystems and Environment 137:192–201.

Zhu, X., M. Burger, T. A. Doane, and R. W. Horwath. 2013. Ammonia oxidation pathways and nitrifier denitrification are significant sources of N_2O and NO under low oxygen availability. Proceedings of the National Academy of Sciences USA 110:6328–6333.

10

Simulating Crop Growth and Biogeochemical Fluxes in Response to Land Management Using the SALUS Model

Bruno Basso and Joe T. Ritchie

The Green Revolution, through the adoption of new crop varieties, irrigation, and agrochemicals, saved about 1 billion people from famine by increasing global food production (FAO 2011). We now recognize that these enormous gains in agricultural production were accompanied by harm to agriculture's natural resource base, jeopardizing our future ability to meet human food, fuel, and fiber needs for a growing population. Earth's population is projected to increase from ~7 billion in 2011 to ~9 billion in 2050. Given the future challenges to food production and environmental integrity, it is imperative that sustainable land management of agricultural production become an important priority for policy makers. Agricultural crop and soil management practices often cause degradation of the environment, especially the quality of ground and surface water and the fertility of agricultural soils. Clearly, a sustainable framework for developing and improving land use for crop production must be based on long-term and broad-based perspectives.

Sustainable land management is the focus of many research programs, ranging from socioeconomic to ecological, since sustainability is an integrated concept with associated challenges. A multiplicity of factors can prevent production systems from being sustainable; the goals set by a sustainable crop production system may be in conflict with one another, and solutions that work in one site or region with a particular soil, climatic, and socioeconomic setting may not be appropriate in others (Robertson and Harwood 2013). On the other hand, with sufficient attention to indicators of sustainability, a number of practices and policies could be implemented to accelerate the adoption of sustainable practices. Indicators to quantify changes in crop production systems over time at different hierarchical levels are needed for evaluating the sustainability of different land management strategies. Indicators should encompass (1) crop productivity, (2) socioeconomic and ecological well-being, and (3) resource availability.

Approaches for improving land management for the sustainability of crop production should be based on reduced chemical inputs, as well as higher resource use efficiency, enhanced nutrient cycling, and integrated pest management. Modeling is necessary to identify the best approaches because field experiments cannot be conducted with sufficient detail in space and time to find the best land management practices for sustainable crop production across diverse environmental settings. Input from long-term crop and soil management experiments, including measurements of crop yields, soil properties, biogeochemical fluxes, and relevant socioeconomic indicators, is critical to develop and test the models.

Simulation models, when suitably tested in reasonably diverse locations over sufficient time periods, provide a useful tool for finding combinations of management strategies to reach the multiple goals required for sustainable crop production. Models can provide land managers and policy makers with ways to extrapolate experimental results from one location to others where soil, landscape, and climate information is available. When biophysical simulation model results are combined with socioeconomic information, a Decision Support System (DSS) can provide management options for maximizing sustainability goals. Decision Support Systems describe a wide range of computer software programs designed to make site-specific recommendations for pest management (Michalski et al. 1983, Beck et al. 1989), farm financial planning (Boggess et al. 1989), and general crop and land management (Plant 1989). Decision Support System software packages have been designed primarily for use by crop consultants and other specialists who work with farmers and policy makers, although some are used directly by farmers. Users provide site-specific information about soil properties, crop type and management, weather conditions, and other data specific to the software. Typically, a given DSS provides a variety of management options to reach desired sustainability goals.

Process-based models of crop growth and development are integral parts of the most effective DSS models and have been developed and used for more than 40 years, since the advent of high-speed computers. During this time, two scientific teams have integrated such models into DSSs, namely, DSSAT (Tsuji et al. 1998) and APSIM (McCown et al. 1996), and both have proven useful for many groups involved in agricultural research and decision making throughout the world. The International Consortium of Agricultural Systems Applications (ICASA) was formed from several modeling groups to promote the efficient and effective use of functional models for problem solving and decision making (Ritchie 1995). Crop models that simulate crop growth, the timing of critical growth stages, and grain yields have added soil and plant carbon and nitrogen dynamics for different climate, soil, and management conditions (e.g., Parton et al. 1988).

Here, we provide a general overview of crop simulation models followed by a concise description of the model Systems Approach for Land Use Sustainability (SALUS) for evaluating the impact of agronomic management on crop yields, carbon (C) and nitrogen (N) dynamics, and environmental performance. We describe key model components and the minimum data required for simulating crop yields under different management practices. The research at the Kellogg Biological Station Long-Term Ecological Research Site (KBS LTER) provides the opportunity to test models of long-term changes in soil carbon, nitrogen leaching, crop yields, and gaseous emissions from soil. Data from KBS LTER also provide an excellent context for illustrating the utility and limitations of crop models, and we use these data to show two examples of model applications: (1) an evaluation of nitrate leaching as affected by nitrogen fertilizer management in a corn (*Zea mays* L.) and alfalfa (*Medicago sativa* L.) rotation and (2) soil carbon dynamics under various tillage systems. We also illustrate spatially connected processes by linking SALUS to digital terrain modeling.

Crop Models

Crop simulation models range from simple to complex. Simple models are often adopted to estimate yield across large land areas based on statistical information related to climate and historical yields and include little detail about the soil–plant system. The more sophisticated physically based models are capable of providing additional details on processes in the soil–plant–atmosphere system, but sophisticated models demand detailed initial environmental and agronomic information that may be unavailable in many situations.

Crop models may be either deterministic or stochastic. Deterministic models provide a specific outcome for a certain set of conditions, with all plants and soil within the simulation space assumed to be uniform. Stochastic models produce outcomes that incorporate uncertainty due to spatial variability of soil properties, temporal variability of weather conditions, abiotic and biotic factors not accounted for in a deterministic model, and uncertainties of model logic and functions. However, stochastic crop models are at an early stage of development and not used in DSSs to our knowledge.

To overcome some of the problems of using deterministic crop models, soils with known spatial variability can be grouped into small homogenous units and the results aggregated to model yield at the whole-field scale. Similarly, running simulations over multiple years with deterministic yields accounts for temporal variability (Basso et al. 2007).

Deterministic crop models can be statistical, mechanistic, or functional (Addiscott and Wagenet 1985, Ritchie and Alagarswamy 2002). Statistical models—fitting a function to observed weather variables and crop regional yield statistics to predict crop yield—were the first crop models used for large-scale yield estimations. Average regional yields were regressed on time to reveal a general trend in crop yields (Thompson 1969; Gage et al. 2015, Chapter 4 in this volume). An example is the upward trend in crop yield over the past several decades due to technological advancements in genetics and management, especially the increased use of fertilizers. Thompson (1986) quantified the impact of climate change and variability on corn yield in five U.S. states using a statistical model. In that study, preseason precipitation (September–June), June temperature, and temperature and rainfall in July and August were closely correlated with corn yield variations from the trend. Recently, Gage et al. (2015, Chapter 4 in this volume) incorporated climate effects into regional yield trends with the use of a Crop Stress Index (CSI). This approach significantly improved predictions of historical yields of corn and soybean.

In general, the results of statistical models cannot be extrapolated to other places and time periods because of variation in soils, landscapes, and weather not included in the population of information from which the statistical relationship was derived. Furthermore, the impact of agricultural technology cannot be extrapolated over space and time. Despite these limitations, statistical models can provide many insights about past yields and historical influences (Gage et al. 2015, Chapter 4 in this volume) and can be used to inform the other kinds of models.

Mechanistic models are based on known physical, chemical, and biological processes occurring in the soil–plant–atmosphere continuum. Soon after computers became available, mechanistic models were developed to simulate photosynthetic processes such as light interception, uptake of carbon dioxide (CO_2), carbon allocation to different plant organs, and loss of CO_2 during respiration, as well as the dynamics of soil water including infiltration, evaporation, drainage, and root uptake.

Mechanistic models describe processes at fine time scales (e.g., photosynthesis and transpiration processes) but a large amount of input information is required to execute them. Uncertainties in some assumptions make mechanistic model outcomes less certain and often make them less useful to those outside of the model development group (Basso et al. 2012a). Mechanistic models are rarely adopted to solve problems; rather, they are often used for academic purposes to gain a better understanding of specific processes and interactions.

Functional models are based on empirical functions that approximate complex processes, such as a crop's interception of energy using plant leaf area (as an indicator of biomass) and radiation use efficiency (as a measure of biomass produced per unit of radiation intercepted). This type of function is relatively simple and usually produces reasonable results when compared to field measurements, although it has uncertainties related to the fraction of biomass partitioned to roots and nonlinear photosynthetic responses to light. Another example is the simulation of potential evapotranspiration using the well-known functional Penman or Priestley–Taylor equations, which have been used successfully for decades although they are highly simplified compared to mechanistic evapotranspiration models.

Functional crop models use simplified equations and logic to partition simulated biomass into various plant organs, which are integrated to estimate total biomass and yield. Functional models primarily use "capacity" concepts to describe the amount of water available to plants as compared to using "instantaneous rate" concepts from soil physics. The difference between the upper and lower limits of soil water-holding capacity determines the amount of water available to plants.

Functional models typically use daily time step inputs for weather and management variables such as precipitation, solar radiation, temperature, irrigation, and fertilizer use. Low data input requirements make these models attractive when detailed data on biophysical processes are lacking. These models, when properly tested, can provide an appropriate level of detail needed for assessing many aspects of crop production. Functional type models are now routinely used in DSSs.

Examples of Models to Simulate Crop Ecosystems

Here, we provide a brief summary of the most widely adopted models to simulate soil organic matter, soil water, and biogeochemical fluxes and how these variables affect crop growth in response to land management.

Soil Organic Matter and Gas Emission Models

One of the most widely used soil organic matter (SOM) models is the CENTURY model developed by Parton et al. (1988) to simulate long-term (10–1000 years) patterns in surface SOM dynamics, plant production, and nutrient cycling (N, phosphorus [P], and sulfur [S]). The model uses a monthly time step with monthly average maximum air temperature (at 2 m height), monthly precipitation, soil texture (sand, silt, and clay content), nutrient and lignin content of dead plant material, and atmospheric and soil inputs of N. Plant material is divided into structural (difficult to decompose) and metabolic (readily decomposable) fractions. Soil organic matter is divided into active, slow, and passive pools. Decomposition of plant material and SOM is a function of soil water and temperature, and is influenced by soil type and the C/N ratio of decomposing material. A complete description of the N and soil C model is presented by Parton et al. (1987). The plant submodel is highly simplified, using only inputs of stored water at planting, precipitation during plant growth, a fixed water-use efficiency, and available soil N. Partitioning of C and N into various plant components is performed using fixed partitioning coefficients. While emphasizing long-term organic matter dynamics, the CENTURY model lacks details important for short-term soil water and crop growth dynamics as well as soil management other than N inputs.

A daily incrementing modification of CENTURY called NGAS-DAYCENT or simply DAYCENT (Parton et al. 1996, 1998, 2001; Del Grosso et al. 2000a, b) simulates trace gas fluxes of nitric oxide (NO), nitrous oxide (N_2O), and dinitrogen (N_2) from soils as well as methane (CH_4) formation and oxidation. The DAYCENT model has been used to simulate national N_2O emissions in the United States from major cropped soil regions (Del Grosso et al. 2006). Soil water calculations are performed at hourly time steps, which may not match other processes simulated at daily time steps (Basso et al. 2010, 2011).

Another mechanistic SOM and gas emission model is the DeNitrification–DeComposition (DNDC) model. The DNDC model has been used for estimating N_2O and CH_4 emissions from agricultural lands (Li 1995, 2000), but it requires substantially more input detail than other models.

While providing much detail about soil greenhouse gas emissions and carbon dynamics, these three models lack detail for estimating crop yield. Thus, they are useful for simulating SOM and soil greenhouse gas dynamics but have limited utility for evaluating the sustainable production of food, fuel, and fiber.

Crop and Soil Water Models

The Decision Support System for Agrotechnology Transfer (DSSAT) (Tsuji et al. 1998) contains a suite of crop models widely used to simulate crop biomass and yield as influenced by weather, soil, crop management, and crop genotype. The primary crop models contained in DSSAT are CROPGRO for major grain legumes, CERES for cereal crops, and SUBSTOR for crops with belowground storage organs. The models were developed with a goal of minimizing the data needed for prediction and control purposes. Simulations are executed on a daily time step using solar radiation, temperatures (maximum and minimum), and precipitation, thereby accounting for day-to-day variation that can be substantial. They are based on empirical functions to estimate the soil water balance (runoff, drainage, evapotranspiration, soil storage) and biomass production. Input needs include soil physical and chemical properties for several depth increments as available in soil surveys. Crop management input needs include date of sowing, plant population, dates and quantities of nutrient and irrigation water applications, photoperiod, and crop genotype. Air temperature and photoperiod during critical phases of development determine plant ontogeny and biomass partitioning, and are based on plant genotype. The DSSAT system has two options for simulating N balance and SOM: the original SOM model (Godwin and Singh 1998) and a modified CENTURY model that operates on a daily time increment and at soil depth increments that conform to DSSAT (Gijsman et al. 2002).

The Environmental Policy Integrated Climate (EPIC) model was originally designed to simulate soil erosion and its effects on soil fertility (Williams et al. 1984). EPIC has now evolved into a comprehensive agro-ecosystem model capable of simulating biomass and yields of crops grown in complex rotations and under diverse management practices such as tillage, irrigation, fertilization, and liming (Williams 1995). The SOM module in EPIC uses processes similar to CENTURY but with daily time increments and several soil depths. The soil water balance submodel is similar to that in DSSAT models.

The Agricultural Production Simulator (APSIM) model is another widely used model (Keating et al. 2003) similar in detail to DSSAT and EPIC. APSIM was developed with a modular structure to allow testing and use of various methods of simulating several components of the soil, plant, and atmosphere system.

Rivington and Koo (2010), in a recent comprehensive meta-analysis of crop modeling for climate change and food security, reported that DSSAT crop models were the models most commonly used by various groups surveyed throughout the world. The report revealed perceived model limitations and made suggestions for model improvements based on user feedback.

Simulation of Crop Yield

Yield simulation in crop models is based on two processes: crop growth and development. The fraction of total biomass partitioned into grain or other harvested biomass is termed the economic yield. Crop simulations thus involve the two-step

Table 10.1. Factors affecting crop growth and development and their sensitivity to water and nitrogen deficits.

	Growth Rate		Development (Duration)	
	Mass	Expansion	Phasic	Morphological
Principal environmental factor affecting the process	Solar radiation	Temperature	Temperature, photoperiod	Temperature
Degree of variation between genotypes	Low	Low	High	Low
Sensitivity to water deficit	Low–Stomata Moderate–Leaf wilting and rolling	High–Vegetative stage Low–Grain filling stage	Low–Delay in vegetative stage	Low–Main stem High–Tillers and branches
Sensitivity to nitrogen deficiency	Low	High	Low	Low–Main stem High–Tillers and branches

Source: Ritchie and Alagarswamy (2002).

process of estimating total biomass using crop growth rate and duration and partitioning that biomass into harvested components. Separating growth and development processes also allows a distinction between sources and sinks of assimilate (i.e., photosynthetically produced carbon) within the various plant organs. A plant can be exposed to source or sink limitation during its growth cycle, where "source" refers to the production of organic matter by photosynthesis, and "sink" refers to the assimilation of that organic matter in tissues. The assimilates are stored in roots or elsewhere if the sink demand is less than source supply, as the aboveground plant parts cannot grow faster than the sink demand. During seed development, stored assimilates become available to augment daily grain fill demand.

Table 10.1 summarizes the environmental factors that influence crop growth and development and the sensitivity of these processes to water and N deficits. In the next sections, we discuss the three major processes—growth, development, and yield and yield components—important in simulating crop yield.

Crop Growth

Net photosynthesis is simulated in functional models using radiation use efficiency (RUE), which assumes that daily biomass production is directly proportional to intercepted photosynthetically active radiation (IPAR), a concept introduced by Monteith (1977). Model simulations need to consider variations in the RUE proportionality constant over the time interval measured (hourly, weekly, or seasonal), the form of biomass measured (aboveground, belowground, or specific plant part), and the type of radiation measured (i.e., total solar or photosynthetically active radiation).

Accurate leaf area index (LAI) estimates are crucial for models based on IPAR. Since LAI is the ratio of plant leaf area to the average ground area covered, it can change dramatically over the growing season until a full plant canopy has developed.

Crop Development

Phasic development describes the duration of different growth phases and the bio-mass partitioning among different plant organs. Morphological development refers to organ development during the plant life cycle. Both are affected by temperature (Table 10.1), as calculated by growing degree-days. Phasic development is also affected by photoperiod and genetics (Table 10.1). Genetic diversity within a crop species enables plants to be adapted to diverse settings in different regions of the world. For example, wheat genotypes are grown from temperate Argentina (latitude 50° S) to Sweden (60° N) and in tropical regions between.

Plant growth rate and duration are equally important in determining potential crop yields; hence, the accuracy of yield simulation models. Record high yields of annual crops are always obtained in cooler environments where there is maximal duration of daylight for plant growth. Warmer climates can equal the total annual yields of the cooler region yields by growing more than one crop per year.

The principal functional approach used to estimate the duration of crop growth is based on thermal time calculation (Gallagher 1979). Thermal time (t_d) is the accumulation of degree-days (i.e., °C d) above a base temperature and is calculated as

$$t_d = \sum_{i=1}^{n} \left(T_a - T_b \right)$$

where T_a is 24-hour daily mean temperature; T_b is the base temperature below which the crop growth ceases; and n is the number of days. T_a is usually approximated by taking the mean of daily maximum and minimum temperatures (Ritchie and NeSmith 1991).

Thermal time to simulate development requires temperature to be measured close to the growing point of the plant. Ritchie and NeSmith (1991) showed that using air temperatures to calculate thermal time and to predict the number of leaf tips and leaf development overpredicted leaf numbers in the CERES corn model, and required correction using a higher phyllochron value (i.e., duration between leaf tip appearances) (Vinocur and Ritchie 2001).

Several crop species are sensitive to photoperiod. In general, plants adapted to grow in shorter day lengths (e.g., corn, sorghum, and soybean) develop more quickly when exposed to shorter days. Plants adapted to grow in longer day lengths (e.g., wheat and barley) grow more quickly when exposed to longer days. In addition to temperature, Ritchie and NeSmith (1991) showed that photoperiod in corn can significantly affect leaf number and the duration of vegetative stages.

Yield and Yield Components

Simulation procedures for yield estimates differ among crop models. One approach is to assume a constant fraction of biomass produced at maturity (i.e., the point of economic yield) or to assume a constant increment of biomass production each day after grain filling starts. Another approach is to separately estimate the yield components (ear number, kernels per ear, and kernel weight).

The simplest level of yield simulation assumes that economic yield is a constant fraction of total aboveground biomass at maturity, known as the harvest index (HI). This index can range from 0.40 to 0.55 for corn. Both the EPIC (Williams et al. 1989) and CROPSYS (Stockle et al. 1994) models are based on HI calculations. Some models estimate corn yields using a constant rate of change in HI after silking (Muchow et al. 1990, Muchow and Sinclair 1991, Sinclair and Muchow 1995). In this case, the rate of change of HI for corn is 0.015 d^{-1} during the entire period of kernel growth. The accuracy of yield simulations by models based on the HI concept depends on the accuracy of simulating total aboveground biomass as well as the stability of HI. Such models are of more limited value in situations where the crop yields are low because of water deficits that constrain HI.

Kernel number (KN) is an important predictor of yield in most cereal crops (Evans 1993), and reflects the irreversible effects of water deficit and nutrient deficiencies that occur around the time of anthesis. Crop models using the KN concept are based on two approaches. A simple approach calculates KN from biomass at anthesis, while a more complex one estimates KN from biomass production during a critical period (around silking in the case of corn). SALUS, for example, uses the simulated stem weight at anthesis to simulate grain number.

The Systems Approach to Land Use Sustainability Model

The Systems Approach to Land Use Sustainability (SALUS) (Basso et al. 2006, 2010) is similar to the DSSAT family of models but is designed to simulate not only yields of crops in rotation, but also soil, water and nutrient dynamics as a function of management strategies over multiple years (Fig. 10.1). SALUS accounts for the effects of rotations, planting dates, plant populations, irrigation and fertilizer applications, and tillage practices. The model simulates daily plant growth and soil processes on a daily time step during the growing season and fallow periods. SALUS contains (1) crop growth modules, (2) SOM and nutrient cycling modules, and (3) soil water balance and temperature modules. The model simulates the effects of climate and management on the water balance, SOM, N and P dynamics, heat balance, plant growth, and plant development. Within the water balance, surface runoff, infiltration, surface evaporation, saturated and unsaturated soil water flow, drainage, root water uptake, soil evaporation, and transpiration are simulated. Soil organic matter decomposition, along with N mineralization and formation of ammonium and nitrate, N immobilization, and gaseous N losses are also simulated.

Crop development in the SALUS model is based on thermal time calculations modified by day length and vernalization. Potential crop growth depends on intercepted light using solar radiation data and simulated LAI, and is reduced by water or nitrogen limitations. The crop growth modules in SALUS are derived from the CERES model originally developed for single-year and monoculture simulations (Ritchie 1998, Ritchie et al. 1998). Phasic development is controlled by temperature and photoperiod and is governed by variety-specific genetic coefficients. The main external inputs required for the crop growth simulations are the genetic coefficients and climate data (daily solar radiation, precipitation, and air temperature).

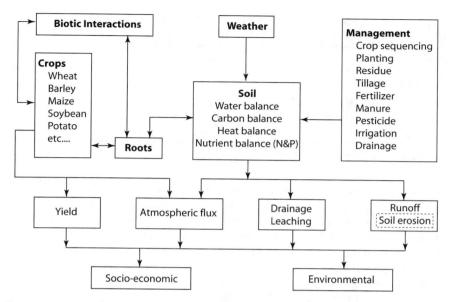

Figure 10.1. Components of the System Approach for Land Use Sustainability (SALUS) model.

The SALUS soil N and SOM modules are derived from CENTURY (Parton et al. 1988) with several new additions and modifications. The model simulates SOM and N mineralization/immobilization from three soil organic carbon pools (active, slow, and passive) that vary in their turnover rates and characteristic C:N ratios (see Paul et al. 2015, Chapter 5 in this volume). There are two crop residue/ fresh organic matter pools (structural and metabolic) for representing recalcitrant and easily decomposable residues, based on residue lignin and N content. A surface active SOM pool associated with the surface residue pools was added to better represent conservation tillage systems and perennial crops. A soil P model incorporates inorganic and organic P dynamics. Inorganic P is divided into labile, active, and stable pools.

The soil water balance module has advanced from the DSSAT models with new improvements in calculating infiltration, drainage, evaporation, and runoff. The time-to-ponding (TP) concept (White and Sully 1989) replaces the previous runoff and infiltration calculations based on the USDA-NRCS runoff curve number. SALUS does not account for impacts of pests, disease, or extreme weather such as hail.

Input data required by SALUS consist of weather, soil and crop management, soil properties, genetic characteristics of the crop, and the site location. When models are to be used in new locations, plant data such as phenology, biomass, and economic yield are needed. To test the soil simulations, information on water content and nitrate concentrations is helpful.

Weather uncertainty is the major source of insecurity and risk in agricultural production. SALUS accounts for weather variability by using several decades of existing weather information. The minimum weather dataset required for SALUS is listed in Table 10.2. Daily totals of rainfall and solar radiation along with the

Table 10.2. The minimum and optional weather datasets required as inputs for crop simulation models.

Minimum	Optional
Daily solar radiation	Daily dewpoint temperature
Daily maximum temperature	Daily wind run
Daily minimum temperature	Daily net radiation
Daily precipitation	Precipitation intensity

Source: Ritchie and Alagarswamy (2002).

maximum and minimum temperature are considered the minimum needed for relatively accurate crop simulation. The main weather element of greatest concern in most agricultural regions is the temporal distribution of rainfall. Solar radiation is the main weather variable for describing the energy available for crop growth and evapotranspiration (ET). Temperature is necessary to simulate crop phenology and to modify growth and ET.

Ideally, weather data should be obtained at a site near the area where the model is to be applied, especially for daily rainfall. Temperature and radiation are more spatially uniform, so the weather station need not be on-site. Most weather stations record rainfall and temperature but not always solar radiation. Accurate solar radiation data can be obtained from NASA (http://power. larc.nasa.gov/cgi-bin/cgiwrap/solar/agro.cgi), although the spatial resolution is given in 1° grid cells. This NASA site also provides all the daily weather data required by DSSAT and SALUS, but with the same spatial resolution issues as with solar radiation.

The minimum soil information required to run crop simulation models such as SALUS is listed in Table 10.3. On-site measurement of soil properties is recommended where possible to validate the model for a specific site. Not all soil input data may be available, in which case soil characteristics such as texture, bulk density, and organic matter content can serve as surrogate measures. However, the lower limit of available soil water and the field capacity or drained upper limit (DUL) water content are often more accurate when measured in the field than when using laboratory measurements of field soil samples. These measurements must be made when field conditions are at or near their lower and upper limits. In well-drained soil, the DUL can best be measured after the profile has been thoroughly wetted and allowed to drain without irrigation until drainage practically stops. The lower limit is best measured during a dry period in the growing season when water content ceases to decline in the root-zone because of a shortage of soil water.

The minimum crop management information required to run crop simulation models such as SALUS is listed in Table 10.4. When irrigation is used, the dates, amounts and mode of application are required. Information on the type, dates, and mode of fertilizer application is necessary to simulate nutrient dynamics, although often model assessments of crop yield assume that nutrient availability is not limiting.

Table 10.3. Soil datasets required as inputs for crop simulation models.

Minimum	Desirable for Specific Applications	Initial Conditions
Lower limit water content at 10- to 20-cm depths	Hydraulic conductivity and water retention curves at 10- to 20-cm depths	Water content at 10- to 20-cm depths
Field capacity soil water content at 10- to 20-cm depths	Runoff curve number	Soil nitrate concentration at 10- to 20-cm depths
Crop rooting depth	Surface albedo	Soil ammonium concentration at 10- to 20-cm depths
Hydraulic conductivity at soil depths that restrict water flow	Soil pH at 10- to 20-cm depths	Soil extractable phosphorus at 10- to 20-cm depths (if phosphorus subroutine is run)
	Soil organic carbon in upper depths	Fresh plant residues or manure amounts and depth of incorporation
	Soil textural characterization for 10- to 20-cm depths	
	Surface water ponding capacity	
	Soil bulk density at 10- to 20-cm depths	
	Groundwater depth bypass flow fraction	

Source: Ritchie and Alagarswamy (2002).

Table 10.4. Crop management datasets required as inputs for crop simulation models.

Minimum	Optional
Crop cultivar characteristics	Row spacing
Planting date and depth	Row direction
Plant population density	Pesticide inputs
Irrigation inputs (date, amount, depth)	Harvest date
Fertilizer inputs (date, amount, type)	
Crop residues or manure inputs (dates, quality, amount)	

Source: Ritchie and Alagarswamy (2002).

The crop variety, genotype, or cultivars also must be specified; cultivars may vary significantly in the duration of developmental phases and in the partitioning of assimilates within the plant. Wheat and corn cultivar information is generally expressed as genetic coefficients, which allow models to simulate crop phenology over a wide range of latitudes and planting times.

Assessment of Biogeochemical Fluxes under Different Management Strategies Using SALUS

Nitrate Leaching Following Manure Application

Organic sources of N are often considered superior to inorganic fertilizers because they decompose slower and promote better soil structure and overall soil quality. However, there has been little field-based research to quantify nitrate leaching when animal manure is applied as the primary source of nutrients in intensive crop production systems. It is possible that when organic sources of N fertilizer are used, nitrate leaching may be greater than when using inorganic N because organic N is converted to inorganic N only slowly, so large quantities of organic N are needed to provide enough N to rapidly growing crop plants during the relatively short time of intense N uptake. More surplus N may then be mineralized and available for leaching at the end of the growing season.

Basso and Ritchie (2005) quantified N leaching from KBS plots receiving large quantities of either animal manure (18 ton ha^{-1} yr^{-1}) or inorganic N fertilizer (120 kg N ha^{-1} yr^{-1}) from January 1994 to December 1999 in a corn–alfalfa rotation. The results were used to validate the ability of SALUS to simulate nitrate leaching. Most of the water drainage occurred early in the season or after harvest and was lower during the growing period of the crop. SALUS provided a reasonable simulation of the amount of water drained and nitrate leached for both manure and inorganic N fertilizer over the 6 years of the study (Table 10.5). The manure plots leached 33% more N as nitrate (NO_3^-) than did the plots treated with inorganic N, illustrating the trade-off between the organic matter benefits of manure and a greater N loss to the environment (Millar and Robertson 2015, Chapter 9 in this volume). Field studies and the validated model results showed that leaching can be substantial if a high quantity of manure is applied to soils in autumn (Basso and Ritchie 2005, Beckwith et al. 1998, Chambers et al. 2000).

Soil Carbon Changes in Cropped and Unmanaged Ecosystems

Soil tillage has contributed significantly to the increase in atmospheric CO_2 that has occurred over the last two centuries (Wilson 1978). Historically, intensive tillage of agricultural soils has led to substantial losses of soil C, ranging from 30 to 50% of preconversion levels (Davidson and Ackerman 1993). These CO_2 losses are related to soil fracturing and opening, which facilitates the movement of CO_2 out of—and oxygen into—the soil (Reicosky 1997, Lal 2004), and especially to the destruction of soil aggregates (e.g., Grandy and Robertson 2006; Paul et al. 2015, Chapter 5 in this volume), which exposes otherwise protected organic matter to microbial attack. Although conventional moldboard plowing buries nearly all plant residue, it leaves the soil in a rough, loose, and open condition, which maximizes CO_2 loss and results in a consistent reduction in SOM. Reduced tillage results in more soil C retention or sequestration, which reduces its atmospheric release (Cole 1996, Paustian et al. 1998, Rasmussen et al. 1998).

Table 10.5. Comparison of SALUS simulations to field measurements for cropping systems with either inorganic nitrogen (N) or manure additions.

Variable	Inorganic N (140 kg ha⁻¹)	Manure (18 ton ha⁻¹)
Biomass (kg ha⁻¹)		
Measured[a]	20,893	21,015
Simulated[a]	21,450	21,932
(RMSE[b])	450	645
Cumulative Nitrate Leaching[c] (kg NO₃-N ha⁻¹)		
Measured	279	367
Simulated	273	362
RMSE[b]	15.7	14.3
Cumulative Drainage[c] (mm)		
Measured	1904	1857
Simulated	1901	1862
RMSE[b]	24.4	54.5

[a]Corn dry biomass harvested for silage (1997).
[b]RMSE = Root Mean Square Error.
[c]Nitrate leaching and drainage were measured and modeled over a 6-year (1994–1999) corn–alfalfa rotation at the W.K. Kellogg Biological Station south of the KBS LTER Main Cropping System Experiment (selected data from Basso and Ritchie 2005).

A major concern among producers is the possibility of yield reductions associated with permanent no-till management compared to conventional tillage (Grandy et al. 2006). Residue cover on the soil surface reflects solar radiation and acts as an insulator, slowing warming of the soils in the spring. This effect is more noticeable in temperate climates with wet and cool springs because high soil water content maintained by residue cover is combined with low incoming energy (Allmaras et al. 1977). Reicosky et al. (1977) reported that on poorly drained soils, corn yields were decreased because poorly drained soils are usually colder in the growing season due to higher water content. When vegetative corn development is delayed by lower soil temperature because of residue cover, yield can be lost due to a shortened growth period. However, residue cover can improve soil water availability by increasing infiltration, protecting the soil surface from erosion, and reducing evaporative losses. Thus, residue cover can improve yields in lower rainfall years and in drier locations (Basso et al. 2006, Bertocco et al. 2008).

SALUS was recently used to simulate soil carbon changes in different land use management practices, including tillage, at the KBS LTER (Senthilkumar et al. 2009; Paul et al. 2015, Chapter 5 in this volume). The simulations of soil C changes obtained using SALUS were consistent with measurements in the Conventional and No-till systems of the KBS Main Cropping System Experiment (Fig. 10.2). The model also simulated the observed large loss of soil C in fertilized, conventionally tilled plots in an adjacent experiment.

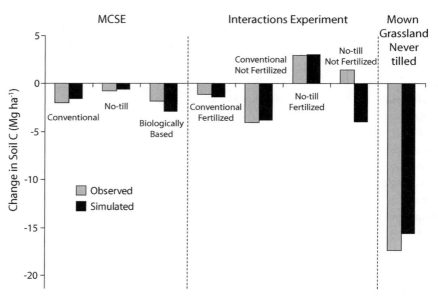

Figure 10.2. Measured and simulated changes in soil carbon after 18–21 years of different management systems at the KBS LTER Main Cropping System Experiment (MCSE) and at the Interactions Experiment, an adjacent continuous corn tillage (conventional vs. no-till) × nitrogen fertilizer (fertilized vs. not fertilized) experiment. Modified from Senthilkumar et al. (2009).

Linking Crop Models with Digital Terrain Analysis for Assessing Spatially Connected Processes

The assessment of soil water spatial patterns is crucial for understanding crop yield variation across the landscape. Soil water within a field is highly variable in space and time as a result of several processes that occur at different scales and because of complex interactions among weather, topography, soil, and vegetation. The effect of topographic convergence and divergence in natural landscapes has a major impact on soil water balance (Moore and Grayson 1991). Without consideration of the terrain characteristics, accurate simulation of soil water balance in entire, nonuniform fields is not possible. Spatial variability of soil water content is often the cause of yield variation over space and time. Accurate estimation of the spatial variability of soil water is also important for other applications including soil erosion, groundwater flow models, and precision agriculture.

The dynamics of soil water balance and crop growth have been extensively modeled to assess the risk associated with uncertainty in water availability (Jones et al. 1993). Soil–plant–atmosphere models often simulate vertical drainage but not lateral movement and water routing across the landscape (Basso 2000).

Existing digital terrain models are able to partition the landscape into a series of interconnected elements to spatially route water flow (Moore et al. 1993, Vertessy et al. 1993). Most digital terrain models fill the depressions in landscapes to provide a continuous flow of water to streams, making their application for agricultural purposes limited. Basso (2000) created a spatial soil water balance model called SALUS-TERRAE that accounts for water pooling in depressions, surface and sub-surface water movements, and the water runoff–runon mechanism occurring on the landscape. SALUS-TERRAE was developed by coupling the Ritchie vertical–soil–water balance model (Ritchie 1998) with TERRAE, a digital terrain model developed by Gallant and Basso (2013). SALUS-TERRAE is a spatial soil water balance model composed of vertical and lateral components of the water balance. The model requires a digital elevation map for partitioning the landscape into a series of interconnected irregular elements. Weather and soil information for the soil water balance simulation is also needed.

SALUS-TERRAE has been applied at a location in Michigan similar to the KBS LTER. Figure 10.3 shows the spatial variability of soil water content across the landscape the day after a rainfall event of 65 mm. SALUS-TERRAE was able to simulate the higher surface ponding capacities in the depression areas. The model performed well when compared to field measurements of soil water content for the entire growing season (Fig. 10.4): the root mean square error (RMSE) between measured and simulated results varied from 0.22 cm to 0.68 cm (Basso 2000, Batchelor et al. 2002). The net surface flow (Fig. 10.5) is the difference between the amount of water leaving each element (runoff) from that running onto the element (runon). The highest value (–5 cm) is observed at summit positions in the landscape since these elements do not have any water running into them. Application of the SALUS-TERRAE model can benefit precision agriculture by being able to select the appropriate management strategy for optimizing management practices across the landscape.

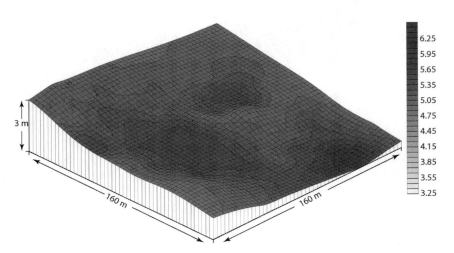

Figure 10.3. Simulated kriged map of soil water content (cm) for the surface (0–26 cm) soil layer using the digital terrain model SALUS-TERRAE in a sandy loam soil in Durand, Michigan. Redrawn from Basso (2000).

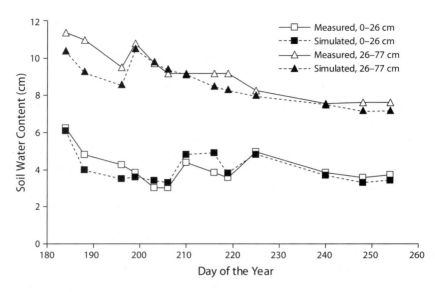

Figure 10.4. Measured and simulated water content for the soil profile (0–26 cm) and (26–77 cm) in the medium elevation zone (upper saddle) for the entire season in Durand, Michigan. Redrawn from Basso (2000).

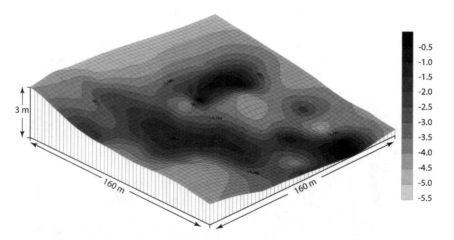

Figure 10.5. Simulated kriged map of net surface flow (cm) calculated as the difference between run-on and run-off using SALUS-TERRAE on a sandy loam soil in Durand, Michigan. Redrawn from Basso (2000).

A current limitation in most crop models is the assumption of uniform plant distribution. Yield variability at the field scale is the norm rather than the exception (Sadler et al. 1994; Basso et al. 2001, 2012b). Visual observations as well as measurements commonly indicate that crops are not uniformly distributed, and therefore assuming they are can be an unrealistic assumption and a significant source of uncertainty in yield simulations. A correction procedure based on the extent of variation in plant stand uniformity or dominant plant density may be necessary. Correction also is required to compensate for yield loss from plants missing in a population; to some extent, neighboring plants can compensate for missing plants because they have more space to intercept light. Pommel and Bonhomme (1998) demonstrated the degree of compensation and losses from irregular stands in corn.

Summary

Simulation models are important for providing producers and policy makers with better decision-making capabilities. By predicting the response of different sustainability indicators to changes in crop management and climate, models can provide much needed information for designing sustainable cropping systems and landscapes. Functional models are particularly useful in that they integrate crop growth and yield with environmental responses such as nitrate leaching, carbon sequestration, erosion, and nitrous oxide emissions. It would be impossible for a single model to address all the issues regarding sustainable crop productivity or meet the goals of every researcher, planner, or policy maker. However, based on the successes of models like DSSAT, CENTURY, SALUS, and EPIC—along with continuing technological improvements—it is reasonable to expect development of more useful Decision Support System models to meet a growing range of demands. SALUS is promising because it has a crop model with several years of testing and is coupled with tested conservative simulations of soil C, N, and P models, allowing users to account for the impact of agronomic management on crop net primary productivity and on the environment. It also has tested capability to simulate climate change impact on production and the environment. The coupling with TERRAE makes SALUS a unique system with the capabilities of simulating the effects of topography and terrain attributes on water routing across the landscape.

References

Addiscott, T. M., and R. J. Wagenet. 1985. Concepts of solute leaching in soils: a review of modeling approaches. European Journal of Soil Science 36:411–424.

Allmaras, R. R., E. A. Hallauer, W. W. Nelson, and S. D. Evans. 1977. Surface energy balance and soil thermal property modifications by tillage induced soil structure. Technical Bulletin 306–1977, Minnesota Agricultural Experiment Station, University of Minnesota, Minneapolis, Minnesota, USA.

Basso, B. 2000. Digital terrain analysis and simulation modeling to assess spatial variability of soil water balance and crop production. Dissertation, Michigan State University, East Lansing, Michigan, USA.

Basso, B., M. Bertocco L. Sartori, and E. C. Martin. 2007. Analyzing the effects of climate variability on spatial pattern of yield in a maize-wheat-soybean rotation. European Journal of Agronomy 26:82–91.

Basso, B., D. Cammarano, A. Troccoli, D. Chen, and J. T. Ritchie. 2010. Long-term wheat response to nitrogen in a rainfed Mediterranean environment: field data and simulation analysis. European Journal of Agronomy 33:132–138.

Basso, B., and J. T. Ritchie. 2005. Impact of compost, manure and inorganic fertilizer on nitrate leaching and yield for a 6-year maize-alfalfa rotation in Michigan. Agriculture, Ecosystems & Environment 108:329–341.

Basso, B., J. T. Ritchie, D. Cammarano, and L. Sartori. 2011. A strategic and tactical management approach to select optimal N fertilizer rates for wheat in a spatially variable field. European Journal of Agronomy 35:215–222.

Basso, B., J. T. Ritchie, P. R. Grace, and L. Sartori. 2006. Simulation of tillage systems impacts on soil biophysical properties using the SALUS model. Italian Journal of Agronomy 4:677–688.

Basso, B., J. T. Ritchie, and J. W. Jones. 2012a. On modeling approaches for effective assessment of hydrology of bioenergy crops: comments on Le at al. (2011) Proc Natl Acad Sci USA 108:15085–15090. European Journal of Agronomy 38:64–65.

Basso, B., J. T. Ritchie, F. J. Pierce, R. P. Braga, and J. W. Jones. 2001. Spatial validation of crop models for precision agriculture. Agricultural Systems 68:97–112.

Basso, B., L. Sartori, D. Cammarano, P. Grace, C. Sorensen, and S. Fountas. 2012b. Environmental and economic evaluation of N fertilizer rates in a maize crop in Italy: a spatial and temporal analysis using crop models. Biosystems Engineering 113:103–111.

Batchelor, W. D., B. Basso, and J. O. Paz. 2002. Examples of strategies to analyze spatial and temporal yield variability using crop models. European Journal of Agronomy 18:141–158.

Beck, H. W., P. Jones, and J. W. Jones. 1989. SOYBUG: an expert system for soybean insect pest management. Agricultural Systems 30:269–286.

Beckwith, C. P., J. Cooper, K. A. Smith, and M. A. Shepherd. 1998. Nitrate leaching loss following application of organic manures to sandy soils in arable cropping. Soil Use and Management 14:123–130.

Bertocco, M., B. Basso, L. Sartori, and E. C. Martin. 2008. Evaluating energy efficiency of site-specific tillage in maize in NE Italy. Bioresource Technology 99:6957–6965.

Boggess, W. G., P. J. van Blokland, and S. D. Moss. 1989. FinARS: a financial analysis review expert systems. Agricultural Systems 31:19–34.

Chambers, B. J., K. A. Smith, and B. F. Pain 2000. Strategies to encourage better use of nitrogen in animal manures. Soil Use and Management 16:157–166.

Cole, C. V. K. 1996. Agricultural options for mitigation of greenhouse gas emissions. Pages 746–771 in R. T. Watson, M. C. Zinyowera, and R. H. Moss, editors. Climate change 1995: impacts, adaptations and mitigation of climate change. Scientific-technical analysis. Contribution of Working Group II to the Second Assessment Report of the Intergovernmental Panel on Climate Change. Cambridge University Press, Cambridge, UK.

Davidson, E. A., and I. L. Ackerman. 1993. Changes in soil carbon inventories following cultivation of previously untilled soils. Biogeochemistry 20:161–193.

Del Grosso, S. J., W. J. Parton, A. R. Mosier, D. S. Ojima, A. E. Kulmala, and S. Phongpan. 2000a. General model for N2O and N2 gas emissions from soils due to denitrification. Global Biogeochemical Cycles 14:1045–1060.

Del Grosso, S. J., W. J. Parton, A. R. Mosier, D. S. Ojima, C. S. Potter, W. Borken, R. Brumme, K. Butterbach-Bahl, P. M. Crill, K. E. Dobbie, and K. A. Smith. 2000b. General CH_4 oxidation model and comparisons of CH_4 oxidation in natural and managed systems. Global Biogeochemical Cycles 14:999–1019.

Del Grosso, S. J., W. J. Parton, A. R. Mosier, D. S. Walsh, D. S. Ojima, and P. E. Thornton. 2006. DAYCENT national scale simulations of N_2O emissions from cropped soils in the USA. Journal of Environmental Quality 35:1451–1460.

Evans, L. T. 1993. Crop evolution, adaptation, and yield. Cambridge University Press, Cambridge, UK.

FAO (Food and Agriculture Organization of the United Nations). 2011. Save and grow: a policymaker's guide to the sustainable intensification of smallholder crop production. FAO, Rome, Italy.

Gage, S. H., J. E. Doll, and G. R. Safir. 2015. A crop stress index to predict climatic effects on row-crop agriculture in the U.S. North Central Region. Pages 77–103 in S. K. Hamilton, J. E. Doll, and G. P. Robertson, editors. The ecology of agricultural ecosystems: long-term research on the path to sustainability. Oxford University Press, New York, New York, USA.

Gallagher, J. N. 1979. Ear development: processes and prospects. Pages 3–9 in J. H. J. Spiertz and T. H. Kramer, editors. Crop physiology and cereal breeding. PUDOC, Wageningen, The Netherlands.

Gallant, J. C., and B. Basso. 2013. Creating a flow-oriented modeling mesh using the stream function. Page 346 in J. Piantadosi, R. S. Anderssen, and J. Boland, editors. Adapting to change: the multiple roles of modelling. 20th International Congress on Modelling and Simulation (MODSIM2013). Adelaide, Australia.

Gijsman, A. J., G. Hoogenboom, W. J. Parton, and P. C. Kerridge. 2002. Modifying DSSAT crop models for low-input agricultural systems using a soil organic matter-residue module from CENTURY. Agronomy Journal 94:462–474.

Godwin, D. C., and U. Singh. 1998. Nitrogen balance and crop response to nitrogen in upland and low-land cropping systems. Pages 55–78 in G. Y. Tsuji, G. Hoogenboom, and P. K. Thornton, editors. Understanding options for agricultural production. Kluwer Academic Publishers, Dordrecht, The Netherlands.

Grandy, A. S., and G. P. Robertson. 2006. Initial cultivation of a temperate-region soil immediately accelerates aggregate turnover and CO_2 and N_2O fluxes. Global Change Biology 12:1507–1520.

Grandy, A. S., G. P. Robertson, and K. D. Thelen. 2006. Do productivity and environmental trade-offs justify periodically cultivating no-till cropping systems? Agronomy Journal 98:1377–1383.

Jones, J. W., W. T. Bowen, W. G. Boggess, and J. T. Ritchie. 1993. Decision support systems for sustainable agriculture. Pages 123–138 in Technologies for sustainable agriculture in the tropics. ASA Special Publication 56. ASA, CSSSA, and SSSA, Madison, Wisconsin, USA.

Keating, B. A., P. S. Carberry, G. L. Hammer, M. E. Probert, M. J. Robertson, D. Holzworth, N. I. Huth, J. N. G. Hargreaves, H. Meinke, Z. Hochman, G. McLean, K. Verburg, V. Snow, J. P. Dimes, M. Silburn, E. Wang, S. Brown, K. L. Bristow, S. Asseng, S. Chapman, R. L. McCown, D. M. Freebairn, and C. J. Smith. 2003. An overview of APSIM, a model designed for farming systems simulation. European Journal of Agronomy 18:267–288.

Lal, R. 2004. Soil carbon sequestration impacts on global climate change and food security. Science 304:1623–1627.

Li, C. 1995. Impact of agricultural practices on soil C storage and N_2O emissions in 6 states in the US. Pages 101–112 in R. Lal, J. Kimble, E. Levine, and B. A. Stewart, editors. Soil management and greenhouse effect, Advances in soil science, CRC Press, Boca Raton, Florida, USA.

Li, C. 2000. Modeling trace gas emissions from agricultural ecosystems. Nutrient Cycling in Agroecosystems 58:259–276.

McCown, R. L., G. L. Hammer, J. N. G. Hargreaves, D. P. Holzworth, and D. M. Freebairn. 1996. APSIM: a novel software system for model development, model testing and simulation in agricultural system research. Agricultural Systems 50:255–271.

Michalski, R. S., J. H. Davis, V. S. Bisht and J. B. Sinclair. 1983. A computer-based advisory system for diagnosing soybean diseases in Illinois. Plant Disease 4:459–463.

Millar, N., and G. P. Robertson. 2015. Nitrogen transfers and transformations in row-crop ecosystems. Pages 213–251 in S. K. Hamilton, J. E. Doll, and G. P. Robertson, editors. The ecology of agricultural ecosystems: long-term research on the path to sustainability. Oxford University Press, New York, New York, USA.

Monteith, J. L. 1977. Climate and the efficiency of crop production in Britain. Philosophical Transactions of the Royal Society of London B 281:277–294.

Moore, I. D., P. E. Gessler, G. A. Nielsen, and G. A. Peterson. 1993. Soil attribute prediction using terrain analysis. Soil Science Society of America Journal 57:443–452.

Moore, I. D., and R. B. Grayson. 1991. Terrain-based catchment partitioning and runoff prediction using vector elevation data. Water Resources Research 27:1177–1191.

Muchow, R. C., and T. R. Sinclair. 1991. Water deficit effects on maize yields modeled under current and "greenhouse" climates. Agronomy Journal 83:1052–1059.

Muchow, R. C., T. R. Sinclair, and J. M. Bennett. 1990. Temperature and solar radiation effects on potential maize yield across locations. Agronomy Journal 82:338–343.

Parton, W. J., M. D. Hartman, D. S. Ojima, and D. S. Schimel. 1998. DAYCENT and its land surface submodel: description and testing. Global and Planetary Change 19:35–48.

Parton, W. J., E. A. Holland, S. J. Del Grosso, M. D. Hartman, R. E. Martin, A. R. Mosier, D. S. Ojima, and D. S. Schimel. 2001. Generalized model for NO_x and N_2O emissions from soils. Journal of Geophysical Research-Atmospheres 106:17403–17419.

Parton, W. J., A. R. Mosier, D. S. Ojima, D. W. Valentine, D. S. Schimel, and K. Weier. 1996. Generalized model for N_2 and N_2O production from nitrification and denitrification. Global Biogeochemical Cycles 10:401–412.

Parton, W. J., D. S. Schimel, C. V. Cole, and D. S. Ojima. 1987. Analysis of factors controlling soil organic matter levels in Great Plains Grasslands. Soil Science Society of America Journal 51:1173–1179.

Parton, W. J., J. W. B. Stewart, and C. V. Cole. 1988. Dynamics of C, N, P, and S in grassland soils: a model. Biogeochemistry 5:109–131.

Paul, E. A., A. Kravchenko, A. S. Grandy, and S. Morris. 2015. Soil organic matter dynamics: controls and management for sustainable ecosystem functioning. Pages 104–134 in S. K. Hamilton, J. E. Doll, and G. P. Robertson, editors. The ecology of agricultural ecosystems: long-term research on the path to sustainability. Oxford University Press, New York, New York, USA.

Paustian, K., E. T. Elliott, and K. Killian. 1998. Modeling soil carbon in relation to management and climate change in some agroecosystems in central North America. Pages 459–471 in R. Lal, J. M. Kimble, R. F. Follett, and B. A. Stewart, editors. Soil processes and the carbon cycle. Advances in Soil Science. CRC Press, Boca Raton, Florida, USA.

Plant, R. E. 1989. An integrated expert decision support system for agricultural management. Agricultural Systems 29:49–66.

Pommel, B., and R. Bonhomme. 1998. Variations in the vegetative and reproductive systems in individual plants of a heterogeneous maize crop. European Journal of Agronomy 8:39–49.

Rasmussen, P. E., K. W. T. Goulding, J. R. Brown, P. R. Grace, H. H. Janzen, and M. Korschens. 1998. Long-term agroecosystem experiments: assessing agricultural sustainability and global change. Science 282:893–896.

Reicosky, D. C. 1997. Tillage-induced CO_2 emission from soil. Nutrient Cycling in Agroecosystems 49:273–285.

Reicosky D. C., D. K. Cassel, R. L. Blevins, W. R. Gill, and G. C. Naderman. 1977. Conservation tillage in the southeast. Journal of Soil Water Conservation 32:13–20.

Ritchie, J. T. 1995. International consortium for agricultural systems applications (ICASA): establishment and purpose. Agricultural Systems 49:329–335.

Ritchie, J. T. 1998. Soil water and plant stress. Pages 41–54 in G. Y. Tsuji, G. Hoogenboom, and P. K. Thorton, editors. Understanding options for agricultural production. Kluwer Academic Publishers, Dordrecht, The Netherlands.

Ritchie, J. T., and G. Alagarswamy. 2002. Overview of crop models for assessment of crop production. Pages 43–68 in O. C. Doering III, J. C. Randolph, J. Southworth, and R. A. Pfeifer, editors. Effects of climate change and variability on agricultural production systems. Kluwer Academic Publishing, Dordrecht, The Netherlands.

Ritchie, J. T., and D. S. NeSmith. 1991. Temperature and crop development. Pages 6–29 in J. T. Ritchie and H. J. Hanks, editors. Modeling plant and soil systems. American Society of Agronomy, Madison, Wisconsin, USA.

Ritchie, J. T., U. Singh, D. C. Godwin, and W. T. Bowen. 1998. Cereal growth, development and yield. Pages 79–98 in G. Y. Tsuji, G. Hoogenboom, and P. K. Thornton, editors. Understanding options for agricultural production. Kluwer Academic Publishers, Dordrecht, The Netherlands.

Rivington, M., and J. Koo. 2010. Report on the meta-analysis of crop modelling for climate change and food security survey. CGIAR Research Program on Climate Change, Agriculture and Food Security, Copenhagen, Denmark. <http://cgspace.cgiar.org/handle/10568/10255>

Robertson, G. P., and R. R. Harwood. 2013. Sustainable agriculture. Pages 111–118 in S. A. Levin, editor. Encyclopedia of biodiversity. Second edition. Volume 1. Academic Press, Waltham, Massachusetts, USA.

Sadler, E. J., P. J. Bauer, and W. J. Busscher. 1994. Spatial corn yield during drought in the SE coastal plain. Pages 365–381 in P. C. Robert, R. H. Rust, and W. E. Larsen, editors. Site-specific management for agricultural systems. Proceedings of Second International Conference, ASA-SSA-CSSA, Minneapolis, Minnesota, USA.

Senthilkumar, S., B. Basso, A. N. Kravchenko, and G. P. Robertson. 2009. Contemporary evidence for soil carbon loss in the U.S. corn belt. Soil Science Society of America Journal 73:2078–2086.

Sinclair, T. R., and R. C. Muchow. 1995. Effects of nitrogen supply on maize yields: I. Modeling physiological responses. Agronomy Journal 87:632–641.

Stockle, C. O., S. A. Martin, and G. S. Campbell. 1994. CropSyst, a cropping systems model: water/nitrogen budgets and crop yield. Agricultural Systems 46:335–339.

Thompson, L. M. 1969. Weather and technology in production of corn in US corn belt. Agronomy Journal 61:453–456.

Thompson, L. M. 1986. Climate change, weather variability, and corn production. Agronomy Journal 78:649–653.

Tsuji, G. Y., G. Hoogenboom, and F. C. Thornton, editors. 1998. Understanding options for agricultural production. Kluwer Academic Publishers, Dordrecht, The Netherlands.

Vertessy, R. A., J. T. Hatton, P. J. O'Shaughnessy, and M. D. A. Jayasuriya. 1993. Predicting water yield from a mountain ash forest catchment using a terrain analysis based catchment model. Journal of Hydrology 150:665–700.

Vinocur, M. G., and J. T. Ritchie. 2001. Maize leaf development biases caused by air-apex temperature differences. Agronomy Journal 93:767–772.

White, I., and M. J. Sully. 1989. Use of hydrological robustness of time-to-incipient-ponding. Soil Science Society of America Journal 53:1343–1346.

Williams, J. R. 1995. The EPIC model. Pages 909–1000 in V. P. Singh, editor. Computer models of watershed hydrology. Water Research Publications, Highlands Ranch, Colorado, USA.

Williams, J. R., C. A. Jones, and P. T. Dyke. 1984. A modeling approach to determining the relationship between erosion and soil productivity. Transactions of the ASAE 27:129–144.

Williams, J. R., C. A. Jones, J. R. Kiniry, and D. A. Spanel. 1989. The EPIC crop growth model. Transactions of the ASAE 32:479–511.

Wilson, A. T. 1978. The explosion of pioneer agriculture: contribution to the global CO_2 increase. Nature 273:40–41.

11

Water Quality and Movement
in Agricultural Landscapes

Stephen K. Hamilton

Water quality refers to the physical, chemical, and biological characteristics of surface waters and groundwaters that determine their suitability for use by humans and aquatic life. Both natural and managed landscapes provide an important ecosystem service by maintaining water quality, which is key for water supply, recreational use, aesthetic values, and biodiversity, including fish and wildlife habitat. The upper U.S. Midwest is a region endowed with abundant groundwater, lakes, and wetlands as a result of a glacial topography and humid climate. Within this region lies the Kellogg Biological Station Long-Term Ecological Research site (KBS LTER) in southwest Michigan, situated in a heterogeneous, largely rural landscape (Fig. 11.1). The region in the vicinity of KBS is ideal for comparative study of how water quality changes as water moves through landscapes and how the transport of nutrients via water movement through watersheds is affected by land use and land cover, human activities, and biogeochemical transformations in surface water bodies. Agricultural influences on water quality are of particular interest in the U.S. Midwest, where nutrient export from farmland to groundwaters, lakes, and streams can lead to high nitrate in drinking water supplies and to surface water eutrophication (i.e., excessive algal and plant growth).

Water-quality work at the KBS LTER has two major goals. The first is to improve our understanding of how water quality changes as water flows across the landscape, including the effects of natural processes as well as changes ascribed to agricultural row-crop management. A second goal is to examine how the movement of water through streams and wetlands may lead to changes in water quality that include retention or removal of nutrients of concern for eutrophication, and to investigate the specific processes responsible for the changes. Although the information presented is largely from the KBS region, it is generally applicable to landscapes with intensive agriculture.

In this chapter, I discuss the main effects of row-crop agriculture on water quality, as illustrated by findings from the KBS LTER and from studies in the KBS

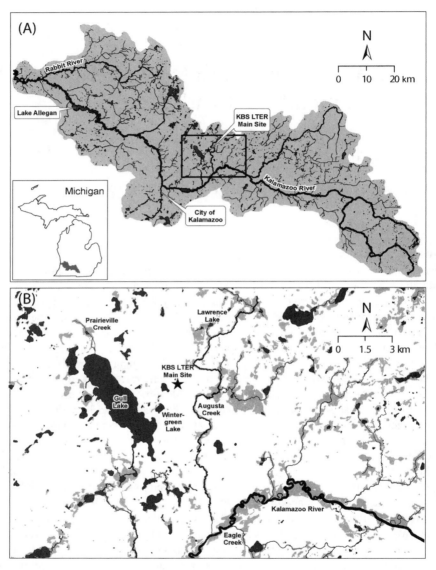

Figure 11.1. Location of the Kellogg Biological Station Long-term Ecological Research site (KBS LTER; within rectangle) in the (A) Kalamazoo River watershed, which includes Kalamazoo, the largest urban area. Enlargement of the rectangle area (B) shows lakes and streams in black and wetlands in gray. Surface-water features referred to in the text are labeled.

region that have shown how abiotic and biotic factors, including human activity, interact to determine water quality. I have organized the chapter by separating the landscape flow path into two distinct parts: (1) water movement from precipitation to soils and groundwater, and (2) receiving surface waters including streams, wetlands, and lakes. Unless noted otherwise, the data presented in this chapter are derived from water samples collected from 1996–2009 using consistent methods as detailed in Hamilton et al. (2001, 2007, 2009). Most of the water samples were collected in nonurban regions of Kalamazoo, Barry, and Allegan counties, within

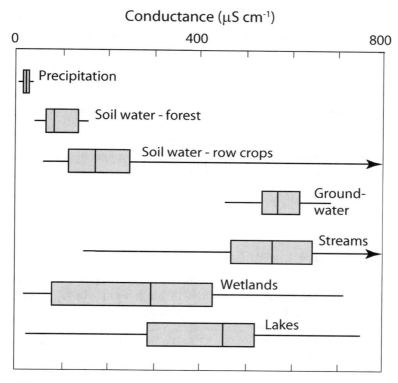

Figure 11.2. Distributions of specific conductance of precipitation, soil waters sampled at 1.2 m depth, groundwater pumped from residential wells, and surface waters. The conductance (corrected to 25°C) reflects the total ionic content. Boxes show the interquartile range and median, and lines show the range (arrows indicate outliers not shown here). Data sources: (i) Precipitation based on 30 years of year-round monitoring by NADP/NTN (2011) (volume-weighted means for 1979–2008; see Table 11.1); (ii) Soil water from forest based on 98 samples taken at three of the KBS LTER Main Cropping System Experiment (MCSE; Table 11.2) Deciduous Forest plots during Mar–Nov from 2000–2003 (Kurzman 2006); (iii) Soil water from row-crop fields based on 128 samples taken at three of the MCSE Conventional plots during Mar–Nov from 2000–2003 (Kurzman 2006); (iv) Groundwater based on 19 pumped well samples collected from 1996–2008; and (v) Streams, wetlands, and lakes based on means for sites (n = 245 streams, 174 wetlands, and 184 lakes), mainly from southwest Michigan except for lakes which were sampled across the Lower Peninsula of Michigan.

the Kalamazoo River watershed (Fig. 11.1) or just north of it in southern Barry County. A rich body of KBS research on the limnology and ecology of local lakes and streams that has been published since the 1970s aided in the interpretation of patterns and processes.

What can be learned from patterns of water chemistry across landscapes? By making simple measurements of specific conductance, we find that the total ion content of water changes dramatically across the landscape, and the patterns observed reflect biogeochemical processes along subsurface and surface flow paths (Fig. 11.2). Ion content increases greatly as water from precipitation passes through soils and the underlying unsaturated zone before reaching groundwater reservoirs. Streams largely reflect the chemistry of their groundwater sources. In contrast, lakes and particularly wetlands tend to have lower and more variable total ion contents, and hence specific conductances, mainly due to variable relative contributions of groundwater and direct precipitation. These changes in water chemistry have implications for the biological availability of nitrogen (N), phosphorus (P), and silicon (Si)—the nutrients that often limit aquatic primary production. Total ion content also affects mineral precipitation reactions—particularly those involving calcium carbonate—that, in turn, affect the transport and fate of pollutants. In this chapter, we evaluate landscape patterns in water quality, including both major ions and key nutrient forms, and consider how the effects of agricultural activities are superimposed on natural patterns and processes.

From Precipitation to Groundwater

Precipitation Amount and Variability

Southern Michigan's climate is continental and is significantly influenced by the Great Lakes. Mean air temperatures at KBS range from −3.8°C in January to 22.9°C in July (1981–2010; NCDC 2013). Mean seasonal cycles of precipitation and Thorthwaite potential evapotranspiration (PET) are typical for this part of North America (Crum et al. 1990, Fig. 11.3). From 1951 to 1980, the mean annual precipitation for KBS was 855 mm, and the difference between precipitation and PET was 124 mm. And during the next 30 years (1981–2010), the mean annual precipitation increased 17% (1005 mm). Total precipitation varies considerably from year to year; since 1996, water levels in a number of local wetlands and lakes near KBS site have varied over 1–2 m (for an example, see Robertson and Hamilton 2015, Chapter 1 in this volume), reflecting this interannual climate variability.

Hydrogeology and Landscape Flow Paths

Southwest Michigan's recently glaciated landscape is distinct because of its high degree of linkage between surface waters and groundwaters and the predominant influence of groundwater discharge on streams, rivers, most lakes, and many wetlands (Grannemann et al. 2008). The landscape geomorphology is largely a reflection of the most recent continental glaciation, when the Saginaw and Lake

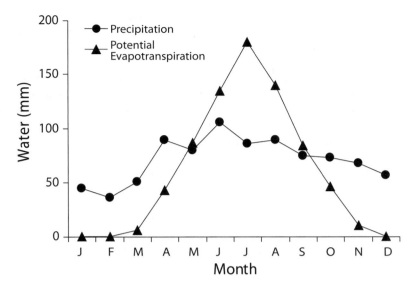

Figure 11.3. Monthly means of precipitation and potential evapotranspiration (Thornthwaite method) based on measurements at KBS from 1951–1980. Redrawn from Crum et al. (1990).

Michigan ice lobes converged in this area (Kincare and Larson 2009). As these lobes retreated beginning around 18,000 years ago, they left till plains and upland end moraines that contain sand and gravel mixed with finer materials. Outwash plains—with medium to coarse sand and gravel derived from the erosion of these moraines—were deposited by the general southward drainage of the glacial melt-waters (Rheaume 1990, Kehew et al. 1996). In the vicinity of KBS, the uncon-solidated glacial till and outwash deposits extend to depths of 15–60 m, with the thinnest deposits found along the Kalamazoo River valley where elevations are lowest (Rheaume 1990). The bedrock lying beneath the glacial deposits is either the Coldwater Shale or, to the northeast, the Marshall Sandstone formation. The glacial deposits are so thick that bedrock outcrops are not found in the area.

Land around KBS drains southward toward the Kalamazoo River, a tributary to Lake Michigan with a drainage basin of 5260 km^2 (Fig. 11.1). The Kalamazoo River watershed is approximately 50% agricultural land (dominated by corn and soybean, with wheat and alfalfa as well), 33% unmanaged terrestrial uplands (mostly secondary deciduous forest and successional old fields), 7% lakes and wet-lands, and 10% urban. Most agricultural land around KBS is naturally well drained and is neither tile drained nor ditched (Schaetzl 2009). Most of the area's abundant lakes and wetlands occupy depressions (kettles) formed by melting residual glacial ice, and many lack surface connections to other water bodies. The Kalamazoo River and its floodplain occupy an incised valley that formed from drainage of glacial meltwaters. Tributary streams reach the river through side valleys that dissect the outwash plains, and often originate in or pass through lakes and wetlands. A sub-stantial fraction of the upland area contains undulating, hummocky terrain that is difficult to assign to a particular stream watershed based on surface topography. Instead, analysis of groundwater levels in domestic water-supply wells can be used to reveal the direction and magnitude of underground flow and can provide a reasonable indication of watershed boundaries (Lusch 2009).

Hydrologic budgets for the Kalamazoo River watershed (Allen et al. 1972) show for a 34-year record (1933–1966) that of the period's 889 mm of annual precipitation, about 580 mm (65%) was returned to the atmosphere by evapotranspiration. Most of the remainder became river runoff, mainly via groundwater flow paths. The annual rate of groundwater recharge by precipitation in the area averaged 229 mm (26% of mean annual precipitation) and occurred mainly during the cooler months of November through May, when evapotranspiration rates are low (Fig. 11.3).

In general, surface runoff is low because glacial deposits in the watershed have high hydraulic conductivity, which facilitates infiltration. Thus, on a landscape scale most precipitation that is not evapotranspired readily infiltrates the soil and percolates to the water table, making groundwater flow paths especially important. Groundwater flow in turn supplies water to lakes, streams, and rivers. The hydrological linkages between ground and surface waters in the KBS LTER landscape resemble those well documented in the North Temperate Lakes LTER site in northern Wisconsin (Webster et al. 2006).

A hydrologic budget has been constructed for Augusta Creek (Fig. 11.1), the tributary of the Kalamazoo River draining ~90 km^2 of land including the eastern side of the KBS LTER sites (Rheaume 1990). For a year of average precipitation (1977: 950 mm), evapotranspiration returned 65% of the annual precipitation to the atmosphere (Fig. 11.4), the same percentage found by Allen et al. (1972) in earlier work on the entire Kalamazoo River watershed. The remaining 35% was discharged as stream runoff, of which an estimated 75% entered as groundwater flow, corresponding to an annual groundwater recharge rate of 248 mm (26% of precipitation in that year). Thus, the area around KBS, which includes the Main Cropping System Experiment (MCSE) of the KBS LTER (Robertson and Hamilton 2015, Chapter 1 in this volume), has a similar groundwater recharge rate to that of the overall landscape of the region.

While such hydrologic budgets provide a long-term water balance, they give no picture of the time scales of water movement through landscapes (Webster et al. 2006). Given the importance of groundwater flow paths and the large volume of groundwater reservoirs, transit times of water through these watersheds are undoubtedly long compared to watersheds in which overland flow is a more important route for water movement. Groundwater dating—using tracers such as industrial chlorofluorocarbon gases—shows that the mean age of groundwater sampled from water-supply wells in recently glaciated landscapes is often several decades (Saad 2008, Rupert 2008, Stewart et al. 2010, Hamilton 2012). Although the mean age of groundwater discharged into streams can be similar to that of water from groundwater wells, streams typically receive convergent groundwater flow paths of widely differing ages (Böhlke 2002). While groundwater dating based on tracers has not been conducted in the vicinity of KBS, a model for the Augusta Creek watershed shows the expected convergence of flow paths of widely varying distances and residence times to deliver water to the stream (Bartholic et al. 2007).

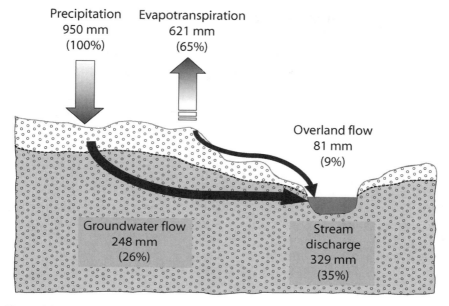

Figure 11.4. Water budget for the Augusta Creek watershed representing a year of average precipitation. Annual fluxes are expressed as equivalent depth of water over the watershed, and as a percentage of the annual precipitation. Shading represents the water table. Data from Rheaume (1990).

Precipitation Chemistry

On account of air pollution, precipitation at KBS is more acidic and enriched in nitrate (NO_3^-) and sulfate (SO_4^{2-}) than it would be otherwise (Table 11.1, Fig. 11.5), although concentrations of these anions and the accompanying acidity have been declining in recent years. Southwest Michigan is within a broad region of moderate to high acid deposition rates that extends across the midwestern and northeastern United States (Driscoll et al. 2001). High atmospheric deposition rates of NO_3^- and SO_4^{2-} in industrialized regions result from anthropogenic N and sulfur (S) oxide emissions, originating primarily from fossil fuel combustion. In addition, anthropogenic activities—particularly those dealing with livestock and fertilized crops—likely increase ammonium (NH_4^+) deposition (Konarik and Aneja 2008). From 1979 to 2010, the mean wet deposition rates of NH_4^+-N, NO_3^--N, and SO_4^{2-} at KBS were 3.09, 3.44, and 20.2 kg ha^{-1} yr^{-1}, respectively. Sources and fluxes of organic N and P in atmospheric deposition have not often been measured and are not well understood. Total (wet + dry, or bulk) deposition rates of N and P were determined by Rheaume (1990) by combining data from two National Oceanic and Atmospheric Administration stations and two U.S. Geological Survey stations in Kalamazoo County. Rheaume estimated total annual bulk deposition rates, which are mostly due to wet deposition, as 14.8 kg N ha^{-1} yr^{-1} for total nitrogen (NO_3^- + NH_4^+ + organic N) and 0.34 kg P ha^{-1} yr^{-1} for total phosphorus (PO_4^{3-} + organic P). Tague (1977) reported a similar atmospheric deposition rate for total P (0.33 kg ha^{-1} yr^{-1}) at KBS during 1974.

Table 11.1. Hydrochemistry of precipitation and groundwater in the vicinity of Kellogg Biological Station.

Measurement[*]	Precipitation		Groundwater		
	USGS[a]	NADP/NTN[b]	Hamilton[c]	USGS -Kalamazoo[d]	WMU-Barry[e]
# dates or sites	14 dates	1240 dates	18 sites	46 sites	170–179 sites
Conductance (μS cm^{-1}; 25°C)	34.0	21.7	556	587	403
pH	4.3	4.47	7.48	7.34	NA
ANC[*] (mg HCO$_3^-$ L^{-1})	~0	~0	318	259	240
Ca^{2+} (mg L^{-1})	0.19	0.21	80.6	81.0	70.0
Mg^{2+} (mg L^{-1})	<0.01	0.04	25.0	25.0	23.0
Na$^+$ (mg L^{-1})	<0.2	0.08	6.40	5.1	9.30
K$^+$ (mg L^{-1})	<0.1	0.02	1.54	1.0	1.60
Cl$^-$ (mg L^{-1})	0.25	0.13	11.0	11.0	0
SO$_4^{2-}$ (mg L^{-1})	3.5	2.25	36.7	32.0	35.00
NO$_3^-$ (mg N L^{-1})	0.61	0.39	1.85	0.19	0
NH$_4^+$ (mg N L^{-1})	0.56	0.35	0.035	0.04	0.06
Si (mg Si L^{-1})	0.0	NA	6.96	1.69	3.04
DOC[*] (mg L^{-1})	NA	NA	1.65	NA	NA
TDN[*] (mg L^{-1})	1.6	NA	NA	0.46	NA
TDP[*] (mg L^{-1})	0.03	NA	0.011	0.01	NA

[*]DOC = dissolved organic carbon. ANC = acid-neutralizing capacity (also known as total alkalinity). TDN = total dissolved N. TDP = total dissolved P.
[a]Rheaume (1990: Tables 7 and 8 in that report); data are means of wet deposition for a precipitation station near the mouth of Gull Creek, sampled for most variables 14 times between November 1986 and September 1987.
[b]National Atmospheric Deposition Program (NRSP-3)/National Trends Network means of annual volume-weighted means in wet deposition for 1979–2010 from the KBS station (MI26), downloaded August 14, 2011. NADP/NTN Coordination Office, Illinois State Water Survey, Champaign, IL.
[c]Mostly residential water-supply wells at homes on KBS property and most were sampled once; Hamilton lab database queried December 2009.
[d]Rheaume (1990: Table 20); data are medians for 46 observation wells located throughout Kalamazoo County. The mean concentration of NO$_3^-$ was 3.64 mg N L^{-1}.
[e]Kehew and Brewer (1992: Table 2); data are medians for 170–179 domestic wells located throughout Barry County (excludes wells judged to be contaminated by anthropogenic NO$_3^-$, NH$_4^+$, or Cl$^-$).

Notably, long-term monitoring of precipitation chemistry at KBS has shown a marked reduction of SO$_4^{2-}$ and NO$_3^-$ wet deposition since the 1980s and a consequent increase in the pH of precipitation (NADP/NTN 2011). Wet deposition of SO$_4^{2-}$ has fallen from a mean annual rate of 28.8 kg ha^{-1} yr^{-1} in the 1980s to a mean annual rate of 12.2 kg ha^{-1} yr^{-1} in recent years (2006–2010). A less marked but significant decrease in NO$_3^-$ wet deposition is also apparent, falling from a mean annual rate of 4.23 kg N ha^{-1} yr^{-1} in the 1980s to 2.51 kg N ha^{-1} yr^{-1} from 2006 to 2010. There is no apparent trend for NH$_4^+$ over this period. Precipitation acidity has decreased greatly in conjunction with SO$_4^{2-}$ and NO$_3^-$ decreases; precipitation pH

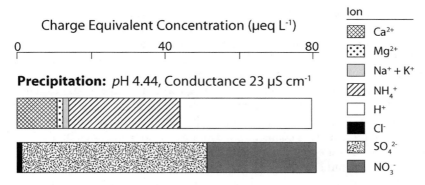

Figure 11.5. Ionic composition of precipitation at KBS, expressed as charge equivalents. Data are volume-weighted means for the 30-year monitoring record (1979–2008; see Table 11.1 for sources).

has increased from a mean of 4.34 in the 1980s to 4.83 from 2006 to 2010, and total acid deposition has fallen from 0.44 to 0.14 kg H^+ ha^{-1} yr^{-1} in the same time frame. Similar trends have been observed throughout the midwestern and eastern United States and likely reflect reduced emissions of S and N oxides due to tighter environmental controls on anthropogenic sources (Driscoll et al. 2001, Kahl et al. 2004).

Natural Chemical Changes as Water Moves through Soils

Water from precipitation changes greatly in chemical composition as it percolates through soils. Some of these changes reflect dissolution of materials at or near the soil surface, but as water percolates through the upper few meters of the soil profile, mineral weathering changes its chemistry further. Because the soils and underlying glacial deposits in the vicinity of KBS are recently formed, they still contain an abundance of readily weathered minerals. Groundwater at KBS begins to acquire its chemical signature within a few meters of the soil surface.

Upland soils in the KBS landscape are generally well-drained loams that developed under deciduous forest (Robertson and Hamilton 2015, Chapter 1 in this volume). The forest was interspersed with smaller areas of oak openings, prairie, and savanna before most areas were converted to agriculture during European settlement in the mid-1800s (Chapman and Brewer 2008). In recent decades, row-crop agriculture has continued on the relatively level outwash plains, while secondary forest has developed on much of the less productive land that was abandoned from agriculture during the 1900s. The most common soil formations are alfisols that cover most of the upland areas, with histosols, mollisols, and entisols occupying lower-lying areas around lakes, wetlands, and streams (Schaetzl 2009). Soil mineralogy on upland soils at KBS is dominated by quartz, K-feldspar, plagioclase, and amphibole (Jin et al. 2008a).

Carbonate minerals (calcite and dolomite) are abundant in the glacial deposits from which KBS soils are derived. Acid precipitation has little effect on most surface water bodies in the region because the acidity is neutralized as water passes through these carbonate-rich soils on its way to groundwater, and because most surface waters receive groundwater inputs that are alkaline. However, in surface soils (0.5–2 m, or deeper in the coarsest soils) where carbonate minerals have been leached since deglaciation and their acid neutralization capacity diminished, acid precipitation could cause various biotic stresses including aluminum toxicity and leaching of calcium to the point where plant growth may experience calcium deficiency (Driscoll et al. 2001). In cropped soils, substantial acidity is generated by nitrification stimulated by N fertilizers and by removal of base cations in harvest, but much of this acidity is counteracted by the application of agricultural lime (calcium carbonate or dolomite) at several year intervals (see below).

As indicated earlier, the hydrochemical changes that occur as water percolates through the first few meters of soil are dramatic (Fig. 11.6). For example, as precipitation percolates through unfertilized soils at KBS, soil solute concentrations increase considerably, primarily due to mineral dissolution (Kurzman 2006). In the upper 1.2 m—the carbonate-leached zone in these soils—silicate mineral weathering produces modest increases in solutes including base cations (calcium [Ca^{2+}], magnesium [Mg^{2+}], sodium [Na^+]), dissolved Si, and carbonate alkalinity (Jin et al. 2008a, b). However, a much larger and abrupt increase in total solutes occurs beneath about 1.5 m as percolating water contacts carbonate minerals. Within the carbonate mineral zone, the soil water reacts with calcite and dolomite to substantially increase concentrations of dissolved Ca^{2+}, Mg^{2+}, and acid neutralizing capacity (ANC, almost entirely due to bicarbonate at these pH values), with a concomitant increase in pH and specific conductance (Hamilton et al. 2007, Jin et al. 2008a). Nitrate also contributes significantly to the total anion composition in the soil waters depicted in Fig. 11.6.

Effects of Agricultural Management on Water Quality

Land management practices—particularly fertilization and cultivation of N-fixing crops—can profoundly influence the chemical composition of percolating waters (Böhlke 2002, Chen and Driscoll 2009). Nitrogen often readily moves as NO_3^- from cropping systems into groundwater, and in well-drained soils groundwater NO_3^- leaching rates beneath fertilized crops are commonly 10–50% of fertilizer N application rates (Böhlke 2002, Raymond et al. 2012). In contrast, P and other contaminants such as metals and pesticides that are less mobile tend to be retained in upland soils, with some notable exceptions such as the herbicide atrazine and its derivatives (Unterreiner and Kehew 2005, Bexfield 2008, Saad 2008). However, P binding by soils eventually becomes saturated with high rates of application, increasing the mobility of P in soils and potentially leading to its export into surface waters (Domagalski and Johnson 2011, Kleinman et al. 2011, Sharpley et al. 2013).

The KBS Main Cropping System Experiment (MCSE; Table 11.2; Robertson and Hamilton 2015, Chapter 1 in this volume) provides an opportunity to investigate how conventional and alternative management of row crops affects the quality of water percolating through soils in comparison with unmanaged (nonagricultural) vegetation at various stages of ecological succession. Such investigations commonly use tension samplers to collect soil water samples (using a vacuum)

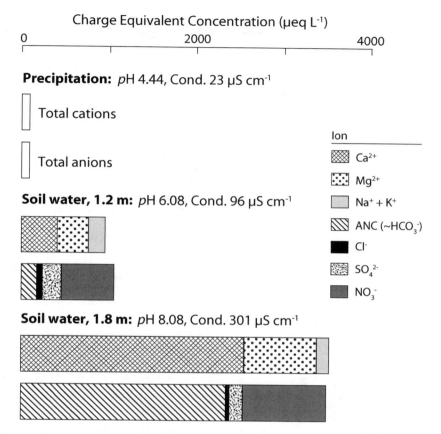

Figure 11.6. Ionic composition, expressed as charge equivalents, of soil water collected with tension samplers at 1.2 and 1.8 m depths, and of precipitation (data from Fig. 11.5). The soil water samples represent hydrochemical changes in precipitation as it percolates through soils without added fertilizer, lime, and organic wastes. Soils at the 1.2-m depth had no detectable carbonate minerals at the three MCSE Deciduous Forest sites, whereas those collected from 1.8-m depth contained abundant carbonate minerals (8 wt % calcite and 9 wt % dolomite: Jin et al. 2008a). Cond. = specific conductance; ANC = acid neutralizing capacity, due almost entirely to bicarbonate (HCO_3^-) alkalinity in these waters. Soil water at 1.2 m based on 98 samples taken at three of the KBS Deciduous Forest sites during Mar–Nov from 2000–2003 (data from Kurzman 2006). Soil water at 1.8 m based on 35 samples from a monolith lysimeter located near the MCSE, within the 3BC2 horizon, collected during Mar–May from 2003–2004 (data from Jin et al. 2008a).

for measuring the concentration of solutes in percolating water leaving the rooting zone, and when combined with water budget estimates of drainage, they provide an estimate of total hydrologic solute loss. A comprehensive analysis of major solute hydrochemistry in soil water collected from tension samplers located just beneath the rooting zone of MCSE systems (Kurzman 2006) shows that the hydrologic loss of solutes varies with vegetation and agronomic management. The total ionic

Table 11.2. Description of the KBS LTER Main Cropping System Experiment (MCSE).[a]

Cropping System/Community	Dominant Growth Form	Management
Annual Cropping Systems		
Conventional (T1)	Herbaceous annual	Prevailing norm for tilled corn–soybean–winter wheat (c–s–w) rotation; standard chemical inputs, chisel-plowed, no cover crops, no manure or compost
No-till (T2)	Herbaceous annual	Prevailing norm for no-till c–s–w rotation; standard chemical inputs, permanent no-till, no cover crops, no manure or compost
Reduced Input (T3)	Herbaceous annual	Biologically based c–s–w rotation managed to reduce synthetic chemical inputs; chisel-plowed, winter cover crop of red clover or annual rye, no manure or compost
Biologically Based (T4)	Herbaceous annual	Biologically based c–s–w rotation managed without synthetic chemical inputs; chisel-plowed, mechanical weed control, winter cover crop of red clover or annual rye, no manure or compost; certified organic
Perennial Cropping Systems		
Alfalfa (T6)	Herbaceous perennial	5- to 6-year rotation with winter wheat as a 1-year break crop
Poplar (T5)	Woody perennial	Hybrid poplar trees on a ca. 10-year harvest cycle, either replanted or coppiced after harvest
Coniferous Forest (CF)	Woody perennial	Planted conifers periodically thinned
Successional and Reference Communities		
Early Successional (T7)	Herbaceous perennial	Historically tilled cropland abandoned in 1988; unmanaged but for annual spring burn to control woody species
Mown Grassland (never tilled) (T8)	Herbaceous perennial	Cleared woodlot (late 1950s) never tilled, unmanaged but for annual fall mowing to control woody species
Mid-successional (SF)	Herbaceous annual + woody perennial	Historically tilled cropland abandoned ca. 1955; unmanaged, with regrowth in transition to forest
Deciduous Forest (DF)	Woody perennial	Late successional native forest never cleared (two sites) or logged once ca. 1900 (one site); unmanaged

[a]Site codes that have been used throughout the project's history are given in parentheses. Systems T1–T7 are replicated within the LTER main site; others are replicated in the surrounding landscape. For further details, see Robertson and Hamilton (2015, Chapter 1 in this volume).

content, as indicated by the conductance of soil solutions, ranged widely across the various systems with the lowest levels beneath the Deciduous Forest and the highest levels beneath the Conventional agriculture systems. Nitrate concentrations were correlated with total dissolved phosphorus (TDP) concentrations in soil water, and higher concentrations of both NO_3^- and TDP were associated with fertilization or cultivation of alfalfa, an N-fixing crop (Fig. 11.7).

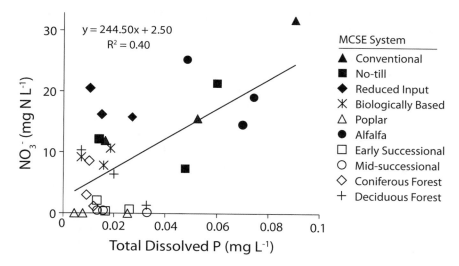

$y = 244.50x + 2.50$
$R^2 = 0.40$

MCSE System

▲ Conventional
■ No-till
◆ Reduced Input
✳ Biologically Based
△ Poplar
● Alfalfa
□ Early Successional
○ Mid-successional
◇ Coniferous Forest
+ Deciduous Forest

Figure 11.7. Total dissolved phosphorus (TDP) and nitrate (NO_3^-) in soil waters collected with tension samplers at 1.2 m depth from MCSE systems. Row-crop systems receiving fertilizer are shown with solid symbols. Each symbol represents the mean of 100–150 samples collected at a particular site from 2000–2003; n = 3 replicate sites for each system. The regression line is fit to all points and is highly significant (P < 0.001). Data from Kurzman (2006).

Although soil water NO_3^- and TDP concentrations are correlated across MCSE systems, these two ions are differentially retained by surface soils, as illustrated by following changes in N:P ratios. The molar N:P ratio of $NO_3^- + NH_4^+$ to TDP in precipitation is ~70–86, based on the NADP/NTN or USGS N data, respectively, assuming TDP is ~0.03 mg L^{-1} (Table 11.1). Soils tend to retain P but lose N by NO_3^- leaching. The molar N:P ratio in soil water beneath the root zone, averaged across all MCSE cropping systems, was 749 based on the ratio of $NO_3^- + NH_4^+$ to TDP concentrations (Kurzman 2006). Thus, the water percolating out of the root zone is enriched in N, largely in the form of NO_3^-, relative to P. This increase in dissolved N:P ratios as precipitation infiltrates the soil reflects not only gains in N but also the tendency for P to be retained in the soil. As a result, soils in the root zone of the MCSE contain much more P relative to N; Robertson et al. (1997) reported a molar N:P in the upper 15 cm of MCSE soils of 0.3, based on extractable inorganic N ($NO_3^- + NH_4^+$) and total P. Given the considerable depth of the unsaturated zone below the root zone (and the depth of soil water sampling), and the predominance of inorganic phosphate in the soil water (>90% of TDP, on average: Kurzman 2006), it is possible that much of the P that is transported downward in percolating water becomes sequestered with minerals below the depth of sampling, but possibly within the reach of deeply rooted plants. This is consistent with the lower TDP concentrations typical of groundwater pumped from water-supply wells (means, 0.011 mg L^{-1} in well waters vs. 0.027 mg L^{-1} in soil waters). This selective retention of P in soils helps to explain why N is the most frequently limiting nutrient in terrestrial ecosystems despite the high N inputs, while P tends to be limiting in groundwater-fed surface waters.

Estimates of NO_3^- leaching under MCSE systems clearly show the contrast between annual row crops and unmanaged perennial vegetation and the role of fertilization and N-fixing crops as sources of NO_3^- (Fig. 11.8). Syswerda et al. (2012) combined measured NO_3^- concentrations in soil water beneath the root zone with modeled soil water export to provide estimates of NO_3^- leached from the root zone over 11 years. Soil water export (drainage) is markedly higher in the Conventional, No-till, and Deciduous Forest systems (Fig. 11.8A) than in the Reduced Input, Biologically Based, Alfalfa, Poplar, and Early and Mid-successional systems, indicating differences in evapotranspiration losses among these systems.

Nitrate leaching fluxes ranged from less than 1 kg N ha^{-1} yr^{-1} in unfertilized Poplar to 62 kg N ha^{-1} yr^{-1} in the Conventional corn–soybean–wheat rotation (Fig. 11.8B). Over 75% of the fertilizer added to the Conventional system was lost as NO_3^- over the 11-year period. Mean annual leaching losses were also high in the No-till, Reduced Input, and Biologically Based annual cropping systems, but progressively and significantly lower in each; the Biologically Based system leached only 19 kg N ha^{-1} yr^{-1}, on average. Alfalfa leached even less N over this period, and most of this occurred during a normal break year when a small unfertilized grain crop was grown prior to reestablishing the alfalfa stand. The lowest rates of NO_3^- leaching were observed in the Early and Mid-successional systems, and in Poplars, none of which were regularly fertilized (the Poplar system received 60 kg N ha^{-1} in its first year). The relatively mature Deciduous Forest leached more NO_3^- than these systems, presumably reflecting low nutrient demand by the forests' steady-state biomass in combination with high drainage rates. Over the 11-year period, volume-weighted mean NO_3^- concentrations in drainage water (Fig. 11.8C) were highest in the Conventional system—above the threshold for acceptable drinking water quality of 10 mg L^{-1} as N, established by the U.S. Environmental Protection Agency—and they approached this threshold in the No-till and Reduced Input systems. Dissolved organic N and NH_4^+ were measured in these soil water samples in earlier work and were found to be relatively minor constituents compared to NO_3^- (Syswerda et al. 2012). These findings corroborate results from other studies (Power et al. 2001), although few studies have included such a diversity of land cover and management regimes at a single location nor for this long a period.

Carbonate mineral dissolution and precipitation in soils can contribute to soil-atmosphere carbon dioxide (CO_2) exchanges (Hamilton et al. 2007). Dissolution of carbonate minerals occurs in reaction with acidity, and can be either a source or a sink for CO_2 depending on whether the reaction occurs with strong acids (e.g., nitric acid) or carbonic acid, respectively. Acidity in agricultural soils is mostly produced by two processes: (1) carbonic acid formed by CO_2 produced when roots and microbes respire, and (2) nitric acid produced during nitrification of NH_4^+ to NO_3^-. Acidification can reduce soil fertility through several mechanisms; therefore, typically farmers of land free of native carbonate minerals periodically add carbonate minerals (agricultural lime) to the soil to counteract acidifying processes.

Hamilton et al. (2007) analyzed the pathways by which carbonate minerals dissolve in soils and groundwater of the KBS region. In particular, they investigated

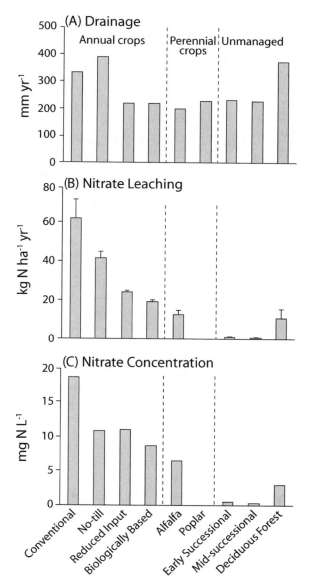

Figure 11.8. Determination of nitrate (NO$_3^-$) leaching losses from MCSE systems. Modeled water drainage (A) was combined with NO$_3^-$ concentrations measured at 1.2 m during periods of drainage to estimate mean annual NO$_3^-$ leaching losses (B) over an 11-year period. Volume-weighted mean NO$_3^-$ concentrations in drainage water (C) were calculated from the total fluxes and drainage over the 11 years. Means of 100-150 water samples from three replicate plots in each system. The 11-year period from 1995–2006 spanned 3.5 full rotations (corn-soybean-wheat) of the annual cropping systems. Data from Syswerda et al. (2012).

how soil microbial processes influence the fate of the carbon in carbonate, dissolved from either native carbonate minerals or from agricultural liming materials. The degree to which percolating water can dissolve carbonate minerals and accumulate Ca^{2+} and Mg^{2+} is controlled by temperature and pH. The pH of soil water is inversely proportional to dissolved free CO_2 from root and microbial respiration and to mineral acidity from precipitation and from internal soil processes. Nitrification of fertilizer-derived N by soil bacteria is an important source of nitric acid in the fertilized systems, and reaction of nitric acid with carbonate minerals can cause their dissolution to switch from a net CO_2 sink to a source as NO_3^- concentrations increase in infiltrating waters. Whether carbonate minerals dissolve in reaction with dissolved free CO_2 or with nitric acid, either reaction yields dissolved Ca^{2+} and Mg^{2+}. In glacial landscapes such as southern Michigan, carbonate mineral dissolution is an important source of these cations when the sum of Ca^{2+} and Mg^{2+} concentrations exceeds ~2 meq L^{-1} (Hamilton et al. 2007); dissolution of other minerals (e.g., silicates) does not generate such high concentrations, although soils elsewhere can contain other significant sources of these cations.

The charge equivalents of Ca^{2+}, Mg^{2+}, HCO_3^-, and NO_3^- in soil solutions yield clues about the relative importance of carbonate mineral dissolution by reaction with carbonic acid vs. nitric acid (Hamilton et al. 2007). Concentrations of Ca^{2+} and Mg^{2+} were significantly and positively correlated with NO_3^- ($r = 0.9$ and 0.8, respectively) across all MCSE systems (Kurzman 2006), reflecting the importance of nitric acid reaction with carbonate minerals, as discussed above. And positive correlations of Ca^{2+} and Mg^{2+} with NO_3^- in soil water have also been observed in other agricultural systems (Böhlke 2002). High NO_3^- concentrations in soil water beneath or emerging from fertilized agricultural fields are associated with unnaturally high concentrations of Ca^{2+} and Mg^{2+} (i.e., well above the charge equivalent of HCO_3^-), likely as a result of the additional dissolution capacity of nitric acid (and possibly sulfuric acid in some cases) compared with carbonic acid (Böhlke 2002, Hamilton et al. 2007).

Groundwater Quality

Groundwater in the vicinity of KBS occurs in unconsolidated glacial deposits as an upper unconfined aquifer, with an underlying semi-confined aquifer in some areas (Allen et al. 1972). The depth of the unsaturated zone (i.e., from the land surface to the water table) is generally <15 m. Groundwater surface gradients tend to follow the land surface, but as noted above, the direction of groundwater flow is not always apparent from surface topography. Nonetheless, groundwater is generally recharged in the upland areas and discharged to the lower-lying lakes, streams, and wetlands. Lateral flow toward streams and rivers is likely to be more important in the outwash plains, while the morainal systems are more likely to be dominated by localized vertical flow systems (Kehew and Brewer 1992). Wetlands and lakes can receive groundwater through-flow, in which groundwater inputs may enter on one end and exit back to the groundwater on the other end (Kehew et al. 1998).

Groundwater flow paths can be complex in glacial deposits because of the presence of interbedded layers of coarse and fine materials, as shown by a detailed study of flow patterns in an outwash plain located ~35 km southwest of KBS (Kehew et al. 1996). For the Augusta Creek watershed near KBS, spatial patterns of groundwater flow have been inferred from maps of the water table in unconfined aquifers, as derived from water-supply well records (Bartholic et al. 2007). In the land surrounding Gull Lake (Fig. 11.1B), the elevation of the water table varies only about 40 m; the water table around the MCSE lies about 30 m higher than the lowest land-surface elevation in the area (along the Kalamazoo River).

Hydrochemical information on local groundwaters is available mainly from analyses of domestic wells and is summarized in Table 11.1 (Allen et al. 1972, Rheaume 1990, Kehew and Brewer 1992). Most domestic wells in the area pump groundwater from relatively shallow depths of <25 m, so the analysis of well water samples provides an indication of the hydrochemistry of groundwaters that would be discharged to lakes and streams. The major ion composition of a well close to the KBS LTER main site is depicted in Fig. 11.9. As discussed above, the dissolution of carbonate minerals is largely responsible for the high ionic strength of ground-water around KBS (Kehew et al. 1996), as it is throughout the lower peninsula of Michigan (Wahrer et al. 1996, Jin et al. 2008b). Compared to the water sampled from the soil profile at a depth of 1.8 m (Fig. 11.6), which is just into the carbonate mineral zone, groundwater has higher concentrations of Ca^{2+}, Mg^{2+} and HCO_3^-, and higher ratios of Mg^{2+} to Ca^{2+} (Jin et al. 2008b, 2009). Groundwater also often has higher SO_4^{2-} concentrations than the soil waters measured at KBS; SO_4^{2-} in ground-water can originate from gypsum ($CaSO_4$) dissolution or from oxidation of sulfide minerals (e.g., pyrite FeS_2) in the glacial deposits (Böhlke 2002).

Local groundwaters can generally be classified as the calcium–magnesium–bicarbonate type due to the dissolution of carbonate minerals. Ferrous iron concentrations may be elevated in groundwaters with little or no dissolved oxygen. These concentrations likely increase during the passage of recharge waters through zones of low redox potential such as lake and wetland sediments (Kehew et al. 1996) or when there is very high organic loading at the soil surface.

In agricultural landscapes, NO_3^- concentrations in domestic wells commonly exceed the drinking-water standard of 10 mg N L^{-1} (Rupert 2008, Puckett et al. 2011), although the KBS well water shown in Fig. 11.9 had a particularly low NO_3^- concentration (13 µg N L^{-1} = 1 µeq L^{-1}) compared to other wells in the area. Chowdhury et al. (2003) analyzed data for 8733 wells in Kalamazoo County (where KBS is located) and found that NO_3^- concentrations exceeded 5 and 10 mg N L^{-1} in 28% and 3% of the wells, respectively; concentrations >5 mg N L^{-1} were considered indicative of surface contamination sources. Rheaume (1990) presented evidence that NO_3^- concentrations in the area have increased substantially over the past few decades, documented the spatial patterns in NO_3^- contamination as of the late 1980s, and formulated an N budget that implicates agricultural fertilizers as the primary cause of the increased contamination of local aquifers. As mentioned above, atmospheric deposition is a substantial source of N loading as well. Groundwater in the deeper layers of the glacial deposits still tends to be lower in NO_3^- (Rheaume 1990, Kehew et al. 1996), which likely reflects its longer turnover time and the relatively recent history of fertilizer use in the region as well as reactions that consume NO_3^- in the groundwater system (Böhlke 2002).

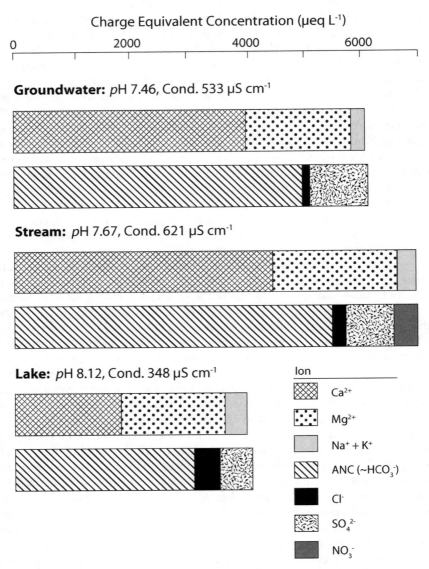

Figure 11.9. Ionic composition, expressed as charge equivalents, of groundwater from KBS wells, a groundwater-fed stream (Prairieville Creek) draining outwash plains northwest of KBS, and the large inland lake (Gull Lake) into which that stream flows. Note the increased scale from Figure 11.6. Cond. = specific conductance; NO_3^- concentrations in groundwater and the lake water were too low to depict. Data are means for 19 groundwater wells with variable sampling dates, 41 sampling dates for the creek, and 86 sampling dates for the lake.

Surface-derived contaminants other than NO_3^- do not yet appear to be an extensive problem in local groundwaters around KBS, and a survey of a wide suite of organic contaminants and trace elements in wells and streams throughout Kalamazoo County revealed little contamination in the rural areas (Rheaume 1990). Hydrogeological characteristics indicate, however, that most of the area surrounding KBS is moderately to highly susceptible to groundwater contamination from surface sources (Rheaume 1990). In the glacial deposits of Barry County, a rural area just north of KBS, Kehew and Brewer (1992) did find evidence of surface-derived contamination from human activity in water samples collected from domestic water-supply wells. About 25% of those wells had not only elevated concentrations of NO_3^- (>1 mg N L^{-1}) but also NH_4^+ (>0.5 mg N L^{-1}) and/or chloride (≥10 mg L^{-1}), the latter likely originating from the use of de-icing salts on roads and probably also the addition of potash (most commonly composed of KCl) to agricultural fields.

Arsenic is a common health concern in well water, including wells around KBS. In Michigan its origin is thought to be largely from native minerals, but mobilization of arsenic could be related to human activities, including increased loading of soluble cations (e.g., Ca^{2+} and Mg^{2+} associated with NO_3^- leaching, as discussed above). Anthropogenic increases in cation loading can cause arsenic to be released from ion exchange sites in soils, potentially resulting in arsenic accumulation downstream in groundwater flow paths (Böhlke 2002).

Surface Waters

Streams and Rivers

Streams and rivers are important in the transformation and retention of nutrients, often reducing the nutrient load in water passing through them (Alexander et al. 2009, Mulholland et al. 2008). Alterations in surface flow paths—by draining wetlands and straightening streams—prevent water from passing through soil and wetland filters that naturally retain nutrients and sediments, and result in degraded water quality downstream (NRC 1995). There has been less wetland drainage in the immediate vicinity of KBS than in most other parts of southern Michigan and the rest of the upper Midwest (Prince 1997) because it is difficult to drain the abundant isolated depression wetlands.

Streams around KBS resemble the groundwater in their total ionic content because they derive most of their annual discharge from groundwater (Fig. 11.2). The stream example depicted in Fig, 11.9 is for Prairieville Creek (Fig. 11.1), a tributary to Gull Lake that originates from a complex of groundwater springs and drains a largely agricultural watershed. Streams with lower conductance than groundwater tend to be outflows from larger lakes and wetlands that capture more direct precipitation that dilutes groundwater inflows. The few streams in the region around KBS that have significantly higher conductance than groundwater tend to be in urban areas and likely receive pollution, particularly Na^+ and Cl^-. Overland flow from storm runoff or snow melt also tends to dilute the total ion content. When groundwater emerges at the surface, it loses dissolved CO_2 by diffusion to the atmosphere. This, as well as warmer summertime temperatures, often causes calcium carbonate to precipitate on underwater surfaces and sediments. Carbonate precipitates are frequently visible in local springs and streams as well as in wetlands and lakes where groundwater inflow rates are high. The influence of carbonate precipitation on streamwater concentrations of Ca^{2+} and acid neutralizing capacity (carbonate alkalinity) is small unless the water resides in a lake or reservoir along the stream network (Szramek and Walter 2004, Reid and Hamilton 2007, Baas 2009).

Most local streams receive considerable groundwater inputs as their valleys cut downward through the glacial outwash plains, with groundwater entering via diffuse seepage and occasional discrete springs along their channels. Stream riparian zones with high rates of groundwater discharge are known to be hotspots of biogeochemical activity due to high biological productivity and the constant delivery of reactive solutes by groundwater flow (McClain et al. 2003). Hedin et al. (1998) analyzed biogeochemical processes at the soil–stream interface in a small tributary of Augusta Creek (Fig. 11.1) to show the changing importance of aerobic and anaerobic microbial processes across a gradient of decreasing oxygen availability. Denitrification, sulfate reduction, and methanogenesis occurred within the anaerobic zones.

The Lotic Intersite Nitrogen Experiment (LINX) investigated N cycling in headwater streams across the United States, including streams in the vicinity of KBS. This experiment used a coordinated set of whole-stream stable isotope additions to reveal how streams act as N processors (Peterson et al. 2001; Webster et al. 2003; Mulholland et al. 2008, 2009; Hall et al. 2009). The first set of LINX studies examined N cycling through food webs and included a stable isotope addition of ^{15}N in NH_4^+ for 6 weeks. Using ^{15}N as a tracer enables the flow of N to be traced through the stream ecosystem. This is because the minute amount of ^{15}N added has little effect on N availability and ^{15}N behaves the same in the N cycle as does ^{14}N, the much more abundant isotope in nature. Results for Eagle Creek, a second-order tributary of the Kalamazoo River southeast of KBS (Fig. 11.1), showed a rapid turnover of dissolved NH_4^+ and NO_3^- in spite of relatively stable concentrations. The distribution of the ^{15}N tracer in stream organic matter and organisms revealed the importance of assimilative uptake of N by heterotrophic bacteria and fungi as well as benthic algae (algae dwelling on submersed stream surfaces) for N uptake into food webs (Hamilton et al. 2001, 2004; Raikow and Hamilton 2001).

A second set of LINX experiments focused on NO_3^- dynamics in headwater streams across the United States and included whole-stream ^{15}N tracer additions in three predominant land-cover types: forest or other natural vegetation, agriculture, and urban/suburban. The Rabbit River, a tributary of the Kalamazoo River (Fig. 11.1), was included in the study. A survey of streams in the Rabbit River watershed showed the highest NO_3^- concentrations in watersheds with the most agriculture (Arango and Tank 2008). And nationally, streams in agricultural watersheds tended to have the highest NO_3^- concentrations, followed by urban/suburban ones. Among the 72 experiments conducted across eight biomes, there were positive relationships between NO_3^- concentrations and the rates of biotic N uptake and of denitrification. However, at high NO_3^- loading, stream N removal did not increase proportionately with increasing NO_3^- loads, suggesting that the capacity of streams to remove NO_3^- can become saturated, which then allows more NO_3^- to move downstream. High NO_3^- loading was correlated with increased stream primary production (mainly by algae), and NO_3^- removal by denitrification was correlated with increased stream respiration (which, in turn, is driven by organic matter supply). The role of stream denitrification as a source of atmospheric nitrous oxide, a potent greenhouse gas, was also elucidated from these ^{15}N tracer experiments, providing evidence that streams and rivers are more important to global nitrous oxide emissions than previously thought (Beaulieu et al. 2008, 2011). These cross-site studies show how stream channels and riparian zones could be better managed to promote higher rates of N removal by denitrification and thereby reduce eutrophication of downstream waters, but with the undesirable side effect of enhanced nitrous oxide emission to the atmosphere.

For Augusta Creek, like many streams in this glacial landscape, the large groundwater inflow and the presence of lakes and wetlands along the stream system result in a stable physicochemical environment compared to streams in other kinds of landscapes (Allen et al. 1972, Fongers 2008). The water level of Augusta Creek has been monitored since October 1964 by the U.S. Geological Survey (Station 04105700), and the long-term mean discharge (1964–1994) near its mouth is 1.28 $m^3 s^{-1}$. Because of the relatively slow movement of groundwater across the local landscape, the water discharged by the creek is likely to reflect the recharge from upland parts of the watershed over the past several decades (Stewart et al. 2010, Tesoriero et al. 2013). Given this time lag, the stream water of Augusta Creek may not be in equilibrium with or reflect changes in the current recharge of upland areas. Bartholic et al. (2007) presented a groundwater flow model that estimates spatial patterns of groundwater inflow and water temperature in Augusta Creek; this model is used to show how large-scale groundwater withdrawals might affect stream discharge and habitat for cool-water fishes. Augusta Creek and most other local streams carry little inorganic material in suspension, but transport significant quantities of bed load, mainly as sand, which may be affected by changes in stream discharge.

Monitoring of NO_3^- concentrations in Augusta Creek by the KBS LTER since 1997 has shown increasing concentrations (Fig. 11.10A). However, when corrected for discharge variation using the concentration–discharge relationship (Fig. 11.10B) and daily discharge records from the USGS station, NO_3^- fluxes appear consistent and stable across 13 years of observation (Fig. 11.10C). Thus, the temporal increase in concentrations is explained by changes in stream discharge rather than by increasing amounts of NO_3^- exported from the watershed. This tendency for little short-term change in stream NO_3^- export is not uncommon and likely reflects the time lags and attenuation associated with groundwater movement through the watershed (Basu et al. 2011, Sprague et al. 2011, Hamilton 2012) and possibly also the delayed release of nitrogen from fertilized soils (Sebilo et al. 2013).

Total P concentrations are low and NO_3^- concentrations can be high in many headwater streams, reflecting the importance of groundwater inflows and the high mobility of NO_3^- relative to P in the soil–groundwater system, as discussed earlier. When N-enriched groundwaters are discharged to surface waters, PO_4^{-3} is further

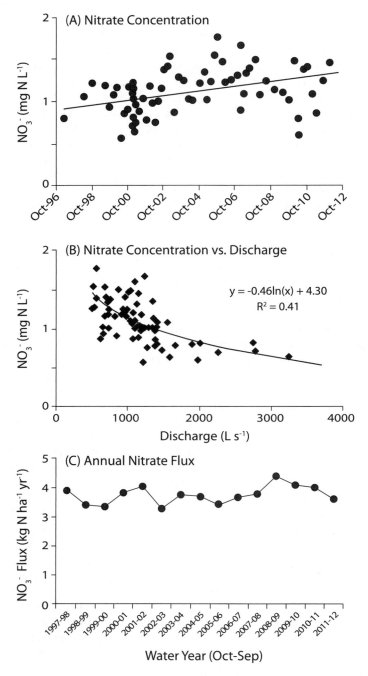

Figure 11.10. Nitrate export in stream water from Augusta Creek: (A) concentrations, (B) NO₃⁻ concentration-discharge relation, and (C) annual NO₃⁻ fluxes in the lower reaches of Augusta Creek, which drains land east of and including part of the KBS LTER main site. Annual fluxes are based on the concentration-discharge relation (64 sampling dates) and daily discharge measured by the U.S. Geological Survey.

removed by co-precipitation with calcite and sedimentation. As a result, primary production in local streams and other groundwater-fed surface waters tends to be strongly limited by the availability of P (Wetzel 1989, Hamilton et al. 2009).

Augusta Creek discharges into the Kalamazoo River, and further downstream in the Kalamazoo River system nutrient concentrations increase considerably. This river system has several reservoirs, and the largest and most downstream reservoir (Lake Allegan; Fig. 11.1) has a history of poor water quality ascribed mainly to eutrophication by high P loading. Reid and Hamilton (2007) presented evidence for shifting patterns of nutrient limitation in these reservoirs, with N, P and Si all potentially important as limiting factors for algal growth, but only at times of low river discharge and reservoir flushing. Phosphorus contributions from the various subwatersheds of the Kalamazoo River and the transport and transformation of P and other elements through the river system were subsequently quantified in greater detail through intensive monitoring and modeling (Baas 2009). This work documented the importance of nonpoint (diffuse) sources of P loading to the river system, including the release of sediment P from a reservoir (Morrow Lake) upstream of the City of Kalamazoo to river water, possibly reflecting a legacy of high sediment loading rates.

Wetlands

Wetlands are generally recognized for their positive effects on water quality, for example, nutrient removal and sediment trapping (NRC 1995). Wetlands are abundant in the landscape around KBS, covering 10% of the four townships in the vicinity (Fig. 11.1). The National Wetland Inventory (U.S. Fish and Wildlife Service) provides maps and information on the general types of wetlands and their spatial patterns (http://www.fws.gov/wetlands/Data/index.html). The present vegetation and, to some extent, the hydrology of local wetlands has been strongly influenced by the history of land use since settlement in the 1830s (Rich 1970). Logging, grazing, farming, construction of ponds and drainage channels, and mining of peat and calcium carbonate were common practices in local wetlands. And the effects of these activities on the geomorphology, hydrology, and vegetation still persist today.

The abundance and diversity of wetlands near KBS make this landscape ideal for comparative studies. The wide range in specific conductance of wetland waters shows that they are the most hydrochemically variable surface waters in the landscape (Fig. 11.2). This is because wetlands have highly variable hydraulic connectivity with groundwater: they span the range from receiving negligible groundwater inputs to having their water budgets dominated by groundwater through-flow (Kehew et al. 1998).

Thobaben and Hamilton (2014) surveyed wetland hydrology across a diverse set of 24 wetlands around KBS. They estimated the relative importance of groundwater as a source of water to the wetlands by comparing solute composition and found—consistent with previous studies in lakes—that dissolved Mg^{2+} is the best tracer of groundwater in this landscape (Wetzel and Otsuki 1974, Stauffer 1985). This is because (1) concentrations are negligible in precipitation but consistently high in groundwater in contact with dolomite minerals in the glacial deposits, (2) biological activity has no detectable effect on Mg^{2+} concentrations, and (3) Mg^{2+}

does not tend to precipitate from solution in wetlands or other surface waters. The study found that the relative importance of groundwater varied from near zero to near 100% of the water supplied to the wetlands, and that this explained much of the variation in plant species composition across wetland sites. Statistical analyses revealed seven wetland plant communities across the 24 wetlands: leatherleaf bogs, bogs, a poor fen, fens, sedge meadow fens, wet swamps, and dry swamps. The environmental variables most strongly related to these community classes were water pH, fraction of groundwater, soil nutrient availability, hydrochemical variables that covaried with pH and fraction of groundwater, and shading by tall woody species. Water levels were of secondary importance.

Biogeochemical processes in aquatic sediments of wetlands and other shallow waters have been studied in detail in connection with the KBS LTER as well as in earlier work focused mainly on lakes (see below). Whitmire and Hamilton (2008) compared rates of anaerobic metabolism (denitrification, iron reduction, sulfate reduction, and methanogenesis) in three groundwater- and three precipitation-fed wetlands. Denitrification was not measurable in these wetlands, all of which had low NO_3^- concentrations. Iron reduction was measurable mainly in precipitation-fed wetlands, while sulfate reduction was only measurable in the groundwater-fed wetlands (groundwater supplies SO_4^{2-}; see Fig. 11.9). Methanogenesis was measurable in all wetlands, with no differences between wetlands with contrasting water sources. In terms of total carbon mineralization by anaerobic metabolism, sulfate reduction and methanogenesis were the most important processes in groundwater- and precipitation-fed sites, respectively.

Whitmire and Hamilton (2005) examined NO_3^- removal in wetland sediments around KBS and found that N removal was linked to S cycling in many of the study sites. Further research in the Hamilton lab revealed that the linkage is biologically mediated and consistent with NO_3^- utilization by S-oxidizing bacteria (Burgin and Hamilton 2008, Payne et al. 2009). While NO_3^- evidently can either be denitrified or reduced to NH_4^+ during S oxidation, only denitrification represents a permanent loss of reactive N from the ecosystem. The factors governing the relative importance of these two reactions remain incompletely understood (Burgin and Hamilton 2007), but we can conclude that there is enough reduced S in these environments for S-oxidizing bacteria to play a significant role in NO_3^- removal via a form of chemolithoautotrophic denitrification. This stands in contrast to what has long been thought (Robertson and Groffman 2015): heterotrophic denitrification (i.e., denitrification coupled to the oxidation of organic matter) is not the only process that can permanently remove N in fresh waters.

Lakes

Lakes are abundant and diverse in the landscape around KBS, and they are an important ecological, aesthetic, and recreational resource. Larger, deeper lakes have been the subject of much study because of their importance to people (Wetzel 2001). Notably, most lakes in the area are shallow (i.e., <2 m deep) and become entirely filled with emergent and floating-leaf aquatic vegetation during the summer; therefore, they would be considered wetlands in the National Wetlands Inventory (Cowardin et al. 1979). Most local lakes are hydraulically connected to adjacent groundwater (Allen et al. 1972, Tague 1977, Rheaume 1990, Kehew et al. 1996), although the relative importance of groundwater and precipitation varies. Lakes fed primarily by groundwater (commonly known as "spring-fed lakes") are the most common, as evident from the distribution of their conductances (Fig. 11.2). As noted earlier, lakes around KBS are similar in their spatial and temporal hydrologic variation to lakes in the vicinity of the North Temperate Lakes (NTL) LTER site (Webster et al. 2006). However, in contrast to lakes at NTL, lakes around KBS contain more alkaline, ion-rich water, reflecting the influence of the abundant carbonate minerals in glacial deposits.

Groundwater-fed lakes around KBS commonly exhibit high rates of calcite precipitation during the summer, when water temperatures warm and aquatic primary production increases the pH through assimilation of CO_2 (Hamilton et al. 2009). Carbonate precipitates often build up in the sediments of such lakes. The effect of calcite precipitation on lake Ca^{2+} concentrations is significant: up to half of the Ca^{2+} in groundwater inputs may precipitate out during the summer (Fig. 11.11).

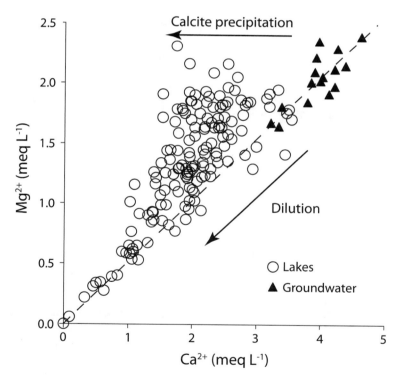

Figure 11.11. Calcium (Ca^{2+}) and magnesium (Mg^{2+}) concentrations in southern Michigan lakes during the summer, and in groundwater from wells at KBS. Assuming similar groundwater composition across the Lower Peninsula, precipitation water inputs would dilute the ion concentrations along the dashed line, whereas calcite precipitation would lower the ratio of Ca^{2+} to Mg^{2+}. Data based on 152 lakes across southern Michigan, most sampled once in the summer from 1996–2008, and 16 groundwater wells with variable sampling dates.

Algal and plant growth in such lakes tends to be strongly limited by P availability (Wetzel 2001, Schindler 2012). Above a pH of ~9, the co-precipitation of phosphate with calcite contributes to the limited availability of P. Sedimentation of this co-precipitated P serves as a negative feedback mechanism for aquatic primary production, potentially ameliorating P-driven eutrophication (Koschel et al. 1983, Hamilton et al. 2009). Calcite precipitation can also act as a negative feedback to aquatic primary production by attenuating light in the water column and by flocculation and consequent sedimentation of algal cells (Koschel et al. 1983). Calcite precipitation and deposition on underwater surfaces impact the ecology and biogeochemistry of lakes in many other ways, for example, by binding trace metals and dissolved organic matter and by smothering biofilms and underwater plant leaves (Kelts and Hsü 1978, Wetzel 2001).

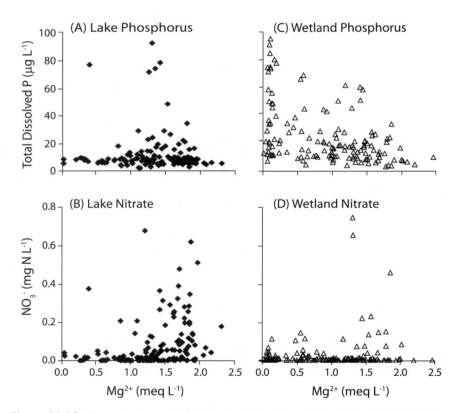

Figure 11.12. Concentrations of total dissolved phosphorus (P) and nitrate (NO$_3^-$) in lakes (A, B) and wetlands (C, D), in relation to the importance of groundwater as indicated by magnesium (Mg^{2+}) concentrations. Groundwater at equilibrium with dolomite tends to contain ~2 meq L^{-1} of Mg^{2+} whereas the concentration of Mg^{2+} in precipitation is negligible (Fig. 11.11). Data based on 152 lakes across southern Michigan, most sampled once in the summer, and 17 wetland sites in the KBS area, many sampled multiple years in May and Oct; sampling was conducted from 1996–2008.

The influence of groundwater inputs on nutrient availability differs between lakes and wetlands in southern Michigan, as can be seen in Fig. 11.12, although the sample set given is biased by the prevalence of groundwater-fed lakes in the area (i.e., those with Mg^{2+} concentrations exceeding about 1.0 meq L^{-1}). Both TDP and NO_3^- are often, but not always, higher in groundwater-fed lakes (Fig. 11.12). Wetlands, especially those fed by precipitation (as indicated by Mg^{2+} < 1 meq L^{-1}), often have TDP concentrations that are high by lake standards, for reasons that are not well understood. In contrast, NO_3^- concentrations tend to be lower in wetlands even where groundwater with high NO_3^- concentrations is an important water source, likely reflecting the high capacity of wetlands for both

Table 11.3. Hydrochemistry of surface waters in the vicinity of Kellogg Biological Station.

Measurement*	Surface waters				
	Augusta Creek[a]	Gull Lake[b]	Wintergreen Lake[c]	Lawrence Lake[d]	Lower Crooked Lake[e]
# of dates	34	42	16	7	18
Conductance (μS cm^{-1}; 25°C)	481	348	343	441	205
pH	8.06	8.12	8.09	7.94	8.53
ANC* (mg HCO$_3^-$ L^{-1})	286	189	182	266	100
Ca^{2+} (mg L^{-1})	70.6	36.8	38.9	63.0	21.0
Mg^{2+} (mg L^{-1})	21.3	21.7	18.6	22.7	9.33
Na$^+$ (mg L^{-1})	6.45	6.30	6.54	5.54	6.98
K$^+$ (mg L^{-1})	1.21	1.24	6.10	2.43	0.47
Cl$^-$ (mg L^{-1})	11.1	11.8	14.2	6.40	17.6
SO$_4^{2-}$ (mg L^{-1})	19.8	23.3	8.74	18.1	4.16
NO$_3^-$ (mg N L^{-1})	1.10	0.20	0.06	0.62	0.02
NH$_4^+$ (mg N L^{-1})	0.019	0.090	0.150	0.183	0.014
Si (mg Si L^{-1})	5.55	3.37	0.84	5.91	0.14
DOC* (mg L^{-1})	4.92	5.65	11.44	4.30	9.24
TDN* (mg L^{-1})	NA	NA	NA	NA	NA
TDP* (mg L^{-1})	0.008	0.007	0.049	0.009	0.007

*DOC = dissolved organic carbon, ANC = acid-neutralizing capacity (also known as total alkalinity), TDN = total dissolved N, TDP = total dissolved P.

[a]Augusta Creek was sampled near its mouth at the Village of Augusta; Hamilton lab database queried December 2009.

[b]Gull Lake data are means of integrated samples of the epilimnion and point samples from the middle of the hypolimnion (Raikow et al. 2004); Hamilton lab database queried December 2009.

[c]Wintergreen Lake data are means of samples taken mostly at its outflow stream, representing the epilimnion when it was stratified; Hamilton lab database queried December 2009.

[d]Lawrence Lake data are means of samples taken at several depths when it was stratified; Hamilton lab database queried December 2009.

[e]Lower Crooked Lake data are means of samples taken near the surface, mostly in May and October when it was not stratified; Hamilton lab database queried December 2009.

denitrification and NO_3^- assimilation by plants and algae (Whitmire and Hamilton 2005). Denitrification can also be an important sink for NO_3^- in groundwater-fed lakes, even though concentrations remain quite high (Bruesewitz et al. 2012, Finlay et al. 2013).

The limnology of three local lakes—Gull, Wintergreen, and Lawrence (Fig. 11.1B)—was widely studied during the 1970s and 1980s, and these lakes remain among the most studied inland water bodies in the world. Table 11.3 presents data on the hydrochemistry of these lakes based on a recent survey; detailed information on the spatial and temporal variability of some of the variables can be found in the earlier studies. These three lakes represent contrasts in morphometry, hydrology, hydrochemistry, and ecology, but all are "hardwater" lakes that mix vertically in the spring and fall and develop thermal stratification during the summer and in ice-forming winters. The hardness of their water reflects the high Ca^{2+} and Mg^{2+} concentrations in groundwater inflows.

Recent studies on Gull Lake include work on invasive zebra mussels (*Dreissena polymorpha*), which thrive in alkaline waters and could potentially colonize most of the lakes in the region. These mussels tend to promote dominance by *Microcystis aeruginosa*, a phytoplanktonic cyanobacterium that often produces a potent toxin (Raikow et al. 2004; Knoll et al. 2008; Bruesewitz et al. 2009; Sarnelle et al. 2005, 2012). This is of particular interest because Gull Lake is only moderately productive (i.e., mesotrophic) and harmful cyanobacterial blooms, often produced by this species, are traditionally associated with eutrophic lakes that have greater levels of P enrichment.

Summary

Landscape-level patterns in water quality in recently glaciated landscapes can be ascribed to a combination of natural and anthropogenic influences. Weathering of minerals—particularly the carbonate minerals calcite and dolomite—produces marked changes in water quality as water from precipitation percolates through the upper 1–2 m of the soil profile, and these changes influence underlying groundwater and all downstream groundwaters and surface waters. Nitrate pollution of infiltrating water is particularly apparent beneath conventionally N-fertilized annual row crops, whereas perennial crops, poplar plantations, and natural successional vegetation leach comparatively little N. In contrast, P tends to be retained in most upland soils, and hence groundwater N:P ratios are high, contributing to the importance of P for limiting aquatic primary production in most groundwater-fed surface waters.

Hydrologic exchanges between surface and groundwater systems are particularly important in the KBS landscape and have numerous ecological ramifications. In nonurban areas, most water is delivered to streams and rivers by groundwater flow. Many lakes also receive a large fraction of their water from groundwater and thus their chemistry resembles that of groundwater, although calcium carbonate precipitation and biotic uptake in surface waters can reduce concentrations of Ca^{2+}, acid-neutralizing capacity, and labile nutrients (N, P, and Si). Elevated concentrations of NO_3^- present a water-quality issue in this region, particularly where groundwater serves as a water supply for people, but passage of water through streams, wetlands, and lakes can remove much of the NO_3^-.

Landscapes similar to KBS are common across the upper U.S. Midwest, as well as in northern Europe. Agricultural row crops, in particular, annual crops that are heavily fertilized, produce notable effects on water quality that extend through water flow paths to groundwater, streams, rivers, wetlands, and lakes. The groundwater flow path introduces a protracted time lag for the movement of contaminants from the land surface to streams and lakes (Hamilton 2012). Urban and residential development further alters water quality in myriad ways, and those effects are superimposed on the water-quality patterns described here. The long-term, landscape-level research described here provides the foundation for understanding natural patterns in water quality and the influence of agricultural activities on groundwater and surface waters; such an understanding is essential for better management of water and water quality for human and ecological benefits.

References

Alexander, R. B., J. K. Bohlke, E. W. Boyer, M. B. David, J. W. Harvey, P. J. Mulholland, S. P. Seitzinger, C. R. Tobias, C. Tonitto, and W. M. Wollheim. 2009. Dynamic modeling of nitrogen losses in river networks unravels the coupled effects of hydrological and biogeochemical processes. Biogeochemistry 93:91–116.

Allen, W. B., J. S. Miller, and W. W. Wood. 1972. Availability of water in Kalamazoo County, southwestern Michigan. Water Supply Paper 1973, U.S. Geological Survey, Washington, DC, USA.

Arango, C. P., and J. L. Tank. 2008. Land use influences the spatio-temporal controls on nitrification and denitrification in headwater streams. Journal of the North American Benthological Society 27:90–107.

Baas, D. G. 2009. Inferring dissolved phosphorus cycling in a TMDL watershed using biogeochemistry and mixed linear models. Dissertation, Michigan State University, East Lansing, Michigan, USA.

Bartholic, J., S. S. Batie, S. Seedang, H. Abbas, S. G. Li, W. Northcott, L. Wang, S. Lacy, S. A. Miller, J. A. Andresen, M. Kaplowitz, M. Branch, J. Asher, J. Shi, and M. Selman. 2007. Restoring the Great Lakes water through the use of water conservation credits and integrated water balance analysis system. Institute of Water Research, Michigan State University, East Lansing, Michigan, USA.

Basu, N. B., S. E. Thompson, and P. S. C. Rao. 2011. Hydrologic and biogeochemical functioning of intensively managed catchments: a synthesis of top-down analyses. Water Resources Research 47:W00J15. doi:10.1029/2011WR010800

Beaulieu, J. J., C. P. Arango, S. K. Hamilton, and J. L. Tank. 2008. The production and emission of nitrous oxide from headwater streams in the Midwestern United States. Global Change Biology 14:878–894.

Beaulieu, J. J., J. L. Tank, S. K. Hamilton, W. M. Wollheim, R. O. Hall, Jr., P. J. Mulholland, B. J. Peterson, L. R. Ashkenas, L. W. Cooper, C. N. Dahm, W. K. Dodds, N. B. Grimm, S. L. Johnson, W. H. Mcdowell, G. C. Poole, H. M. Valett, C. P. Arango, M. J. Bernot, A. J. Burgin, C. L. Crenshaw, A. M. Helton, L. T. Johnson, J. M. O'Brien, J. D. Potter, R. W. Sheibley, D. J. Sobota, and S. M. Thomas. 2011. Nitrous oxide emission from denitrification in stream and river networks. Proceedings of the National Academy of Sciences USA 108:214–219.

Bexfield, L. M. 2008. Decadal-scale changes of pesticides in ground water of the United States, 1993-2003. Journal of Environmental Quality 37:S226–S239.

Böhlke, J.-K. 2002. Groundwater recharge and agricultural contamination. Hydrogeology Journal 10:153–179.

Bruesewitz, D. A., J. L. Tank, and S. K. Hamilton. 2009. Seasonal effects of zebra mussels (*Dreissena polymorpha*) on littoral nitrogen transformation rates in Gull Lake, Michigan, U.S.A. Freshwater Biology 54:1427–1443.

Bruesewitz, D. A., J. L. Tank, and S. K. Hamilton. 2012. Incorporating spatial variation of nitrification and denitrification rates into whole-lake nitrogen dynamics. Journal of Geophysical Research—Biogeosciences 117:G00N07. doi:10.1029/2012JG002006

Burgin, A. J., and S. K. Hamilton. 2007. Have we overemphasized the role of denitrification in aquatic ecosystems? A review of nitrate removal pathways. Frontiers in Ecology and the Environment 5:89–96.

Burgin, A. J., and S. K. Hamilton. 2008. NO_3^- driven SO_4^{2-} production in freshwater ecosystems: implications for N and S cycling. Ecosystems 11:908–922.

Chapman, K. A., and R. Brewer. 2008. Prairie and savanna in southern Lower Michigan: history, classification, ecology. Michigan Botanist 47:1–48.

Chen, X., and C. T. Driscoll. 2009. Watershed land use controls on chemical inputs to Lake Ontario embayments. Journal of Environmental Quality 38:2084–2095.

Chowdhury, S. H., A. E. Kehew, and R. N. Passero. 2003. Correlation between nitrate contamination and ground water pollution potential. Ground Water 41:735–745.

Cowardin, L. M., V. Carter, F. C. Golet, and E. T. LaRoe. 1979. Classification of wetlands and deepwater habitats of the United States. FWS/OBS-79/31, U.S. Fish and Wildlife Service, Washington, DC, USA.

Crum, J. R., G. P. Robertson, and F. Nurenberger. 1990. Long-term climate trends and agricultural productivity in Southwestern Michigan. Pages 53–58 in D. Greenland and L. W. Swift, editors. Climate variability and ecosystem response. U.S. Department of Agriculture, U.S. Forest Service, Southeastern Forest Experiment Station, Asheville, North Carolina, USA.

Domagalski, J. L., and H. M. Johnson. 2011. Subsurface transport of orthophosphate in five agricultural watersheds, USA. Journal of Hydrology 409:157–171.

Driscoll, C. T., G. B. Lawrence, A. J. Bulger, T. J. Butler, C. S. Cronan, C. Eagar, K. F. Lambert, G. E. Likens, J. L. Stoddard, and K. C. Weathers. 2001. Acidic deposition in the Northeastern United States: sources and inputs, ecosystem effects, and management strategies. BioScience 51:180–198.

Finlay, J. C., G. E. Small, and R. W. Sterner. 2013. Human influences on nitrogen removal in lakes. Science 342:247–250.

Fongers, D. 2008. Kalamazoo River Watershed hydrologic study. Michigan Department of Environmental Quality, East Lansing, Michigan, USA.

Grannemann, N. G., R. J. Hunt, J. R. Nicholas, T. E. Reilly, and T. C. Winter. 2008. The importance of ground water in the Great Lakes Region. Water-Resources Investigations Report 00–4008, U.S. Geological Survey, Washington, DC, USA.

Hall, R. O., J. L. Tank, D. J. Sobota, P. J. Mulholland, J. M. O'Brien, W. K. Dodds, J. R. Webster, H. M. Valett, G. C. Poole, B. J. Peterson, J. L. Meyer, W. H. McDowell, S. L. Johnson, S. K. Hamilton, N. B. Grimm, S. V. Gregory, C. N. Dahm, L. W. Cooper, L. R. Ashkenas, S. M. Thomas, R. W. Sheibley, J. D. Potter, B. R. Niederlehner, L. T. Johnson, A. M. Helton, C. M. Crenshaw, A. J. Burgin, M. J. Bernot, J. J. Beaulieu, and C. P. Arango. 2009. Nitrate removal in stream ecosystems measured by N-15 addition experiments: total uptake. Limnology and Oceanography 54:653–665.

Hamilton, S. K. 2012. Biogeochemical time lags may delay responses of streams to ecological restoration. Freshwater Biology 57 (Suppl. s1):43–57.

Hamilton, S. K., D. A. Bruesewitz, G. P. Horst, and O. Sarnelle. 2009. Biogenic calcite-phosphorus precipitation as a negative feedback to lake eutrophication. Canadian Journal of Fisheries and Aquatic Sciences 66:343–350.

Hamilton, S. K., A. L. Kurzman, C. Arango, L. Jin, and G. P. Robertson. 2007. Evidence for carbon sequestration by agricultural liming. Global Biogeochemical Cycles 21:GB2021.

Hamilton, S. K., J. L. Tank, D. F. Raikow, E. Siler, N. J. Dorn, and N. Leonard. 2004. The role of in-stream vs. allochthonous N in stream food webs: modeling the results of a nitrogen isotope addition experiment. Journal of the North American Benthological Society 23:429–448.

Hamilton, S. K., J. L. Tank, D. F. Raikow, W. M. Wollheim, B. J. Peterson, and J. R. Webster. 2001. Nitrogen uptake and transformation in a midwestern U.S. stream: a stable isotope enrichment study. Biogeochemistry 54:297–340.

Hedin, L. O., J. C. von Fischer, N. E. Ostrom, B. P. Kennedy, M. G. Brown, and G. P. Robertson. 1998. Thermodynamic constraints on nitrogen transformation and other biogeochemical processes at soil-stream interfaces. Ecology 79:684–703.

Jin, L., S. K. Hamilton, and L. M. Walter. 2008a. Mineral weathering rates in glacial drift soils (SW Michigan, U.S.A.): new constraints from seasonal sampling of waters and gases at soil monoliths. Chemical Geology 249:129–154.

Jin, L., N. Ogrinc, S. K. Hamilton, K. Szramek, T. Kanduc, and L. M. Walter. 2009. Inorganic carbon isotope systematics in soil profiles undergoing silicate and carbonate weathering (Southern Michigan, USA). Chemical Geology 264:139–153.

Jin, L., E. Williams, K. Szramek, L. M. Walter, and S. K. Hamilton. 2008b. Silicate and carbonate mineral weathering in soil profiles developed on Pleistocene glacial drift (Michigan, USA): mass balances based on soil water geochemistry. Geochemica Acta 72:1027–1042.

Kahl, J. S., J. L. Stoddard, R. Haeuber, S. G. Paulsen, R. Birnbaum, F. A. Deviney, J. R. Webb, D. R. DeWalle, W. E. Sharpe, C. T. Driscoll, A. T. Herlihy, J. H. Kellogg, P. S. Murdoch, K. Roy, K. E. Webster, and N. S. Urquhart. 2004. Have U.S. surface waters responded to the 1900 Clean Air Act Amendments? Environmental Science and Technology 38:484A–490A.

Kehew, A. E., and M. K. Brewer. 1992. Groundwater quality variations in glacial drift and bedrock aquifers, Barry County, Michigan, USA. Environmental Geology and Water Sciences 20:105–115.

Kehew, A. E., R. N. Passero, R. V. Krishnamurthy, C. K. Lovett, M. A. Betts, and B. A. Dayharsh. 1998. Hydrogeochemical interaction between a wetland and an unconfined glacial drift aquifer, southwestern Michigan. Ground Water 36:849–856.

Kehew, A. E., W. T. Straw, W. K. Steinman, P. G. Barrese, G. Passarella, and W. Peng. 1996. Ground-water quality and flow in a shallow glaciofluvial aquifer impacted by agricultural contamination. Ground Water 34:491–500.

Kelts, K., and K. J. Hsu. 1978. Freshwater carbonate sedimentation. Pages 295–323 in A. Lerman, editor. Lakes: chemistry, geology, physics. Springer-Verlag, New York, New York, USA.

Kincare, K., and G. J. Larson. 2009. Evolution of the Great Lakes. Pages 174–190 in R. J. Schaetzl, J. Darden, and D. Brandt, editors. Michigan geography and geology. Pearson Custom Publishing, New York, New York, USA.

Kleinman, P. J. A., A. N. Sharpley, A. R. Buda, R. W. McDowell, and A. L. Allen. 2011. Soil controls of phosphorus in runoff: management barriers and opportunities. Canadian Journal of Soil Science 91:329–338.

Knoll, L. B., O. Sarnelle, S. K. Hamilton, C. E. H. Kissman, A. E. Wilson, J. B. Rose, and M. R. Morgan. 2008. Invasive zebra mussels (*Dreissena polymorpha*) increase cyanobacterial toxin concentrations in low-nutrient lakes. Canadian Journal of Fisheries and Aquatic Sciences 65:448–455.

Konarik, S., and V. P. Aneja. 2008. Trends in agricultural ammonia emissions and ammonium concentrations in precipitation over the Southeast and Midwest United States. Atmospheric Environment 42:3238–3252.

Koschel, R., J. Benndorf, G. Proft, and F. Recknagel. 1983. Calcite precipitation as a natural control mechanism of eutrophication. Archiv fur Hydrobiologie 98:380–408.

Kurzman, A. L. 2006. Changes in major solute chemistry as water infiltrates soils: comparisons between managed agroecosystems and unmanaged vegetation. Dissertation, Michigan State University, East Lansing, Michigan, USA.

Lusch, D. P. 2009. Groundwater and karst. Pages 223–248 in R. J. Schaetzl, J. Darden, and D. Brandt, editors. Michigan geography and geology. Pearson Custom Publishing, New York, New York, USA.

McClain, M. E., E. W. Boyer, C. L. Dent, S. E. Gergel, N. B. Grimm, P. M. Groffman, S. C. Hart, J. W. Harvey, C. A. Johnston, E. Mayorga, W. H. McDowell, and G. Pinay. 2003. Biogeochemical hot spots and hot moments at the interface of terrestrial and aquatic ecosystems. Ecosystems 6:301–312.

Mulholland, P. J., R. O. Hall, D. J. Sobota, W. K. Dodds, S. E. G. Findlay, N. B. Grimm, S. K. Hamilton, W. H. McDowell, J. M. O'Brien, J. L. Tank, L. R. Ashkenas, L. W. Cooper, C. N. Dahm, S. V. Gregory, S. L. Johnson, J. L. Meyer, B. J. Peterson, G. C. Poole, H. M. Valett, J. R. Webster, C. P. Arango, J. J. Beaulieu, M. J. Bernot, A. J. Burgin, C. L. Crenshaw, A. M. Helton, L. T. Johnson, B. R. Niederlehner, J. D. Potter, R. W. Sheibley, and S. M. Thomas. 2009. Nitrate removal in stream ecosystems measured by N-15 addition experiments: denitrification. Limnology and Oceanography 54:666–680.

Mulholland, P. J., A. M. Helton, G. C. Poole, R. O. Hall, S. K. Hamilton, B. J. Peterson, J. L. Tank, L. R. Ashkenas, L. W. Cooper, C. N. Dahm, W. K. Dodds, S. Findlay, S. V. Gregory, and N. B. Grimm. 2008. Stream denitrification across biomes and its response to anthropogenic nitrate loading. Nature 452:202–206.

NADP/NTN (National Atmospheric Deposition Program/National Trends Network). 2011. National Atmospheric Deposition Program (NRSP-3)/National Trends Network means of annual volume-weighted means for 1979–2010 form the KBS station (MI26). NADP/NTN Coordination Office, Illinois State Water Survey, Champaign, Illinois 61820, USA.

NCDC (National Climate Data Center). 2013. Summary of monthly normal 1981–2010. Gull Lake Biology Station, MI US. Accessed at http://www.ncdc.noaa.gov/cdo-web/ search on 17 Jan 2013.

NRC (National Research Council). 1995. Wetlands: characteristics and boundaries. National Academies Press, Washington, DC, USA.

Payne, E. K., A. J. Burgin, and S. K. Hamilton. 2009. Sediment nitrate manipulation using porewater equilibrators reveals potential for N and S coupling in freshwaters. Aquatic Microbial Ecology 54:233–241.

Peterson, B. J., W. M. Wollheim, P. J. Mulholland, J. R. Webster, J. L. Meyer, J. L. Tank, E. Marti, W. B. Bowden, H. M. Valett, A. E. Hershey, W. H. McDowell, W. K. Dodds,

S. K. Hamilton, S. Gregory, and D. D. Morrall. 2001. Control of nitrogen export from watersheds by headwater streams. Science 292:86–90.

Power, J. F., R. Wiese, and D. Flowerday. 2001. Managing farming systems for nitrate control: a research review from management systems evaluation areas. Journal of Environmental Quality 30:1866–1880.

Prince, H. 1997. Wetlands of the American Midwest: a historical geography of changing attitudes. University of Chicago Press, Chicago, Illinois, USA.

Puckett, L. J., A. J. Tesoriero, and N. M. Dubrovsky, 2011. Nitrogen contamination of surficial aquifers—A growing legacy. Environmental Science and Technology 45:839–844.

Raikow, D. F., and S. K. Hamilton. 2001. Bivalve diets in a Midwestern U.S. stream: a stable isotope enrichment study. Limnology and Oceanography 46:514–522.

Raikow, D. F., O. Sarnelle, A. E. Wilson, and S. K. Hamilton. 2004. Dominance of the noxious cyanobacterium *Microcystis aeruginosa* in low-nutrient lakes is associated with exotic zebra mussels. Limnology and Oceanography 49:482–487.

Raymond, P. A., M. B. David, and J. E. Saiers. 2012. The impact of fertilization and hydrology on nitrate fluxes from Mississippi watersheds. Current Opinion in Environmental Sustainability 4:212–218.

Reid, N. J., and S. K. Hamilton. 2007. Controls on algal abundance in a eutrophic river with varying degrees of impoundment (Kalamazoo River, Michigan, USA). Lake and Reservoir Management 23:219–230.

Rheaume, S. J. 1990. Geohydrology and water quality of Kalamazoo County, Michigan, 1986–88. Water-Resources Investigations Report 90–4028, U.S. Geological Survey, Washington, DC, USA.

Rich, P. H. 1970. Post-settlement influences upon a Southern Michigan marl lake. The Michigan Botanist 9:3–9.

Robertson, G. P. and P. M. Groffman. 2015. Nitrogen transformations. Pages 421–426 in E. A. Paul, editor. Soil microbiology, ecology and biochemistry. Fourth edition. Academic Press, Burlington, Massachusetts, USA.

Robertson, G. P., and S. K. Hamilton. 2015. Long-term ecological research at the Kellogg Biological Station LTER Site: conceptual and experimental framework. Pages 1–32 in S. K. Hamilton, J. E. Doll, and G. P. Robertson, editors. The ecology of agricultural ecosystems: long-term research on the path to sustainability. Oxford University Press, New York, New York, USA.

Robertson, G. P., K. M. Klingensmith, M. J. Klug, E. A. Paul, J. R. Crum, and B. G. Ellis. 1997. Soil resources, microbial activity, and primary production across an agricultural ecosystem. Ecological Applications 7:158–170.

Rupert, M. G. 2008. Decadal-scale changes of nitrate in ground water of the United States, 1988–2004. Journal of Environmental Quality 37:S240–S248.

Saad, D. A. 2008. Agriculture-related trends in groundwater quality of the glacial deposits aquifer, central Wisconsin. Journal of Environmental Quality 37:S240–S248.

Sarnelle, O., A. E. Wilson, S. K. Hamilton, L. B. Knoll, and D. F. Raikow. 2005. Complex interactions between the zebra mussel, *Dreissena polymorpha*, and the harmful phytoplankter, *Microcystis aeruginosa*. Limnology and Oceanography 50:896–904.

Sarnelle, O., J. D. White, G. P. Horst, and S. K. Hamilton. 2012. Phosphorus addition reverses the positive effect of zebra mussels (*Dreissena polymorpha*) on the toxic cyanobacterium, *Microcystis aeruginosa*. Water Research 46:3471–3478.

Schaetzl, R. J. 2009. Soils. Pages 315–329 in R. J. Schaetzl, J. Darden, and D. Brandt, editors. Michigan geography and geology. Pearson Custom Publishing, New York, New York, USA.

Schindler, D.W. 2012. The dilemma of controlling cultural eutrophication in lakes. Proceedings of the Royal Society B. doi:10.1098/rspb.2012.1032

Sebilo, M., B. Mayer, B. Nicolardot, G. Pinay, and A. Mariotti. 2013. Long-term fate of nitrate fertilizer in agricultural soils. Proceedings of the National Academy of Sciences USA 110:18185–18189.

Sharpley, A., H. P. Jarvie, A. Buda, L. May, B. Spears, and P. Kleinman. 2013. Phosphorus legacy: Overcoming the effects of past management practices to mitigate future water quality impairment. Journal of Environmental Quality 42: 1308–1326.

Sprague, L. A., R. M. Hirsch, and B. T. Aulenbach. 2011. Nitrate in the Mississippi River and its tributaries, 1980 to 2008: Are we making progress? Environmental Science and Technology 45:7209–7216.

Stauffer, R. E. 1985. Use of solute tracers released by weathering to estimate groundwater inflow to seepage lakes. Environmental Science and Technology 19:405–411.

Stewart, M. K., U. Morgenstern, and J. J. McDonnell. 2010. Truncation of stream residence time: How the use of stable isotopes has skewed our concept of streamwater age and origin. Hydrological Processes 24:1646–1659.

Syswerda, S. P., B. Basso, S. K. Hamilton, J. B. Tausig, and G. P. Robertson. 2012. Long-term nitrate loss along an agricultural intensity gradient in the Upper Midwest USA. Agriculture, Ecosystems and Environment 149:10–19.

Szramek, K., and L. M. Walter. 2004. Impact of carbonate precipitation on riverine inorganic carbon mass transport from a mid-continent, forested watershed. Aquatic Geochemistry 10:99–137.

Tague, D. F. 1977. The hydrologic and total phosphorus budgets of Gull Lake, Michigan. Thesis, Michigan State University, East Lansing, Michigan, USA.

Tesoriero, A. J., J. H. Duff, D. A. Saad, N. E. Spahr, and D. M. Wolock. 2013. Vulnerability of streams to legacy nitrate sources. Environmental Science and Technology 47:3623–3629.

Thobaben, E. T., and S. K. Hamilton. 2014. The relative importance of groundwater and its ecological implications in diverse glacial wetlands. American Midland Naturalist 172:205–218.

Unterreiner, G. A., and A. E. Kehew. 2005. Spatial and temporal distribution of herbicides and herbicide degradates in a shallow glacial drift aquifer/surface water system, south-central Michigan. Ground Water Monitoring and Remediation 25:87–95.

Wahrer, M. A., D. T. Long, and R. W. Lee. 1996. Selected geochemical characteristics of ground water from the glaciofluvial aquifer in the Central Lower Peninsula of Michigan. Water-Resources Investigations Report 94–4017, U.S. Geological Survey, Washington, DC, USA.

Webster, J. R., P. J. Mulholland, J. L. Tank, H. M. Valett, W. R. Dodds, B. J. Peterson, W. B. Bowden, C. N. Dahm, S. Findlay, S. V. Gregory, N. B. Grimm, S. K. Hamilton, S. L. Johnson, E. Marti, W. B. McDowell, J. L. Meyer, D. D. Morrall, S. A. Thomas, and W. M. Wollheim. 2003. Factors affecting ammonium uptake in streams—an inter-biome perspective. Freshwater Biology 48:1329–1352.

Webster, K. E., C. J. Bowser, M. P. Anderson, and J. D. Lenters. 2006. Understanding the lake-groundwater system: Just follow the water. Pages 19–48 in J. J. Magnuson, T. K. Kratz, and B. J. Benson, editors. Long-term dynamics of lakes in the landscape. Oxford University Press, New York, New York, USA.

Wetzel, R. G. 1989. Wetland and littoral interfaces of lakes: productivity and nutrient regulation in the Lawrence Lake ecosystem. Freshwater Wetlands and Wildlife, U.S. Department of Energy Symposium Series 61:283–302.

Wetzel, R. G. 2001. Limnology: lake and river ecosystems. Academic Press, San Diego, California, USA.

Wetzel, R. G., and A. Otsuki. 1974. Allochthonous organic carbon of a marl lake. Archiv fur Hydrobiologie 73:31–56.

Whitmire, S. L., and S. K. Hamilton. 2005. Rapid removal of nitrate and sulfate by freshwater wetland sediments. Journal of Environmental Quality 34:2062–2071.

Whitmire, S. L., and S. K. Hamilton. 2008. Rates of anaerobic microbial metabolism in wetlands of divergent hydrology on a glacial landscape. Wetlands 28:703–714.

12

Mitigation of Greenhouse Gases in Agricultural Ecosystems

Ilya Gelfand and G. Philip Robertson

Modern cropping systems use substantial amounts of fossil energy in the form of fertilizers, pesticides, and fuel for field operations. An important environmental consequence of this use is the emission of greenhouse gases (GHGs) to the atmosphere, from sources both direct and indirect. Direct sources include fossil fuel used for tillage and other field operations as well as GHGs produced and consumed by microbes in cropped soils. Indirect sources include fossil energy used off-site to produce fertilizers and other agronomic inputs, as well as GHGs produced by microbes in noncropped sites that receive nutrients escaped from cropped fields. Row-crop agriculture can thus be either a net source or sink of GHGs, with the balance (net emission or uptake) influenced greatly by management practices.

All three of the major biogenic GHGs are affected by agriculture: carbon dioxide (CO_2), methane (CH_4), and nitrous oxide (N_2O). Not including postharvest activities or land-use conversion caused by agricultural expansion, agriculture is responsible for 10–14% of total global anthropogenic GHG emissions (Barker et al. 2007, Smith et al. 2007). This includes ~84% of anthropogenic N_2O emissions and ~53% of anthropogenic CH_4 emissions (Robertson 2004). The manufacture of agrochemicals adds another 0.6–1.5% to the global total (Vermeulen et al. 2012).

Most agricultural CO_2 emissions are from land conversion and fossil fuel use. Methane emissions associated with agriculture are from enteric fermentation by ruminant animals such as cattle, cultivated rice soils, animal wastes, and agricultural biomass burning. In addition, land conversion to agriculture substantially reduces microbial CH_4 oxidation in soil, thereby attenuating an important CH_4 sink and effectively increasing CH_4 in the atmosphere. Nitrous oxide emissions from agriculture are produced mostly from nitrogenous fertilizers, with lesser contributions from animal wastes and biomass burning.

The global importance of GHG fluxes from established cropping systems and their sensitivity to management make agriculture an attractive sector for mitigation measures. And because many of these fluxes are interdependent and sensitive to the same management practices (though often differentially sensitive), there are many opportunities to manage them together. In fact, because management practices

Table 12.1. Description of the KBS LTER Main Cropping System Experiment (MCSE).[a]

Cropping System/Community	Dominant Growth Form	Management
Annual Cropping Systems		
Conventional (T1)	Herbaceous annual	Prevailing norm for tilled corn–soybean–winter wheat (c–s–w) rotation; standard chemical inputs, chisel-plowed, no cover crops, no manure or compost
No-till (T2)	Herbaceous annual	Prevailing norm for no-till c–s–w rotation; standard chemical inputs, permanent no-till, no cover crops, no manure or compost
Reduced Input (T3)	Herbaceous annual	Biologically based c–s–w rotation managed to reduce synthetic chemical inputs; chisel-plowed, winter cover crop of red clover or annual rye, no manure or compost
Biologically Based (T4)	Herbaceous annual	Biologically based c–s–w rotation managed without synthetic chemical inputs; chisel-plowed, mechanical weed control, winter cover crop of red clover or annual rye, no manure or compost; certified organic
Perennial Cropping Systems		
Alfalfa (T6)	Herbaceous perennial	5- to 6-year rotation with winter wheat as a 1-year break crop
Poplar (T5)	Woody perennial	Hybrid poplar trees on a ca. 10-year harvest cycle, either replanted or coppiced after harvest
Coniferous Forest (CF)	Woody perennial	Planted conifers periodically thinned
Successional and Reference Communities		
Early Successional (T7)	Herbaceous perennial	Historically tilled cropland abandoned in 1988; unmanaged but for annual spring burn to control woody species
Mown Grassland (never tilled) (T8)	Herbaceous perennial	Cleared woodlot (late 1950s) never tilled, unmanaged but for annual fall mowing to control woody species
Mid-successional (SF)	Herbaceous annual + woody perennial	Historically tilled cropland abandoned ca. 1955; unmanaged, with regrowth in transition to forest
Deciduous Forest (DF)	Woody perennial	Late successional native forest never cleared (two sites) or logged once ca. 1900 (one site); unmanaged

[a]Site codes that have been used throughout the project's history are given in parentheses. Systems T1–T7 are replicated within the LTER main site; others are replicated in the surrounding landscape. For further details, see Robertson and Hamilton (2015, Chapter 1 in this volume).

produce different effects on GHG fluxes, it is especially important to consider them together, that is, to take a systems approach toward their understanding and management (Robertson 2014).

In this chapter, we describe an ecosystems approach to documenting changes in GHG fluxes in intensive row-crop agriculture. We draw, in particular, on results from the Kellogg Biological Station Long-Term Ecological Research site (KBS LTER), where GHG fluxes have been studied in the Main Cropping System Experiment (MCSE; Table 12.1; Robertson and Hamilton 2015, Chapter 1 in this volume) since 1989. We discuss the value of long-term comparisons of different cropping systems in determining the potential for management practices to contribute to or mitigate GHG fluxes. We end with consideration of the GHG implications of crop production not only for grain but also for cellulosic biomass, which is anticipated to become increasingly important in a future that includes cellulosic biofuels.

Row-Crop Agriculture and GHG Mitigation

Historically, agricultural impacts on atmospheric chemistry have been dominated by land-use change. Since the late eighteenth century, conversion of forests and grasslands to cropland has resulted in emissions of CO_2 to the atmosphere on the order of 130 to 170 Pg C (Wilson 1978, Sauerbeck 2001), mostly due to immediate biomass burning and subsequent soil carbon (C) oxidation. Global CO_2 emissions from deforestation today amount to ~1.5 Pg C yr^{-1} (Canadell et al. 2007).

In few established croplands today are GHG emissions dominated by soil C oxidation. Rather, emissions now are dominated by CO_2 from fossil fuel combustion during farm operations; CO_2 produced during the manufacture and transport of fertilizers, pesticides, and other agricultural inputs; N_2O emitted when nitrogen (N) fertilizers are applied to soil; and CH_4 emitted during flooded conditions in lowland rice. In most of the world's established agricultural soils (except drained wetlands), soil C is either stable or, if managed appropriately, increasing, though this trend could be reversed by a warming climate (Senthilkumar et al. 2009; Paul et al. 2015, Chapter 5 in this volume).

The need for mitigation of agricultural GHG emissions becomes especially important in light of the agricultural intensification yet required to feed an increasing and more affluent world population (Tilman et al. 2011, Mueller et al. 2012). Although intensification to date has improved yields on existing farmland and thereby fed more people at a lower per-capita GHG cost (i.e., at a lower GHG cost per unit yield) (Burney et al. 2010), the efficiency gained has not been sufficient to halt the increase in GHG emissions from agriculture.

Growing demands for biofuel feedstocks could further increase agriculture's GHG footprint: over the next several decades, millions of hectares will likely be converted to biofuel cropping systems that will consume fuel and fertilizer and could—if not carefully managed—exacerbate rather than alleviate atmospheric GHG loading (Melillo et al. 2009). Bioenergy cropping systems correctly implemented, on the other hand, provide a substantial opportunity for mitigating anthropogenic GHG contributions as well as providing other environmental benefits (Robertson et al. 2008, NRC 2009, Tilman et al. 2009).

In light of the expectation for worldwide expansion and intensification of agriculture in the coming decades, it seems crucial to pursue opportunities for reducing the GHG contributions of agricultural crop production. Many such opportunities are available, particularly in the areas of soil C conservation (CAST 2011) and better N management (Robertson and Vitousek 2009). Through the strategic adoption of agronomic practices known to attenuate GHG emissions (e.g., Millar et al. 2010), agriculture could contribute significantly to climate change mitigation.

Long-term research such as that conducted at the KBS LTER has a particularly important contribution to make in climate change mitigation because of the variable nature and slow rate of change for many agricultural GHG fluxes. While some emissions are sudden, such as biomass burning during land clearing, and others are episodic but easily quantified, such as fuel used during agronomic operations, others can be difficult to reliably estimate based on short-term observations because they change very slowly or are temporally variable. Changes in soil C sequestration, for example, are normally too gradual to detect on an annual basis: a change of 50 g C m^{-2} (a typical annual gain after conversion to no-till management) cannot be distinguished in 1 year against a spatially variable background pool of 5000 g C m^{-2}. Long-term research provides the time necessary to document such changes; detecting an increase of 500 g C m^{-2} over 10 years is much more tractable (Kravchenko and Robertson 2011).

Similarly, changes in N$_2$O emissions are difficult to detect against a background of high temporal variability. Nitrous oxide emissions from soils are notoriously variable and unpredictable: fluxes can change an order of magnitude within a single day (e.g., Ambus and Robertson 1998, Barton et al. 2008) in response to a variety of environmental drivers. Long-term N$_2$O research provides the large set of measurements and hence the statistical power needed to assess differences among agronomic systems and practices against an otherwise confusing backdrop of short-term variability.

Providing a Common Basis for Systemwide Comparisons

The Concept of Global Warming Impact (GWI)

Greenhouse gases vary greatly in radiative forcing and residence time in the atmosphere, so it is not enough to know that one system stores more soil C but liberates more N$_2$O than another system that oxidizes more CH$_4$: a reference is needed to appropriately weight the effect of different gases on the atmosphere's capacity to hold heat. The Global Warming Potential (GWP; IPCC 2001) index satisfies this need. The GWP is a combined measure of the radiative forcing of a given GHG based on its physical capacity to absorb infrared radiation, its current concentration in the atmosphere, and its atmospheric lifetime.

By convention, CO_2 has a GWP of 1; the GWPs of all other gases are expressed relative to this. Because GHGs have different atmospheric lifetimes, their GWPs change differentially after emission—for example, the GWP for a quantity of N_2O emitted today is higher than it will be a century from now, when less of it will remain in the atmosphere. Methane, with its briefer atmospheric lifetime (~12 years vs. 114 years for N_2O), will have a correspondingly smaller impact a century after emission. To provide a common means for comparison, the IPCC has identified 100 years as an appropriate standard time horizon for comparing mitigation options (Forster et al. 2007). Methane has a 100-year GWP of 23 and N_2O, 298. Manufactured halocarbons with atmospheric lifetimes of millennia can have 100-year GWPs greater than 10,000 (Prinn 2004).

We use the term GWI to refer to the effect of a given activity or group of activities on the atmosphere's heat-trapping capacity. Both GWP and GWI are measured in CO_2 equivalents (CO_2e). By way of example, a cropping practice that releases 1 g m^{-2} of CO_2 has a GWI of 1 g CO_2e m^{-2}, and a practice that releases 1 g m^{-2} of N_2O has a GWI of 298 g CO_2e m^{-2}; the GWI of both practices combined would be 299 g CO_2e m^{-2}. Thus, management practices that affect N_2O fluxes can disproportionately influence climate forcing relative to practices that affect fluxes of CO_2.

GWI in Practice

The literature is rich with estimates for GWIs of individual cropping activities. These include the effects of tillage on soil C sequestration (e.g., no-till management increases soil organic C; Paul et al. 2015, Chapter 5 in this volume); the amount of CO_2 emitted by the manufacture, transport, and application of agrochemicals; and the amount of N_2O emitted from fertilized fields as a function of the rate, timing, and formulation of N fertilizer (Millar and Robertson 2015, Chapter 9 in this volume). Still rare, however, are full-cost accountings of entire cropping systems or farms, in which GWIs from all significant sources are tallied to provide a systemwide net GWI.

Cropping systems with a net positive GWI are net emitters of GHGs and thus drivers of anthropogenic climate change, whereas systems with a net negative GWI mitigate climate change. Important to realize, however, is that any system or practice with a GWI lower than that which is currently the norm will represent mitigation relative to business as usual—even if the GWI of the new system or practice remains positive. Equally important is the notion that only by placing GWIs for different practices in an ecosystem context can the net benefits of any change be assessed. No-till practices, for example, will save fuel and store more soil C relative to conventional tillage, but the need for additional herbicide use has a C cost that will offset some of the fuel savings and soil C gain, and in some soils no-till practices may increase N_2O emissions (van Kessel et al. 2013).

Results from a full-cost analysis of GWI in the MCSE (Table 12.2) illustrate both tradeoffs and synergies. In one of the first whole-system analyses of the contribution of different GHGs to agriculture's GWI, Robertson et al. (2000) showed that the GWI of MCSE cropping systems differed markedly—and for different reasons.

Table 12.2. Global Warming Impacts for the first decade (1989–1999) of the MCSE.[a]

System	Global Warming Impact (GWI)[b] (g CO_2e m^{-2} yr^{-1})						
	Soil C	N Fertilizer	Lime	Fuel	N_2O	CH_4	Net GWI
Annual Crops (corn–soybean–wheat rotation)							
Conventional	0	27	23	16	52	−4	114
No-till	−110	27	34	12	56	−5	14
Reduced Input	−40	9	19	20	60	−5	63
Biologically Based	−29	0	0	19	56	−5	41
Perennial Crops							
Alfalfa	−161	0	80	8	59	−6	−20
Poplar	−117	5	0	2	10	−5	−105
Successional Communities							
Early Successional	−220	0	0	0	15	−6	−211
Mid-successional	−32	0	0	0	16	−15	−31
Mown Grassland (never tilled)	0	0	0	0	18	−17	1
Deciduous Forest	0	0	0	0	21	−25	−4

[a]See Table 12.1 for a description of systems. All systems are replicated (n = 3–6).
[b]Net GWI is determined as the sum of GWI components: soil carbon (C) sequestration, agronomic inputs of nitrogen (N) fertilizer, lime and fuel, and GHG exchanges of nitrous oxide (N_2O) and methane (CH_4) with the atmosphere. Units are carbon dioxide equivalents (CO_2e; g m^{-1} yr^{-1}) based on IPCC conversion factors (IPCC 2007). Negative values indicate net climate change mitigation potential.
Source: Robertson et al. (2000).

Net GWIs over a 9-year period (Table 12.2) ranged from 114 g CO_2e m^{-2} yr^{-1} (net emission) in the conventionally managed corn–soybean–wheat rotation to −211 g CO_2e m^{-2} yr^{-1} (net mitigation) in the Early Successional community abandoned from agriculture 9 years earlier. Net GWIs also differed substantially among the annual cropping systems: net GWI was low in the No-till system (14 g CO_2e m^{-2} yr^{-1}) and intermediate in the Reduced Input and Biologically Based systems (63 and 41 g CO_2e m^{-2} yr^{-1}, respectively), suggesting the potential for substantial mitigation relative to the Conventional management.

Close analysis shows the source of these differences. While in all the annual crops, N_2O production was the largest single source of GWI, in the No-till system soil C storage more than offset the GWI of N_2O emissions, although additional contributions from N fertilizer manufacture, lime (calcium and magnesium carbonate) application, and fuel use kept GWI in the No-till system positive (Table 12.2). And although not enough C was stored in the Reduced Input and Biologically Based systems to offset N_2O production, savings from lower N fertilizer and lime use helped to reduce their net GWI to about half that of the Conventional system.

The hybrid Poplar system's combination of low N_2O emissions and enhanced soil C accumulation over 9 years resulted in a substantial mitigation capacity of −105 g CO_2e m^{-2} yr^{-1} (Table 12.2). Although Alfalfa, the other perennial system evaluated, also had substantial soil C accumulation, much of this was offset by agricultural lime applications and high N_2O emissions—both related to alfalfa's N fixation capacity. Nitrogen fixation provides inorganic N to nitrifying bacteria, which in turn provide nitrate (NO_3^-) to denitrifiers, and both nitrifiers and denitrifiers produce N_2O (Ostrom et al. 2010). Nitrifiers also produce acidity, increasing the need for lime application. As a result, alfalfa possessed only a modest net mitigation capacity in spite of high rates of soil C sequestration.

Results from the full-cost analysis of GWI in the MCSE (Table 12.2) also suggest an eventual diminution of the Early Successional community's strong mitigation potential. Older successional communities had a substantially higher net GWI (though still negative), primarily because of lower soil C accumulation (Table 12.2). For example, in the late successional Deciduous Forest net soil C accumulation was nil, and although CH_4 oxidation was significant, it was largely offset by N_2O emissions, leading to an overall GWI close to zero.

Gelfand et al. (2013) extended the GWI analysis of the MCSE by an additional decade, and although results showed similar trends, there were two important differences (Table 12.3). First, Hamilton et al. (2007) found that lime contributions to GWI are likely far less than calculated earlier due to how lime is dissolved in these soils. Dissolution by strong acids such as nitric (HNO_3) leads to immediate CO_2 release—as was assumed in the earlier analysis. However, dissolution by carbonic acid (H_2CO_3)—a weak acid existing in equilibrium with dissolved CO_2—leads to net CO_2 capture by the soil solution and its hydrologic export as bicarbonate (HCO_3^-), which resides in the groundwater system for long periods. Thus, the net GWI in KBS soils, where dissolution by the two reactions tends to occur in about equal proportions, is likely nil. And second, a more recent and deeper soil C sampling (Syswerda et al. 2011) showed that soil C sequestered by the hybrid Poplar system was largely lost during reestablishment after harvesting, when for ~2 years soils were warmer and moister as a result of greater insolation and reduced transpiration due to lack of canopy cover. These results revise but do not substantially alter the original study's conclusion that different cropping practices contribute differentially to a given cropping system's GWI, and they illustrate how a long-term systems approach is necessary to fully partition the benefits and liabilities of specific management systems.

Biofuel and Energy Flux Considerations

Neither Robertson et al. (2000) nor others (e.g., Mosier et al. 2004) considered the end use of the biomass produced by cropping systems in their calculations of GWI—all harvested biomass was assumed to be oxidized to CO_2, thereby providing no further mitigation capacity. If, on the other hand, harvested biomass is used for energy that would otherwise be provided by fossil fuel, then an additional mitigation credit must be added to the GWI of the cropping system that produced it, so long as additional GHGs are not produced elsewhere by land cleared to offset a potential loss of food production (Searchinger et al. 2008, 2009). For example, the MCSE Poplar system discussed above would gain an additional mitigation credit of ~319 g CO_2e m^{-2} yr^{-1} were those trees grown on previously unforested land not

Table 12.3. GWIs over two decades of the MCSE.[a]

System	Global Warming Impact (GWI, g CO_2e m^{-2} yr^{-1})										
	Soil C	N Fertilizer	Lime	P	K	Seed	Pest	Fuel	N_2O	CH_4	Net GWI Revised
Annual Crops (corn–soybean–wheat rotation)											
Conventional	0 (31)	33	3	0.4	1.3	7.3	7.1	13	37 (6)	–1 (0)	101 (32)
No-till	–122 (31)	33	4	0.3	1.3	7.0	15.5	9	39 (3)	–1 (0)	–14 (31)
Reduced Input	–92 (122)	11	2	0.2	1.0	7.9	4.3	21	35 (2)	–1 (0)	–11 (122)
Biologically Based	–183 (31)	0	0	0	0	7.9	0	20	32 (3)	–1 (0)	–124 (31)
Perennial Crops											
Alfalfa	–122 (92)	0	14	0.3	4	6	3	11	46 (4)	–1 (0)	–39 (92)
Poplar	61 (53)	3	0	0	0	0	2	1	6 (1)	–1 (0)	73 (53)
Successional Communities											
Early Successional	–397 (31)	0	0	0	0	0	0	0	11 (1)	–1 (0)	–387 (31)
Mid-successional	–214 (275)	0	0	0	0	0	0	0	16 (3)	–3 (1)	–201 (275)
Mown Grassland (never tilled)	0	0	0	0	0	0	0	0	11 (2)	–4 (0)	7 (2)
Deciduous Forest	0	0	0	0	0	0	0	0	12 (2)	–5 (1)	7 (2)

[a]See Table 12.2 for further explanation. Mean values (±SE) are based on data from 1989 to 2009, except mean values of N_2O and CH_4 are based on data from 1991 to 2011.
Source: Revised from data in Gelfand et al. (2013).

now used for food crops and were the harvested biomass used to make biofuel that offsets fossil fuel use (Gelfand et al. 2013).

Energy balance can also provide a common basis for comparing the GWIs of different cropping systems because of the interconnection between GHG emissions and energy use (West and Marland 2002). An accurate estimate of agricultural energy efficiency—the ratio of useable energy in the end products to the energy used for production (Gelfand et al. 2010)—can, in addition to illustrating GWI differences, provide insights into how society can meet food and fuel security needs most energy efficiently. Energy efficiency can be calculated using energy balance tools (e.g., Kim and Dale 2003), and becomes especially useful when assessing the potential for bioenergy crops to offset fossil fuel use.

Table 12.4 shows the large range in annual energy inputs and food energy outputs for the MCSE annual cropping systems (Gelfand et al. 2010). While the Conventional system produced more than 10 times the energy in food than was used in farming (72.7 vs. 7.1 GJ ha^{-1} yr^{-1}), the No-till system produced even more energy (78.5 GJ ha^{-1} yr^{-1}) and at two-thirds of the energy input (4.9 GJ ha^{-1} yr^{-1}), for a net energy efficiency (energy output:input ratio) of 16, far higher than that of the Conventional (10). High energy costs of tillage account for most of the difference. Gelfand et al. (2010) also showed that the energy efficiency for food production was always higher than for liquid fuel production from the same crops, even when crop residues were to be used for fuel. However, this analysis assumes that food is produced for direct human consumption; allocating a portion of food crops to support livestock for meat and dairy production would change the energy balance because of the inherent inefficiency of energy transfer through food chains.

The Importance of System Boundaries for GWI Comparisons

A full accounting of the GWI or energy balance of an agricultural ecosystem requires a clear definition of the boundaries that meet the purpose and needs of the analysis. Inclusion of solar energy inputs, for example, would make fossil fuel

Table 12.4. Crop yields and energy balances for the annual cropping systems of the MCSE from 1989 to 2007.[a]

Annual Cropping System	Crop Yield (Mg ha^{-1} yr^{-1})			Crop Rotation Energy Balance (GJ ha^{-1} yr^{-1})		
	Corn	Wheat	Soybean	Farming Energy Inputs	Food Energy Output[b]	Net Energy Efficiency[c]
Conventional	5.9	3.5	2.3	7.1	72.7	10
No-till	6.3	3.7	2.7	4.9	78.5	16
Reduced Input	5.2	3.1	2.6	5.2	66.9	13
Biologically Based	4.1	2.1	2.4	4.8	53.1	11

[a]See Table 12.1 for a description of systems.
[b]Food produced for direct human consumption (i.e., not via livestock production for human consumption).
[c]Net Energy Efficiency calculated as output to input ratio.
Source: Gelfand et al. (2010).

energy inputs insignificant even though they are often the component of greatest interest. Energy in human labor and machinery manufacture (Hülsbergen et al. 2001) could similarly be included, but these inputs do not significantly differ among production-scale farming systems regardless of the final products (Pimentel and Patzek 2005). Thus, if one is calculating an energy balance to determine the most energy-efficient system, only significant sources of manageable energy need be included, that is, energy inputs that differ and are affected by various management options. Such comparisons assume that differences outside the farm gate are negligible, that is, that the energy costs of labor inputs, farm implements, and storing or transporting crop yields are identical or sufficiently similar to be an insignificant part of the overall system budget. This makes analyses more tractable, as measurements of fluxes and pools at the farm scale are relatively straightforward.

Thus, the choice of system boundaries should be explicit and based on the needs of the study. As for nutrient budgets or biogeochemical cycles (Robertson 1982), boundaries should be expanded only as far as necessary to encompass the fluxes relevant to the question under study. In a comparative analysis of biofuel cropping systems, for example, it make sense to expand the boundary to include the cost of transporting harvested grain and cellulosic biomass, as does inclusion of the fate of grain ethanol end-products such as dry distillers grain.

Components of GHG Balances in Cropping Systems

The primary purpose of an agricultural GHG balance is to track the exchanges of GHGs between cropping systems and the atmosphere. Figure 12.1 summarizes major fluxes between these two pools. The cropping system contains three main compartments: agricultural inputs that cost CO_2e to manufacture and transport, GHG production and consumption by soil microbes, and CO_2 captured by the cropping system and ultimately emitted in consumption of the harvested biomass. All three compartments are interrelated and influenced by management decisions.

The GWI of a given system can be studied using a mass-balance approach, which accounts for fluxes into and out of the system and provides estimates of change in the pool of interest—ultimately resulting in GHG exchanges (expressed as CO_2e) with the atmosphere:

$$\frac{dX}{dt} = Flux\,In\,X(t) - Flux\,Out\,X(t)$$

where X is the pool of interest, and *Flux In* and *Flux Out* are the sum of all measured and estimated fluxes into or out of the studied system over a given time period t. Although t is usually annualized, when processes involve different time scales, it is important that t be appropriately normalized, such as over the length of a rotation. A comparison of a 1-year continuous corn rotation to a 3-year corn–soybean–wheat rotation, for example, should be performed over at least one 3-year period to capture different crop effects, and preferably more in order to capture climatic variation. The same is true for other periodic management practices as well; for example, if

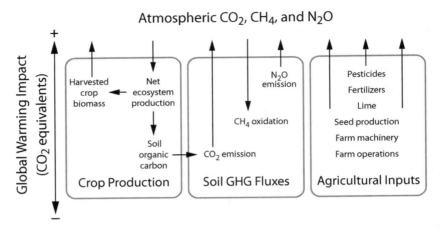

Figure 12.1. Conceptual diagram of Global Warming Impact (GWI) components in agricultural cropping systems. Arrows indicate the flux of carbon dioxide (CO_2), methane (CH_4), and nitrous oxide (N_2O) between cropping systems and the atmosphere. Atmospheric exchanges are CO_2 unless noted otherwise.

a no-till system is plowed every few years to solve a management problem or reap mineralizable N benefits, then t needs to span one or more of these tillage cycles.

Agricultural Inputs

Management decisions have a strong influence on the magnitude of CO_2e emissions associated with agricultural inputs including seed production, agrochemicals, and fuel use in farm operations. For example, the MCSE Conventional system, which is tilled, emits 35% more CO_2 from fuel use than does the No-till system (Fig. 12.2). Although the No-till system lacks soil preparation, the additional herbicides and energy required at planting (because the soil is more resistant than had it been plowed) partly offset the CO_2e savings associated with reduced fuel use by not tilling (Fig. 12.2). Likewise, synthetic N fertilizer can be a large source of CO_2e because of CO_2 emitted during its manufacture (Table 12.2), but this cost is avoided in alfalfa, which acquires its N from the atmosphere through biological N fixation. However, this savings is almost entirely offset by the CO_2e costs of alfalfa's increased agricultural lime and potassium (K) requirements. Thus, overall CO_2e emissions of the Alfalfa system are ~60% of those of the No-till and Conventional systems, despite the absence of N fertilizer use (Fig. 12.2).

The CO_2e cost of producing agricultural lime (0.04 g CO_2e kg^{-1}; West and Marland 2002) is independent of its fate. As noted earlier, Hamilton et al. (2007) estimate that CO_2 emissions from agricultural lime applied to KBS soils are fully offset by CO_2 capture when at least 50% of the lime is dissolved by carbonic acid rather than by a strong mineral acid. Nitric acid in agricultural soils is largely produced by nitrifying bacteria that produce 2 moles of H$^+$ for every mole of ammonium oxidized to NO_3^- (Robertson and Groffman 2015), and this can be

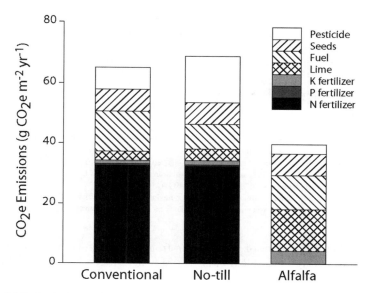

Figure 12.2. Emissions of carbon dioxide equivalents (CO_2e) from agricultural fuel and chemical inputs in the Conventional and No-till corn-soybean-wheat and the Alfalfa systems of the Main Cropping System Experiment (MCSE). Emissions from N fertilizer represent production costs only, not the resultant emission of nitrous oxide after fertilizer application.

an important source of strong acid in fertilized soils. The relative significance of these reactions with liming materials in most soils and environmental settings is not well known.

Pesticides have high CO_2e production costs (4–5 kg CO_2e kg^{-1}) but a disproportionately low impact on ecosystem CO_2e fluxes because usually only a few grams of active ingredient are applied per hectare; they thus represent only ~10% of total agricultural inputs, except in the No-till system, where they represent ~20% of total inputs (Fig. 12.2). Seeds have a larger impact on GWI due to their high production costs and seeding rates of ~20, 70, and 170 kg ha^{-1} for corn, soybean, and wheat, respectively (Gelfand et al. 2013). Estimates of GWI for seeds vary widely, however, depending on how seed production costs are estimated: West and Marland (2002) used a dollar value method to estimate a cost of 0.25 kg CO_2e kg^{-1} soybean seed (Table 12.5); and Sheehan et al. (1998) estimate a cost of 2.62 kg CO_2e kg^{-1} soybean seed based on 150% of the soybean energy content of 23.8 MJ kg^{-1} (Rathke et al. 2007). Based on average actual production costs for irrigated soybean, we estimate a cost of 0.31 kg CO_2e kg^{-1} soybean seed (Table 12.5; West and Marland 2002).

Other inputs not common to KBS cropping systems can also have significant GHG costs. Most notable among these is irrigation. Pumped irrigation uses energy to move water from lower landscape positions or groundwater to the crop, and the electricity or diesel used for this can readily become the dominant component of the GWI of irrigated systems (Mosier et al. 2005). Irrigation scheduling

Table 12.5. Estimation of the GHG cost of producing 1 kg of soybean seeds using three different approaches.

Approach	GHG Cost (kg CO_2e kg^{-1} seeds)
U.S. dollar value[a]	0.25
Energy content[b]	2.62
Direct estimation[c]	0.31

[a]From West and Marland (2002).
[b]Assumes all energy for soybean production is derived from fossil diesel with an energy content of 36.4 MJ L^{-1}; the energy content of soybean seeds is 23.8 MJ kg^{-1}; CO_2 emission from burning fossil diesel is 2.67 kg CO_2 L^{-1}.
[c]Based on CO_2e emissions from irrigated soybean production (239.9 kg C ha^{-1} yr^{-1}; West and Marland 2002) and average U.S. soybean yield (2.8 Mg ha^{-1}; http://www.nass.usda.gov/).

can also affect the amount of NO_3^- driven from the root zone into surface water and groundwater systems (Gehl et al. 2005), where it can be denitrified to N_2O (Beaulieu et al. 2011).

Nitrous Oxide and Methane Fluxes

Soil N_2O emissions are directly related to soil N availability and therefore to N fertilization and N fixation. In KBS LTER systems, those with high soil N availability—either from fertilizer inputs (e.g., in the Conventional and No-till systems) or from N fixed by leguminous cover crops (e.g., by red clover in the Reduced Input and Biologically Based systems) or by the primary crops themselves (e.g., soybean and alfalfa)—showed higher N_2O emissions than did systems with lower N inputs and availability (Fig. 12.3). This is a common finding in the N_2O literature (see Robertson and Vitousek 2009, Millar et al. 2010); in fact, global GHG inventories for agricultural N_2O emissions are largely based on a simple percentage of national fertilizer N inputs (IPCC 2006). Higher N_2O emissions in crops with lower N availability (i.e., wheat vs. soybeans, Fig. 12.3) suggest, however, that not only N availability but also specific crop (i.e., rotation type) may have an effect.

Nitrous oxide emissions appear to be especially high where N fertilization exceeds crop N requirements. McSwiney and Robertson (2005) found a nonlinear, exponentially increasing N_2O emission rate from KBS soils in continuous corn as fertilization levels increased beyond the point required for maximum yield. Others have since found similar responses (Grant et al. 2006, Ma et al. 2010, Millar et al. 2010, Hoben et al. 2011), suggesting that mitigation efforts directed toward more precise fertilizer use may have greater payoffs than those estimated by inventory methods based on a simple percentage of inputs. Millar et al. (2010, 2012, 2013) incorporated this relationship into C market incentives that can compensate farmers for more conservative N fertilizer use, which in theory is a promising way to promote fertilizer conservation in general, with both climate and water quality (Hamilton 2015, Chapter 11 in this volume) benefits.

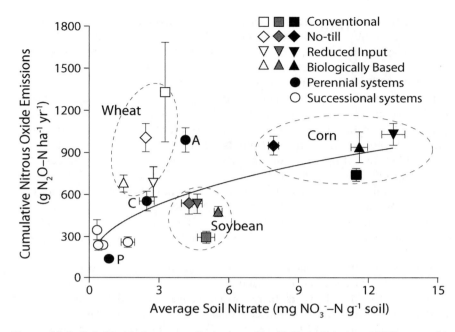

Figure 12.3. Relationship between soil nitrous oxide (N_2O) emissions and KCl-extractable nitrate (NO_3^-) in near-surface soils, fit with a power regression ($y = 358.9 \times (1 - e^{(0.37)})$, $R^2 = 0.58$, $p < 0.001$; the wheat portions of Conventional and No-till systems are not included in the regression). Annual systems are circled by crop; perennial systems are labeled as A = Alfalfa, P = Poplar, and C = Coniferous Forest.

Nitrous oxide is also emitted from aquatic systems that drain agricultural watersheds. Considerable NO_3^- is lost from intensively fertilized fields (Syswerda et al. 2012, Hamilton 2015, Chapter 11 in this volume), and based on watershed mass balances, most of this NO_3^- appears to be denitrified to N_2O and N_2. A recent cross-site study of stream N cycling that includes the broader watershed around KBS (Beaulieu et al. 2008, 2011) suggests that streams and rivers play a particularly important role in N transformations, and may be responsible for a surprising proportion of global anthropogenic N_2O emissions.

Methane is consumed by—rather than emitted from—most field crop systems other than flooded rice. In most well-aerated soils, more CH_4 is oxidized to CO_2 by methanotrophic bacteria than is produced by methanogenic bacteria. This means that soil methanotrophs also consume atmospheric CH_4, helping to attenuate atmospheric concentrations that would otherwise build at even higher rates than are occurring today. Methane oxidation by soil methanotrophs is estimated to consume around 30 Tg yr^{-1}. Although this is only ~5% of the total global CH_4 flux (Forster et al. 2007), it is close to the rate at which CH_4 is accumulating in the atmosphere (37 Tg yr^{-1}), suggesting that were consumption reduced—or intensified—atmospheric concentrations might be likewise affected.

Conversion of forest and grassland soils to agriculture reduces rates of soil CH_4 oxidation by 80–90% (Mosier et al. 1991, Smith et al. 2000, Del Grosso et al. 2000). In the MCSE, CH_4 oxidation in the Conventional system is about 20% of the rate in the Deciduous Forest (Robertson et al. 2000, Suwanwaree and Robertson 2005). Much of this suppression appears to stem from greater N availability in cropped soils rather than N fertilizer per se or tillage-induced changes in soil structure: oxidation was equally low in the unfertilized Biologically Based system, and fertilizing Deciduous Forest plots immediately reduces oxidation rates for the period that inorganic N pools are elevated, while tilling them has no discernible effect (Fig. 12.4; Suwanwaree and Robertson 2005). Gulledge and Schimel (1998) showed that much of the effect of N appears related to the competitive inhibition of CH_4 oxidation enzymes by ammonium ions. A longer time period of measurements of GHG fluxes from KBS soils shows, however, some recovery of CH_4 oxidation in the Biologically Based, Alfalfa, and Early successional systems 20 years after establishment (Table 12.6), despite relatively high N availability.

Nitrogen availability alone also does not explain the very slow recovery of CH_4 oxidation rates in abandoned cropland or in cropland converted to unfertilized perennial crops. After 10 years, there was no recovery of oxidation rates in either the Poplar system or in the Early Successional community (Robertson et al. 2000)—two systems in which soil NO_3^- levels and NO_3^- leaching rates are vanishingly low

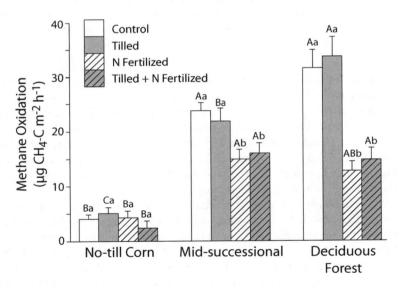

Figure 12.4. The reduction of methane (CH_4) oxidation upon soil disturbance and ammonium nitrate fertilization (100 kg N ha⁻¹) in the No-till (planted in corn), Mid-successional, and Deciduous Forest systems of the MCSE. Vertical bars are standard errors of the mean (SE, n = 3 sites × 7 sampling dates). Different uppercase and lowercase letters represent significant treatment differences ($p < 0.05$) among and within sites, respectively. Modified from Suwanwaree and Robertson (2005).

Table 12.6. Nitrous oxide (N_2O) and methane (CH_4) fluxes and GWIs from 1991 to 2010 of the MCSE.[*]

System	GHG Flux[†]		GWI	
	N_2O-N	CH_4-C	N_2O	CH_4
	(g ha^{-1} d^{-1})		(g CO$_2$e m^{-2} y^{-1})	
Annual Crops (corn–soybean–wheat rotation)				
Conventional	2.15 (0.33)[a]	−0.69 (0.09)[a]	36.6 (5.6)	−0.8 (0.1)
No-till	2.27 (0.15)[a]	−0.65 (0.06)[a]	38.6 (2.5)	−0.7 (0.1)
Reduced Input	2.06 (0.13)[a]	−0.57 (0.05)[a]	35.0 (2.3)	−0.6 (0.1)
Biologically Based	1.91 (0.15)[a]	−0.85 (0.03)[b]	32.5 (2.6)	−1.0 (0.0)
Perennial Crops				
Alfalfa	2.72 (0.24)[a]	−0.87 (0.08)[b]	46.16 (4.2)	−1.0 (0.1)
Poplar	0.38 (0.04)[b]	−0.81 (0.05)[a,b]	6.4 (0.6)	−0.9 (0.1)
Successional Community				
Early Successional	0.66 (0.05)[c]	−0.89 (0.05)[b]	11.2 (0.9)	−1.0 (0.1)

[*]Statistically significant differences (ANOVA repeated measures, $p < 0.05$) are indicated by different letters within columns. GHG fluxes are based on untransformed values and GWIs are carbon dioxide equivalents (CO$_2$e), calculated using a 100-year time horizon (IPCC 2007), and all are expressed as mean (±SE, $n = 4$ replicates).
[†]Soil GHG fluxes were sampled April–December, 1991–2010. Positive values indicate emission to the atmosphere; negative values are uptake.
Source: Gelfand et al. (2013).

(Syswerda et al. 2012). After 20 years of abandonment, however, CH_4 oxidation does begin to recover slightly in these systems (Table 12.6; Gelfand et al. 2013). Measurements in our Mid-successional community suggest that it takes 50 years or more for CH_4 oxidation to exceed 50% of preconversion rates (Robertson et al. 2000, Suwanwaree and Robertson 2005).

Why does CH_4 consumption take so long to recover to preconversion levels? Part of the explanation may be related to methanotroph community composition and, in particular, methanotroph diversity (Gulledge et al. 1997). Levine et al. (2011) found substantially higher methanotroph diversity in MCSE systems with higher oxidation rates, suggesting that microbial community composition (see Schmidt and Waldron 2015, Chapter 6 in this volume) may matter for CH_4 oxidation in the same way that it matters for N_2O production via denitrification (Cavigelli and Robertson 2000, 2001; Schmidt and Waldron 2015, Chapter 6 in this volume).

Soil CH_4 oxidation is not known to be affected by any existing agronomic practice; it is as low in the MCSE No-till and Reduced Input systems and in various organic systems of the Living Field Lab Experiment (Robertson and Hamilton 2015, Chapter 1 in this volume; Snapp et al. 2015, Chapter 15 in this volume) as it is in the fertilized Conventional system (Suwanwaree 2003). Alternative agronomic practices that increase the capacity for CH_4 oxidation could have the potential for significant GWI mitigation (Gelfand et al. 2013).

Crop Carbon Dioxide Capture

Crop production goals, that is, crop productivity and how its biomass is used, have great influence on the GWI of agricultural systems. Net Ecosystem Productivity (NEP; Fig. 12.1) is the net annual uptake of CO_2 from the atmosphere by the plant–soil system, defined as gross primary production (total CO_2 uptake) less ecosystem respiration (total CO_2 produced) (Randerson et al. 2002). Net Ecosystem Productivity thus represents the overall C balance, with a few caveats (Chapin et al. 2006). In long-established annual cropping systems, NEP is typically zero—as much CO_2 is respired as is captured annually. Although not all the biomass may be consumed in the year produced—a portion of the crop residue, for example, may persist as soil organic matter (SOM) C for decades or centuries (see Paul et al. 2015, Chapter 5 in this volume)—for ecosystems at C-balance equilibrium an equivalent amount of older SOM C may be respired. Thus on an annual basis, as much CO_2 will leave the ecosystem as enters.

Cropping systems recently converted from forests or grasslands have a negative NEP, annually releasing more CO_2 to the atmosphere than they capture. More CO_2 is respired than fixed because the original vegetation left on-site, including roots, will decompose and long-stored SOM will be rapidly oxidized when tillage breaks up soil aggregates and exposes protected C to microbial attack (Grandy and Robertson 2006, 2007; Paul et al. 2015, Chapter 5 in this volume). In most temperate regions, the SOM content of recently converted soils will approach a new steady-state equilibrium at 40–60% of original levels in 40–60 years (Paul et al. 1997).

Conversely, cropping systems that are gaining C have a positive NEP. In annual cropping systems, this occurs when SOM accumulates with the adoption of no-till cultivation or cover crops. When left untilled, soil aggregates that form around small particles of organic matter are more stable and protect the organic matter from microbial oxidation (Six et al. 2000, Grandy et al. 2006)—thereby allowing soil C pools to rebuild to some proportion of their original C content (West and Post 2002). At KBS, rates of soil C gain in the No-till system are typical of gains elsewhere in the Midwest (West and Post 2002): ~33 g C m^{-2} yr^{-1} in the Ap horizon, with no significant change in deeper layers to 1 m (Syswerda et al. 2011).

Even in tilled soils, cover crops can build SOM quickly—in the unfertilized Biologically Based system, C was sequestered in the surface soil (A/Ap horizon) at ~50 g C m^{-2} yr^{-1} over the first 12 years of establishment (Syswerda et al. 2011). The mechanisms underlying cover crop gains are not yet clear, but may be related to the greater polyphenolic content in legume residue that could slow its decomposition (Palm and Sánchez 1991). Chemical protection may also be occurring in the Early Successional community, where in addition to the cessation of tillage, plant residue diversity and perennial roots help to explain C sequestration rates in the surface soil of >100 g C m^{-2} yr^{-1} over the first 12 years of abandonment (Syswerda et al. 2011).

Perennial crops provide an additional soil C advantage by having permanent deep roots, which both sequester C in long-lived belowground tissue and produce exudates that microbes at depth can transform into recalcitrant organic matter (Wickings et al. 2012). In the Early Successional community, soils below the Ap horizon accumulated ~33 g C m^{-2} yr^{-1} over the first 12 years of establishment, and SOM also increased at depth in the Alfalfa and Poplar systems during this time (Syswerda et al. 2011).

Climate Change Mitigation through Sustainable Biofuel Production

Projections of reduced fossil fuel availability and growing concerns about the environmental impacts of fossil fuel use have stimulated interest in renewable energy sources from agricultural crops (Robertson et al. 2008, Tilman et al. 2009), which would result in the concomitant expansion of cropland to satisfy new production demands (Field et al. 2008, Feng and Babcock 2010). Biofuels produced from crops could provide climate benefits by offsetting fossil fuel use. Offsets are created when fuels produced from harvested crop biomass are used instead of fossil fuels. A fossil fuel CO_2 offset credit is equivalent to the amount of CO_2 in the avoided fossil fuel use. A full cost accounting or life cycle analysis is necessary to determine the net amount of fossil fuel CO_2 actually avoided: feedstocks can greatly differ in their net C balance (Fargione et al 2010, Gelfand et al. 2013), and calculations must include both the direct C debt accrued from creating a new biofuel cropping system (Fargione et al. 2008, Gelfand et al. 2011) as well as the indirect debt created by the need to clear land elsewhere to replace lost food production (Searchinger et al. 2008). Moreover, crop residue removed to produce biofuel is residue that would otherwise have contributed to maintain or build soil C (Wilhelm et al. 2007), such that its removal can be a net GWI cost as foregone soil C sequestration. Although the mitigation impact of a biofuel cropping system can be substantial, benefits depend entirely on where and how and which crops are grown. Two examples from KBS serve to illustrate this point: one based on the use of existing cropping systems for biofuel production, and the other on the conversion of former cropland enrolled for 22 years in the USDA's Conservation Reserve Program (CRP).

The GWI of Established Biofuel Crops

Currently, most U.S. biofuel production is ethanol made from corn grain. A small amount of biodiesel is produced from soybean and other oil seed crops. In other parts of the world, sugarcane and oil seed crops such as palm oil are used to produce biofuels, and in the future cellulosic ethanol will likely be produced from agricultural wastes and residues, perennial grasses, and woody vegetation (NRC 2009). Future fuels will likely also include butanol, alkanes, and other so-called drop-in hydrocarbons, and biomass is also likely to be combusted directly to produce electricity and heat, avoiding some of the energy loss associated with biorefining and with burning fuel in internal combustion engines of individual automobiles.

Over the next several decades, then, agricultural biomass will increasingly be used as feedstocks to produce a variety of energy sources. This will place unprecedented demands on croplands globally; in the United States, agricultural biomass needs are expected to approach 700 Tg (Perlack et al. 2005, NRC 2009), which could take as much as 90 million ha (222 million acres) of additional cropland (CAST 2011, Robertson et al. 2011)—about half as much U.S. land as we use today for all annual crops. Impacts on the biogeochemistry (Robertson et al. 2011) and biodiversity (Fletcher et al. 2011) of agricultural landscapes are likely to be correspondingly high. The climate change implications of these impacts make it all the more important that policy and landowner decisions be based on accurate GWI assessments.

Gelfand et al. (2013) used 20 years of observations from the MCSE to analyze the life-cycle C balances of systems that could potentially be harvested for use as biofuel feedstocks. For the two annual crop systems evaluated—the Conventional and No-till systems—they assumed grain was used for grain-based ethanol (corn, wheat) or biodiesel (soybean), and that 60% of wheat straw was used for cellulosic ethanol. No residue was removed from the corn or soybean portions of the rotations in order to protect existing SOM stores (NRC 2009). Three perennial cropping systems provided biomass for cellulosic ethanol—Alfalfa, Poplar, and the Early Successional community, which was either fertilized or unfertilized.

Resulting GHG balances (Fig. 12.5B) show a negative (net mitigating) GWI for all biofuel cropping systems. Fossil fuel offset credits were greatest in the Alfalfa and fertilized Early Successional communities and lowest in the more intensively managed systems. The differences were related to both yield and management. For an example, high yields of the No-till system were balanced by relatively high management inputs, which decreased total fossil fuel offset credits. On the other hand, cellulosic biomass produced in the less productive Early Successional system lacked significant management inputs and therefore provided more fossil fuel offset credits (Fig. 12.5A). Credits for the Early Successional community would be substantially higher were technology developed to improve harvest efficiency for perennial grasses, now only 55% (Monti et al. 2009). Nevertheless, the Early Successional community still exhibited the highest net mitigation potential with a GWI of about -851 g CO_2e m^{-2} yr^{-1}, while the more productive No-till system was only fourth, with a net GWI of -397 g CO_2e m^{-2} yr^{-1}. Alfalfa was intermediate to these with a mitigation potential of about -605 g CO_2e m^{-2} y^{-1} because of the high GWI cost of increased N_2O emissions and lower SOC accumulation (Fig. 12.5B). The net mitigation potential of the Poplar system was low, owing to the lack of net soil organic C gain over its rotation including the subsequent break period. Fertilizing the Early Successional community increased its productivity and thus its fossil fuel offset by ~35%, though net GWI remained basically unchanged due to the greater CO_2e cost of the fertilizer N and increased soil N_2O emissions associated with fertilization. Nevertheless, by increasing productivity with no net change in GWI, N fertilization would reduce the amount of land needed to produce a given amount of biofuel feedstock.

The boundary of this analysis includes the full life cycle of biofuel and fossil fuel production. Expanding the boundary to include indirect land-use effects could change GWIs significantly for the worse. More specifically, the GWI of these systems will be significantly less mitigating if biofuel crops were to displace

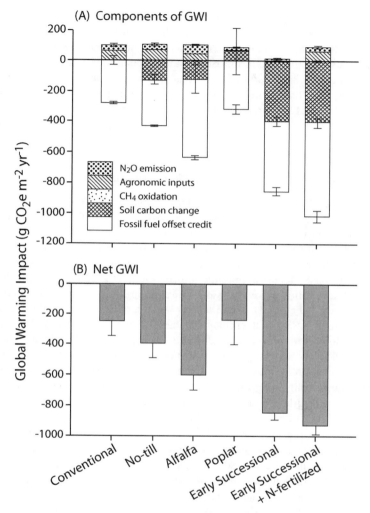

Figure 12.5. Components of Global Warming Impact (GWI, A) and the net impact (B) for agricultural and successional ecosystems in the MCSE, if harvested for cellulosic biofuel feedstock production. Error bars represent standard error (n = 6). Conventional and No-till are in a corn-soybean-wheat rotation. Redrawn from Gelfand et al. (2013).

food crops that must then be grown elsewhere on land not otherwise in agricultural production—what C markets call leakage. To avoid leakage, biofuel crops could be grown on marginal land, that is, land not now used for food crops or grazing. This could also avoid the ethical issue of food vs. fuel when feedstocks are grown on arable cropland.

Perennial grasses are particularly well suited for such marginal lands—after establishment, they require no agronomic attention other than harvest and per-haps low rates of fertilization, and thus should have few environmental liabilities.

Moreover, the right mixture of grasses could provide habitat for beneficial insects as well as for birds and other wildlife, providing additional environmental benefits especially if the marginal land were otherwise degraded due to prior management. Using land-use databases and the EPIC model (Zhang et al. 2010) to scale KBS results for the fertilized Early Successional community to a 10-state U.S. North Central region, Gelfand et al. (2013) estimated that marginal lands could produce at least 21×10^9 L of biofuel annually, or about 25% of the 80 billion L 2022 target legislated for advanced biofuels by the U.S. Energy Independence and Security Act of 2007.

The GWI of Land-Use Conversion for Biofuel Production

In 2014, about 10 million ha of former U.S. cropland were enrolled in the USDA Conservation Reserve Program (CRP) (USDA-FSA 2014). Converting these conservation plantings—most commonly in grassland vegetation—back to cropland risks the release of substantial amounts of stored soil organic C, effectively creating a C debt that models suggest could wipe out the benefits of up to 48 years of subsequent grain-based feedstock production (Fargione et al. 2008). Actual measurements of C debt following conversion, however, are not yet available, and theory suggests that the debt could be significantly less than this with careful management.

In 2009 three KBS fields enrolled in the CRP program since 1987 were converted from long-term brome grass (*Bromus inermis*) to no-till soybean as a recommended break crop prior to growing various cellulosic feedstocks. The advantage of soybeans as a break crop is that glyphosate-tolerant soybeans can be sprayed multiple times during the growing season to kill any remnants of the preexisting vegetation (brome grass, in this case). A CO_2 eddy covariance tower was placed in each field and in an unconverted CRP reference field (Zenone et al. 2011). Eddy covariance towers measure net ecosystem CO_2 flux by observing CO_2 concentrations and the movement of air between the atmosphere and the plant canopy at intervals of one-tenth of a second, allowing estimation of CO_2 fluxes that are then summed over a 30-minute period to provide half-hour snapshots of net ecosystem C gain and loss. Summing the half-hour snapshots over days and weeks provides, ultimately, the annual NEP of the studied ecosystem. In this way, total soil C change can be inferred long before it can be measured directly with soil sampling.

Figure 12.6A shows seasonal patterns of NEP in the converted and reference CRP systems during the year of conversion. Net Ecosystem Productivity was negative in both systems at the beginning of the year, reflecting net emissions of CO_2 as soil respiration exceeded wintertime photosynthesis by brome grass, which was nil. The negative fluxes turned positive beginning in the spring (around Day 100) as brome grass CO_2 fixation began to exceed total respiration. The CRP reference system continued to gain CO_2 until ca. Day 220, when brome grass senescence in the fall led to reduced photosynthesis, and respiration again dominated the CO_2 flux. By the end of the year, however, the cumulative NEP was still positive (above the origin in Fig. 12.6A), indicating net sequestration of CO_2 within the ecosystem. In the CRP converted system, on the other hand, an herbicide application around Day 120 interrupted CO_2 fixation by the brome grass, and the system continued to lose more

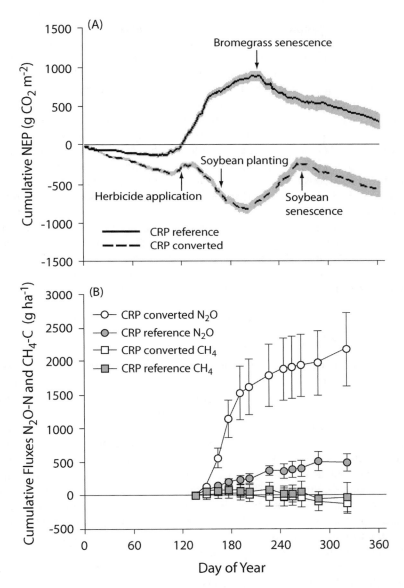

Figure 12.6. Cumulative fluxes of greenhouse gases from Conservation Reserve Program (CRP) grasslands converted to no-till soybean crops. A) Average cumulative net ecosystem productivity (NEP) during 2009 for the CRP reference field (top solid line) and converted field (bottom dashed line). Positive values indicate net CO_2 sequestration. Shaded area represents the standard deviation of cumulative NEP. B) Average net cumulative fluxes of N_2O (circles) and CH_4 (squares) at the study sites over the same period. Error bars represent standard errors (n = 3 replicate fields for CRP converted and n = 4 replicates within one field for CRP reference). Redrawn from Gelfand et al. (2011).

CO_2 than it gained until around Day 200, when net photosynthesis by the recently planted soybeans exceeded the respiration of the herbicide-treated grasses. Once the soybeans senesced around Day 260, respiration again dominated the system's CO_2 flux, and the cumulative NEP remained negative (i.e., net CO_2 release) until the end of the year, by which time some 500 g CO_2 m^{-2} had been emitted by the system.

Overall, during the first year of the conversion study, converted fields lost ~520 g CO_2 m^{-2}, mostly from the decomposition of killed grasses and soil C oxidation. This compares to a gain of ~300 g CO_2 m^{-2} by the reference field, which sequestered C into belowground biomass and SOM (Zenone et al. 2011, Gelfand et al. 2011).

Combining eddy covariance results with the other major sources of GWI in the system—farming inputs and N_2O and CH_4 fluxes, in particular—provides a measure of net GWI analogous to other, less continuous methods. N_2O emissions were also substantially higher in the converted sites (Fig. 12.6B), contributing to a total GWI or C debt of 68±7 Mg CO_2e m^{-2} (Gelfand et al. 2011). This measured C debt (from no-till conversion of CRP fields to agricultural production) is substantial but stands at the lower end of previously modeled estimates of 75–305 Mg CO_2e m^{-2} (Fargione et al. 2006, Searchinger 2008). No-till continuous corn or corn–soybean rotations, when used for grain ethanol production, could repay this C debt in 29–40 years, which is somewhat shorter than previously estimated (Fargione et al. 2008).

Summary

Intensively managed crop production systems contribute substantially to anthropogenic climate change, but changing how systems are managed could mitigate their impact. GWI analyses provide a measure for comparing the climate benefits and costs of different management practices and, by summation, of entire cropping systems. Major components of GWI include land-use change (where appropriate), farming inputs (fuel, fertilizers, pesticides), soil C change, and fluxes of the non-CO_2 GHGs N_2O and CH_4. Nitrous oxide emissions represent the largest GWI in the MCSE annual cropping systems, mainly stemming from high fertilizer inputs but also from the cultivation of N-fixing crops. Improved N management thus represents one of the largest potentials for the mitigation of agricultural GHG emissions. Soil organic C gain represents an equally large mitigation potential where soils could be managed to sequester C via no-till management, cover crops, and the cultivation of perennial crops. Perennial, cellulosic biofuel crops offer substantial climate change mitigation potential so long as their production does not cause food crops with a higher GWI to be planted elsewhere.

References

Ambus, P., and G. P. Robertson. 1998. Automated near-continuous measurement of carbon dioxide and nitrous oxide fluxes from soil. Soil Science Society of America Journal 62:394–400.

Barker, T., I. Bashmakov, L. Bernstein, J. E. Bogner, P. R. Bosch, R. Dave, O. R. Davidson, B. S. Fisher, S. Gupta, K. Halsnæs, G. J. Geij, S. Kahn Riveiro, S. Kobayashi,

M. D. Levine, D. L. Martino, O. Masera, B. Metz, L. A. Meyer, G. J. Nabuurs, A. Najam, N. Nakicenovic, H.-H. Rogner, J. Roy, J. Sathaye, R. Schock, P. Shukla, R. E. H. Sims, P. Smith, D. A. Tirpak, D. Urge-Vorsatz, and D. Zhou. 2007. Technical Summary. Pages 25–93 in B. Metz, O. R. Davidson, P. R. Bosch, R. Dave, and L. A. Meyer, editors. Climate change 2007: Mitigation. Contribution of Working Group III to the Fourth Assessment Report of the Intergovernmental Panel on Climate Change. Cambridge University Press, New York, New York, USA.

Barton, L., R. Kiese, D. Gatter, K. Butterbach-Bahl, R. Buck, C. Hinz, and D. V. Murphy. 2008. Nitrous oxide emissions from a cropped soil in a semi-arid climate. Global Change Biology 14:177–192.

Beaulieu, J. J., C. P. Arango, S. K. Hamilton, and J. L. Tank. 2008. The production and emission of nitrous oxide from headwater streams in the Midwestern United States. Global Change Biology 14:878–894.

Beaulieu, J. J., J. L. Tank, S. K. Hamilton, W. M. Wollheim, R. O. Hall, Jr., P. J. Mulholland, B. J. Peterson, L. R. Ashkenas, L. W. Cooper, C. N. Dahm, W. K. Dodds, N. B. Grimm, S. L. Johnson, W. H. McDowell, G. C. Poole, H. M. Valett, C. P. Arango, M. J. Bernot, A. J. Burgin, C. L. Crenshaw, A. M. Helton, L. T. Johnson, J. M. O'Brien, J. D. Potter, R. W. Sheibley, D. J. Sobota, and S. M. Thomas. 2011. Nitrous oxide emission from denitrification in stream and river networks. Proceedings of the National Academy of Sciences USA 108:214–219.

Burney, J. A., S. J. Davis, and D. B. Lobell. 2010. Greenhouse gas mitigation by agricultural intensification. Proceedings of the National Academy of Sciences USA 107:12052–12057.

Canadell, J. G., C. Le Quere, M. R. Raupach, C. B. Field, E. T. Buitenhuis, P. Ciais, T. J. Conway, R. A. Gillett, R. A. Houghton, and G. Marland. 2007. Contributions to accelerating atmospheric CO_2 growth from economic activity, carbon intensity, and efficiency of natural sinks. Proceedings of the National Academy of Sciences USA 104:18866–18870.

CAST (Council for Agricultural Science and Technology). 2011. Carbon sequestration and greenhouse gas fluxes in agriculture: challenges and opportunities. Task Force Report No.142, CAST, Ames, Iowa, USA.

Cavigelli, M. A., and G. P. Robertson. 2000. The functional significance of denitrifier community composition in a terrestrial ecosystem. Ecology 81:1402–1414.

Cavigelli, M. A., and G. P. Robertson. 2001. Role of denitrifier diversity in rates of nitrous oxide consumption in a terrestrial ecosystem. Soil Biology & Biochemistry 33:297–310.

Chapin, F. S., G. M. Woodwell, J. T. Randerson, E. B. Rastetter, G. M. Lovett, D. D. Baldocchi, D. A. Clark, M. E. Harmon, D. S. Schimel, R. Valentini, C. Wirth, J. D. Aber, J. J. Cole, M. L. Goulden, J. W. Harden, H. Heimann, R. W. Howarth, P. A. Matson, A. D. McGuire, J. M. Melillo, H. A. Mooney, J. C. Neff, R. A. Houghton, M. L. Pace, M. G. Ryan, S. W. Running, O. E. Sala, W. H. Schlesinger, and E. D. Schulze. 2006. Reconciling carbon-cycle concepts, terminology, and methods. Ecosystems 9:1041–1050.

Del Grosso, S. J., W. J. Parton, A. R. Mosier, D. S. Ojima, C. S. Potter, W. Borken, R. Brumme, K. Butterbach-Bahl, P. M. Crill, K. E. Dobbie, and K. A. Smith. 2000. General CH_4 oxidation model and comparisons of CH_4 oxidation in natural and managed systems. Global Biogeochemical Cycles 14:999–1019.

Fargione, J., J. Hill, D. Tilman, S. Polasky, and P. Hawthorne. 2008. Land clearing and the biofuel carbon debt. Science 319:1235–1237.

Fargione, J. E., R. J. Plevin, and J. D. Hill. 2010. The ecological impact of biofuels. Annual Review Ecology, Evolution, and Systematics 41:351–377.

Feng, H., and B. A. Babcock. 2010. Impacts of ethanol on planted acreage in market equilibrium. American Journal of Agricultural Economics 92:789–802.

Field, C. B., J. E. Campbell, and D. B. Lobell. 2008. Biomass energy: the scale of the potential resource. Trends in Ecology & Evolution 23:65–72.

Fletcher Jr., R. J., B. A. Robertson, J. Evans, J. R. R. Alavalapati, P. J. Doran, and D. J. Schemske. 2011. Biodiversity conservation in the era of biofuels: risks and opportunities. Frontiers in Ecology and the Environment 9:161–168.

Forster, P., P. Ramsaswamy, T. Artaxo, T. Bernsten, R. Betts, D. W. Fahey, J. Haywood, J. Lean, D. C. Lowe, G. Myhre, J. Nganga, R. Prinn, G. Raga, M. Schultz, and R. Van Dorland. 2007. Changes in atmospheric constituents and in radiative forcing. Pages 129–234 in D. Solomon, D. Qin, M. Manning, Z. Chen, K. B. Marquis, M. Averyt, M. Tignor, and H. L. Miller, editors. Climate change 2007: The physical science basis. Contribution of Working Group I to the Fourth Assessment of the Intergovernmental Panel on Climate Change. Cambridge University Press, Cambridge, UK.

Gehl, R. J., J. P. Schmidt, L. R. Stone, A. J. Schlegel, and G. A. Clark. 2005. In situ measurements of nitrate leaching implicate poor nitrogen and irrigation management on sandy soils. Journal of Environmental Quality 34:2243–2254.

Gelfand, I., R. Sahajpal, X. Zhang, R. C. Izaurralde, K. L. Gross, and G. P. Robertson. 2013. Sustainable bioenergy production from marginal lands in the US Midwest. Nature 493:514–517.

Gelfand, I., S. S. Snapp, and G. P. Robertson. 2010. Energy efficiency of conventional, organic, and alternative cropping systems for food and fuel at a site in the U.S. Midwest. Environmental Science and Technology 44:4006–4011.

Gelfand, I., T. Zenone, P. Jasrotia, J. Chen, S. K. Hamilton, and G. P. Robertson. 2011. Carbon debt of Conservation Reserve Program (CRP) grasslands converted to bioenergy production. Proceedings of the National Academy of Sciences USA 108:13864–13869.

Grandy, A. S., and G. P. Robertson. 2006. Aggregation and organic matter protection following tillage of a previously uncultivated soil. Soil Science Society of America Journal 70:1398–1406.

Grandy, A. S., and G. P. Robertson. 2007. Land-use intensity effects on soil organic carbon accumulation rates and mechanisms. Ecosystems 10:58–73.

Grandy, A. S., G. P. Robertson, and K. D. Thelen. 2006. Do productivity and environmental trade-offs justify periodically cultivating no-till cropping systems? Agronomy Journal 98:1377–1383.

Grant, R. F., E. Pattey, T. W. Goddard, L. M. Kryzanowski, and H. Puurveen. 2006. Modeling the effects of fertilizer application rate on nitrous oxide emissions. Soil Science Society of America Journal 70:235–248.

Gulledge, J., A. P. Doyle, and J. P. Schimel. 1997. Different NH4+-inhibition patterns of soil CH_4 consumption: a result of distinct CH_4-oxidizer populations across sites? Soil Biology and Biochemistry 29:13–21.

Gulledge, J., and J. P. Schimel. 1998. Low-concentration kinetics of atmospheric CH_4 oxidation in soil and mechanism of NH_4^+ inhibition. Applied and Environmental Microbiology 64:4291–4298.

Hamilton, S. K. 2015. Water quality and movement in agricultural landscapes. Pages 275–309 in S. K. Hamilton, J. E. Doll, and G. P. Robertson, editors. The ecology of agricultural ecosystems: long-term research on the path to sustainability. Oxford University Press, New York, New York, USA.

Hamilton, S. K., A. L. Kurzman, C. Arango, L. Jin, and G. P. Robertson. 2007. Evidence for carbon sequestration by agricultural liming. Global Biogeochemical Cycles 21:GB2021.

Hoben, J. P., R. J. Gehl, N. Millar, P. R. Grace, and G. P. Robertson. 2011. Nonlinear nitrous oxide (N_2O) response to nitrogen fertilizer in on-farm corn crops of the US Midwest. Global Change Biology 17:1140–1152.

Hülsbergen, K.-J., B. Feil, S. Biermann, G.-W. Rathke, W.-D. Kalk, and W. Diepenbrock. 2001. A method of energy balancing in crop production and its application in a long-term fertilizer trial. Agriculture, Ecosystems & Environment 86:303–321.

IPCC (Intergovernmental Panel on Climate Change). 2001. Climate change 2001: the scientific basis. Contribution of Working Group I to the Third Assessment Report of the Intergovernmental Panel on Climate Change, J. T. Houghton, Y. Ding, D. J. Griggs, M. Noguer, P. J. Van der Linden, X. Dai, K. Maskell, and C. A. Johnson, editors. Cambridge University Press, New York, New York, USA.

IPCC (Intergovernmental Panel on Climate Change). 2006. N_2O emissions from managed soils, and CO_2 emissions from lime and urea application Pages 11.11–11.54 in H. S. Eggleston, L. Buendia, K. Miwa, T. Ngara, and K. Tanabe, editors. 2006 IPCC Guidelines for National Greenhouse Gas Inventories. IGES (Institute for Global Environmental Strategies), Hayama, Japan.

Kim, S., and B. E. Dale. 2003. Cumulative energy and global warming impact from the production of biomass for biobased products. Journal of Industrial Ecology 7:147–162.

Kravchenko, A. N., and G. P. Robertson. 2011. Whole-profile soil carbon stocks: the danger of assuming too much from analyses of too little. Soil Science Society of America Journal 75:235–240.

Levine, U., T. K. Teal, G. P. Robertson, and T. M. Schmidt. 2011. Agriculture's impact on microbial diversity and associated fluxes of carbon dioxide and methane. The ISME Journal 5:1683–1691.

Ma, B. L., T. Y. Wu, N. Tremblay, W. Deen, M. J. Morrison, N. B. McLaughlin, E. G. Gregorich, and G. Stewart. 2010. Nitrous oxide fluxes from corn fields: on-farm assessment of the amount and timing of nitrogen fertilizer. Global Change Biology 16:156–170.

McSwiney, C. P., and G. P. Robertson. 2005. Nonlinear response of N_2O flux to incremental fertilizer addition in a continuous maize (*Zea mays* L.) cropping system. Global Change Biology 11:1712–1719.

Melillo, J. M., J. M. Reilly, D. W. Kicklighter, A. C. Gurge, T. W. Cronin, S. Paltsev, B. S. Felzer, X. Wang, A. P. Sokolov, and C. A. Schlosser. 2009. Indirect emissions from biofuels: how important? Science 326:1397–1399.

Millar, N., and G. P. Robertson. 2015. Agricultural nitrogen: boon and bane. Pages 213–251 in S. K. Hamilton, J. E. Doll, and G. P. Robertson, editors. The ecology of agricultural ecosystems: long-term research on the path to sustainability. Oxford University Press, New York, New York, USA.

Millar, N., G. P. Robertson, A. Diamant, R. J. Gehl, P. R. Grace, and J. P. Hoben. 2012. Methodology for quantifying nitrous oxide (N_2O) emissions reductions by reducing nitrogen fertilizer use on agricultural crops. American Carbon Registry, Winrock International, Little Rock, Arkansas, USA.

Millar, N., G. P. Robertson, A. Diamant, R. J. Gehl, P. R. Grace, and J. P. Hoben. 2013. Quantifying N_2O emissions reductions in US agricultural crops through N fertilizer rate reduction. Verified Carbon Standard, Washington, DC, USA.

Millar, N., G. P. Robertson, P. R. Grace, R. J. Gehl, and J. P. Hoben. 2010. Nitrogen fertilizer management for nitrous oxide (N_2O) mitigation in intensive corn (Maize) production: an emissions reduction protocol for US Midwest agriculture. Mitigation and Adaptation Strategies for Global Change 15:185–204.

Monti, A., S. Fazio, and G. Venturi. 2009. The discrepancy between plot and field yields: harvest and storage losses of switchgrass. Biomass & Bioenergy 33:841–847.

Mosier, A., R. Wassmann, L. Verchot, J. King, and C. Palm. 2004. Methane and nitrogen oxide fluxes in tropical agricultural soils: sources, sinks and mechanisms. Environment, Development and Sustainability 6:11–49.

Mosier, A. R., A. D. Halvorson, G. A. Peterson, G. P. Robertson, and L. Sherrod. 2005. Measurement of net global warming potential in three agroecosystems. Nutrient Cycling in Agroecosystems 72:67–76.

Mosier, A. R., D. Schimel, D. Valentine, K. Bronson, and W. Parton. 1991. Methane and nitrous oxide fluxes in native, fertilized and cultivated grasslands. Nature 350:330–332.

Mueller, N. D., J. S. Gerber, M. Johnston, D. K. Ray, N. Ramankutty, and J. A. Foley. 2012. Closing yield gaps through nutrient and water management. Nature 490:254–257.

NRC (National Research Council). 2009. Liquid transportation fuels from coal and biomass: technological status, costs, and environmental impacts. National Academies Press, Washington, DC, USA.

Ostrom, N. E., R. Sutka, P. H. Ostrom, A. S. Grandy, K. H. Huizinga, H. Gandhi, J. C. von Fisher, and G. P. Robertson. 2010. Isotopologue data reveal bacterial denitrification as the primary source of N_2O during a high flux event following cultivation of a native temperate grassland. Soil Biology & Biochemistry 42:499–506.

Palm, C. A., and P. A. Sánchez. 1991. Nitrogen release from the leaves of some tropical legumes as affected by their lignin and polyphenolic contents. Soil Biology and Biochemistry 23:83–88.

Paul, E. A., A. Kravchenko, A. S. Grandy, and S. Morris. 2015. Soil organic matter dynamics: controls and management for ecosystem functioning. Pages 104–134 in S. K. Hamilton, J. E. Doll, and G. P. Robertson, editors. The ecology of agricultural ecosystems: long-term research on the path to sustainability. Oxford University Press, New York, New York, USA.

Paul, E. A., K. Paustian, E. T. Elliott, and C. V. Cole, editors. 1997. Soil organic matter in temperate agroecosystems: long-term experiments in North America. CRC Press, Boca Raton, Florida, USA.

Perlack, R. D., L. L. Wright, A. F. Turhollow, R. L. Graham, B. J. Stokes, and D. C. Erbach. 2005. Biomass as feedstock for a bioenergy and bioproducts industry: the technical feasibility of a billion-ton annual supply. Technical Report DOE/GO-102005–2135, U.S. Department of Energy, Washington, DC, USA.

Pimentel, D., and T. Patzek. 2005. Ethanol production using corn, switchgrass, and wood; biodiesel production using soybean and sunflower. Natural Resources Research 14:65–76.

Prinn, R. G. 2004. Non-CO_2 greenhouse gases. Pages 205–216 in C. B. Field and M. R. Raupach, editors. The global carbon cycle. Island Press, Washington, DC, USA.

Randerson, J. T., F. S. Chapin III, J. W. Harden, J. C. Neff, and M. E. Harmon. 2002. Net ecosystem production: a comprehensive measure of net carbon accumulation by ecosystems. Ecological Applications 12:937–947.

Rathke, G.-W., B. J. Wienhold, W. W. Wilhelm, and W. Diepenbrock. 2007. Tillage and rotation effect on corn-soybean energy balances in eastern Nebraska. Soil & Tillage Research 97:60–70.

Robertson, G. P. 1982. Regional nitrogen budgets: approaches and problems. Plant and Soil 67:73–79.

Robertson, G. P. 2004. Abatement of nitrous oxide, methane, and the other non-CO_2 greenhouse gases: the need for a systems approach. Pages 493–506 in C. B. Field and M. R. Raupach, editors. The global carbon cycle. Island Press, Washington, DC, USA.

Robertson, G. P. 2014. Soil greenhouse gas emissions and their mitigation. Pages 185–196 in N. K. Van Alfen, editor. The encyclopedia of agriculture and food systems. Academic Press, San Diego, California, USA.

Robertson, G. P., V. H. Dale, O. C. Doering, S. P. Hamburg, J. M. Melillo, M. M. Wander, W. J. Parton, P. R. Adler, J. N. Barney, R. M. Cruse, C. S. Duke, P. M. Fearnside, R. F. Follett, H. K. Gibbs, J. Goldemberg, D. J. Miadenoff, D. Ojima, M. W. Palmer, A. Sharpley, L. Wallace, K. C. Weathers, J. A. Wiens, and W. W. Wilhelm. 2008. Sustainable biofuels redux. Science 322:49–50.

Robertson, G. P., and P. M. Groffman. 2015. Nitrogen transformations. Pages 421–426 in E. A. Paul, editor. Soil microbiology, ecology, and biochemistry. Fourth edition. Academic Press, Burlington, Massachusetts, USA.

Robertson, G. P., and S. K. Hamilton. 2015. Long-term ecological research at the Kellogg Biological Station LTER Site: conceptual and experimental framework. Pages 1–32 in S. K. Hamilton, J. E. Doll, and G. P. Robertson, editors. The ecology of agricultural ecosystems: long-term research on the path to sustainability. Oxford University Press, New York, New York, USA.

Robertson, G. P., S. K. Hamilton, S. J. Del Grosso, and W. J. Parton. 2011. The biogeochemistry of bioenergy landscapes: carbon, nitrogen, and water considerations. Ecological Applications 21:1055–1067.

Robertson, G. P., E. A. Paul, and R. R. Harwood. 2000. Greenhouse gases in intensive agriculture: contributions of individual gases to the radiative forcing of the atmosphere. Science 289:1922–1925.

Robertson, G. P., and P. M. Vitousek. 2009. Nitrogen in agriculture: balancing the cost of an essential resource. Annual Review of Environment and Resources 34:97–125.

Sauerbeck, D. R. 2001. CO_2 emissions and C sequestration by agriculture perspectives and limitations. Nutrient Cycling in Agroecosystems 60:253–266.

Schmidt, T. M., and C. Waldron. 2015. Microbial diversity in agricultural soils and its relation to ecosystem function. Pages 135–157 in S. K. Hamilton, J. E. Doll, and G. P. Robertson, editors. The ecology of agricultural ecosystems: long-term research on the path to sustainability. Oxford University Press, New York, New York, USA.

Searchinger, T., R. Heimlich, R. A. Houghton, F. Dong, A. Elobeid, J. Fabiosa, S. Tokgoz, D. Hayes, and T.-H. Yu. 2008. Use of U.S. croplands for biofuels increases greenhouse gases through emissions from land-use change. Science 319:1238–1240.

Searchinger, T. D., S. P. Hamburg, J. Melillo, W. L. Chameides, P. Havlik, D. M. Kammen, G. E. Likens, R. N. Lubowski, M. Obersteiner, M. Oppenheimer, G. P. Robertson, W. H. Schlesinger, and G. D. Tilman. 2009. Fixing a critical climate accounting error. Science 326:527–528.

Senthilkumar, S., B. Basso, A. N. Kravchenko, and G. P. Robertson. 2009. Contemporary evidence for soil carbon loss in the U.S. corn belt. Soil Science Society of America Journal 73:2078–2086.

Sheehan, J., V. Camobreco, J. Duffield, M. Graboski, and H. Shapouri. 1998. Life cycle inventory of biodiesel and petroleum diesel for use in an urban bus. National Renewable Energy Laboratory, Golden, Colorado, USA.

Six, J., E. T. Elliott, and K. Paustian. 2000. Soil macroaggregate turnover and microaggregate formation: a mechanism for C sequestration under no-tillage agriculture. Soil Biology & Biochemistry 32:2099–2103.

Smith, K. A., K. E. Dobbie, B. C. Ball, L. R. Bakken, B. K. Situala, S. Hansen, and R. Brumme. 2000. Oxidation of atmospheric methane in Northern European soils, comparison with other ecosystems, and uncertainties in the global terrestrial sink. Global Change Biology 6:791–803.

Smith, P., D. Martino, Z. Cai, D. Gwary, H. Janzen, P. Kumar, B. McCarl, S. Ogle, F. O'Mara, C. Rice, B. Scholes, and O. Sirotenko. 2007. Agriculture. Pages 498–540 in B. Metz, O. R. Davidson, P. R. Bosch, R. Dave, and L. A. Meyer, editors. Climate Change 2007: Mitigation. Contribution of Working Group III to the Fourth Assessment Report of the Intergovernmental Panel on Climate Change. Cambridge University Press, Cambridge, UK, and New York, New York, USA.

Snapp, S. S., R. G. Smith, and G. P. Robertson. 2015. Designing cropping systems for eco-system services. Pages 378–408 in S. K. Hamilton, J. E. Doll, and G. P. Robertson, editors. The ecology of agricultural ecosystems: long-term research on the path to sustainability. Oxford University Press, New York, New York, USA.

Suwanwaree, P. 2003. Methane oxidation in terrestrial ecosystems: patterns and effects of disturbance. Dissertation, Michigan State University, East Lansing, Michigan, USA.

Suwanwaree, P., and G. P. Robertson. 2005. Methane oxidation in forest, successional, and no-till agricultural ecosystems: effects of nitrogen and soil disturbance. Soil Science Society of America Journal 69:1722–1729.

Syswerda, S. P., B. Basso, S. K. Hamilton, J. B. Tausig, and G. P. Robertson. 2012. Long-term nitrate loss along an agricultural intensity gradient in the Upper Midwest USA. Agriculture, Ecosystems & Environment 149:10–19.

Syswerda, S. P., A. T. Corbin, D. L. Mokma, A. N. Kravchenko, and G. P. Robertson. 2011. Agricultural management and soil carbon storage in surface vs. deep layers. Soil Science Society of America Journal 75:92–101.

Tilman, D., R. Socolow, J. A. Foley, J. Hill, E. Larson, L. Lynd, S. Pacala, J. Reilly, T. Searchinger, and C. Somerville. 2009. Beneficial biofuels—the food, energy, and environment trilemma. Science 325:270.

Tilman, D., C. Balzer, J. Hill, and B. L. Befort. 2011. Global food demand and the sustainable intensification of agriculture. Proceedings of the National Academy of Sciences USA 108:20260–20264.

USDA-FSA (U.S. Department of Agriculture-Farm Service Agency). 2014. Conservation programs. Statistics, CRP contract summary and statistics. Washington, DC, USA. http://www.fsa.usda.gov/FSA/webapp?area=home&subject=copr&topic=crp-st

van Kessel, C., R. Venterea, J. Six, M. A. Adviento-Borbe, B. Linquist, and K. J. van Groenigen. 2013. Climate, duration, and N placement determine N_2O emissions in reduced tillage systems: a meta-analysis. Global Change Biology 19:33–44.

Vermeulen, S. J., B. M. Campbell, and J. S. I. Ingram. 2012. Climate change and food systems. Annual Review of Environment and Resources 37:195–222.

West, T. O., and G. Marland. 2002. A synthesis of carbon sequestration, carbon emissions, and net carbon flux in agriculture: comparing tillage practices in the United States. Agriculture, Ecosystems & Environment 91:217–232.

West, T. O., and W. M. Post. 2002. Soil organic carbon sequestration rates by tillage and crop rotation: a global data analysis. Soil Science Society of America Journal 66:1930–1946.

Wickings, K., A. S. Grandy, S. C. Reed, and C. C. Cleveland. 2012. The origin of litter chemical complexity during decomposition. Ecology Letters. 15:1180–1188.

Wilhelm, W. W., J. M. F. Johnson, D. L. Karlen, and D. T. Lightle. 2007. Corn stover to sustain soil organic carbon further constrains biomass supply. Agronomy Journal 99:1665–1667.

Wilson, A. T. 1978. The explosion of pioneer agriculture: contribution to the global CO_2 increase. Nature 273:40–41.

Zenone, T., J. Chen, M. W. Deal, B. Wilske, P. Jasrotia, J. Xu, A. K. Bhardwaj, S. K. Hamilton, and G. P. Robertson. 2011. CO_2 fluxes of transitional bioenergy crops:

effect of land conversion during the first year of cultivation. Global Change Biology-Bioenergy 3:401–412.

Zhang, X., R. C. Izaurralde, D. Manowitz, T. O. West, W. M. Post, A. M. Thompson, V. P. Bandaru, J. Nichols, and J. R. Williams. 2010. An integrative modeling framework to evaluate the productivity and sustainability of biofuel crop production systems. Global Change Biology Bioenergy 2:258–277.

13

Farmer Decisions about Adopting Environmentally Beneficial Practices

Scott M. Swinton, Natalie Rector,
G. Philip Robertson, Christina B. Jolejole-Foreman,
and Frank Lupi

Farmers hugely influence the mix of ecosystem services that rural landscapes provide. Their management choices about crop and livestock production practices affect services linked to water, soil, climate, and wild species. Apart from cropland and pastures, farmers also control woodlots, wetlands, and meadows that can keep groundwater clean, actively mitigate greenhouse gas (GHG) emissions, and provide habitat for beneficial insects (Power 2010, Swinton et al. 2007; Swinton et al. 2015, Chapter 3 in this volume). Given that farmers have such influence over rural ecosystems, it is important to ask how they decide whether and how much to adopt environmentally beneficial practices.

These questions about farmer behavior take us inside the Social System section of the Kellogg Biological Station Long-Term Ecological Research (KBS LTER) conceptual model (Fig. 1.4 in Robertson and Hamilton 2015, Chapter 1 in this volume). The model shows that humans respond to flows of ecosystem services as well as to other drivers of change. Yet human behavior is tremendously variable, and farmers are no exception. Their individual objectives and how they experience external stimuli affect how they respond. For professional farmers, income generation is a major objective. They experience all sorts of ecosystem services, but they are in business to produce and sell provisioning services such as food, fiber, and bioenergy products. Incentives, rules, perceptions, personal values, and social norms (Chen et al. 2009) all shape how they manage agricultural ecosystems.

In this chapter, we examine patterns in U.S. agriculture over the past century to understand how present-day patterns evolved. We then draw on research with crop farmers about their decisions to adopt agricultural practices that provide environmental benefits. We close by reviewing the means to encourage greater environmental stewardship both within current U.S. legal structures and beyond.

Managed Ecosystems and Human Impacts

The impact of human domination of Earth's ecosystems is well documented (Vitousek et al. 1997). Our large ecological footprint owes much to the effects of intentional ecosystem management (Farber et al. 2006). Among managed ecosystems, agricultural systems cover the largest area, with estimates ranging from 25 to 38% of Earth's land area (Wood et al. 2000, Millennium Ecosystem Assessment 2005), and arguably have the greatest environmental impact (Robertson and Swinton 2005). Not only does agriculture compose over half of the land area that is not either desert or permafrost, but also agricultural systems are increasing in area. Indeed, the prospect of a growing reliance on biofuels is likely to drive greater global growth in cultivated land area than even the ~20% growth that Tilman et al. (2001) predicted by 2050.

Agricultural impacts on global ecosystem services are significant. Smith et al. (2007) estimate that 10–14% of total global GHG emissions originate from agriculture, and that does not include land-use change. Land-use change, mostly associated with deforestation for agriculture, is responsible for another 12–17%. Watershed biogeochemical models supported by empirical evidence suggest that agriculture is responsible for over 70% of the phosphorus and nitrogen carried by the Mississippi River to the hypoxic zone of the Gulf of Mexico (Alexander et al. 2008), and similar dead zones exist in other coastal regions around the world (Diaz and Rosenberg 2008). Groundwater reserves that serve drinking water wells and recharge surface streams have been significantly contaminated by agriculture during the twentieth century (Böhlke 2002). On the plus side, carbon reserves in U.S. agricultural soils are estimated to have risen during 1982–1997 due to the replacement of moldboard plowing by conservation tillage practices (i.e., reduced or no tillage), land retirement from agriculture, and reduced use of bare-soil fallow periods (Eve et al. 2002). The magnitude of changes in ecosystem services across air, water, and land indicates the importance of agricultural management effects on services at global, regional, and local scales. The recent evolution of U.S. agriculture helps to explain these environmental impacts.

Trends in United States Agriculture

The past half century of U.S. agriculture has seen rising economic efficiency at producing marketed products. Particularly striking has been the trend of rising productivity (Fig. 13.1). During 1948–2001, the 1.9% annual increase in total factor productivity permitted the real value of agricultural output to rise steadily without the use of additional inputs (Ball et al. 1997, Dumagan and Ball 2009). For example, fertilizer and pesticide use have remained essentially constant

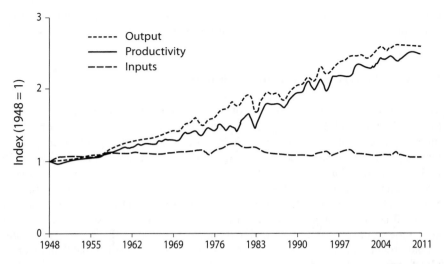

Figure 13.1. Changes in agricultural output, inputs, and total factor productivity in the United States, 1948–2011. Total factor productivity refers to gains in output that were not embodied in added inputs. Redrawn from ERS (2012a).

since 1980 (Gardner 2002). Rises in productivity combined with farm consolidation have contributed to both a sharp growth in the number of consumers supported by each U.S. farmer and some of the world's lowest food expenditures as a proportion of income.

Rising agricultural productivity has, however, been accompanied by environmental harm. In particular, large quantities of agrochemicals applied to farm fields miss their targets (Snapp et al. 2010), escaping to cause environmental damage elsewhere. To farmers, these wasted input costs are offset by the value of increased output. But the cost to society is large, as documented for pesticides (Paul et al. 2002) and nitrogen (Secchi et al. 2007) in ground and surface waters, including hypoxia in the Gulf of Mexico (Alexander et al. 2008). Agricultural mechanization has been another driver of productivity gains. Yet the removal of field borders to facilitate mechanized farming has resulted in fewer field edges and reduced biodiversity in areas of highly productive farmland (Meehan et al. 2011).

Evidence of environmental damage from agriculture, especially to water quality, led to a series of U.S. environmental programs for agriculture during the 1980s and 1990s. For croplands, these programs either paid to remove environmentally sensitive land from crop production (e.g., the Conservation Reserve Program [CRP]) or shared farmer costs of improving environmental performance (e.g., Environmental Quality Incentives Program [EQIP], Wildlife Habitat Incentives Program [WHIP], and Conservation Reserve Enhancement Program [CREP]).

The potential for agriculture to provide beneficial ecosystem services (Daily 1997) was increasingly recognized during the first decade of the 2000s. Managing agriculture as an ecosystem means not only managing for marketed products but also for socially valued ecosystem services (Antle and Capalbo 2002, Robertson et al. 2004, Farber et al. 2006, Swinton et al. 2006, 2007; Swinton et al. 2015, Chapter 3 in this volume). Examples of such services include farm management for GHG mitigation via carbon sequestration and reduced greenhouse gas emissions (Robertson 2004), water-quality improvement via reduced nutrient and pesticide leaching and runoff (Hamilton 2015, Chapter 11 in this volume), and biodiversity habitat that enables enhanced biocontrol of agricultural pests by natural enemies and crop pollination by wild pollinators (Landis et al. 2008; Landis and Gage 2015, Chapter 8 in this volume).

Some environmentally beneficial agricultural practices have low and uneven levels of adoption by U.S. farmers. For example, more complex crop rotations are known to provide environmental benefits related to pest protection and nutrient conservation. Yet in 1997, 53% of U.S. corn (*Zea mays* L.) and soybean (*Glycine max* L.) acreage was in a simple 2-year corn–soybean rotation, with only 10% in a rotation that included small grains (Padgitt et al. 2000). Only 31% of U.S. corn farmers practiced soil testing in 2002 (Christensen 2002). In California, by 2007 the adoption of organic farming remained under 2% of farmers and less than 1% of the state's agricultural value; the number of California farmers becoming organic-certified was nearly offset by the number allowing their certification to lapse (Serra et al. 2008).

Agricultural research over several decades, including 20 years at KBS LTER, has identified clear environmental benefits of rotations of corn, soybean, and wheat (*Triticum aestivum* L.) with a winter cover crop and reduced fertilizer application. Nonetheless, the 2008 Crop Management and Environmental Stewardship Survey linked to the KBS LTER (Jolejole 2009, Ma et al. 2012) found that Michigan corn and soybean farmers devoted only 8% of their land to wheat and 5% to winter cover crops, while only 22% reported applying fertilizer at rates below those recommended by university extension and only 21% applied pesticides at rates below label recommendations (Ma et al. 2012).

Some environmentally beneficial practices have been readily adopted by farmers. In the same 2008 Michigan farm survey, 82% of farmers reported that they practiced reduced tillage (compared to moldboard plow), including 55% who practiced no-till in some years (Jolejole 2009). Eighty-seven percent also reported scouting for insect pests to guide pesticide decisions. At the national scale, U.S. conservation land set aside through the CRP was fully enrolled shortly after its inception in 1985 (ERS 2012b). And a high enrollment rate has persisted in spite of increasingly stringent environmental criteria. Since EQIP was created in 1996, it too has seen more farmer interest in environmental cost share programs than its budget or acreage caps allow.

Notwithstanding the success of these programs, the combination of national and local patterns of low adoption of many (but not all) environmentally beneficial cropping practices raises the questions: Why are rates of adoption of environmentally beneficial farming practices not higher? Why are some practices adopted but not others?

Drivers of Farmer Adoption of Environmental Technologies

For a new agricultural technology about which farmers are knowledgeable, the determinants of its adoption fall into two basic categories: barriers and incentives. As Nowak (1992) observed,

> Farmers do not adopt production technologies for two basic reasons: they are either unable or unwilling. These reasons are not mutually exclusive. Farmers can be able yet unwilling, willing but unable, and, of course, both unwilling and unable. (p. 14)

The barriers and incentives paradigm offers a compelling explanation for much of observed farmer behavior with respect to environmental stewardship. A national study based on 2001–2003 data found that when farmers see a conservation technology as advantageous and not costly to adopt, adoption can proceed rapidly (Lambert et al. 2006). For example, adoption of seed-embodied conservation technologies like herbicide tolerance and transgenes that encode for the *Bacillus thuringiensis* (Bt) toxin reached high levels in just a few years. The rapid signups and overenrollment of CRP fit this model because lease payments were a clear incentive for which there were no barriers other than knowledge of the program and time available to apply for it.

Technologies embodied in equipment and other capital goods, on the other hand, tend to face high cost as a barrier to adoption. As a result, attractive but capital-intensive technologies are adopted more slowly. They tend to be more quickly adopted by large-scale farmers who can spread fixed costs over more land and may be able to hire staff with the necessary skills. For example, conservation tillage has been widely adopted in the United States (Lambert et al. 2006), but its adoption was much slower than improved seeds because it became cost-effective to adopt only when the time came to replace equipment (Krause and Black 1995).

Uncertainty can be another barrier to farm technology adoption. Farmers may be reluctant to invest when significant uncertainty (including the uncertain costs of learning by trial and error) accompanies an investment in a new technology. For example, free-stall dairy barns offer improved manure handling as well as operational efficiencies, but they were adopted slowly on account of uncertainty about future returns on investment (Purvis et al. 1995). Organic farming technologies have been adopted slowly largely because of the time lag to certification and a degree of management complexity that can make future earnings uncertain (Musshoff and Hirschauer 2008).

Demands on management time can also be a barrier to technology adoption, especially for small and part-time farmers (Lambert et al. 2006). In contrast, full-time farmers are more likely to invest the management time or hire a specialized employee who can do so.

National Trends in Adoption of Cropping Practices Used at KBS LTER

Adoption of the conservation technologies included in the KBS LTER Main Cropping System Experiment (MCSE) row-crop systems (Robertson and Hamilton 2015, Chapter 1 in this volume) is high in some areas of the United States but low in others. The Agricultural Resource Management Survey—begun by the USDA in 1996 and most recently conducted in 2010—offers online data summaries for tracking adoption of crop production practices (ERS 2012c)

Although over 80% of U.S. row crops are grown in rotations (Wallander 2013), there has been a trend toward simplified crop rotations that focus on the most profitable crops. In 1996–2006, the most recent decade for which preceding crop data are available, the corn–soybean rotation was increasingly adopted at the expense of both continuous corn and rotations with small grains. Over that period, corn following small grains dropped from 10% to less than 8% of U.S. corn acreage. By contrast, corn following soybean expanded from 54 to 60% of corn area. That pattern appears to have continued in the early 2000s, as corn and soybean displaced over 10 million acres of wheat and hay, in response to market price signals (Jekanowski and Vocke 2013).

Winter cover crops such as clover and rye are planted in the fall following harvest and plowed under prior to the following summer's primary crop. Winter cover crops are more common after soybean than after corn, mainly due to an earlier soybean harvest that provides a longer planting time in the fall. In 1997 winter cover crops were grown on 1% of U.S. corn land and 5% of U.S. soybean land (Padgitt et al. 2000). As of 2010, 3 to 7% of U.S. farms planted cover crops, but the area remained small, roughly 1% of cropland (Wallander 2013).

By contrast, acreage under conservation tillage increased from 26 to 41% between 1990 and 2004 (Sandretto and Payne 2006). This is likely linked to the availability and rapid adoption of herbicide-tolerant crop varieties that simplify herbicide decisions and reduce the need for tillage to control weeds. Between 1996 and 2006, the percentage of U.S. soybean acreage planted with genetically modified, herbicide-resistant seed rose from 7 to 97% (ERS 2012c). Over that same period, the mean number of tillage operations in soybean fell from three to one, and mean number of herbicide applications from three to two.

Corn farmers consistently applied nitrogen fertilizer to 96–99% of planted acres during 1996–2010 at annual average rates that grew from 151 to 160 kg N ha^{-1} yr^{-1} from 2000 to 2010 (134 to 143 lb acre^{-1} yr^{-1}) (ERS 2012c). Over that same period, the percentage of corn land area that underwent soil nitrogen testing rose from 21% in 1996 to 28% in 2005, and then fell back to 22% in 2010. Plant tissue nutrient testing crept upward from 2 to 4% of planted area. Using the same USDA survey data, Ribaudo et al. (2011) found that only 35% of U.S. corn land met best management practice norms for rate, timing, and method of nitrogen fertilizer application in 2006.

In summary, the adoption of conservation technologies like those used at KBS LTER has occurred in the case of conservation tillage practices and (to a lesser extent) small grains in the crop rotation, but not for cover crops or reduced nitrogen rates (Table 13.1). Given the documented ecosystem service benefits from all these technologies, it is important to ask what impedes adoption of the full set of them.

Attitudes toward Adopting Conservation Practices

Direct questioning of farmers can shed light on the motives behind their adoption patterns. We gathered both qualitative and quantitative data on how Michigan corn

Table 13.1. KBS LTER technologies and adoption trends among U.S. corn and soybean farmers.

Technology	U.S. Land-Area Trend among Corn/ Soybean Farmers	Source
Small grain in crop rotation	Declined in corn rotations from 10% to 8% in 1995–2005; wheat area down 2000–2002 to 2010–2012	ERS (2012c)[a]; Jekanowski and Vocke (2013)
Conservation tillage	Rose from 26% to 41% in 1990–2004	Sandretto and Payne (2006)
Fertilizer-reduced rates	Nitrogen use on corn stable, but excessive rates declined from 41% to 35% in 2001–2005	Ribaudo et al. (2011)
Soil nitrogen testing on corn	Trended 21, 28, and 22% in 1996, 2005, and 2010, respectively	ERS (2012c)[a]
Cover crop	Declined from 5% to 2% of soybean land in 1997–2006	Padgitt et al. (2000), C. Greene (personal communication)

[a]Data retrieved by Swinton from the Agricultural Resource Management Survey online database, but these fields not accessible in online tailored reports tool (ERS 2012c).

and soybean farmers view various conservation practices. Qualitative data came from 39 full-time corn and soybean farmers interviewed in six focus groups held in south-central and central Michigan during February and March 2007. Three of the 39 were organic farmers. In 2006 the participants had farmed between 273–2750 acres. Focus group participants were recruited by Michigan State University Extension agricultural educators and were financially compensated for their participation. The farmers completed short questionnaires about their farms, current management, and attitudes toward specific conservation practices. After discussing their views about these practices, they took part in a series of experimental auctions that were designed to reveal what it would cost them to adopt various conservation practices.

The quantitative data come from the 2008 Crop Management and Environmental Stewardship Survey, a statistically representative survey of Michigan corn and soybean farmers, described previously (Swinton et al. 2015, Chapter 3 in this volume). Farmers were asked specific questions about current farming practices and their attitudes toward conservation. They were also asked hypothetical questions about adopting new practices, and their willingness to adopt was used to estimate the potential supply of ecosystem services in exchange for payments.

Some conservation practices of interest to ecological researchers had already been adopted by the farmers surveyed. Figure 13.2 ranks 11 practices and the percentage of farmers currently using them. Two practices were used by over 80% of the farmers. These included reduced tillage (as compared to moldboard plow) and scouting for pests to guide pesticide decisions. What did these practices have in common? Both either saved labor (for tillage and pesticide application operations) or input costs (pesticide and fuel) without reducing expected crop revenue. They were largely viewed as win-win choices, helping both the environment and farm profitability.

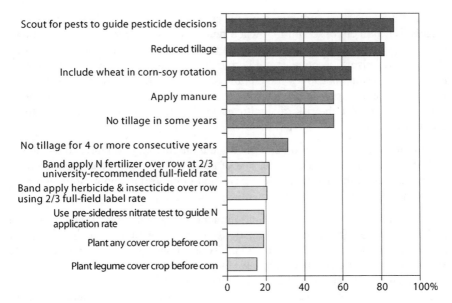

Figure 13.2. Percentage of Michigan corn-soybean farmers reporting current use of selected practices (n = 1408).

A second group of conservation practices were used by 55–65% of the farmers (Fig. 13.2). These included no-till in some years (but not continuously), applying manure, and including wheat in rotation with corn and soybean crops. What made this second category of practices slightly less attractive? No-till planting reduces weed control options (particularly for farmers not using glyphosate-tolerant crops, still a notable number at the time of the survey). Applying manure can compete for time with other farming tasks. As for why a third of respondents opted not to include any wheat in the corn–soybean rotation, three explanations came from the focus group participants: (1) wheat can be less profitable than corn or soybean, (2) wheat diseases in Michigan can reduce yield and grain quality, and (3) demand for white winter wheat had declined at local grain elevators. The common element among these three common but not ubiquitous practices is that under certain circumstances, all had the potential for reduced revenues, higher costs, or greater labor demand during busy periods.

A third group of conservation practices appealed to less than a third of the focus group farmers (Fig. 13.2). These practices were viewed by many as fundamentally problematic for one reason or another. Winter cover crops were widely perceived to delay or complicate spring planting by two mechanisms: (1) the need to allow cover crops sufficient time to accumulate spring biomass without reducing soil moisture excessively, and (2) the need to kill the cover crop prior to crop planting. The pre-sidedress nitrate test, a just-in-time test to estimate nitrogen fertilizer needs after the corn is up and growing, also confronted the timeliness problem—soil must be sent to a laboratory for analysis, which can delay fertilizer decisions and make equipment scheduling more difficult, especially when weather is poor for field work. As for continuous no-till cultivation (for more than 4 years consecutively), farmers observed that when fields went several years without tillage, they were invaded by perennial weeds that were difficult to control with herbicides. Reducing fertilizer, insecticide, and herbicide use was viewed by most farmers as having two serious negatives: potentially sacrificing crop yield and boosting the risk of herbicide-resistant insects and weeds. The least attractive practices were perceived to involve the risk of significant income loss. In the words of one focus group participant,

> "I think if you're going to get a good yield on anything you have to use the full rate of fertilizer and the full rate of pesticide . . . . Otherwise you're not getting the profits that you should have, and I don't think you're doing a service to the future crops you're planting either."

Many farmers expressed their commitment to environmental stewardship, but typically they saw it in a trade-off relationship with profitability and gave a higher priority to profitability and business viability. Said one focus group participant, "I always try to choose practices that have environmental benefits but if it's going to cause me to lose money then I can't take that choice."

Farmers' willingness to adopt stewardship practices was also influenced by their perception of how much they would benefit directly. Such benefits might be monetary, such as a greater profit margin or higher future land value, or nonmonetary, such as safer groundwater for family use. The survey respondents were asked to consider six environmental benefits from conservation agriculture and to rate the relative importance of these services "to me" on a three-point Likert scale of (1) highly important to me, (2) somewhat important to me, and (3) unimportant to me. Benefits included less global warming, less pesticide risk, less phosphorus and nitrate pollution, more soil conservation, and more soil organic matter. A parallel set of questions followed asking respondents to rate importance "to society" instead of "to me."

Upon taking differences between their answers for the relative importance "to me" versus "to society," paired difference t-tests revealed clear statistical differences ($p < 0.01$). Figure 13.3 shows that farmers rated soil organic matter, soil conservation, and reduced nitrate leaching as much more important to themselves than to society. To a lesser degree, they also found less phosphorus runoff and less pesticide risk to be more beneficial to themselves than to society at large. In contrast, they found reduced global warming to be much more important to society than to themselves. These responses conform to the economic distinction between private and public goods. The first three benefits are largely private: benefits to soil organic matter and soil conservation contribute directly to crop productivity. Reduced nitrate leaching protects the quality of groundwater, which most Michigan farms rely on for drinking, and it keeps fertilizer nitrogen in the field where it can contribute to crop productivity. Less markedly, the survey respondents also found themselves to benefit more than society from reduced phosphorus runoff and reduced pesticide risk to human health; both help farmers as well as neighbors in the region. By contrast, reduced global warming was clearly viewed as more beneficial to society at large, which is characteristic of a public good.

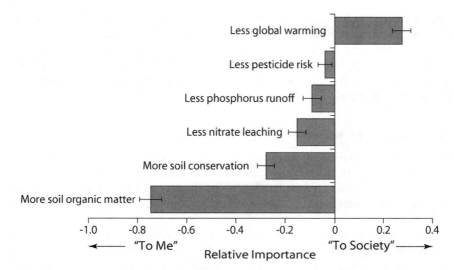

Figure 13.3. Michigan corn-soybean farmer ratings of the importance of the environmental benefits "to me" minus importance of the benefits "to society," in 2008 (n=1443). Error bars = 2 standard errors based on paired difference t-test. Negative values imply the service was rated more important "to me" than "to society"; positive values indicate the converse. Redrawn from Robertson et al. (2014).

Both survey respondents and focus group participants expressed familiarity with the management practices presented in the questionnaire, so knowledge of the practices was present. Those practices that they had adopted, or were willing to adopt, were ones that offered clear private benefits. Conservation tillage practices are a case in point. Reduced tillage was adopted by 82% of Michigan corn and soybean farmers surveyed, with 65% adopting no-till in some years. According to Sandretto and Payne (2006), reduced tillage lowers expenses for labor, fuel, and equipment and may improve yields. Among the four MCSE annual crop systems, reduced tillage is represented both by the Conventional chisel-till system and by the No-till system, the latter being both the most profitable (Swinton et al. 2015, Chapter 3 in this volume) and the provider of lowest greenhouse gas emissions (Gelfand and Robertson 2015, Chapter 12 in this volume). So conservation tillage appears to offer farmers private profitability benefits at the same time that it provides beneficial environmental externalities.

Most of the farmers surveyed and interviewed reported that environmental traits of the potential cropping systems were secondary to profitability traits. This ranking explains farmers' reluctance to reduce herbicide use by using more costly mechanical weed control or to reduce nitrogen fertilizer use by substituting more costly winter cover crops—patterns that are also evident at the national scale.

Even when farmers desire to adopt environmental technologies, barriers can impede them from doing so. Of the technologies evaluated in the 2008 Crop Management and Environmental Stewardship Survey, reduced tillage, especially no-till, has the greatest capital requirements because it requires special equipment.

Prior research has found that the practice of no-till cultivation was delayed by the normal capital replacement process because it requires purchase of an expensive no-till planter (Krause and Black 1995). Despite this financial hurdle, adoption of conservation tillage has been expanding over the past 15 years. By contrast, reducing fertilizer rates and planting cover crops are generally accessible technologies that most farmers have not elected to practice.

Incentives and Payments for Ecosystem Services

In focus group interviews, farmers made it clear that maintaining profitability is a necessary condition when it comes to deciding which crop management technologies to adopt. Hence, cropping practices that offer public benefits but impose private costs were acceptable to most participants only in exchange for a payment.

The next logical question, then, is what payments would be required to get farmers to adopt practices with greater public benefits but also higher private costs? Farmer focus group participants were invited to participate in three rounds of experimental auctions to elicit their willingness to adopt progressively more costly stewardship practices. The experimental auctions were modeled on USDA CRP procurement auctions in which the low bidder receives the payment contract. However, the experimental auctions differed from the CRP process in two important ways: (1) farmers were presented with pre-set bid amounts and had to decide whether or not to accept them and, if so, on how many acres, and (2) the lowest bidders were enrolled in the program and paid the amount per acre of the second lowest bid. This approach provides participants an incentive to truthfully reveal the minimum amount that they would be willing to accept because no single bid controls both the payment awarded and whether enrollment is successful (Harrison and List 2004, Milgrom and Weber 1982).

Farmers were asked to consider a basic corn–soybean cropping system (System A). Considering their own costs and environmental views, they were asked to determine how much they would need to be paid to replace System A with one of the four lower input systems shown in Table 13.2. All the alternative systems (B to E) included reduced tillage, pre-sidedress nitrate testing on corn, and split nitrogen applications. Beyond these, Systems C to E each added a new level of stewardship: winter cover crop (C, D, E), wheat in rotation with corn and soybean (D, E), and agrochemicals applied at two-thirds of the normal recommended rate (university rate for fertilizer and pesticide label rate for pesticides). In each case, participants were invited to specify the number of acres they would be willing to supply from their own farms at the bid offered.

Experimental auction responses (Table 13.3) showed a clear willingness to adopt environmental stewardship practices if the price is right. As the systems become more complicated with higher direct and opportunity costs, (1) fewer farmers were willing to participate and (2) the average payment the farmers would need to receive increased. In particular, farmers would require higher

Table 13.2. Alternative cropping systems offered to six farmer focus groups, Michigan, 2007.

Cropping System	A	B	C	D	E
Tillage	Mixed	Reduced	Reduced	Reduced	Reduced
Fertilizer timing	At planting	Split	Split	Split	Split
Nitrate test on corn[a]	No	Yes	Yes	Yes	Yes
Winter cover crop	No	No	Yes	Yes	Yes
Rotation	Corn–soybean	Corn–soybean	Corn–soybean	Corn–soybean–wheat	Corn–soybean–wheat
Mineral fertilizer rate	Full	Full	Full	Full	2/3 recommended
Pesticide rate	Label	Label	Label	Label	2/3 label[b]

[a]Pre-sidedress nitrate test, which requires split nitrogen fertilizer application after corn is growing.
[b]Full rate added within rows, mechanical cultivation between rows.

Table 13.3. Farmer willingness to implement low-input cropping practices in exchange for payment.[a]

Cropping System	B	C	D	E
Rotation/Management	Corn–soybean	Corn–soybean, winter cover	Corn–soybean–wheat, winter cover	Corn–soybean–wheat, winter cover agrochemicals at 2/3 rates
Participated (%)	90	85	72	59
Average payment offered if participated ($US)	37	57	44	71
Average acres offered	1315	1203	947	877
Average acres offered if participated	1470	1436	1274	1353

[a]Results are compared to Cropping System A in Table 13.2 and are based on an experimental auction involving 39 Michigan farmers in six 2007 focus groups.

payments to grow wheat and a winter cover crop. Beyond that, they would require yet a higher payment to reduce agrochemical rates from university extension recommendations (F. Lupi, unpublished data). Admittedly, the small sample size calls for caution in drawing quantitative inferences. However, the subsequent 2008 Crop Management and Environmental Stewardship Survey of 1688 corn–soybean farmers found a similar pattern: although farmers' willingness to consider enrolling in a payment-for environmental-services program was determined by their environmental attitudes, experience, education and equipment owned, the amount of land they would enroll depended more on the payment level and other income-related factors that would compensate the costs of participation (Ma et al. 2012).

What We Can Conclude about Incentives for Ecosystem Service?

Farmer adoption of conservation practices depends on awareness, attitudes, barriers, and incentives. The low-input practices studied at KBS LTER offer documented environmental benefits, including greenhouse gas mitigation (the permanent No-till, Reduced Input, and Biologically Based systems) and reduced nitrate leaching (the Reduced Input and Biologically Based systems). Michigan farmers have been adopting conservation tillage practices, as have farmers nationally. However, management that includes permanent no-till, rotation with small grains, winter cover crops, and reduced agrochemical input rates has not been widely adopted, which is also consistent with national patterns. Evidently, where a management option is win-win for both private profitability and the environment at large, farmers will adopt the practice. But where there are trade-offs that affect profitability, most farmers are reluctant to shoulder what they perceive as a private burden for the benefit of the public at large.

In focus group interviews and a statewide survey, Michigan farmers expressed familiarity with the conservation practices used at KBS LTER, but they were generally inclined to adopt only those practices that are profitable. Farmers generally believed they should be compensated to undertake practices that benefited a wider public than the farm. In experimental auctions and a subsequent mail survey, they expressed willingness to adopt low-input cropping practices in exchange for payments that would increase with the complexity and cost of the practices to be undertaken.

Similar findings internationally have led to an explosion of interest in payments for environmental services (Pagiola et al. 2002, Lipper et al. 2009). The ideal for sustainable financing of such projects is that they emerge from markets between willing buyers and sellers. However, designing such an exchange for agricultural ecosystem services can be extremely demanding, even when external start-up funding is involved (Bohlen et al. 2009). Alternatively, government programs can offer payments, although in the past, budgetary and political limitations have constrained U.S. programs such as the Conservation Stewardship Program and EQIP. Moreover, EQIP is not a true payment for ecosystem services program because its payments share input costs; they do not pay for ecosystem service outcomes per se.

Although funding payment for ecosystem service programs in the United States may be politically difficult to expand through existing farm bill mechanisms (Batie 2009), there exist alternative avenues for inducing farmers to adopt costly practices that generate wider environmental benefits. Tradable pollution permits have been inspired by the cost-effectiveness of tradable emissions permits within the Clean Air Act cap on sulfur dioxide from U.S. electrical power plants. Similar cap and trade programs have been proposed for water-borne nutrients (Hoag and Hughes-Popp 1997, Stephenson et al. 1999, Ribaudo et al. 2011, Millar and Robertson 2015, chapter 9 in this volume); greenhouse gas emissions (Konyar 2001, Post et al. 2004); and nitrous oxide abatement (Millar et al. 2010, Ribaudo et al. 2011). Such programs potentially offer farmers a market-based incentive to offer ecosystem services. However, tradable permit programs for water pollution and GHG emissions have failed to gain traction because they lack legally binding caps on emissions needed to motivate significant payment rates.

Taxes on polluting inputs are another incentive avenue. Agrochemical input taxes have been implemented at low levels in many U.S. states. While these taxes generate revenues often directed toward financing more environmental improvements, they have not led to widespread adoption of reduced agrochemical use. The literature suggests that tax rates would have to be very high in order to trigger significant reductions in agrochemical application rates (Swinton and Clark 1994, Claassen and Horan 2001, Ribaudo et al. 2011).

The adoption of environmental stewardship practices can also be made a precondition for farmers to gain access to desirable opportunities. In the government sector, conservation compliance is already required for farmers to access many farm subsidy programs, and those requirements could be expanded (Ribaudo et al. 2011). In the private sector, a number of large food companies have mandated certain management practices in the name of corporate social responsibility (Maloni and Brown 2006). For example, McDonald's requires that contracted poultry farmers meet certain animal welfare requirements. By making market access contingent on the adoption of environmental practices, the incentive for farmer adoption is tied to the value of being in that market.

To put this recent incentive research into a broader perspective, U.S. farmer attitudes and adoption behavior emerge from a legal institutional setting under which farmers have broad latitude to use their land, so long as there is no directly traceable harm done to someone else. This U.S. legal precedent evolved from English common law during a time when no scientific basis existed to demonstrate links between farm input use and outcomes that were unimaginable at that time, such as hypoxia in the Gulf of Mexico or changes in the global climate. Even when those links are acknowledged, the nonpoint source nature of much agricultural pollution impedes tracing outcomes to individual sources. Moreover, much of the damage from agricultural pollution results from the combined effect of individual contributions, making it difficult to tie aggregate impacts to individual actions.

Property rights hinge on the relationship between the person and the property. The judicial interpretation of that relationship has evolved over time (Williams 1998, Merrill and Smith 2001). Mainly in nonagricultural settings has it come to provide greater protection for the interests of persons other than the property owner. Nuisance law does recognize certain property rights of neighbors, but the law has yet to recognize the attenuation of agricultural property rights based on nonpoint source pollution. This matters both to decisions about agricultural production practices and to resultant environmental quality.

The assignment of property rights affects the very definition of production costs (Coase 1960, Norris et al. 2008, Schmid 2004). Because most U.S. farmers hold the implicit right to allow excess nutrients and greenhouse gases to move into off-farm water and air, they expect to be compensated for internalizing these disposal costs. This definition of property rights is coming under challenge from a view based on relations among members of society (Singer 2000). Kling

(2011) has argued that in a populous world with greater scientific understanding of off-farm emission effects, property rights should change so that farming is subject to the same "polluter pays" principle as industry. Were that to occur, farmers would be responsible for pollution mitigation costs for which they now expect to be compensated. Given that most of the conservation technologies discussed in this chapter would abate such pollution, voluntary adoption would be more likely.

Apart from policy and market incentive programs, technological change offers another potential avenue for a greater voluntary provision of ecosystem services from agriculture. Conservation tillage in association with herbicide-tolerant, genetically modified crops has been adopted in the United States chiefly because farmers find it to be efficient and profitable. Arguably, this phenomenon has provided important benefits via both reduced greenhouse gas emissions and reduced pesticide runoff from farm fields (NRC 2010). However, these two benefits have come with the potential for risks associated with gene release as well as perceived health risks (Uzogara 2000), which has led to the banning of genetically modified crops in Europe. The development of more win-win technologies that are profitable to farmers while offering public benefits remains possible. Incentives to generate such technologies could be enhanced by payment programs that would induce innovation of environmental technologies for agriculture (Swinton and Casey 1999) or by changes in property rights that hold farmers responsible for reducing the release of excess agrochemicals and greenhouse gases from agricultural activities (Norris et al. 2008).

Summary

Agricultural ecosystems are managed directly for human benefit. Farmers make decisions with the knowledge and resources they command to meet their goals in a complex, risky setting. Working ecosystems like agriculture are managed chiefly to provide farm income, while producing food, fiber, and biofuels to meet human needs. During the twentieth century, U.S. farmers became increasingly efficient at producing food and fuel through more reliance on agrochemical inputs. Recent calls for a rebalanced, more diverse mix of ecosystem services from agriculture raise a fundamental question: What will induce farmers to adopt more environmentally beneficial practices? By what avenues will they balance food, fiber, and fuel production with ecosystem services like carbon sequestration, improved water quality, and functional biodiversity?

Farmer adoption of new management practices depends on awareness, attitudes, available resources, and incentives. Research with Michigan farmers indicates that they are largely aware of the low-input systems studied at KBS LTER. Yet few row-crop farmers have chosen to adopt these systems in their entirety. Focus group interviews, experimental auctions, and a statewide mail survey suggest that farmer reluctance to adopt low-input practices stems from a perception of lower profitability and higher labor requirements. While no-till farming with conventional fertilization was profitable and attractive for many farmers, reduced chemical inputs appealed only in the presence of special incentives such as an organic price premium. However, farmer focus group participants and mail survey respondents expressed willingness to adopt low-input practices in exchange for payments for ecosystem services. Apart from existing government cost-share programs like the USDA Environmental Quality Incentives Program, there appear to be opportunities for payments that could compensate farmers for providing added ecosystem services, such as global warming mitigation and water-quality improvements. Payment programs or changes in legal responsibility for agricultural pollution will likely be necessary to create incentives for technological innovation with environmental benefits.

References

Alexander, R. B., R. A. Smith, G. E. Schwarz, E. W. Boyer, J. V. Nolan, and J. W. Brakebill. 2008. Differences in phosphorus and nitrogen delivery in the Gulf of Mexico from the Mississippi River Basin. Environmental Science and Technology 42:822–830.

Antle, J. M., and S. M. Capalbo. 2002. Agriculture as a managed ecosystem: policy implications. Journal of Agricultural and Resource Economics 27:1–15.

Ball, V. E., J. Bureau, R. Nehring, and A. Somwaru. 1997. Agricultural productivity revisited. American Journal of Agricultural Economics 79:1045–1063.

Batie, S. S. 2009. Green payments and the US Farm Bill: information and policy challenges. Frontiers in Ecology and the Environment 7:380–388.

Bohlen, P. J., S. Lynch, L. Shabman, M. Clark, S. Shukla, and H. Swain. 2009. Paying for environmental services from agricultural lands: an example from the northern Everglades. Frontiers in Ecology and the Environment 7:46–55.

Böhlke, J.-K. 2002. Groundwater recharge and agricultural contamination. Hydrogeology Journal 10:153–179.

Chen, X., F. Lupi, G. He, and J. Liu. 2009. Linking social norms to efficient conservation investment in payments for ecosystem services. Proceedings of the National Academy of Sciences USA 106:11812–11817.

Christensen, L. A. 2002. Soil, nutrient, and water management systems used in U.S. corn production. U.S. Department of Agriculture, Economic Research Service, Washington, DC, USA.

Claassen, R., and R. Horan. 2001. Uniform and non-uniform second-best input taxes: the significance of market price effects on efficiency and equity. Environmental and Resource Economics 19:1–22.

Coase, R. 1960. On the problems of social cost. Journal of Law and Economics 3:1–44.

Daily, G. C., editor. 1997. Nature's services: social dependence on natural ecosystems. Island Press, Washington, DC, USA.

Diaz, R. J., and R. Rosenberg. 2008. Spreading dead zones and consequences for marine ecosystems. Science 321:926–929.

Dumagan, J. C., and V. E. Ball. 2009. Decomposing growth in revenues and costs into price, quantity, and total factor productivity contributions. Applied Economics 41:2943–2953.

ERS (Economic Research Service). 2012a. Agricultural productivity in the U.S. U.S. Department of Agriculture, Washington, DC, USA. <http://www.ers.usda.gov/data-products/agricultural-productivity-in-the-us/findings,-documentation,-and-methods.aspx> Accessed January 4, 2014.

ERS (Economic Research Service). 2012b. Conservation programs. U.S. Department of Agriculture, Washington, DC, USA. <http://www.ers.usda.gov/topics/natural-resources-environment/conservation-programs/background.aspx> Accessed July 10, 2012.

ERS (Economic Research Service). 2012c. ARMS farm financial and crop production practices. U.S. Department of Agriculture, Washington, DC, USA. <http://www.ers.usda.
gov/data-products/arms-farm-financial-and-crop-production-practices/documentation.
aspx> Accessed March 7, 2012.

Eve, M. D., M. Sperow, K. Paustian, and R. F. Follett. 2002. National-scale estimation of changes in soil carbon stocks on agricultural lands. Environmental Pollution
116:431–438.

Farber, S., R. Costanza, D. L. Childers, J. Erickson, K. L. Gross, M. Grove, C. S. Hopkinson,
J. Kahn, S. Pincetl, A. Troy, P. Warren, and M. A. Wilson. 2006. Linking ecology and
economics for ecosystem management. BioScience 56:121–133.

Gardner, B. L. 2002. American agriculture in the twentieth century. Harvard University
Press, Cambridge, Massachusetts, USA.

Gelfand, I., and G. P. Robertson. 2015. Mitigation of greenhouse gas emissions in agricultural ecosystems. Pages 310–339 in S. K. Hamilton, J. E. Doll, and G. P. Robertson,
editors. The ecology of agricultural ecosystems: long-term research on the path to sustainability. Oxford University Press, New York, New York, USA.

Hamilton, S. K. 2015. Water quality and movement in agricultural landscapes. Pages 275–
309 in S. K. Hamilton, J. E. Doll, and G. P. Robertson, editors. The ecology of agricultural ecosystems: long-term research on the path to sustainability. Oxford University
Press, New York, New York, USA.

Harrison, G. W., and J. A. List. 2004. Field experiments. Journal of Economic Literature
42:1009–1055.

Hoag, D. L., and J. S. Hughes-Popp. 1997. Theory and practice of pollution credit trading in
water quality management. Review of Agricultural Economics 19:252–262.

Jekanowski, M., and G. Vocke. 2013. Crop outlook reflects near-term prices and longer
term market trends. Amber Waves. <http://www.ers.usda.gov/amber-waves/2013-june/
crop-outlook-reflects-near-term-prices-and-longer-term-market-trends.aspx> Accessed
November 24, 2013.

Jolejole, M. C. B. 2009. Trade-offs, incentives, and the supply of ecosystem services from
cropland. Thesis, Michigan State University, East Lansing, Michigan, USA.

Kling, C. L. 2011. Economic incentives to improve water quality in agricultural landscapes: some new variations on old ideas. American Journal of Agricultural Economics
93:297–309.

Konyar, K. 2001. Assessing the role of US agriculture in reducing greenhouse gas emissions
and generating additional environmental benefits. Ecological Economics 38:85–103.

Krause, M. A., and J. R. Black. 1995. Optimal adoption strategies for no-till technology in
Michigan. Review of Agricultural Economics 17:299–310.

Lambert, D. H., P. Sullivan, R. Claassen, and L. Foreman. 2006. Conservation-compatible
practices and programs: Who participates? U.S. Department of Agriculture, Economic
Research Service, Washington, DC, USA.

Landis, D. A., and S. H. Gage. 2015. Arthropod diversity and pest suppression in agricultural
landscapes. Pages 188–212 in S. K. Hamilton, J. E. Doll, and G. P. Robertson, editors.
The ecology of agricultural ecosystems: long-term research on the path to sustainability. Oxford University Press, New York, New York, USA.

Landis, D. A., M. M. Gardiner, W. van der Werf, and S. M. Swinton. 2008. Increasing corn for
biofuel production reduces biocontrol services in agricultural landscapes. Proceedings
of the National Academy of Sciences USA 105:20552–20557.

Lipper, L., T. Sakuyama, R. Stringer, and D. Zilberman, editors. 2009. Payment for environmental services in agricultural landscapes: economic policies and poverty reduction in
developing countries. Springer, Rome, Italy.

Ma, S., S. M. Swinton, F. Lupi, and C. B. Jolejole-Foreman. 2012. Farmers' willingness to participate in Payment-for-Environmental-Services programmes. Journal of Agricultural Economics 63:604–626.

Maloni, M. J., and M.E. Brown. 2006. Corporate social responsibility in the supply chain: an application in the food industry. Journal of Business Ethics 68:35–52.

Meehan, T. D., B. P. Werling, D. A. Landis, and C. Gratton. 2011. Agricultural landscape simplification and insecticide use in the Midwestern United States. Proceedings of the National Academy of Sciences USA 108:11500–11505.

Merrill, T. W., and H. E. Smith. 2001. What happened to property in law and economics? Yale Law Journal 111:357–398.

Milgrom, P. R., and R. J. Weber. 1982. A theory of auctions and competitive bidding. Econometrica 50:1089–1122.

Millar, N., G. P. Robertson, P. R. Grace, R. J. Gehl, and J. P. Hoben. 2010. Nitrogen fertilizer management for nitrous oxide ($N2O$) mitigation in intensive corn (Maize) production: an emissions reduction protocol for US Midwest agriculture. Mitigation and Adaptation Strategies for Global Change 15:185–204.

Millennium Ecosystem Assessment. 2005. Ecosystems and human well-being: synthesis. Island Press, Washington, DC, USA.

Musshoff, O., and N. Hirschauer. 2008. Adoption of organic farming in Germany and Austria: an integrative dynamic investment perspective. Agricultural Economics 39:135–145.

Norris, P. E., D. B. Schweikhardt, and E. A. Scorsone. 2008. The instituted nature of market information. Chapter 14 in S. S. Batie and N. Mercuro, editors. Alternative institutional structures: evolution and impact. Routledge, London, UK.

Nowak, P. 1992. Why farmers adopt production technology. Journal of Soil and Water Conservation 47:14–16.

NRC (National Research Council). 2010. Toward sustainable agricultural systems in the 21st century. National Academies Press, Washington, DC, USA.

Padgitt, M., D. Newton, R. Penn, and C. Sandretto. 2000. Production practices for major crops in U.S. agriculture, 1990–97. Statistical Bulletin No. SB969, Economic Research Service, U.S. Department of Agriculture, Washington, DC, USA.

Pagiola, S., J. Bishop, and N. Landell-Mills, editors. 2002. Selling forest environmental services: market-based mechanisms for conservation and development. Earthscan, London, UK.

Paul, C. J. M., V. E. Ball, R. G. Felthoven, A. H. Grube, and R. F. Nehring. 2002. Effective costs and chemical use in United States agricultural production: Using the environment as a "free" input. American Journal of Agricultural Economics 84:902–915.

Post, W. M., R. C. Izaurralde, J. D. Jastrow, B. A. McCarl, J. E. Amonette, V. L. Bailey, P. M. Jardine, T. O. West, and J. Zhou. 2004. Enhancement of carbon sequestration in US soils. BioScience 54:895–908.

Power, A. G. 2010. Ecosystem services and agriculture: tradeoffs and synergies. Philosophical Transactions of the Royal Society B: Biological Sciences 365:2959–2971.

Purvis, A., W. G. Boggess, C. B. Moss, and J. S. Holt. 1995. Technology adoption decisions under irreversibility and uncertainty: an "ex ante" approach. American Journal of Agricultural Economics 77:541–551.

Ribaudo, M., J. A. Delgado, L. Hansen, M. Livingston, R. Mosheim, and J. Williamson. 2011. Nitrogen in agricultural systems: implications for conservation policy. U.S. Department of Agriculture, Economic Research Service, Washington, DC, USA.

Robertson, G. P. 2004. Abatement of nitrous oxide, methane, and the other non-CO_2 greenhouse gases: the need for a systems approach. Pages 493–506 in C. B. Field and M. R. Raupach, editors. The global carbon cycle. Island Press, Washington, DC, USA.

Robertson, G. P., J. C. Broome, E. A. Chornesky, J. R. Frankenberger, P. Johnson, M. Lipson, J. A. Miranowski, E. D. Owens, D. Pimentel, and L. A. Thrupp. 2004. Rethinking the vision for environmental research in US agriculture. BioScience 54:61–65.

Robertson, G. P., and S. K. Hamilton. 2015. Long-term ecological research at the Kellogg Biological Station LTER Site: conceptual and experimental framework. Pages 1–32 in S. K. Hamilton, J. E. Doll, and G. P. Robertson, editors. The ecology of agricultural ecosystems: long-term research on the path to sustainability. Oxford University Press, New York, New York, USA.

Robertson, G. P., and S. M. Swinton. 2005. Reconciling agricultural productivity and environmental integrity: a grand challenge for agriculture. Frontiers in Ecology and the Environment 3:38–46.

Sandretto, C., and J. Payne. 2006. Soil management and conservation. Pages 117–128 in K. Wiebe and N. Gollehon, editors. Agricultural resources and environmental indicators. Economic Research Service, U.S. Department of Agriculture, Washington, DC, USA.

Schmid, A. A. 2004. Conflict and cooperation: institutional and behavioral economics. Blackwell, Oxford, UK.

Secchi, S., P. W. Gassman, M. Jha, L. Kurkalova, H. H. Feng, T. Campbell, and C. L. Kling. 2007. The cost of cleaner water: assessing agricultural pollution reduction at the watershed scale. Journal of Soil and Water Conservation 62:10–21.

Serra, L., K. Klonsky, R. Strochlic, S. Brodt, and R. Molinar. 2008. Factors associated with deregistration among organic farmers in California. University of California Press, Davis, California, USA.

Singer, J. W. 2000. Property and social relations: from title to entitlement. Pages 3–20 in C. Geisler and G. Deneker, editors. Property and values: alternatives to public and private ownership. Island Press, Washington, DC, USA.

Smith, P., D. Martino, Z. Cai, D. Gwary, H. Janzen, P. Kumar, B. McCarl, S. Ogle, F. O'Mara, C. Rice, B. Scholes, and O. Sirotenko. 2007. Agriculture. Pages 498–540 in B. Metz, O. R. Davidson, P. R. Bosch, R. Dave, and L. A. Meyer, editors. Climate Change 2007: Mitigation. Contribution of Working Group III to the Fourth Assessment Report of the Intergovernmental Panel on Climate Change. Cambridge University Press, Cambridge, UK, and New York, New York, USA.

Snapp, S. S., L. E. Gentry, and R. R. Harwood. 2010. Management intensity—not biodiversity—the driver of ecosystem services in a long-term row crop experiment. Agriculture, Ecosystems and Environment 138:242–248.

Stephenson, K., L. Shabman, and L. L. Geyer. 1999. Toward an effective watershed-based effluent allowance trading system: identifying the statutory and regulatory barriers to implementation. The Environmental Lawyer 5:775–815.

Swinton, S. M., and F. Casey. 1999. From adoption to innovation of environmental technologies. Pages 351–359 in F. Casey, A. Schmitz, S. Swinton, and D. Zilberman, editors. Flexible incentives for the adoption of environmental technologies in agriculture. Kluwer Academic Press, Boston, Massachusetts, USA.

Swinton, S. M., and D. M. Clark. 1994. Farm-level evaluation of alternative policies to reduce nitrate leaching from midwest agriculture. Agriculture and Resource Economics Review 23:66–74.

Swinton, S.M., C. Jolejole-Foreman, F. Lupi, S. Ma, W. Zhang, and H. Chen. 2015. Economic value of ecosystem services from agriculture. Pages 54–76 in S. K. Hamilton, J. E. Doll, and G. P. Robertson, editors. The ecology of agricultural ecosystems: long-term research on the path to sustainability. Oxford University Press, New York, New York, USA.

Swinton, S. M., F. Lupi, G. P. Robertson, and S. K. Hamilton. 2007. Ecosystem services and agriculture: cultivating agricultural ecosystems for diverse benefits. Ecological Economics 64:245–252.

Swinton, S. M., F. Lupi, G. P. Robertson, and D. A. Landis. 2006. Ecosystem services from agriculture: looking beyond the usual suspects. American Journal of Agricultural Economics 88:1160–1166.

Tilman, D., J. Fargione, B. Wolff, C. D'Antonio, A. Dobson, R. Howarth, D. Schindler, W. H. Schlesinger, D. Simberloff, and D. Swackhamer. 2001. Forecasting agriculturally driven global environmental change. Science 292:281–284.

Uzogara, S. G. 2000. The impact of genetic modification of human foods in the 21st century: a review. Biotechnology Advances 18:179–206.

Vitousek, P. M., H. A. Mooney, J. Lubchenco, and J. M. Melillo. 1997. Human domination of earth's ecosystems. Science 277:494–499.

Wallander, S. 2013. While crop rotations are common, cover crops remain rare. Amber Waves. <http://www.ers.usda.gov/amber-waves/2013-march/while-crop-rotations-are-common,-cover-crops-remain-rare.aspx > Accessed November 24, 2013.

Williams, J. 1998. The rhetoric of property. Iowa Law Review 83:277–361.

Wood, S., K. Sebastian, and S. J. Scherr. 2000. Pilot Analysis of Global Ecosystems (PAGE): agroecosystems. International Food Policy Research Institute and World Resources Institute, Washington, DC, USA.

14

Acoustic Observations in Agricultural Landscapes

Stuart H. Gage, Wooyeong Joo, Eric P. Kasten,
Jordan Fox, and Subir Biswas

Sounds emanating from various habitats and ecosystems provide a rich source of information to interpret ecological phenomena at multiple scales. Acoustic communication is a fundamental property of many animals for breeding and defending territories (Fletcher 1953, Peterson and Dorcas 1994), and acoustic signatures can be used to measure the spatial and temporal distributions of vocal organisms in ecosystems (Kroodsma and Miller 1996, Pijanowski et al. 2011a). Ecosystem sounds, in general, create a *soundscape*, made up of acoustic periodicities and frequencies emitted in aggregate from an ecosystem's biophysical entities (Schafer 1977, Truax 1984, 1999, Qi et al. 2008). Soundscapes can be partitioned into anthropogenic, biological, and physical sources (Napoletano 2004, Qi et al. 2008, Pijanowski et al. 2011a, Joo et al. 2011). They can then be further subdivided to provide valuable insights into the ecology of vocal organisms and their habitats, including their diversity and abundance, as well as phenological events such as seasonal arrivals, dates of reproduction, and breeding communication behavior.

Interpreting animal sounds has been used for many years to survey vocal organisms. For example, the U.S. North American Breeding Bird Survey—one of the largest long-term, national-scale avian observation programs—has been conducted for over 30 years by observers using both auditory and visual cues (Bystrak 1981, Robbins et al. 1986). The North American Amphibian Monitoring Program identifies amphibian species primarily by listening for their calls at night (Weir and Mossman 2005). Recent advances in sensor networks enable the large-scale, automated collection of acoustic signals in natural areas (Estrin et al. 2003, Porter et al. 2005, Pijanowski et al. 2011b, Aide et al. 2013, Ospina et al. 2013). The systematic and synchronous collection of sound recordings at multiple locations, combined with ancillary measurements such as light, temperature, and humidity, can produce an enormous volume of ecologically relevant information.

Soundscape information has the potential to increase our understanding of ecosystem change if sampled over appropriate time intervals (Truax 1984, Wrightson 2000, Sueur 2008). The analysis of entire soundscapes may also produce valuable information about the dynamics of interactions among ecological systems in heterogeneous landscapes (Carles et al. 1999). Further, rapid analysis enables the timely delivery of important environmental information to natural resource managers and can promote public involvement through public access to information about nearby and distant environments.

Automated, distributed acoustic measurements via sensor networks provide additional benefits to ecology and the environmental science community. First, analysis of observations collected through continuous monitoring at fixed sites can reveal spatiotemporal patterns that cannot be captured using site-by-site observations (Gage et al. 2004, Gage and Axel 2013). By monitoring soundscapes continuously from fixed locations, acoustic information can reveal ecosystem change over scales of days to years (Truax 1984). Second, because acoustic monitoring systems can simultaneously monitor multiple locations, acoustic variances can be compared to environmental heterogeneity (Thompson et al. 2001, Michener et al. 2001, West et al. 2001). Third, microphones can collect data from all directions simultaneously despite visual obstructions such as trees or buildings, and at all times of day including night. Finally, recording technology can operate in the field unattended, thereby allowing observations to be made without the interference generated by human presence (West et al. 2001).

Here, we illustrate the use of older recording technology (tape recorders with a clock used in 2005) and the subsequent development of wireless monitoring technology (sensor–transmitter–receiver used in 2007) to measure the soundscape, transmit the sound to a remote computer, and analyze it to understand the spatial and temporal variability of sounds emanating from ecosystems in the Kellogg Biological Station Long-Term Ecological Research site (KBS LTER). In particular, we describe the design, development, and deployment of an automated acoustic recording system and then its application to examine ecological phenomena in a complex agricultural landscape.

Soundscape Taxonomy

The sounds emanating from an ecosystem can be treated as the transmission of signals that carry information (Shannon 1948, Raisbeck 1964). The organism or force generating a sound acts as the encoder and transmitter of a signal that travels through a medium such as air or water. An organism receives the acoustic signal and then registers and decodes it into information.

Acoustic signals can be generally classified as either natural or human-induced sounds (Schafer 1977). Krause (1998) called the natural sounds *biophony*. Napoletano (2004) further classified soundscapes as biological, geophysical, or anthropogenic (Fig. 14.1). Biological sounds can be *intentional* or *incidental* signals. Intentional signals are produced by organisms that wish to communicate information such as mating or distress calls. Incidental signals may contain useful

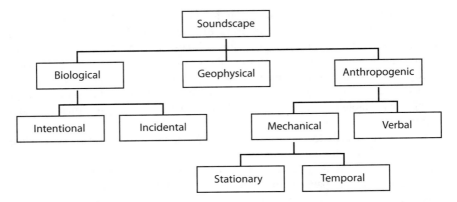

Figure 14.1. An acoustic taxonomic schema of biological, geophysical, and anthropogenic sounds. Anthropogenic verbal communication can also be considered biological, although here they have been categorized separately for clarity.

information but are not dispatched for the explicit purpose of communication. Similarly, human-induced sounds can be *verbal* or *mechanical*. Verbal signals are those produced by the human voice (e.g., talking, shouting, or singing). Conversely, mechanical signals are produced by machinery and technology. Mechanical signals can be *stationary* or *temporal*. Stationary sounds are those that impose themselves on the ambient soundscape indefinitely (e.g., turbulence from ventilation fans), while temporal sounds include noises that impinge on the soundscape for a limited period (e.g., occasional automobile or train traffic). Each of these components occurs at a range of frequencies that can be used to quantify the soundscape. Importantly, this schema differentiates between sounds produced by vocalizing animals and those produced by machines.

Soundscape Analysis

Acoustic diversity metrics attempt to measure and quantify the number and type of signal sources in the soundscape (Slabbekoorn and Peet 2003, Katti and Warren 2004, Warren et al. 2006). Organisms make selective use of acoustic frequencies when attempting to communicate information such as mating potential, territory size, and the presence of predators (Narins 1995, Catchpole and Slater 1995). Essentially, each vocalizing species can develop a dynamic niche by modulating the temporal periodicity and frequency of its respective signals to unused portions of the soundscape in order to avoid competition for spectral or temporal resources (Narins 1995). This suggests that soundscape diversity might be a sensitive index for identifying ecological change. Dale and Beyeler (2001) identified the value and characteristics of ecological indices, among which by their criteria would include environmental acoustics. Derived metrics from the soundscape are potentially valuable as ecological indices because they can provide predictable measures of ecosystem stress that can be used to interpret and measure both ecological and anthropogenic disturbances.

A soundscape consists of a complex of specific sounds (e.g., birdsong, flowing water, train whistle) of varying intensity depending on the source and the distance from the sensor. These sounds can be used as signatures since they are repeatable. A soundscape can also consist of sounds that occur at different frequencies (e.g., birdsong at higher frequencies, mechanical sounds at lower frequencies). Quantifying either type of signal is difficult since multiple organisms may sing simultaneously and their frequencies may overlap, making signature identification difficult. On the other hand, some animals have simple sounds (e.g., spring peeper frogs, *Pseudacris crucifer*) that can be readily quantified. Some organisms signal at low frequencies (e.g., American crow, *Corvus brachyrhynchos*), and thus signal at frequencies similar to those of some mechanical sounds, introducing exceptions to the idea that biological sound can be separated from mechanical sound based on frequency analysis. These problems have prompted research into pattern recognition to characterize entities in the soundscape (e.g., Reynolds and Rose 1995, Anderson et al. 1996, Acevedo et al. 2009, Ranjard and Ross 2008, Brandes 2008, Waddle et al. 2009, Kasten et al. 2010).

To date, most research on soundscapes has focused on understanding acoustic characteristics based on descriptive and qualitative analysis of sounds (Schafer 1977, Krause 1998). Quantitative methods to analyze the soundscape using frequency extraction have been developed by Napoletano (2004) and Qi et al. (2008) using spectrograms. Analysis begins with the creation of a spectrogram, which is a time-varying spectral representation of an acoustic signal that can be visualized, with frequency (in hertz or Hz) of an acoustic signal on the y-axis and time on the x-axis (Haykin 1991). To analyze a spectrogram image produced from a sound recording, one can transform the recording to an image that can then be divided into intervals (e.g., 1-kHz intervals) using image analysis software. The power level represented by the pixel values in each interval can then be summed to provide a value for each frequency interval. This enables the signal power in each frequency interval to be quantified. A more efficient method is to compute the total Power Spectral Density (PSD in watts Hz^{-1}) (Welch 1967) for each 1-kHz interval. This method requires less computational time and eliminates the need to produce spectrogram images and subsequently apply image analysis techniques to quantify the number of pixels in each frequency interval. The specifics for these latter computations are described in Kasten et al. (2012).

Soundscape Index

Napoletano (2004) found that mechanical sounds (anthrophony) mostly occur at low frequencies (1–2 kHz), whereas most biological sounds (biophony) are prevalent above 2 kHz. Geophony (e.g., wind and rain) typically occurs across the entire soundscape spectrum. We developed a Normalized Difference Soundscape Index (NDSI) to separate biophony from anthrophony:

$$\text{NDSI} = \frac{(\beta - \alpha)}{(\beta + \alpha)}$$

where a and β represent the amount of acoustic energy in the biophony (2–11 kHz) and anthrophony (1–2 kHz) frequency domains, respectively. The value of NDSI can range between –1 (pure anthrophony) to 1 (pure biophony). The index has exceptions: some vocalizing organisms such as large birds and amphibians can produce sound within the anthropogenic frequency range, and geophony as well as anthrophony (loud engines) can obscure sounds across the entire spectrum. However, the overall patterns in the soundscape represented by this index can be used to characterize the soundscape both spatially and over time (Gage and Axel 2013).

An Automated Acoustic Recording System

Prior to the development of automated acoustic recording technology, acoustic observations of birds and amphibians were made by visiting a habitat, listening for signals, interpreting them, and then recording their occurrence. By 2000 emergent technologies such as web cameras connected to the Internet were being used to transmit visual observations and the capacity to make automated acoustic observations quickly followed. However, file size associated with acoustic observation can be large and the use of wireless technology to transmit large files remains challenging.

Soundscape recording at KBS LTER was initiated in 2001 using a desktop computer in a field shed at the Main Cropping System Experiment (MCSE; Table 14.1) (Robertson and Hamilton 2015, Chapter 1 in this volume). The computer was programmed to record, capture, and transmit recordings via the Internet to a remote server on the Michigan State University (MSU) campus. A microphone on the outside of the shed captured recordings at regular intervals.

Use of Tape Recorders to Assess Biodiversity and Acoustic Variability in KBS Habitats

Digitizing and quantifying sound recordings provide both a measure of the changing patterns of the soundscape in an ecosystem as well as the identification of vocal species and a characterization of changes in biodiversity of vocal species communities at various temporal and spatial scales (Qi et al. 2008, Joo et al. 2011). At KBS, we recorded sounds in five MCSE communities—Alfalfa, Poplar, Coniferous Forest, Mid-successional, and Deciduous Forest—from May 18 to July 15, 2005 (Fig. 14.2). These early observations of the soundscape were made using an analog cassette tape recording unit that contained a clock to start and stop the recording (Sangean VersaCorder®, C. Crane Co.) and an omni-directional boundary microphone (Model 330-3020, Radio Shack Corp.). These units were placed at each recording site and were set to turn on and off six times in a 24-hour period. The 6095 recordings collected using this technology were converted from analog to digital to enable quantitative analysis. These observations are available at http://lter.kbs.msu.edu/datasets/127.

Table 14.1. Description of the KBS LTER Main Cropping System Experiment (MCSE).[a]

Cropping System/ Community	Dominant Growth Form	Management
Annual Cropping Systems		
Conventional (T1)	Herbaceous annual	Prevailing norm for tilled corn–soybean–winter wheat (c–s–w) rotation; standard chemical inputs, chisel-plowed, no cover crops, no manure or compost
No-till (T2)	Herbaceous annual	Prevailing norm for no-till c–s–w rotation; standard chemical inputs, permanent no-till, no cover crops, no manure or compost
Reduced Input (T3)	Herbaceous annual	Biologically based c–s–w rotation managed to reduce synthetic chemical inputs; chisel-plowed, winter cover crop of red clover or annual rye, no manure or compost
Biologically Based (T4)	Herbaceous annual	Biologically based c–s–w rotation managed without synthetic chemical inputs; chisel-plowed, mechanical weed control, winter cover crop of red clover or annual rye, no manure or compost; certified organic
Perennial Cropping Systems		
Alfalfa (T6)	Herbaceous perennial	5- to 6-year rotation with winter wheat as a 1-year break crop
Poplar (T5)	Woody perennial	Hybrid poplar trees on a ca. 10-year harvest cycle, either replanted or coppiced after harvest
Coniferous Forest (CF)	Woody perennial	Planted conifers periodically thinned
Successional and Reference Communities		
Early Successional (T7)	Herbaceous perennial	Historically tilled cropland abandoned in 1988; unmanaged but for annual spring burn to control woody species
Mown Grassland (never tilled) (T8)	Herbaceous perennial	Cleared woodlot (late 1950s) never tilled, unmanaged but for annual fall mowing to control woody species
Mid-successional (SF)	Herbaceous annual + woody perennial	Historically tilled cropland abandoned ca. 1955; unmanaged, with regrowth in transition to forest
Deciduous Forest (DF)	Woody perennial	Late successional native forest never cleared (two sites) or logged once ca. 1900 (one site); unmanaged

[a]Site codes that have been used throughout the project's history are given in parentheses. Systems T1–T7 are replicated within the LTER main site; others are replicated in the surrounding landscape. For further details, see Robertson and Hamilton (2015, Chapter 1 in this volume).

The number of bird species present and the number of their vocalizations were measured by listening to the acoustic samples and identifying each species' songs and calls. Additionally, a bird species diversity index was calculated (Shannon and Weaver 1963, Magurran 1988, Blair 1996, Hobson et al. 2002).

We recorded 43 bird species and 881 vocal activities (songs and calls) during the 2-month recording period. Species richness and the number of vocalizations varied

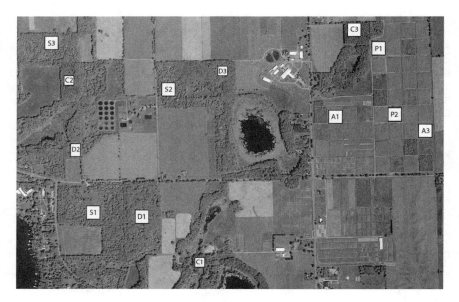

Figure 14.2. Sampling locations at the KBS LTER Main Cropping System Experiment (MCSE) where acoustic tape recorders were deployed from May 18 to July 15, 2005 and set to record six times per day. A = Alfalfa, P = Poplar, C = Coniferous Forest, S = Mid-successional, D = Deciduous Forest.

greatly among the sampling sites (Table 14.2). Fewer bird species were detected in the Alfalfa and Poplar systems than in the Coniferous Forest, Deciduous Forest, and Mid-successional communities. Bird species richness was positively correlated with the number of bird vocalizations (Fig. 14.3). A maximum frequency range of 3 kHz occurred in the Coniferous Forest and Deciduous Forest sites, whereas the maximum frequency range in the Alfalfa, Poplar, and Mid-successional community sites reached 5 kHz, showing that in these recordings, the overall acoustic frequency was lower in forests than in the agricultural (open habitat) communities (Table 14.2).

Use of Wireless and Wired Technology to Record, Transmit, and Interpret Acoustic Observations

Acoustic Sensor Technology Development

We designed and developed an Automated Acoustic Observatory System (AAOS) in 2007 utilizing a low-power sensor platform, a local server, wireless technology, and a remote server. Figure 14.4 illustrates the conceptual framework for placing automated acoustic recorders in remote locations for month-long or longer periods, making automated sound recordings at intervals of minutes to hours, with periodic transmissions to a remote server. The AAOS consists of four components (Fig. 14.4): (A) acoustic recorders in the field that record sounds at frequent

Table 14.2. Avian species identified by listening to digital recordings in KBS LTER Main Cropping System Experiment (MCSE) locations.[a]

Site[b]	Dominant Species (call density)	Species Richness	Number of Vocalizations	Shannon–Wiener Index	Frequency with Maximum Acoustic Power (kHz)
A1	Song sparrow (0.68)	9	34	1.64	5
A2	Song sparrow (0.58)	13	59	2.10	3
P1	Indigo bunting (0.65)	11	27	2.09	5
P2	Song sparrow (0.69)	11	38	1.97	3
C1	Red-winged blackbird (1.00)	16	80	2.31	3
C2	Tufted titmouse (0.42)	14	46	2.46	3
C3	Northern cardinal (0.53)	9	30	1.69	3
S1	Song sparrow (0.59)	25	146	2.64	5
S2	Brown thrasher (0.41)	17	53	2.48	5
S3	Northern cardinal (0.45)	17	77	2.47	3
D1	Scarlet tanager (0.6)	12	73	2.09	3
D2	Baltimore oriole (0.44)	21	120	2.65	3
D3	Eastern wood-pewee (0.31)	25	98	2.91	3

[a]The dominant species was determined based on call density (i.e., the number of vocalizations for that species divided by the total number of recordings). The normalized acoustic power density was generated in each frequency range by slicing every 1000 Hz from 0 to 11,000 Hz. The right column is the most powerful frequency recorded at each site.
[b]A = Alfalfa, P = Poplar, C = Coniferous Forest, S = Mid-successional, and D = Deciduous Forest systems of the MCSE; numbers refer to replicate locations.

intervals, (B) a wireless router to send acoustic recordings around tall vegetation (e.g., poplar trees), (C) a local server to receive the recordings via wireless communication and store the sound recordings locally, and (D) a regional server where the sound recordings are received via the Internet from the local server and analyzed. The recordings, results from the analysis of them (normalized soundscape power by frequency interval), and computed soundscape indices from these values are then placed in a sound library (E) that can be accessed simultaneously by users. The characteristics of this digital acoustic library are described further in Kasten et al. (2012).

Early efforts using autonomous acoustic recorders in the field identified power as the factor that limited recordings to short periods until the advent of low-power processors. Wireless technology allowed the deployment of distributed acoustic sensors, powered by a 12-V battery charged by solar panels, that collect sound recordings frequently (e.g., 30-minute intervals for 30-second durations) and transmit the recorded sounds to a local server for subsequent transfer to a remote server via the Internet. The acoustic sensor platform was designed and developed based on the Crossbow Stargate processor (Crossbow 2006). This processor operated using Linux and required relatively low power (~3 watt). The hardware components of the sensor platform (Fig. 14.5) included

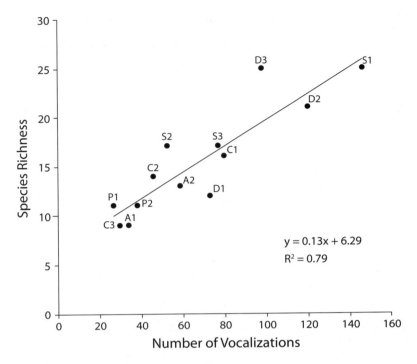

Figure 14.3. The relationship between avian species richness and the number of vocalizations identified from the acoustic samples. The letter and number near each point refers to an MCSE system/replicate number in Figure 14.2.

a processor, a power supply to convert 12-V battery input to 5-V output, an acoustic sensor (microphone), a web camera, a USB hub for additional sensors, a 2-GB flash card for local storage, a wireless communication card (802.11b), and a waterproof case. Power was supplied via a 12-V deep cycle battery charged using an 18-W solar panel.

We programmed the acoustic sensor to capture a 30-second acoustic sample at 30-minute intervals, and to transmit the recording in WAV format to a local server. The local server received approximately 100 MB of audio recordings each day from each recorder. Recordings archived on the local server were subsequently transmitted daily to the web-based digital acoustic library hosted on a laboratory-based server. These recordings are available at http://lter.kbs.msu.edu/datasets/127.

The observatory integrates acoustic sensor technology, wireless networks, and ecological applications using sound recordings from the field. This new assessment tool for ecology and environmental science provides significant opportunities to measure and interpret acoustic signals at relevant spatial and temporal scales (Kasten et al. 2012).

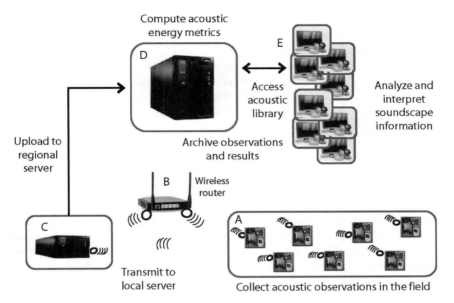

Figure 14.4. The Automated Acoustic Recording System consists of four components: A) acoustic recorders in the field that record sounds at frequent intervals and transmit them using wireless technology, B) one or more wireless routers to relay recordings around tall vegetation, C) a local server to receive the recordings via wireless communication and store them, and D) a regional server where the recordings are received via internet from the local server. The recordings are processed in the regional server and placed in a sound library (E) where they can be accessed.

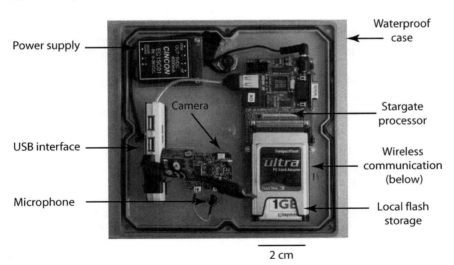

Figure 14.5. The acoustic recorder hardware configuration including a processor, wireless communication, storage, USB interface, power supply, camera, microphone, and a weather-proof enclosure.

We used the AAOS to characterize four of the MCSE communities during June 2007 to investigate the stability and efficiency of the recorders and the integrity and quality of recordings, as well as to determine differences in soundscapes among the communities. The placement of the acoustic recorders, the position of the local server (line power and internet access), and the location of wireless routers used to transmit acoustic signals around tall vegetation to the local server are given at http://lter.kbs.msu.edu/datasets/127. Automated acoustic recorders were placed in Conventional, No-till, Poplar, and Early Successional systems. Winter wheat had been planted in the Conventional and No-till systems the previous fall. A recorder was deployed in three replicates of each system for a total of 12 recorders. The automated recording system collected 11,977 sound recordings, of which 11,777 recordings were valid. Table 14.3 shows the sensor code, location where each sensor was placed, the number of recordings, statistics for total acoustic energy measured by each acoustic sensor, and the percentage of successful recordings. Two sensors malfunctioned during the month-long test (MS04, MS17), whereas others did not record for the entire time due to communication issues or battery failure. The maximum possible number of recordings was 1440 for the month (48 × 30). The Stargate-based acoustic sensors performed adequately during their first field deployment at KBS LTER, although several communication and battery

Table 14.3. Details on the deployment of acoustic sensors in the KBS LTER MCSE.[a]

Sensor Code	Location	Number of Recordings	Total Acoustic Energy (watts kHz^{-1})		Recording Success (%)
			Mean	Standard Deviation	
MS02	Poplar	930	1.50	0.25	65
MS03	Wheat[b]	1374	1.59	0.26	95
MS04	Poplar	[c]M	M	M	M
MS05	Wheat[b]	1151	1.53	0.25	80
MS06	Wheat[b]	1438	1.62	0.28	99
MS07	Early Successional	1426	1.54	0.24	99
MS09	Wheat[b]	1271	1.66	0.23	88
MS11	Early Successional	1151	1.70	0.28	80
MS12	Wheat[b]	1284	1.68	0.28	89
MS13	Wheat[b]	862	1.65	0.23	60
MS15	Early Successional	1090	1.57	0.31	76
MS17	Wheat[b]	M	M	M	M

[a]Details include sensor code, the system where each sensor was located, the number of recordings made during June 2007, statistics for total acoustic energy, and the success of each sensor to record 48 times per day for 30 days.
[b]Wheat includes both Conventional and No-till systems.
[c]M is missing observations due to faulty wireless transmission to server.

drawdown issues were identified that were later resolved. A primary issue was transmission through dense vegetation to the local server. To solve this problem, we placed two wireless routers in strategic positions to avoid tall vegetation between the sensor and local server.

Interpretation of Spatial and Temporal Change in Acoustic Observations

The analytical component of the AAOS automatically computes Power Spectral Density (PSD) values (Welsh 1967) for each of 10 frequency intervals between 1 and 11 kHz. These values were normalized (0–1 range) so that soundscape energy could be compared between locations. In addition, acoustic indices were developed from these values. One index, the Normalized Difference Soundscape Index (NDSI), was used to examine spatial and temporal variability of the KBS LTER soundscape (Kasten et al. 2012, Gage and Axel 2013). The mean NDSI was positive in all systems, indicating that biophony dominated the soundscape everywhere. The means (±standard errors) of the NDSI for the winter wheat (Conventional and No-till combined), Poplar, and Early Successional communities were 0.52 ± 0.01, 0.79 ± 0.01, 0.52 ± 0.08, respectively. Poplar had the highest mean NDSI among the three habitat types and was significantly different from the winter wheat and Early Successional systems ($F = 348.81$, $p < 0.001$), indicating that the Poplar system was more dominated by biological sounds compared to the other communities.

Although overall mean NDSI values are informative, acoustic energy patterns (expressed as watts kHz^{-1}) vary depending on the source of the sound as well as the time of day and the season. The acoustic frequencies and patterns of the frequencies may provide insight into ecological phenomena. The patterns of acoustic energy (watts kHz^{-1}) in each system are shown for four different acoustic frequency intervals (1–2 kHz; 2–3 kHz; 3–4 kHz; and 4–5 kHz) in Fig. 14.6. Both human activity (anthrophony) and some other organisms signal at lower frequencies (e.g., some amphibians, larger birds). Note the precipitous change in acoustic energy at 1–2 kHz (Fig. 14.6A) in all three systems at dawn and dusk. Also note the rise and fall in acoustic energy that can be attributable to human activity during daylight hours (08:00–20:00 h), especially in open areas (wheat and Early Successional systems) compared to Poplar where sound is buffered by vegetation. At the next highest frequency interval (2–3 kHz; Fig. 14.6B), we observe moderately high levels of soundscape energy in all systems during nighttime. At this frequency, the Early Successional community has the highest acoustic energy during the daytime compared to wheat or Poplar. In the next frequency interval (3–4 kHz; Fig. 14.6C), there is a precipitous rise in acoustic energy at dawn (0530 h) and a sharp decline at dusk (2100 h). It is within this frequency range that many species of birds signal. Relatively high levels of energy due to birdsong are sustained during the day. Although the energy in the frequency range of 4–5 kHz (Fig.14.6D) is less than that in the lower frequencies (Figs. 14.6A–C), energy is higher at night in wheat and Poplar and relatively constant in the Early Successional community.

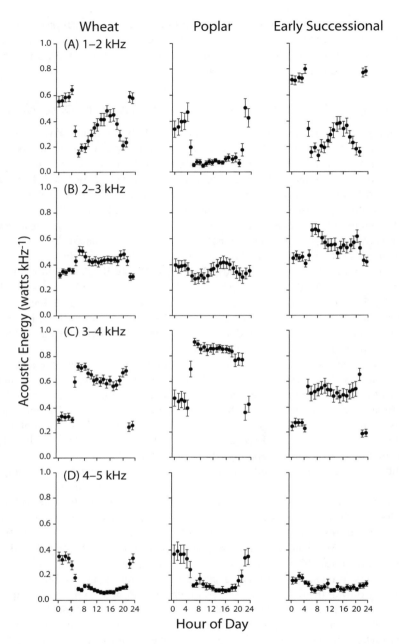

Figure 14.6. The patterns of acoustic energy (watts kHz^{-1}) in each of the examined MCSE systems (Table 14.3) are shown for four different acoustic frequency intervals: A) 1–2 kHz, B) 2–3 kHz, C) 3–4 kHz, and D) 4–5 kHz. Wheat includes both Conventional and No-till systems.

Broader Adoption of Acoustic Technology for Ecological Monitoring

The Biophony Grid Portal

Soundscape research at KBS LTER resulted in a pilot grid computing initiative, led by the National Center for Supercomputer Applications (NCSA) (Butler et al. 2006). The Biophony Grid Portal was developed to demonstrate the potential of grid computing to enhance collaboration and sharing within and external to the national LTER community, based on a large acoustic dataset and algorithms developed by KBS LTER researchers to identify entities in the soundscape. The grid utilities were developed at NCSA, the access system was developed by the LTER Network Office (LNO), and the digital data were located on a grid-enabled server at MSU.

The Biophony Grid Portal was designed to allow an investigator to identify an entity in a subset of a digital sound archive from a set of available locations. The recordings on the MSU server were linked to metadata on the LNO server. An investigator could log onto the grid and access the Biophony Grid Portal via the Internet. The investigator could then select the entity sound signature from a list of recognized entities (e.g., train whistle, chipping sparrow, etc.), together with a location and a range of dates to search. Based on location and the date range, the subset of sound recordings was retrieved from the MSU server and transmitted to the NCSA High Performance Computer (HPC). Results were provided via the Biophony Grid Portal where the investigator could listen to the entity signature, examine the soundscape spectrograms, listen to the sounds, and retrieve a table of signature match probabilities based on the recordings examined.

The grid computing infrastructure contributed to LTER Network–level synthetic science. Scalability of solutions has emerged as an increasingly significant issue, and grid technologies are an important approach to addressing and solving large-scale data and analytical requirements (Butler et al. 2006).

Current Technology

The application of automated soundscape recording, and subsequent storage, analysis, and interpretation have advanced considerably over the past decade. Today, recording technologies are available commercially (e.g., http://www.wildlifeacoustics.com) and new models and acoustic sensor innovations are under way. Digital libraries to archive, analyze, and access acoustic observations have been developed (Villanueva-Rivera and Pijanowski 2012, Kasten et al. 2012, Aide et al. 2013); sound pattern recognition applications have evolved (Kasten et al. 2010, Acevedo et al. 2009, Aide et al. 2013, Ospina et al. 2013); and acoustic indices have been further developed (e.g., Sueur et al. 2008, Joo et al. 2011). The importance of the soundscape as an ecological attribute has been acknowledged (Pijanowski et al. 2011a), and a research plan has been devised to apply the principles of soundscape ecology to monitoring ecological phenomena across landscapes (Pijanowski et al.

2011b). In addition, ecology journals have dedicated issues to soundscape ecology and ecological acoustics (e.g., see *Landscape Ecology* [2011] 26; *Ecological Informatics* [2013] 21).

Summary

There is rising interest in using sound as an ecological attribute that can be monitored and analyzed to provide information about ecological phenomena. Acoustic sensors can further advance ecological science by allowing researchers to capture observations in locations and times that are not easily accessible or feasible, and at time scales that were previously impractical to accommodate. Acoustic sensors will provide new knowledge about organisms and further our understanding of human activities that cause environmental disturbance. The commercialization of programmable acoustic sensor platforms that can be deployed for months with little maintenance will revolutionize how we listen to and interpret our environment. Although there are still constraints to developing a real-time acoustic sensor network system (e.g., power consumption and wireless communication distances), progress in sensor system development will enable biologists to measure and observe complex ecological attributes at detailed spatial and temporal scales, and potentially to forecast changes in ecosystems at regional scales (NRC 2001, Porter et al. 2005, Joo 2009, Joo et al. 2011).

The AAOS has been tested in an operations framework at KBS LTER and now has been expanded to other locations. This web-enabled system has been developed (http://www.real.msu.edu) to accommodate a large number of sensor observations (Kasten et al. 2012) and includes >1,000,000 recordings in 20 soundscape projects ranging in location from Alaska to Australia. The infrastructure developed for this soundscape application will readily fit into a scalable cyber-infrastructure schema such as cloud computing for large-scale acoustic observation networks. New applications using commercially available automated acoustic sensors coupled with digital libraries, remote access systems, and pattern recognition technologies have enabled rapid advances in the large-scale observation and interpretation of soundscapes and their attributes.

References

Acevedo, M. A., C. J. Corrada-Bravo, H. Corrada-Bravo, L. J. Villamnueva-Rivera, and T. M. Aide. 2009. Automated classification of bird and amphibian calls using machine learning: a comparison of methods. Ecological Informatics 4:206–214.

Aide, T. M., C. Corrada-Bravo, M. Campos-Cerqueira, C. Milan, G. Vega, and R. Alvarez. 2013. Real-time bioacoustics monitoring and automated species identification. *PeerJ* 1:e103. doi:10.7717/peerj.103

Anderson, S. E., A. S. Dave, and D. Margoliash. 1996. Template-based automatic recognition of birdsong syllables from continuous recordings. Journal of the Acoustical Society of America 100:1209–1219.

Blair, R. B. 1996. Land use and avian species diversity along an urban gradient. Ecological Applications 6:506–519.

Brandes, T. S. 2008. Automated sound recording and analysis techniques for bird surveys and conservation. Bird Conservation International 18:S163–S173.

Butler, R., M. Servilla, S. Gage, J. Basney, V. Welch, B. Baker, T. Fleury, P. Duda, D. Gehrig, M. Bletzinger, J. Tao, and D. M. Freemon. 2006. Cyberinfrastructure for the analysis of ecological acoustic sensor data: a use case study in grid deployment. Cluster Computing 10:301–310.

Bystrak, D. 1981. The North American breeding bird survey. Studies in Avian Biology 6:34–41.

Carles, J. L., I. López Barrio, and J. V. de Lucio. 1999. Sound influence on landscape values. Landscape and Urban Planning 43:191–200.

Catchpole, C., and P. J. B. Slater. 1995. Bird song: biological themes and variations. Cambridge University Press, New York, New York, USA.

Crossbow. 2006. Stargate developer's guide. Crossbow Technologies, Inc., San Jose, California, USA.

Dale, V. H., and S. C. Beyeler. 2001. Challenges in the development and use of ecological indicators. Ecological Indicators 1:3–10.

Estrin, D., W. Michener, G. Bonito, and W. Bonito. 2003. Environmental cyberinfrastructure needs for distributed sensor networks. Report from a National Science Foundation sponsored workshop. Scripps Institute of Oceanography, La Jolla, California, USA.

Fletcher, H. 1953. Speech and hearing in communication. Van Nostrand, New York, New York, USA.

Gage, S. H., P. Ummadi, A. Shortridge, J. Qi, and P. Jella. 2004. Using GIS to develop a network of acoustic environmental sensors. ESRI International User Conference 2004, San Diego, California, USA.

Gage, S. H., and A. C. Axel. 2013. Visualization of temporal change in soundscape power of a Michigan lake habitat over a 4-year period. Ecological Informatics. <http://dx.doi.org/10.1016/j.ecoinf.2013.11.004>

Haykin, S., editor. 1991. Advances in spectrum analysis and array processing. Prentice Hall, Upper Saddle River, New Jersey, USA.

Hobson, K. A., R. S. Rempel, H. Greenwood, B. Turnbull, and S. L. Van Wilgenburg. 2002. Acoustic surveys of birds using electronic recordings: new potential from an omnidirectional microphone system. Wildlife Society Bulletin 30:709–720.

Joo, W. 2009. Environmental acoustics as an ecological variable to understand the dynamics of ecosystems. Dissertation, Michigan State University, East Lansing, Michigan, USA.

Joo, W., S. H. Gage, and E. P. Kasten. 2011. Analysis and interpretation of variability in soundscapes along an urban-rural gradient. Landscape and Urban Planning 103:259–276.

Kasten, E. P., S. H. Gage, W. Joo, and J. Fox. 2012. The remote environmental assessment laboratory's acoustic library: an archive for studying soundscape ecology. Ecological Informatics 12:50–67.

Kasten, E. P., P. K. McKinley, and S. H. Gage. 2010. Ensemble extraction for classification and detection of bird species. Ecological Informatics 5:153–166.

Katti, M., and P. S. Warren. 2004. Tits, noise, and urban bioacoustics. Trends in Ecology and Evolution 19:109–110.

Krause, B. 1998. Into a wild sanctuary: a life in music and natural sound. Heyday Books, Berkeley, California, USA.

Kroodsma, D. E., and E. H. Miller, editors. 1996. Ecology and evolution of acoustic communication in birds. Comstock Publishing Associates, Ithaca, New York, USA.

Magurran, A. E. 1988. Ecological diversity and its measurement. Princeton University Press, Princeton, New Jersey, USA.

Michener, W. K., T. Baerwald, P. Firth, M. A. Palmer, J. Rosenberger, E. Sandlin, and H. Zimmerman. 2001. Defining and unraveling biocomplexity. BioScience 51:1018–1023.

Napoletano, B. M. 2004. Measurement, quantification and interpretation of acoustic signals within and ecological context. Thesis, Michigan State University, East Lansing, Michigan, USA.

Narins, P. M. 1995. Frog communication. Scientific American 273:62–67.

NRC (National Research Council). 2001. Grand challenges in environmental sciences. National Academy Press, Washington, DC, USA.

Ospina, O. E., L. J. Villanueva-Rivera, C. J. Corrada-Bravo, and T. M. Aide. 2013. Variable response of anuran calling activity to daily precipitation and temperature: implications for climate change. Ecosphere 4:47. <http://dx.doi.org/10.1890/ES12–00258.1>

Peterson, C. R., and M. E. Dorcas. 1994. Automated data acquisition. Pages 47–56 in W. R. Heyer, M. A. Donnelly, R. W. McDiarmid, L. C. Hayek, and M. S. Foster, editors. Measuring and monitoring biological diversity: standard methods for amphibians. Smithsonian Institution Press, Washington, DC, USA.

Pijanowski, B. C., A. Farina, S. H. Gage, S. L. Dumyahn, and B. L. Krause. 2011b. What is soundscape ecology? An introduction and overview of an emerging new science. Landscape Ecology 26:1213–1232.

Pijanowski, B. C., L. J. Villanueva-Rivera, S. L. Dumyahn, A. Farina, B. L. Krause, B. M. Napoletano, S. H. Gage, and N. Pieretti. 2011a. Soundscape ecology: the science of sound in the landscape. BioScience 61:203–216.

Porter, J., P. Arzberger, H.-W. Braun, P. Bryant, S. Gage, T. Hansen, P. Hanson, C.-C. Lin, F.-P. Lin, T. Kratz, W. Michener, S. Shapiro, and T. Williams. 2005. Wireless sensor networks for ecology. BioScience 55:561–572.

Qi, J., S. H. Gage, W. Joo, B. Napoletano, and S. Biswas. 2008. Soundscape characteristics of an environment: a new ecological indicator of ecosystem health. Pages 201–211 in W. Ji, editor. Wetland and water resource modeling and assessment: a watershed perspective. CRC Press, Boca Raton, Florida, USA.

Raisbeck, G. 1964. Information theory: an introduction for scientists and engineers. M.I.T. Press, Cambridge, Massachusetts, USA.

Ranjard, L., and H. A. Ross. 2008. Unsupervised bird song syllable classification using evolving neural networks. Journal of the Acoustical Society of America 123:4358–4368.

Reynolds, D. A., and R. C. Rose. 1995. Robust text-independent speaker identification using Gaussian mixture speaker models. IEEE Transactions on Speech and Audio Processing 3:72–83.

Robbins, C. S., D. Bystrak, and P. H. Geissler. 1986. The breeding bird survey: its first fifteen years, 1965–1979. Resource Publication No. 156, U.S. Fish and Wildlife Service, Washington, DC, USA.

Robertson, G. P., and S. K. Hamilton. 2015. Long-term ecological research at the Kellogg Biological Station LTER Site: conceptual and experimental framework. Pages 1–32 in S. K. Hamilton, J. E. Doll, and G. P. Robertson, editors. The ecology of agricultural ecosystems: long-term research on the path to sustainability. Oxford University Press, New York, New York, USA.

Schafer, R. M. 1977. The soundscape: our sonic environment and the tuning of the world. Destiny Books, Rochester, Vermont, USA.

Shannon, C. E. 1948. The mathematical theory of communication. Bell System Technical Journal 27:379–423, 623–656.

Shannon, C. E., and W. Weaver. 1963. The mathematical theory of communication. University of Illinois Press, Urbana, Illinois, USA.

Slabbekoorn, H., and M. Peet. 2003. Birds sing at a higher pitch in urban noise. Nature 424:267.

Sueur, J., S. Pavoine, O. Hamerlynck, and S. Duvail. 2008. Rapid acoustic survey for biodiversity appraisal. PLoS ONE 3(12):e4065.

Thompson, J. N., O. J. Reichman, P. J. Morin, G. A. Polis, M. E. Power, R. W. Sterner, C. A. Couch, L. Gough, R. Holt, D. U. Hooper, F. Keesing, C. R. Lovell, B. T. Milne, M. C. Molles, D. W. Roberts, and S. Y. Strauss. 2001. Frontiers of ecology. BioScience 51:15–24.

Truax, B. 1984. Acoustic communication. Ablex Publishing, Norwood, New Jersey, USA.

Truax, B., editor. 1999. Handbook for acoustic ecology. Second edition. Cambridge Street Publishing, Vancouver, British Columbia, Canada.

Villanueva-Rivera, L. J., and B. C. Pijanowski. 2012. Pumilio: a web-based management system for ecological recordings. Bulletin of the Ecological Society of America 93:71–81.

Waddle, J. H., T. F. Thigpen, and B. M. Glorioso. 2009. Efficacy of automatic vocalization recognition software for anuran monitoring. Herpetological Conservation and Biology 4:384–388.

Warren, P. S., M. Katti, M. Ermann, and A. Brazel. 2006. Urban bioacoustics: it's not just noise. Animal Behavior 71:491–502.

Weir, L. A., and M. J. Mossman. 2005. North American Amphibian Monitoring Program (NAAMP). Pages 307–313 in M. Lannoo, editor. Amphibian declines: the conservation status of United States species. University of California Press, Berkeley, California, USA.

Welch, P. 1967. The use of fast Fourier transform for the estimation of power spectra: a method based on time averaging over short, modified periodograms. IEEE Transactions on Audio and Electroacoustics 15:70–73.

West, B. W., P. G. Flikkema, T. Sisk, and G. W. Koch. 2001. Wireless sensor networks for dense spatio-temporal monitoring of the environment: a case for integrated circuit, system, and network design. Proceedings of IEEE CAS Workshop on Wireless Communications and Networking. Notre Dame, Indiana, USA. Institute of Electrical and Electronics Engineers, New York, New York, USA. 6 pages.

Wrightson, K. 2000. An introduction to acoustic ecology. Soundscape 1:10–13.

15

Designing Cropping Systems for Ecosystem Services

Sieglinde S. Snapp, Richard G. Smith, and
G. Philip Robertson

Almost all intensive row-crop ecosystems depend on external chemical inputs such as nitrogen fertilizers and pesticides for their high yields. This dependency has had important consequences for the environment and for ecosystem services (MEA 2005; Swinton et al. 2015a, Chapter 3 in this volume) that underpin these systems' long-term sustainability (Robertson and Swinton 2005). Well-known consequences of chemical inputs include biodiversity loss, reductions in water quality, increased greenhouse gas emissions, and degradation of soil resources (Matson et al. 1997, Tilman et al. 2002, Robertson et al. 2004). Agriculture is also facing mounting challenges in the form of climate change and increasing fossil fuel costs—all in the context of an increasing global demand for food over the coming decades. These challenges call for a reevaluation of modern agricultural practices and the supporting and provisioning services that agroecosystems provide.

Several key questions must be answered before we can hope to ensure the long-term sustainability of row-crop production systems and their associated ecosystem services. First, to what degree can agricultural systems be designed to be regenerative (Pearson 2007), enhancing supporting and regulating services so that nutrients and other resources are conserved within the system and any external inputs are used most efficiently? Second, what factors determine agricultural resilience and the capacity for agriculture to maintain productivity in the face of external stressors (Snapp 2008)?

The future of agriculture depends on our ability to understand both the efficient use of natural resources and the ecological principles that promote agroecosystem resilience and stability. While the production of food, fiber, and fuel will remain core goals of farming systems, the provision of other ecosystem services will become increasingly important. Measures to enhance agricultural yield will be evaluated with greater attention to potential trade-offs among other ecosystem services (Syswerda and Robertson 2014).

Experimental investigations at the Kellogg Biological Station Long-Term Ecological Research site (KBS LTER) provide insights into the productivity of different row-crop production systems vis-à-vis ecosystem services such as carbon (C) sequestration, nutrient cycling, pest protection, and energy efficiency, and consequent impacts on soil and water resources. Such insights could be valuable for guiding agriculture through the coming challenges of feeding, clothing, and powering a growing global population with finite natural resources and uncertain trajectories of environmental change.

In this chapter, we report on key agroecosystem performance and ecosystem service indicators measured in KBS cropping system experiments. We include results from both the Main Cropping System Experiment (MCSE) as well as the Living Field Laboratory Experiment (LFL) (Robertson and Hamilton 2015, Chapter 1 in this volume). Alternative cropping systems in these experiments include reduced soil disturbance and complex mixtures of grain crops and winter cover crops such as red clover (*Trifolium pretense* L.) and annual rye (*Secale cereal* L.) that are grown in between grain crops to conserve resources and increase soil fertility. Cover crops are included in many of the KBS LTER agricultural systems because cover crops promote a host of supporting ecosystem services related to soil organic matter, nutrient cycling, water quality, and soil conservation (Snapp et al. 2005, MEA 2005).

Quantifying the benefits and potential trade-offs associated with cover crops and other agricultural practices that maintain or enhance ecosystem services will provide insights for policy makers, land managers, and agricultural advisors. Over two decades of experimentation at KBS provide a unique opportunity to test how a variety of practices—reducing external inputs, no-tillage production, and enhancing plant and residue diversity—affect grain production and ecosystem services, including supporting and regulating ecosystem services.

Agriculture in Michigan

Historical trends in southern Michigan, the location of the KBS LTER, show that agrarian systems have changed dramatically over the last two centuries (Gage et al. 2015, Chapter 4 in this volume). The anthropological record across the upper U.S. Midwest suggests a mosaic of land uses prior to European settlement. Highly diverse horticultural systems were practiced in specific locales, intermixed with low-intensity forest and grassland management (Rudy et al. 2008). In the early to mid-1800s, Americans of European descent moved westward from New England and began to clear forests and drain wetlands for corn (*Zea mays* L.) and small grains including oats (*Avena sativa* L.) and wheat (*Triticum aestivum* L.) (Gray 1996). In the early 1900s, market opportunities expanded for a wide range of horticultural crops and livestock products. New crops such as alfalfa (*Medicago sativa* L.) and soybean (*Glycine max* L.) were promoted by the emerging university extension service, and a diverse suite of crops was grown throughout Michigan (Rudy et al. 2008). A century later, low-diversity row-crop systems supported by agrochemical use now dominate the landscape, and mixed cropping systems—in which livestock are supported mainly or solely by crops grown on-farm—are rare.

Current policies support the production of inexpensive food and affect the livelihoods of many rural communities. To remain economically viable, farmers necessarily focus on maximizing cash crop production (Swinton et al. 2015b, Chapter 13 in this volume). Corn, soybean, and wheat are grown over wide areas due to superior biological traits—including rapid growth, effective resource acquisition, and an ability to translate inputs into grain yield. Changing the portfolio of crops grown and better integrating crops and livestock to improve the delivery of ecosystem services face considerable challenges, in part because the contemporary socioeconomic context and infrastructure reinforce current farming systems.

Further, in recent decades U.S. subsidy policies have favored corn, soybean, and wheat among the few crops that are targeted for direct price support payments. In this environment, it is not surprising that an increasingly narrow range of crops are grown, with less production of forage and hay crops, small grains, minor legumes, or specialty oil and grain crops. At the same time, awareness is growing that reliance on these few crops has environmental costs and increases the vulnerability of significant portions of the food production system.

Long-Term Agricultural Ecosystem Experiments at KBS LTER

The ecological principles that underpin the functioning of natural ecosystems apply to agroecosystems as well, and KBS LTER cropping system experiments are designed to elucidate key processes, population and community-level dynamics, and interactions. The overarching research goal of the KBS LTER is to test the hypothesis that biological processes can substitute for chemical inputs without sacrificing high yields (Robertson and Hamilton 2015, Chapter 1 in this volume). The Main Cropping System Experiment (MCSE) was established in 1989 to provide a practical range of model systems across which these key ecological attributes could be intensively examined over long time periods. The LFL (Sánchez et al. 2004) was established in 1993 to extend MCSE findings to a broader range of farm-relevant cropping systems and, in particular, to separate crop diversity as a distinct factor from crop management. The LFL includes systems with a wider range of crop diversity and nitrogen (N) sources than does the MCSE. Properties that differ among various MCSE and LFL systems include perenniality (the duration of living cover), plant diversity (the number of species in a rotation), types and quantities of fossil fuel vs. biologically derived inputs (including energy), and management-related disturbances (tillage regime).

Resource gradients and disturbances from fire, flooding, tillage, and pests are common regulators of productivity and resource flux in agricultural ecosystems across the world. Processes such as biological N fixation and organic matter accumulation are at the foundation of traditional agriculture (Greenland and Nye 1959), including the bush-fallow or swidden agriculture that was historically practiced across North America prior to European contact (Sylvester and Gutmann 2008). For thousands of years, farmers have used soil disturbance and burning as primary means to enhance nutrient availability in synchrony with crop demand and to suppress weeds. Native vegetation was used as part of long "bush-fallow" crop rotations to allow successional processes to rebuild soil fertility.

Successional processes can be used in a sustainable manner to produce food and other agricultural products in contemporary agricultural systems, given sufficient time and access to land. However, these resources are often in short supply due to economic and population pressures, and modern agricultural systems rely instead on intensive energy and chemical inputs (Tilman et al. 2002). As a consequence, there has been limited attention paid to the underlying ecological processes that mediate agricultural production and control the pools and fluxes of C and nutrients in agricultural soils.

The KBS Main Cropping System Experiment

The Main Cropping System Experiment (MCSE) evaluates agricultural row-crop systems that vary in management intensity (Table 15.1); see Robertson and Hamilton (2015, Chapter 1 in this volume) for a full description. In brief, we compare four management strategies for a corn–soybean–winter wheat rotation: (1) the Conventional system uses fertilizer and herbicide inputs, and conventional tillage as recommended by Michigan State University Extension; (2) the No-till system uses conventional management but for permanent no-till soil management; (3) the Reduced Input system uses biologically based management, including winter cover crops, to reduce synthetic chemical inputs to ~one-third of those used in the Conventional system; and (4) the Biologically Based system uses biologically based management to eliminate synthetic chemical inputs altogether. No systems receive manure or compost. The Reduced Input and Biologically Based systems include cover crops of red clover interseeded in wheat in the spring, and annual rye planted after corn harvest in the fall. Wide row spacing and mechanical cultivation are used to control weeds in these two systems. A fifth system, Alfalfa, is managed conventionally as a continuous forage crop, replanted on a ~6-year schedule following a break year in a grain crop.

The Reduced Input and Biologically Based systems model alternative agriculture practices that make up a small but active sector of U.S. agriculture (Swinton et al. 2015b, Chapter 13 in this volume). For example, the Biologically Based system simulates organic management practices and is USDA-certified organic, although it is unconventionally organic in that neither manure nor compost are used as inputs. The total acreage of certified organic cropland in the U.S. has increased more than 6-fold from 1992 to 2008 and made up 0.7% (1.1 million ha) of total U.S. cropland in 2008 (ERS 2011). The Reduced Input system includes integrated nutrient and pest management practices such as closely monitoring nutrient availability and pest and beneficial insect populations, which allows external inputs to be reduced by about two-thirds in this system. Herbicide application was banded within the crop row until the shift was made in 2011 to broadcast application, and N fertilizer is applied at lower rates. This reduction in N fertilizer use is significant in light of the fact that in many cropping systems fertilizer is frequently applied in excess of crop demand (Robertson 1997, Gardner and Drinkwater 2009). The effects of

Table 15.1. Description of row cropping systems of the KBS LTER Main Cropping System Experiment (MCSE) and Living Field Lab Experiment (LFL).[a]

Experiment/System	Crop Rotation	Management
Main Cropping System Experiment (MCSE)		
Conventional (T1)	Corn–soybean–winter wheat	Prevailing norm for tilled corn–soybean–winter wheat rotation (c–s–w); standard chemical inputs, chisel-plowed, no cover crop, no manure or compost
No-till (T2)	Corn–soybean–winter wheat	Prevailing norm for no-till c–s–w; standard chemical inputs, permanent no-till, no cover crop, no manure or compost
Reduced Input (T3)	Corn–soybean–winter wheat	Biologically based c–s–w managed to reduce synthetic chemical inputs; chisel-plowed, winter cover crop of red clover or annual rye, no manure or compost
Biologically Based (T4)	Corn–soybean–winter wheat	Biologically based c–s–w managed without synthetic chemical inputs; chisel-plowed, mechanical weed control, winter cover crop of red clover or annual rye, no manure or compost; certified organic
Alfalfa (T6)	Alfalfa	5- to 6-year rotation with wheat as a 1-year break crop
Living Field Lab Experiment (LFL)		
Organic	One species—corn Two species—corn with cover crop Three species—corn, corn, soybean, winter wheat Six species—three species rotation with three cover crop species	Biologically based without synthetic chemical inputs; five entry points in annual rotation: continuous corn and each of corn, corn, soybean, wheat (c–c–s–w); winter cover crop(s) of crimson clover in two species rotation and of crimson clover, annual rye, and red clover in six species rotation; chisel-plowed; certified organic practices; dairy compost; mechanical weed control
Integrated Conventional	Same as organic above	Biologically based with reduced synthetic chemical inputs; same crop rotations as organic above; chisel-plowed; targeted application of herbicides and N fertilizer; mechanical weed control

[a]Site codes that have been used throughout the project's history are given in parentheses. For further details, see Robertson and Hamilton (2015, Chapter 1 in this volume).

excess fertilizer N include enhanced production of the greenhouse gas nitrous oxide (McSwiney and Robertson 2005, Hoben et al. 2011), promotion of invasive species (Davis et al. 2000), accelerated changes in some soil organic matter pools (Grandy et al. 2008), and contamination of ground and surface waters with attendant eutrophication (Hamilton 2015, Chapter 11 in this volume; Millar and Robertson 2015, Chapter 9 in this volume).

Prior to the 1980s, row-crop agriculture was heavily reliant on tillage to prepare the soil for planting and to manage weeds. Now a significant proportion of row-crop land is under no-till production (Horowitz et al. 2010), due in part to the use of herbicide-resistant crop varieties. No-till production practices allow farmers to reduce the number of soil-disturbing equipment passes and attendant potential for surface erosion. Benefits of no-till include a decreased requirement for fossil fuel, reduced loss of soil, nutrients, and pesticides in runoff water, better soil water infiltration and water-holding capacity, and a more stable environment for soil organisms. In the MCSE No-till system, the number of equipment passes has been reduced by 26%—from 8.4 to 6.2 per year—as compared to the Conventional system (Gelfand et al. 2010). Reduced soil disturbance has led to greater soil C accumulation in the top 5 cm of soil in the No-till system (3.6 kg C m^{-2}) compared to the Conventional system (3.2 kg C m^{-2}) (Syswerda et al. 2011). The Biologically Based system also led to a greater soil C accumulation (3.8 kg C m^{-2}), similar to the No-till system, but as discussed in more detail below, this occurred despite frequent soil disturbance from plowing and rotary hoeing for mechanical weed control.

One of the sustainability principles evaluated in the MCSE is the role of plant diversity in agroecosystem performance, including net primary productivity (NPP), nutrient retention, and ecosystem stability. Generally, positive relationships have been shown among diversity, NPP, and other ecosystem services in grasslands and other low nutrient, semi-managed, and natural systems (e.g., Hector et al. 1999; Tilman et al. 2001; Hooper et al. 2005, 2012). In row-crop systems, biodiversity is generally a function of crop rotation, intercropping, and inclusion of accessory crops such as winter cover crops. In both long-term (Syswerda et al. 2011, 2012) and shorter-term comparisons (Drinkwater et al. 1998, Maeder et al. 2002), more diverse organic systems have accumulated more soil organic matter and leached less nitrate than paired systems under conventional management.

However, such comparisons cannot distinguish between the effects of plant diversity per se and other management practices that differ among the experimental systems. As a consequence, diversity has rarely been studied as a discrete factor (Gross et al. 2015, Chapter 7 in this volume). The LFL and Biodiversity experiments (Smith et al. 2008; Robertson and Hamilton 2015, Chapter 1 in this volume; Gross et al. 2015, Chapter 7 in this volume) were established at KBS to more explicitly investigate the effects of plant diversity and rotational complexity in biologically based cropped ecosystems (Sánchez et al. 2004, Snapp et al. 2010a).

The KBS Living Field Lab Experiment

The LFL was designed with input from a farmer advisory group and has played an important role, especially in outreach at KBS. The aim of the LFL is to test farm-relevant combinations of intensively managed systems where crop diversity can be examined as a separate factor from other management factors (Sánchez et al. 2004). This allows comparison of common crop sequences such as continuous corn vs. more diverse corn rotations. Management factors include nutrient sources (combinations of conventional fertilizer and composted manure) and weed control (conventional herbicide inputs vs. mechanical cultivation).

The factorial, split-plot design of the LFL includes management regime (Organic vs. Integrated Conventional) as a main plot system and plant biodiversity (comparing one, two, three, and six plant species) as subplot treatments. The Organic system relies on certified organic practices including the application of dairy compost for fertility and tillage for weed management, while the Integrated Conventional system uses herbicides at rates one-third of conventional rates and composted manure and synthetic N fertilizers at rates two-thirds of conventional (Sánchez et al. 2002, 2004). Five entry points for cropping system diversity are included within each management system, where one plot is continuous corn and the other four plots represent each phase of a 4-year rotation of corn–corn–soybean–wheat. Cover crops are grown on half of each plot (red clover, crimson clover [*Trifolium incarnatum* L.] and annual ryegrass); the other half is winter fallowed, with some limited plant cover provided by the presence of winter annual weeds (Smith and Gross 2006a).

It is experimentally challenging to manipulate crop diversity in isolation from management, as management practices are typically "bundled" and therefore multifunctional (Snapp et al. 2010a). In conventional management, for example, reliance on chemical inputs for pest control and nutrient supply allows simplification of the system to a few highly productive species. In contrast, biologically based management, including organic management, commonly relies on a mixture of species that promote internal processes such as N fixation, mineralization, and pest suppression (Lowrance et al. 1984, Drinkwater and Snapp 2007).

Agronomic Lessons from the KBS MCSE

Productivity

Agronomic productivity in the MCSE annual crops over 17 years (1989–2007) has shown consistent responses to management. The No-till system has the overall highest average annual grain yield across all three crops at 4.2 Mg ha^{-1} and the Biologically Based system the lowest at 2.9 Mg ha^{-1} (Table 15.2; Gelfand et al. 2010). Grain yield in the cereal crops—corn and wheat—has been substantially reduced under biologically based management, compared to no-till. In contrast, soybean yields have not been reduced under biologically based management compared to conventional. Although soil moisture and weed pressure likely contribute to low yields in some years, insufficient N supply appears to be the key factor influencing biologically based corn and wheat production. This is supported by the success of soybeans, which provide their own N via biological N fixation, and by low levels of soil inorganic N (nitrate [NO$_3^-$] and ammonium [NH$_4^+$]) in soils of the Biologically Based system during other parts of the rotation. For example, at midseason in the corn phase of the Biologically Based system, soils contain 17 mg N kg^{-1}, on average, as compared to 29 mg N kg^{-1} at midseason in soils of the Conventional system (Millar and Robertson, 2015, Chapter 9 in this volume). Nitrogen deficiency in organic systems has been observed in other agroecosystem experiments (Cavigelli et al. 2008), and it is typical of agriculturally converted grassland areas where minimal agricultural inputs are applied (Smith et al. 2008). Biological management for the MCSE relies on N fixation from legumes in the rotation, with no supplementation from manure or compost.

Table 15.2. Annual average crop grain yield and energy balances for KBS LTER MCSE annual cropping systems over the period 1989–2007.

System	Crop Yield				Crop Rotation Energy Balance[a]			Energy Efficiency Output:Input Ratio
	Corn	Wheat	Soybean	System	Farming Energy Inputs	Food Energy Output	Net Energy Gain	
	(Mg ha⁻¹ yr⁻¹)				(GJ ha⁻¹ yr⁻¹)			
Conventional	5.90	3.54	2.33	3.92	7.1	72.7	65.6	10
No-till	6.25	3.74	2.65	4.21	4.9	78.5	73.6	16
Reduced Input	5.23	3.09	2.57	3.63	5.2	66.9	61.7	13
Biologically Based	4.08	2.05	2.48	2.87	4.8	53.1	48.3	11
Alfalfa				6.85	5.5	26.1	20.6	5

[a]Energy balance of systems is based on actual farm management operations and inputs (from Gelfand et al. 2010). Food energy output is for direct human consumption except in the case of alfalfa, which is fed to livestock.

Year-to-year yield variability has been high in all MCSE systems (Smith et al. 2007). This is not surprising, given that annual precipitation has ranged from 60 to 110 cm per year over this period. Precipitation is historically evenly distributed throughout the growing season at KBS, but over the last two decades dry spells have commonly occurred during critical crop development stages, and well-drained KBS loam soils have a limited ability to buffer midseason drought because of their relatively low moisture holding capacity (150 mm to 1 m; Crum and Collins 1995). Climatic predictions for Michigan as for the Midwest call for a lower frequency but increasing severity of precipitation events (Schoof et al. 2010; Gage et al. 2015, Chapter 4 in this volume).

The MCSE experimental design for annual crops—with one crop rotation phase present per year—allows management effects on interannual yield variability to be tested for each crop for a different set of years. Smith et al. (2007) analyzed temporal variability by calculating the coefficient of variation for interannual grain yield to show that corn yield variability was not influenced by management system. In wheat, however, variability followed the ranking No-till (coefficient of variation, CV = 0.22) < Conventional (0.35) < Reduced Input (0.40) < Biologically Based (0.50). Soybean yield variability overall was lower, and the response to management was similar to that of wheat: No-till (CV = 0.18) = Conventional (0.18) < Reduced Input (0.25) < Biologically Based (0.34). The lower temporal variability in No-till and Conventional systems suggests that intensive use of agricultural chemicals can mitigate the impacts of weather variability—whether due to weeds, nutrient supply, timely access to fields, or some other less obvious factor. This finding stands in contrast to a long-term field experiment in Pennsylvania showing that organic production systems maintained yields in low rainfall years (Lotter et al. 2003). Improvements in soil organic matter and water storage have been proposed as key processes supporting gains in yield stability under organic management (El-Hage Scialabba and Müller-Lindenlauf 2010). However, MCSE data provide no evidence that Reduced Input and Biologically Based systems resulted in greater cropping system resistance to changing weather patterns.

Soil Carbon

Over the first 10 years of the MCSE, the No-till system rapidly accumulated soil C relative to the Conventional system; by 2001 soil C was 8.5 kg m^{-2} to a 1-m depth. This was 23% higher than in the Conventional system, which held 6.9 kg C m^{-2} (Fig. 15.1A). Surprisingly, over the same time period, a 20% increase of soil C in the Biologically Based system also occurred, even though this system relies on frequent soil disturbance to manage weeds (Fig. 15.1B). This suggests that the soil C that accumulates under biological management is physically stable and persistent. In contrast, soil C accumulation under no-till management is evidently vulnerable

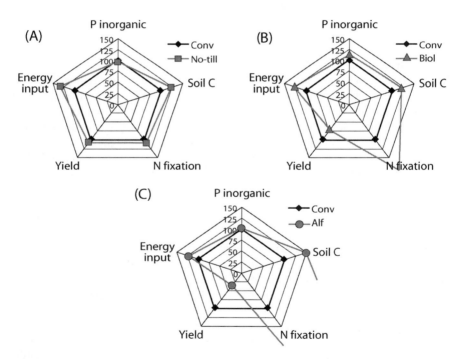

Figure 15.1. Ecosystem services from the KBS LTER Main Cropping System Experiment (MCSE). Results are presented as percent change relative to the Conventional (Conv) system in radial graphs for (A) No-till, (B) Biologically Based (Biol), and (C) Alfalfa (Alf). Values for the Conventional system used as 100%: soil inorganic phosphorus 30 mg P kg^{-1}; soil C 3.2 kg m^{-1}; biological N fixation 27 kg N ha^{-1} yr^{-1}; grain yield 3.92 Mg ha^{-1}, with alfalfa forage biomass valued at 1/5th grain; and energy inputs 7.1 GJ ha^{-1} yr^{-1}. Note that biological N fixation for alfalfa extends beyond the range of the figure.

to loss with physical disturbance; Grandy and Robertson (2006) found rapid C loss following a single initial tillage of plots in the MCSE Mown Grassland (never tilled) community. That C gains under biologically based management occurred despite tillage implicates that other factors, such as crop diversity, residue quality, or a longer annual crop duration, may lead to more persistent soil C accumulation than under no-till. Combining these factors with no-till is an intriguing possibility for building soil C even faster than no-till or biologically based management alone.

Research on soil C sequestration in other long-term agroecosystem experiments has been consistent with KBS MCSE findings. Biodiverse cropping systems have been shown to be generally associated with soil C gains if N fertilizer inputs are nil or minimal (Drinkwater et al. 1998, Russell et al. 2009). More important than diversity may be the duration of living cover. The perennial crop systems in the MCSE have accumulated more C than have the annual crop systems. Carbon gains in the Alfalfa system, for example, had increased by 50% 12 years after establishment (Grandy and Robertson 2007). This may be attributed, in part, to the long growing season for C fixation in alfalfa, which has ~70 more growing days than the 115-day corn growing season and ~30 more days than corn interseeded with a winter cover crop. No tillage is conducted in the Alfalfa system, and the combination of C inputs from roots, continuous cover, and lack of disturbance has led to substantial gains in soil C (Fig. 15.1).

Energy Efficiency

Evaluating the performance of different management systems is a challenge when inputs and outputs vary considerably. Substantial amounts of fossil energy are consumed in common management inputs and practices, including fertilizer, pesticides, and field operations conducted by labor-saving machinery. Organic farming is often assumed to require less energy because of the absence of synthetic chemicals and fertilizers (Pimentel et al. 2005), even though field operations can also be energy-intensive. Forage production and conservation tillage systems are moderately intensive types of agriculture. High economic yields tend to be associated with energy-intensive, conventional agriculture, and questions arise regarding the associated trade-offs. Is a system with low-energy input more efficient if outputs are also low? An assessment of energy balance for the whole cropping system is one way to evaluate these trade-offs (Hülsbergen et al. 2001, Gelfand et al. 2010).

In the MCSE, annual farming energy inputs varied from 4.8 GJ ha^{-1} in the Biologically Based system to 7.1 GJ ha^{-1} in the Conventional system (Table 15.2; Gelfand et al. 2010). Energy inputs were generally lower than previously reported for conventional management in long-term row crop trials in Pennsylvania (Pimentel et al. 2005) and in Central Europe (Maeder et al. 2002). This may be a reflection of the recommended management practices for Michigan field crop production, which do not rely on manure amendments or high fertilization rates (Gelfand et al. 2010). Energy outputs were evaluated in terms of food produced for direct human consumption, or indirect consumption in the case of alfalfa, where energy outputs were based on meat produced when harvested biomass is used as ruminant livestock feed (Table 15.2). The No-till system was the most efficient grain production system, with net energy gains of 74 GJ ha^{-1} yr^{-1}. This was due to a combination of high productivity and moderate energy usage. Although the Biologically Based system had similar energy input as the No-till system (4.8 vs. 4.9 GJ ha^{-1} yr^{-1}, respectively), the net energy gain was substantially lower, owing to lower yields of corn and wheat. Consequently, energy efficiency (output:input) among the annual cropping systems followed the order No-till (16) > Reduced Input (13) > Biologically Based (11) > Conventional (10) (Table 15.2; Gelfand et al. 2010).

Greenhouse Gas Mitigation

The MCSE systems have also provided insight into understanding and mitigating impacts of management intensity on greenhouse gas fluxes. Greenhouse gas exchanges between soils and the atmosphere—nitrous oxide (N_2O), carbon dioxide (CO_2), and methane (CH_4)—have been evaluated for all MCSE systems (Robertson et al. 2000; Gelfand and Robertson 2015, Chapter 12 in this volume). Overall, row-crop production in the MCSE increases greenhouse gas emissions from soils through enhanced N_2O emissions and diminished CH_4 consumption, as has been shown in many other systems as well (Robertson and Vitousek 2009; Gelfand and Robertson 2015, Chapter 12 in this volume).

Within a particular cropping system, N fertilization rate is the best single predictor of N_2O emissions (Millar and Robertson 2015, Chapter 9 in this volume) and N_2O emission rates increase exponentially above a certain fertilization threshold, presumably where crop N needs are saturated (McSwiney and Robertson 2005, Ma et al. 2010, Hoben et al. 2011). These observations underscore the importance of applying N fertilizer at a dose that matches crop requirements. This principle can be challenging to implement in a rain-fed environment as crop growth—and thus requirements for N—vary from year to year with precipitation. However, because ~50% of crop N needs are typically met by N mineralization from soil organic matter (Robertson 1997), also itself a precipitation-dependent process (Robertson and Paul 2000), crop response to fertilizer rate tends to be stable from year to year for a given location. At KBS, application of N fertilizer above the crop optimum has been shown to be associated with 2-fold higher emissions of N_2O (McSwiney and Robertson 2005) with little if any yield benefit. This suggests that widespread adoption of more conservative N fertilizer rates could significantly reduce U.S. N_2O emissions (Millar et al. 2010, Grace et al. 2011).

Water Quality

Water quality is an important attribute of cropping system performance—water leaving the system carries sediments and chemicals that can pollute surface and groundwater far from the point of origin. While many components of water quality are measured at KBS LTER (Hamilton 2015, Chapter 11 in this volume), the loss of nitrate by leaching into infiltrating water provides a reasonable sentinel for solute loss in general, and is important in its own right as a contributor to indirect N_2O fluxes downstream (Beaulieu et al. 2011), to human health via groundwater drinking water supplies (Powlson et al. 2008), and to coastal eutrophication (Diaz and Rosenberg 2008).

That nitrate loss differs among MCSE cropping systems provides another metric for gauging differences in their delivery of ecosystem services. To the extent that N conservation can be considered an ecosystem service, then, the system with the lowest nitrate loss (either absolute or relative to yield) can be considered a greater service provider. While the system with the greatest loss could conversely be viewed as the greater disservice provider, comparisons are more straightforward if put in terms of positive services (Swinton et al. 2015a, Chapter 3 in this volume).

By this metric, then, for the annual cropping systems of the MCSE, in absolute terms the Biologically Based system provided the most nitrate conservation, with average leaching losses of 19 kg NO_3^--N ha^{-1} yr^{-1} over an 11-year (1995–2006) period (Syswerda et al. 2012). This compares to the Conventional system's average loss of 62 kg NO_3^--N ha^{-1} yr^{-1}. The No-till and Reduced Input systems were intermediate to these at 42 and 24 kg NO_3^--N ha^{-1} yr^{-1}, respectively. In relative yield-scaled terms, the differences were smaller but the rankings identical: 18, 11, 7.3, and 7.2 kg NO_3^--N Mg^{-1} yield for Conventional, No-till, Reduced Input, and Biologically Based systems, respectively.

Pest Suppression

An important regulating ecosystem service that can be affected by agricultural management practices is suppression of pests. Weeds are a particularly important group of pests because they reduce crop quality by competing for soil nutrients, water, and light, and by interfering with harvest. Agricultural practices such as tillage, crop rotation, fertilizer and herbicide application, and cover crop use can affect weed populations directly by causing seedling mortality, by inhibiting or promoting seed germination, and by changing weed nutrient status (Liebman et al. 2001). Management practices can also affect weeds indirectly by altering weed–crop competitive relationships (Liebman and Davis 2000, Ryan et al. 2010) or through effects on seed predator populations (Menalled et al. 2007). Management additionally influences soil processes, such as feedbacks with soil biota that reduce weed survival and fitness (Li and Kremer 2000, Davis and Renner 2007).

Weed population data have been collected regularly in the MCSE systems (Gross et al. 2015, Chapter 7 in this volume). The most recent syntheses of these data indicate that the four annual row-crop systems differ in terms of capacity for weed suppression (Davis et al. 2005; Gross et al. 2015, Chapter 7 in this volume). In general, the Biologically Based system is the least weed suppressive (i.e., has more weeds), with weed biomass varying across years from 48 to 148 g m^{-2} compared to the other three systems, where biomass has ranged from less than 3 to over 50 g m^{-2} (Davis et al. 2005). The lack of herbicide use and reduced crop productivity in the Biologically Based system have likely contributed to this system's weed pressure.

An additional factor contributing to weed suppression could be weed seed predator populations, which have also differed by system (Menalled et al. 2007). Populations of seed-predating carabid ground beetles (Coleoptera: Carabidae), which are sensitive to soil disturbance, were over three times higher and seed predation rates over two times higher in the No-till system compared to the Biologically Based system. Taken together, these results are consistent with herbicide use and soil disturbance as key determinants of weed suppression services. However, given that both herbicide use and soil disturbance are associated with a host of potential ecosystem disservices, there is a clear need for research into alternative weed management practices that improve pest suppression services without incurring significant trade-offs in the form of soil or water degradation.

Ecosystem Service Trade-offs

Three of the more important ecosystem services associated with different management systems—yield, nitrate loss, and soil C gain—are summarized in Table 15.3.

Table 15.3. Evaluation of yield reductions and environmental gains associated with alternative systems relative to conventional management.[a]

Experiment/ System	Average Crop Yield[b] (kg grain ha^{-1})	Nitrate Leaching Loss[c] (kg NO$_3$-N ha^{-1} yr^{-1})	Soil C Gain[d] (kg C ha^{-1} yr^{-1})	Yield Trade-off— Nitrate Mitigation (kg NO$_3$-N kg^{-1} grain)[e]	Yield Trade-off— Soil C Accumulation (kg C kg^{-1} grain)[f]
Main Cropping System Experiment (MCSE): Corn–Soybean–Wheat					
Conventional	3511	62	0		
No-till	3853	42	330	–0.06	0.96
Reduced Input	3597	24	200	–0.44	2.33
Biologically Based	2765	19	500	0.06	–0.67
Living Field Lab Experiment (LFL): Continuous Corn					
Integrated Conventional	6420	74	80		
Organic	5050	32	900	0.03	0.59

[a]All values expressed on an annual basis, based on yield reductions or enhancements associated with alternative management relative to conventional management vs. reductions in nitrate (NO$_3$-) leached and gains in soil carbon (C).
[b]MCSE grain yield average for corn–soybean–wheat rotations from 1996 to 2007 (Syswerda and Robertson 2014); LFL continuous corn grain yield average from 1994 to 2000 (Snapp et al. 2010a).
[c]MCSE leaching losses monitored from 1995 to 2006 (Syswerda et al. 2012); LFL leaching losses monitored from 1994 to 2000 (Snapp et al. 2010a).
[d]MCSE soil C gain in the A/Ap horizon from 1989 to 2001 (Syswerda et al. 2011); LFL soil C gain in the 0- to 20-cm horizon from 1993 to 2008 (Snapp et al. 2010a).
[e]Trade-off relative to conventional management, where the change in leached nitrate-N is reported as a ratio to change in grain yield. A negative value implies less N leaching per unit of change in yield, indicating a desirable trade-off with respect to that ecosystem disservice.
[f]Trade-off relative to conventional management, where the change in soil C sequestered is reported as a ratio to change in grain yield. A positive value implies more C sequestration per unit of change in yield, indicating a desirable trade-off with respect to that ecosystem service.

In the MCSE, relative to the Conventional system, the Reduced Input system stands out for its ability to reduce nitrate loss and accumulate soil C while maintaining high grain yields. While the Biologically Based system has provided greater nitrate and C conservation benefits, yields have been substantially lower, resulting in a significant yield vs. ecosystem service trade-off. Likewise in the LFL, ecosystem service benefits in the Organic system come at the cost of a significant yield penalty (Table 15.3). The extent to which these trade-offs might be acceptable will depend on many factors, including the goals and economic position of land managers. Public interests as embedded in policy will also need to assess trade-offs among services, weighing the relative and absolute value of different services to society.

Trade-offs also exist between crop yield and greenhouse gas mitigation services, which can be evaluated as reductions in the global warming impact (GWI). The MCSE No-till system combines high yields with a high soil C sequestration potential, and even in the face of higher chemical use, the No-till system had a low net GWI (-14 g CO_2e m^{-2} yr^{-1}) as compared to the Conventional system (101 g CO_2e m^{-2} yr^{-1}; Gelfand and Robertson 2015, Chapter 12 in this volume). In both systems, N fertilizer use contributed similarly to GWI, as did the enhanced liming requirement associated with fertilizer use. Other long-term experiments have shown that N fertilization is associated with decreased exchangeable calcium, magnesium, and potassium levels, and lower cation exchange capacity, leading to increased requirements for liming (Liu et al. 1997). The direct relationship of N fertilization with liming requirements, as well as the role of liming in the overall GWI of agricultural systems, was evaluated at KBS and for agricultural row-crop systems in general by Hamilton et al. (2007).

The Biologically Based system also provides greenhouse gas mitigation services. This system uses no N fertilizer and sequesters considerable soil C over the long term (note that no C gains were observed initially; Robertson et al. 2000). However, the moderate grain yields associated with the Biologically Based system reduce its net energy gain (Table 15.2) and raise its greenhouse gas intensity (g CO_2e per unit yield). Low productivity would also be expected to jeopardize profitability, which is a precondition for economic sustainability. The current profitability of the Biologically Based system depends to a considerable extent on a market premium, such as those typically paid for organic products (Chavas et al. 2009, Jolejole 2009).

Overall, the costs and benefits associated with biodiversity and other alternative practices are complex. The interaction of management practice and cropping system diversity is a primary focus of the LFL, discussed in the next section.

Agronomic Insights from the KBS LFL

Productivity

On a global basis, there is a growing requirement for grain for food and livestock feed. The quantity of grain produced is the most important service provided by field crops, and it is a key determinant of profitability on many farms. The type of grain produced and its market price are important, but the foundation to generating profit is sufficient production, and crop species vary in their ability to deliver this. Corn provides a biological advantage over other major grain crops such as soybean and wheat because of its greater efficiency at transforming sunlight into grain. This is shown by a broad-stroke comparison of LFL grain yields on a whole system basis. Figure 15.2 shows that a continuous corn rotation produced 5.7 Mg ha^{-1} yr^{-1} of grain, on average, over 4 years, whereas over the same period a corn–corn–soybean rotation produced 4.6 Mg ha^{-1} yr^{-1} (means for Organic and Integrated Conventional systems). Cumulative grain production over this period was 20.2 Mg ha^{-1} in the Organic one-species system (continuous corn), 17.1 Mg ha^{-1} in the Organic three-species (rotation crop sequence), 25.7 Mg ha^{-1} in the Integrated Conventional one-species system, and 19.7 Mg ha^{-1} in the Integrated Conventional three-species system.

On a whole system basis, the low-diversity continuous corn system produced the most grain. However, a higher market price for soybean and wheat will in many cases compensate for the moderate yield potential of rotated crops compared to corn. This will be especially true in locales without high corn subsides.

Nevertheless, LFL crop diversity enhanced corn grain yield: corn in rotation was almost always associated with higher yields compared to continuous corn (Fig. 15.2). Although diversity imparted via cover crops was not associated with higher corn grain yield in the Integrated Conventional system, a positive trend was observed in the rotated Organic system. No biodiversity effect of cover crops was

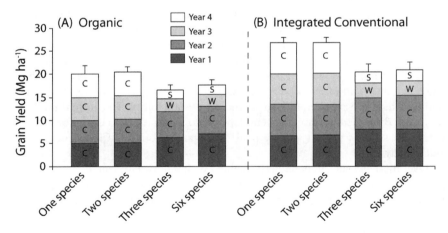

Figure 15.2. Grain yields in the various diversity treatments of the Living Field Lab (means and standard deviations for four replicate plots). A) Annual grain yield average in a four-year rotation sequence under Organic management at the Living Field Lab, where C=corn, W=wheat and S=soybean. The cropping systems include one species (continuous C), two-species (continuous C with a winter annual cover crop), three-species (rotated C-C-S-W), and six species (C-C-S-W with winter cover crops red clover, annual ryegrass and crimson clover). B) Annual grain yield average under Integrated Conventional management; cropping systems as described for Organic.

observed in soybean or wheat. The modest effect of biodiversity on corn productivity may have been influenced by N fertilization: species diversity is known to enhance overall productivity in infertile plant communities (Tilman et al., 2001), but would have a minimal effect in a nutrient-enriched environment such as fertilized corn. Also surprisingly, as for corn in the MCSE, diversity had no effect on the variability of grain yield over time: the coefficient of variation over 12 years was 37% in both continuous corn and in the diversified systems for both the Organic and Integrated Conventional systems (Snapp et al. 2010a).

Management other than rotational diversity also influences grain yield. The average grain yield in monoculture corn in the Organic system was 5.0 Mg ha^{-1} yr^{-1}, 22% lower than monoculture corn in the Integrated Conventional system (Snapp et al. 2010a). Similarly, a 22-year European trial showed that organic management was associated with yields ~20% lower than conventional across a range of crop species (Maeder et al. 2002). In the Mid-Atlantic region of the U. S., Cavigelli et al. (2008) documented >30% yield reductions in organic vs. conventionally managed crops. This yield reduction is not surprising, as management intensity and reliance on external inputs are generally associated with high crop yields. In the adjacent MCSE systems, corn grain yields in the Conventional and No-till systems were higher than in the Reduced Input and Biologically Based systems over the period 1989–2007 (Table 15.2), although overall grain yield of the corn–soybean–wheat rotation in the Reduced Input system equaled that of the Conventional system when averaged over 1996–2007 (Table 15.3).

Moderate yield reductions under organic management are typically compensated for by market premiums. A profitability analysis of a long-term trial in Wisconsin showed an 85 to 110% increase in profit for organically managed grain crop systems when organic price premiums were included (Chavas et al. 2009). Jolejole (2009) found a similar result in an analysis of the MCSE systems: the Biologically Based system was more profitable than the Conventional system when assigned premium prices; otherwise, lower yields and higher labor and cover crop costs offset savings in chemical use.

Grain yield in LFL systems varied markedly with year (Snapp et al. 2010a). Under organic management, where N supply is often limiting, the highest corn yield was obtained >58% of the time in the six-species system (with the legumes red clover and crimson clover). Dry summer conditions in lower-yielding years and the well-drained nature of the site may explain why diversity did not support high corn yields in water-deficient years (Snapp et al. 2010a). Weed competition has also been markedly variable over time and may have contributed to a low corn yield response in some years, despite management designed to control weeds (see below; Smith and Gross 2006a).

Soil Carbon and Phosphorus

Soil organic matter and fertility are key supporting services in agriculture. Management and diversity both affected LFL soil resources. Organic management maintained inorganic phosphorus and enhanced soil organic C by 52%, compared to initial values at the onset of the experiment (Table 15.4). Integrated conventional

Table 15.4. Soil characteristics and significant differences by treatment in the Living Field Lab systems.[a]

Management/ Diversity	Soil C Content (kg m^{-2})	Total N (mg N kg^{-1})	C/N Ratio	Phosphorus[a] (mg P kg^{-1})	Potassium[a] (mg K kg^{-1})	Calcium[a] (mg Ca kg^{-1})
Integrated Conventional						
One species	2.7	0.85	9.43	29.8	94.4	1288
Two species	2.8	0.88	9.22	21.9	79.9	972
Three species	2.8	0.90	9.27	25.4	64.5	1074
Six species	3.3	0.94	10.22	28.0	68.1	1126
Organic						
One species	3.9	1.15	11.09	50.9	94.1	1299
Two species	3.9	1.17	9.91	53.4	125.5	1491
Three species	4.1	1.20	10.96	44.1	101.5	1395
Six species	3.7	1.10	11.07	40.7	117.2	1443
Analysis of Variance (ANOVA) *P*-value						
Management (M)	<0.0001	<0.0001	<0.003	<0.0001	<0.001	0.014
Diversity (D)	NS	NS	NS	0.04	NS	NS
M × D	NS	NS	NS	NS	NS	NS

[a]Soils sampled 0- to 20-cm depth in April 2008. Phosphorus, potassium, and calcium extracted using the Mehlich III method. NS = Not significant ($\alpha = 0.05$).
Source: Adapted from Snapp et al. (2010b).

production, on the other hand, did not maintain soil phosphorus and had almost no discernable effect on soil organic C. Cropping system diversity was not associated with enhanced soil C in the LFL. Processes influencing C sequestration are complex; organic corn rotations have been associated with both declines (Studdert and Echeverria 2000) and accumulations (Russell et al. 2009) of soil C.

Water Quality

Nitrate leaching, an indicator of water quality, was measured in the LFL by installing gravimetric water samplers in cover crop plots. Nitrate loss was pronounced under corn production, primarily in the spring months prior to planting corn (Sánchez et al., 2004; Snapp et al. 2010a). Farmers manage for high soil N availability at the onset of growth, particularly for N-demanding crops such as corn. Soil amended with compost or other N sources that are high in C content (such as cover crop residues) may be an important means for managing N, using temporary immobilization to reduce the spring inorganic N pool (McSwiney et al. 2010). Systems receiving fertilizer N—whether synthetic or organic—require especially careful management during the spring when intense rainfall events and limited plant growth lead to high leaching potential.

The Organic system leached 32 kg NO_3^--N ha^{-1} yr^{-1}, which was about half as much as the Integrated Conventional system (74 kg NO_3^--N ha^{-1} yr^{-1}), as shown by gravimetric lysimeters monitored from 1994 to 2000 (Snapp et al. 2010a;

Table 15.3). Organic management relied on compost and cover crops for N supply. These C-rich nutrient sources may have temporarily immobilized inorganic N and reduced N loss. However, immobilization of N can also reduce crop yields, depending on the competitiveness of plant roots for N, and how fast N is turned over during microbial assimilation (Drinkwater and Snapp 2007).

Pest Suppression

The effects of crop rotation and management system (Integrated Conventional vs. Organic) on weed suppression services in the LFL experiment were investigated from 2001 to 2004 (Smith and Gross 2006a). Over that time, and similar to what has been observed in the MCSE (Davis et al. 2005; Gross et al. 2015, Chapter 7 in this volume), weed biomass was more than 10 times higher in the Organic compared to the Integrated Conventional system. In addition to total weed biomass, the composition of the weed community also differed between the two management systems, with smooth crabgrass (*Digitaria ischaemum* [Schreb.] Schreb. ex Muhl.) and Carolina horsenettle (*Solanum carolinense* L.) dominating the Integrated Conventional system, and common ragweed (*Ambrosia artemisiifolia* L.) and common lambsquarters (*Chenopodium album* L.) dominating the Organic system.

In contrast to the overriding influence of management system, crop rotation per se did not affect weed biomass. Crop rotation did, however, interact with management system to affect the interannual variability of the composition and structure of the weed community. Compared to conventionally managed crops and the Organic continuous corn system, weed community composition in the organic rotation was significantly more variable from one year to the next (Smith and Gross 2006a). This result is likely due, in part, to greater annual changes in the composition of weed species added to the seed bank (Smith and Gross 2006b), and suggests that an important regulating ecosystem service that crop rotation and diversification provide is to reduce the likelihood of developing a consistent weed community that is resistant to other weed management practices applied to the cropping system.

Ecosystem Service Trade-offs

As compared to the Integrated Conventional system, reduced inputs in the Organic system resulted in both yield reduction and enhanced ecosystem services in the form of soil C accretion and lower nitrate leaching losses (Table 15.3). To explicitly explore this trade-off, yield reductions in the Organic two-species and six-species systems were calculated relative to yield in the Integrated Conventional two-species system—chosen as a baseline system that followed all recommended integrated practices. We simultaneously evaluated soil C gains and the extent of nitrate leached in the Organic vs. Integrated Conventional systems over a 6-year period. Finally, the yield reduction was expressed in relation to soil C gained and nitrate leaching reduced (Table 15.3).

The Organic system produced 1370 kg ha^{-1} yr^{-1} less grain than the Integrated Conventional system, but resulted in substantial gains in soil C (820 kg ha^{-1} yr^{-1}) and reduced nitrate leaching losses by 42 kg N ha^{-1} yr^{-1} (Table 15.3). Increasing crop diversity in the LFL Organic system did not further improve these biogeochemical benefits (Snapp et al. 2010b).

In the MCSE, the gains in soil C with Biologically Based (Organic) management were higher than those obtained under No-till management (Table 15.3), but less than those associated with set asides or planting a perennial crop (Grandy and Robertson 2007, Piñeiro et al. 2009). The 50% reduction of nitrate leaching in the Organic system at the LFL was a significant achievement.

Designing Sustainable Agricultural Systems

KBS research highlights some of the trade-offs involved in developing row-crop systems that are more sustainable—more profitable, more environmentally benign, and more socially acceptable: in short, those that deliver a more desirable mix of ecosystem services (Robertson and Harwood 2013, Syswerda and Robertson 2014). Evidence from KBS and elsewhere (Robertson et al. 2007; Snapp et al. 2010a, b) shows that all services cannot be maximized simultaneously in agricultural systems; consequently, it is necessary to set priorities. Presently, priorities are set largely by markets and government policies that incentivize production and allow environmental costs to be externalized to society as a whole. Understanding trade-offs, especially with respect to yields, is an essential first step for incentivizing additional services. To the extent that most farmers' first priority is staying in business (Swinton et al. 2015b, Chapter 13 in this volume), the cost of providing any service that reduces farm profitability must be borne by society. This is particularly true for those services perceived primarily as a public good—greenhouse gas mitigation and water quality, for example. For services perceived to have local value—soil C storage as it affects soil fertility and crop diversity as it affects pest suppression, for example—costs are more willingly borne by the farmers (Swinton et al. 2015b, Chapter 13 in this volume). For example, biologically based row-crop management is shown at KBS LTER to improve soil C storage and water quality, but at the expense of reduced yields in cereals. Making up the yield difference represents the cost of providing these services.

The relationships between yield and other ecosystem services are complex. Management practices are commonly bundled within systems, so it can be difficult to prescribe one practice alone. In the MCSE No-till system, for example, the benefits of low soil disturbance come with a need for greater herbicide use. Within the Reduced Input and Biologically Based systems, the duration of living plant cover is high, providing water quality and soil C benefits, but soil disturbance is also high. These systems also have enhanced rotational diversity that promotes biological N fixation, reducing the requirement for N fertilizer inputs, but more fossil fuel is required to plow under the cover crop, kill weeds, and enhance residue contact with soil to promote the mineralization and release of nutrients in concert with crop demand.

Conservation Agriculture

Conservation agriculture is a broad term that considers many of the system design principles evaluated here (Govaerts et al. 2009). This management approach shifts the emphasis from conservation tillage practices to principles-based management that emphasizes: (1) reduction in tillage to ensure disturbance remains below a set percentage, (2) sufficient retention of residues to provide cover and protect soil from erosive forces, and (3) diversified crop rotations that mitigate against pest problems and ensure a mixture of residue qualities and heterogeneous root system inputs belowground.

The success of permanent no-till management at the MCSE is illustrated by the No-till system's high grain yields (~8% more than Conventional management) and soil C gains (Table 15.3). It is notable that no-till was implemented with no increase in fertilizer N inputs and only a modest increase in herbicide inputs compared to conventional management. Energy inputs are substantially lower with reduced reliance on tillage (e.g., 4.9 GJ ha^{-1} yr^{-1} in the MCSE No-till system compared to 7.1 GJ ha^{-1} yr^{-1} in the Conventional system, Table 15.2), despite modest increases in herbicide use in No-till. Declining fossil fuel supplies and high energy costs are important arguments for the adoption of conservation tillage equipment and practices.

Published studies that have evaluated crop diversification and conservation tillage have shown that only the combination of rotational diversity and reduced tillage is an effective means to enhance soil C and N over the long term (West and Post 2002, Govaerts et al. 2009). Overall, the MCSE showed that permanent no-till management of a corn–soybean–wheat rotation was an effective means to modestly improve a broad range of ecosystem services. It is important to note that this was the only alternative practice that enhanced grain yield relative to conventional management, which may in large part explain the broad adoption of conservation tillage practices. The reduction in number of field operations, leading to reduced fuel and time requirements, may also play a significant role in farmer adoption. However, adoption has not been universal nor, where it has been adopted, is it usually permanent (Horowitz et al. 2010). This is particularly so among farmers who produce on heavy (clayey) soils, or without ready access to herbicides. Another constraint to continued and future adoption of no-till is the increasing number of weed species that are evolving resistance to the primary herbicides used in no-till cropping systems (i.e., glyphosate), which threaten to reduce the longer-term viability of no-till production practices (Johnson et al. 2009) or increase the use of older herbicide chemistries that have greater potential for nontarget impacts and ecosystem disservices (Mortensen et al. 2012).

A first step in promoting conservation agriculture might include improved knowledge about farmer adoption of management practices such as tillage and extended cover. Policies that support innovative conservation practices might include instruments that mitigate risk associated with adoption and adaptive research—these could go far toward enhancing the adoption of conservation agriculture (Feinerman et al. 1992). Innovative research will be required to support adoption by organic farmers and smallholder farmers in developing countries, groups that have thus far been left behind by the conservation agriculture movement (Giller et al. 2009). Long-term research on different types of conservation practices, and associated ecosystem services, is also urgently required. There is a tremendous variety of tillage equipment and integrated practices that can be pursued in combination with manure, cover crops, and rotational crop sequences; all are expected to influence the ecosystem services that are generated.

Organic Agriculture

The principles of organic (biologically based) management are closely aligned with a "semi-closed" system that mainly relies on biological processes to regenerate soil resources and support the growth of healthy plants and animals (Pearson 2007). The duration and diversity of active plant growth and the synchronization of N availability with plant N demand are important features of biologically based management. Following these principles in the MCSE Biologically Based system resulted in enhanced biological N fixation (almost 2-fold higher than in the Conventional system, Fig. 15.1) and soil C gains (25% more than in the Conventional system) that were slow to accrue but occurred in spite of more frequent soil disturbance. Evidence also exists that available soil P has been maintained despite low P inputs (Fig. 15.1). Energy inputs were low in the Biologically Based system compared to Conventional (4.8 GJ ha^{-1} yr^{-1} vs. 7.1 GJ ha^{-1} yr^{-1} in Conventional, Table 15.2). Biologically based crops used no fossil fuel–derived agrochemicals other than fuel for field operations, so total farming energy inputs were equivalent to those of the No-till system.

The yield reductions of cereals observed in the Biologically Based system could be considered a worthwhile trade-off for enhanced ecosystem services, although yield trade-off estimates for nitrate leaching and soil C sequestration in the MCSE indicate that gains in those ecosystem services are negated by the loss in yield (Table 15.3). There are also additional costs incurred for this system and for the Reduced Input system, including labor, tillage, and cover crop establishment, offset somewhat by the lower costs associated with reduced pesticide and fertilizer use. The energy balance conducted for the MCSE systems reflects the lower yield of cereals and net change in inputs associated with the Biologically Based system (Table 15.2).

A significant challenge associated with biologically based management is the labor and land investment in growing cover crops that fix N and build soil C (Drinkwater and Snapp 2007). Not only does this incur seed and management costs for a plant that provides little or no cash value, it also involves opportunity costs. That is, planting diverse crops can infringe on the window of time and resources required to grow higher value crops. For example, cover crops enhance ecosystem services by providing soil cover and active rooting throughout the year. However, planting a summer cash crop is necessarily delayed by the need to first kill and plow under the cover crop. It is also important to allow cover crop residues time to decompose, and the result is an even later planting window. This reduces the length of the growing season, and may require planting shorter season varieties that can have lower yields. This is one likely cause of the lower corn and wheat yields observed in the MCSE Reduced Input and Biologically Based systems.

Opportunity costs are particularly acute challenges for farmers operating in short growing season environments, such as the temperate U.S. Great Lakes region and unimodal rainfall systems in other parts of the world. Farmers have developed systems that maximize use of the biophysical environment (e.g., light, temperature, and moisture) to produce marketable crops. In locations where there is a longer growing season, such as in the southeastern United States, farmers develop double cropping systems with two cash crops per season, rather than following a cash crop with a biology-promoting cover crop (Cavigelli et al. 2008). This reflects the reward structure of current policies, and indeed is a requirement for farm survival in many socioeconomic environments. Overall, the costs and benefits associated with biodiversity are complex and interact with management practices.

Organic row-crop production makes up a very small proportion of midwestern agriculture, but organic acreage is increasing rapidly, even in the absence of policy support for broader adoption (Dimitri and Greene 2002). This would seem to suggest that organic row crops could be promoted in the United States without radically altering policy instruments or incentives. However, we note that as the supply of organic products increases, prices are expected to decline, reducing the premium that now compensates for lower yields.

On the other hand, the Reduced Input system has yields much closer to Conventional, and shares most of the environmental benefits of the Biologically Based system. Cropping systems based on the substitution of biological management for most rather than all chemical inputs could be an attractive hybrid system that optimizes yield and services. Such a system could be widely adopted with proper incentives for farmers' providing desirable services.

Perennial Vegetation

The cumulative effect of perennial vegetation in agroecosystems is dramatic, particularly belowground. In a Russian study, substantial soil C gains to 2 m were observed in grassland and forage systems compared to annual cropping systems (Mikhailova et al. 2000). Perennial legume and grass plantings have been shown to improve soil organic C by 35 to 58% compared to annuals (Bremer et al. 1994), which is comparable to the 45% increase in soil C we observed in our MCSE continuous Alfalfa system (Fig. 15.1).

Nitrogen leaching losses in perennial vegetation vary, influenced by species growth patterns and the importance of biological N fixation. Farming system management is also important, including reactive N inputs and harvest operations. In the MCSE Early Successional community, very low levels of inorganic N have been observed in soil, which is consistent with tight N cycling (Robertson et al. 2000). Alfalfa is intensively managed compared to this Early Successional system, with biomass harvested three times per year, on average, and lime and fertilizers other than N are applied as needed. Nitrogen leaching from the alfalfa system was lower than from any of the annual cropping systems (Syswerda et al. 2012), as has been shown in other shorter-term studies (Randall et al. 1997) despite high levels of biological N fixation. Nitrous oxide production, on the other hand, was as high from the Alfalfa system as from any of the annual cropping systems (Gelfand and Robertson 2015, Chapter 12 in this volume, Millar and Robertson 2015, Chapter 9 in this volume). Randall and colleagues proposed alfalfa plantings as a means to reduce tile drainage nitrate pollution across the U.S. Midwest. A review by Ledgard (2001) found that grazed grassland-legume mixed systems, with minimal fertilizer inputs, were associated with modest N losses from denitrification (6 kg N ha^{-1}) and leaching (23 kg N ha^{-1}).

Generally, perennial crops are confined to marginal farming areas with steep terrain, variable topography, or shallow, infertile soils. In the Midwest—and throughout temperate agricultural regions—alfalfa is the most important perennial crop. In southwest Michigan alfalfa has been grown on ~15% of agricultural land since the 1930s (Sylvester and Gutman 2008). This is presumably due to the species' high-quality residues and high productivity, together with access to ready markets provided by the state's dairy industry. Alfalfa is adapted to a broad range of environments, and there are varieties that can be grown under intensive irrigated and fertilized management.

Perennial vegetation is currently the fastest means for capturing C and reducing farm nitrate losses on a significant scale, and it thus could profoundly improve the delivery of ecosystem services generated from agriculture. Perennial crops grown for cellulosic biofuel on lands now not suited for food crops offer a major opportunity for agriculture to contribute to climate stabilization, soil and N conservation, pest suppression, and other ecosystem services including societal benefits such as national fuel security (Robertson et al. 2008, 2011). Planting perennial forages on land now used for food crops would also provide benefits, but promoting this for broad adaption would require radical changes in agricultural policies and marketing systems.

Cover crop integration is a small step in the direction of perennializing grain crops, and cover crop use could be promoted with existing agricultural policy instruments. An example of wide-scale cover crop adoption is available from the south of Sweden where row-crop farmers were paid to grow cover crops, resulting in a reduction of nitrate leaching (Kirchmann et al. 2002). Another way forward may be to develop grains such as wheat or sorghum that have a perennial life cycle (DeHaan et al. 2005, Glover et al. 2010). Both of these approaches to perennialization deserve support through agricultural policies that promote cover crop integration for incremental improvements in existing systems and research to develop a portfolio of perennial crops that could help to further diversify rural landscapes.

Summary

Row crops can be managed to deliver ecosystem services in addition to yield. KBS LTER results illustrate the delivery of services by current farming systems as well as the principles that can be used to design future systems to enhance the delivery of these and other services. Documented services other than yield include (1) soil C accretion with its positive effects on soil fertility, water-holding capacity, and climate stabilization; (2) nitrate conservation that improves water quality by reducing leaching of nitrate into ground and surface waters; (3) greenhouse gas mitigation including N_2O abatement and lower CO_2 emissions; (4) energy efficiency that saves fuel; and (5) pest suppression that reduces pesticide use.

No current agricultural system can maximize the delivery of all services—almost always the delivery of one service affects the potential delivery of others. Thus, trade-offs must be considered. The most important trade-off for farmers is profitability—the opportunity cost of providing a particular service. No-till management, for example, builds soil C, reduces fuel use, and reduces nitrate leaching, but it can compress the spring planting period and requires specialized equipment. Biologically based management conserves nitrogen, builds soil C, and reduces pesticide loading, but it has associated costs of timely labor requirements and yield reductions not recoverable without the price premiums provided by organic certification.

Key findings from over 20 years of KBS LTER row-crop research on ecosystem services include:

1. Crop (rotational) diversity—planting legumes, in particular—provides enhanced opportunities for biological N fixation and pest regulation.
2. Planting a forage or cover crop makes an annual row-crop system more perennial. Such plantings extend the duration of living cover, support soil C sequestration, reduce N losses, and can lower reliance on chemical inputs.
3. Reducing soil disturbance through conservation tillage enhances soil C sequestration, energy efficiency, and crop yield.

Economic trade-offs were also documented:

1. Compared with conventional crop management, direct costs associated with alternative management practices depend on requirements for extra inputs, labor, and associated costs, as well as any reduction in requirements for chemical inputs.
2. Indirect opportunity costs associated with cover crops include allowing time for biological decomposition to occur, which often requires late plantings of cash crops after cover crops, with associated yield penalties, particularly for corn.
3. Diversification is associated with lower production where moderate-yield crops or cover crops are substituted for high-yield crops such as corn, although economic returns can still be high from moderate-yield crops depending on product prices.

In summary, organic and reduced input management systems deliver more ecosystem services (Table 15.3). However, there is an apparent yield penalty, as shown for organic-managed corn relative to conventional in both the LFL and MCSE. In contrast, conservation tillage is consistently associated with high corn yields relative to conventional management. Further, biologically based soybean yields were consistently high. A simplistic approach to evaluating the yield trade-offs of environmental services provided by these management systems is to examine yield trends over time relative to annual estimates of soil C gain and nitrate leaching

(Table 15.3). In doing so, we find that nitrate leaching can be reduced substantially, albeit with a yield penalty for organic crops, or no penalty in the case of No-till and Reduced Input. Other disservices such as herbicide leaching have not been evaluated here. Given the suite of services considered, no-till is attractive—providing gains in crop yields in conjunction with soil C sequestration and reductions in nitrate leaching. However, the environmental services obtained from no-till are not in themselves sufficient to maximize services and minimize disservices from agriculture. Fine-tuning of N fertilizer application to take into account N fixation and other N inputs is also a promising approach that has many environmental advantages. Adopting a stepwise process would involve initial reliance on N fertilizer adjustments and conservation tillage, followed by more transformative types of alternative management that rely on the principles of diversity and perenniality.

References

Beaulieu, J. J., J. L. Tank, S. K. Hamilton, W. M. Wollheim, R. O. Hall, Jr., P. J. Mulholland, B. J. Peterson, L. R. Ashkenas, L. W. Cooper, C. N. Dahm, W. K. Dodds, N. B. Grimm, S. L. Johnson, W. H. McDowell, G. C. Poole, H. M. Valett, C. P. Arango, M. J. Bernot, A. J. Burgin, C. L. Crenshaw, A. M. Helton, L. T. Johnson, J. M. O'Brien, J. D. Potter, R. W. Sheibley, D. J. Sobota, and S. M. Thomas. 2011. Nitrous oxide emission from denitrification in stream and river networks. Proceedings of the National Academy of Sciences USA 108:214–219.

Bremer, E., H. H. Janzen, and A. M. Johnson. 1994. Sensitivity of total light fraction and mineralizable organic matter to management practices in a Lethbridge soil. Canadian Journal of Soil Science 74:131–138.

Cavigelli, M., J. R. Teasdale, and A. E. Conklin. 2008. Long-term agronomic performance of organic and conventional field crops in the Mid-Atlantic region. Agronomy Journal 100:785–794.

Chavas, J.-P., J. L. Posner, and J. L. Hedtcke. 2009. Organic and conventional production systems in the Wisconsin Integrated Cropping Systems Trial: II. Economic and risk analysis 1993–2006. Agronomy Journal 101:288–295.

Crum, J. R., and H. P. Collins. 1995. KBS soils. Kellogg Biological Station Long-Term Ecological Research, Michigan State University, Hickory Corners, MI. <http://lter.kbs.msu.edu/research/site-description-and-maps/soil-description>

Davis, A. S., K. Renner, and K. L. Gross. 2005. Weed seedbank and community shifts in a long-term cropping systems experiment. Weed Science 53:296–306.

Davis, A. S., and K. A. Renner. 2007. Influence of seed depth and pathogens on fatal germination of velvetleaf (*Abutilon theophrasti*) and giant foxtail (*Setaria faberi*). Weed Science 55:30–35.

Davis, M. A., J. P. Grime, and K. Thompson. 2000. Fluctuating resources in plant communities: a general theory of invasibility. Journal of Ecology 88:528–534.

DeHaan, L. R., D. L. Van Tassel, and S. T. Cox. 2005. Perennial grain crops: a synthesis of ecology and plant breeding. Renewable Agriculture and Food Systems 20:5–14.

Diaz, R. J., and R. Rosenberg. 2008. Spreading dead zones and consequences for marine ecosystems. Science 321:926–929.

Dimitri, C., and C. Greene. 2002. Recent growth patterns in the U.S. organic foods market. Agriculture Information Bulletin Number 777, U.S. Department of Agriculture, Economic Research Service, Washington, DC, USA.

Drinkwater, L. E., and S. S. Snapp. 2007. Nutrients in agroecosystems: rethinking the management paradigm. Advances in Agronomy 92:163–186.

Drinkwater, L. E., P. Wagoner, and M. Sarrantonio. 1998. Legume-based cropping systems have reduced carbon and nitrogen losses. Nature 396:262–265.

El-Hage Scialabba, N., and M. Müller-Lindenlauf. 2010. Organic agriculture and climate change. Renewable Agriculture and Food Systems 25:158–169.

ERS (Economic Research Service). 2011. Organic production. Dataset (Tables 2 and 3). U.S. Department of Agriculture (USDA), Washington DC, USA. <http://www.ers.usda.gov/Data/Organic/> Accessed May25, 2011.

Feinerman, E., J. A. Herriges, and D. Holtkamp. 1992. Crop insurance as a mechanism for reducing pesticide usage: a representative farm analysis. Review of Agricultural Economics 14:169–186.

Gage, S. H., J. E. Doll, and G. R. Safir. 2015. A crop stress index to predict climatic effects on row-crop agriculture in the U.S. North Central Region. Pages 77–103 in S. K. Hamilton, J. E. Doll, and G. P. Robertson, editors. The ecology of agricultural ecosystems: long-term research on the path to sustainability. Oxford University Press, New York, New York, USA.

Gardner, J. B., and L. E. Drinkwater. 2009. The fate of nitrogen in grain cropping systems: a meta-analysis of ^{15}N field experiments. Ecological Applications 19:2167–2184.

Gelfand, I., and G. P. Robertson. 2015. Mitigation of greenhouse gas emissions in agricultural ecosystems. Pages 310–339 in S. K. Hamilton, J. E. Doll, and G. P. Robertson, editors. The ecology of agricultural ecosystems: long-term research on the path to sustainability. Oxford University Press, New York, New York, USA.

Gelfand, I., S. S. Snapp, and G. P. Robertson. 2010. Energy efficiency of conventional, organic, and alternative cropping systems for food and fuel at a site in the U.S. Midwest. Environmental Science and Technology 44:4006–4011.

Giller, K., E. Witter, M. Corbeels, and P. Titonell. 2009. Conservation agriculture and smallholder farming in Africa: the heretic's view. Field Crops Research 114:23–34.

Glover, J. D., J. P. Reganold, L. W. Bell, J. Borevitz, E. C. Brummer, E. S. Buckler, C. M. Cox, T. S. Cox, T. E. Crews, S. W. Culman, L. R. DeHaan, D. Ericksson, B. S. Gill, J. Holland, F. Hu, B. S. Hulke, A. M. H. Ibrahim, W. Jackson, S. S. Jones, S. C. Murray, A. H. Paterson, E. Ploschuk, E. J. Sacks, S. Snapp, D. Tao, D. L. Van Tassel, L. J. Wade, D. L. Wyse, and Y. Xu. 2010. Increased food and ecosystem security via perennial grains. Science 328:1638–1639.

Govaerts, B., N. Verhulst, A. Castellanos-Navarrete, K. D. Sayer, J. Dixon, and L. Dendooven. 2009. Conservation agriculture and soil carbon sequestration: between myth and farmer reality. Critical Reviews in Plant Science 28:97–122.

Grace, P., G. P. Robertson, N. Millar, M. Colunga-Garcia, B. Basso, S. H. Gage, and J. Hoben. 2011. The contribution of maize cropping in the Midwest USA to global warming: a regional estimate. Agricultural Systems 104:292–296.

Grandy, A. S., and G. P. Robertson. 2006. Initial cultivation of a temperate-region soil immediately accelerates aggregate turnover and CO_2 and N_2O fluxes. Global Change Biology 12:1507–1520.

Grandy, A. S., and G. P. Robertson. 2007. Land-use intensity effects on soil organic carbon accumulation rates and mechanisms. Ecosystems 10:58–73.

Grandy, A. S., R. L. Sinsabaugh, J. C. Neff, M. Stursova, and D. R. Zak. 2008. Nitrogen deposition effects on soil organic matter chemistry are linked to variation in enzymes, ecosystems and size fractions. Biogeochemistry 91:37–49.

Gray, S. 1996. The Yankee West: community life on the Michigan frontier. The University of North Carolina Press, Chapel Hill, North Carolina, USA.

Greenland, D. J., and P. H. Nye. 1959. Increases in the carbon and nitrogen contents of tropical soils under natural fallows. Journal of Soil Science 10:284–299.

Gross, K. L., S. Emery, A. S. Davis, R. G. Smith, and T. M. P. Robinson. 2015. Plant community dynamics in agricultural and successional fields. Pages 158–187 in S. K. Hamilton, J. E. Doll, and G. P. Robertson, editors. The ecology of agricultural ecosystems: long-term research on the path to sustainability. Oxford University Press, New York, New York, USA.

Hamilton, S. K. 2015. Water quality and movement in agricultural landscapes. Pages 275–309 in S. K. Hamilton, J. E. Doll, and G. P. Robertson, editors. The ecology of agricultural ecosystems: long-term research on the path to sustainability. Oxford University Press, New York, New York, USA.

Hamilton, S. K., A. L. Kurzman, C. Arango, L. Jin, and G. P. Robertson. 2007. Evidence for carbon sequestration by agricultural liming. Global Biogeochemical Cycles 21:GB2021.

Hector, A., B. Schmid, C. Beierkuhnlein, M. C. Caldeira, M. Diemer, P. G. Dimitrakopoulos, J. A. Finn, H. Freitas, P. S. Giller, J. Good, R. F. Harris, P. Hogberg, K. Huss-Danell, J. Joshi, A. Jumpponen, C. Korner, P. W. Leadley, M. Loreau, A. Minns, C. P. H. Mulder, G. O'Donovan, S. J. Otway, J. S. Pereira, A. Prinz, D. J. Read, M. Scherer-Lorenzen, E.-D. Schulze, A.-S. D. Siamantziouras, E. M. Spehn, A. C. Terry, A. Y. Troumbis, F. I. Woodward, S. Yachi, and J. H. Lawton. 1999. Plant diversity and productivity experiments in European grasslands. Science 286:1123–1127.

Hoben, J. P., R. J. Gehl, N. Millar, P. R. Grace, and G. P. Robertson. 2011. Nonlinear nitrous oxide (N_2O) response to nitrogen fertilizer in on-farm corn crops of the US Midwest. Global Change Biology 17:1140–1152.

Hooper, D. U., E. C. Adair, B. J. Cardinale, J. E. K. Byrnes, B. A. Hungate, K. L. Matulich, A. Gonzalez, J. E. Duffy, L. Gamfeldt, and M. I. O'Connor. 2012. A global synthesis reveals biodiversity loss as a major driver of ecosystem change. Nature 486:105–108.

Hooper, D. U., F. S. Chapin, J. J. Ewel, A. Hector, P. Inchausti, S. Lavorel, J. H. Lawton, D. M. Lodge, M. Loreau, S. Naeem, B. Schmid, H. Setala, A. J. Symstad, J. Vandermeer, and D. A. Wardle. 2005. Effects of biodiversity on ecosystem functioning: a consensus of current knowledge. Ecological Monographs 75:3–35.

Horowitz, J., R. Ebel, and K. Ueda. 2010. "No-till" farming is a growing practice. Economic Information Bulletin Number 70, U.S. Department of Agriculture, Economic Research Service, Washington, DC, USA.

Hülsbergen, K.-J., B. Feil, S. Biermann, G.-W. Rathke, W.-D. Kalk, and W. Diepenbrock. 2001. A method of energy balancing in crop production and its application in a long-term fertilizer trial. Agriculture, Ecosystems and Environment 86:303–321.

Johnson, W. G., V. M. Davis, G. R. Kruger, and S. C. Weller. 2009. Influence of glyphosate-resistant cropping systems on weed species shifts and glyphosate-resistant weed populations. European Journal of Agronomy 31:162–172.

Jolejole, M. C. B. 2009. Trade-offs, incentives, and the supply of ecosystem services from cropland. Thesis, Michigan State University, East Lansing, Michigan, USA.

Kirchmann, H., A. E. J. Johnston, and L. F. Bergström. 2002. Possibilities for reducing nitrate leaching from agricultural land. Ambio 31:404–408.

Ledgard, S. F. 2001. Nitrogen cycling in low input legume-based agriculture, with emphasis on legume/grass pastures. Plant and Soil 228:43–59.

Li, J. M., and R. J. Kremer. 2000. Rhizobacteria associated with weed seedlings in different cropping systems. Weed Science 48:734–741.

Liebman, M., and A. S. Davis. 2000. Integration of soil, crop and weed management in low-external-input farming systems. Weed Research 40:27–47.

Liebman, M., C. L. Molher, and C. P. Staver. 2001. Ecological management of agricultural weeds. Cambridge University Press, Cambridge, UK.

Liu, Y., D. A. Laird, and P. Barak. 1997. Dynamics of fixed and exchangeable NH_4 and K in soils under long-term fertility management. Soil Science Society America Journal 61:310–314.

Lotter, D., R. Seidel, and W. Liebhardt. 2003. The performance of organic and conventional cropping systems in an extreme climate year. American Journal of Alternative Agriculture 18:146–154.

Lowrance, R., B. R. Stinner, and G. J. House, editors. 1984. Agricultural ecosystems: unifying concepts. John Wiley & Sons, New York, New York, USA.

Ma, B. L., T. Y. Wu, N. Tremblay, W. Deen, M. J. Morrison, N. B. McLaughlin, E. G. Gregorich, and G. Stewart. 2010. Nitrous oxide fluxes from corn fields: on-farm assessment of the amount and timing of nitrogen fertilizer. Global Change Biology 16:156–170.

Maeder, P., A. Fliessbach, D. Dubois, L. Gunst, P. Fried, and U. Niggli. 2002. Soil fertility and biodiversity in organic farming. Science 2002:1694–1697.

Matson, P. A., W. J. Parton, A. G. Power, and M. J. Swift. 1997. Agricultural intensification and ecosystem properties. Science 277:504–509.

McSwiney, C. P., and G. P. Robertson. 2005. Nonlinear response of N_2O flux to incremental fertilizer addition in a continuous maize (*Zea mays* sp.) cropping system. Global Change Biology 11:1712–1719.

McSwiney, C. P., S. S. Snapp, and L. E. Gentry. 2010. Use of N immobilization to tighten the N cycle in conventional agroecosystems. Ecological Applications 20:648–662.

MEA (Millenium Ecosystem Assessment). 2005. Our human planet: summary for decision-makers. Island Press, Washington, DC, USA.

Menalled, F. D., R. G. Smith, J. T. Dauer, and T. B. Fox. 2007. Impact of agricultural management systems on carabid beetle communities and weed seed predation. Agriculture, Ecosystems and Environment 118:49–54.

Mikhailova, E. A., R. B. Bryant, I. I. Vassenev, S. J. Schwager, and C. J. Post. 2000. Cultivation effects on soil carbon and nitrogen contents at depth in a Russian chernozem. Soil Science Society of America Journal 64:738–745.

Millar, N., and G. P. Robertson. 2015. Nitrogen transfers and transformations in row-crop ecosystems. Pages 213–251 in S. K. Hamilton, J. E. Doll, and G. P. Robertson, editors. The ecology of agricultural ecosystems: long-term research on the path to sustainability. Oxford University Press, New York, New York, USA.

Millar, N., G. P. Robertson, P. R. Grace, R. J. Gehl, and J. P. Hoben. 2010. Nitrogen fertilizer management for nitrous oxide (N_2O) mitigation in intensive corn (Maize) production: an emissions reduction protocol for US Midwest agriculture. Mitigation and Adaptation Strategies for Global Change 15:185–204.

Mortensen, D. A., J. F. Egan, B. D. Maxwell, M. R. Ryan, and R. G. Smith. 2012. Navigating a critical juncture for sustainable weed management. BioScience 62:75–84.

Pearson, C. J. 2007. Regenerative, semi-closed systems: a priority for twenty-first-century agriculture. BioScience 57:409–418.

Pimentel, D., P. Hepperly, J. Hanson, D. Dougds, and R. Seidel. 2005. Environmental, energetic, and economic comparisons of organic and conventional farming systems. BioScience 55:573–582.

Piñeiro, G., E. G. Jobbágy, J. Baker, B. C. Murray, and R. B. Jackson. 2009. Set-asides can be better climate investment than corn ethanol. Ecological Applications 19:277–282.

Powlson, D. S., T. M. Addiscott, N. Benjamin, K. G. Cassman, T. M. de Kok, H. van Grinsven, J.-L. L'hirondel, A. A. Avery, and C. van Kessel. 2008. When does nitrate become a risk for humans? Journal of Environmental Quality 37:291–295.

Randall, G. W., D. R. Huggins, M. P. Russelle, D. J. Fuchs, W. W. Nelson, and J. L. Anderson. 1997. Nitrate losses through subsurface tile drainage in conservation reserve program, alfalfa, and row crop systems. Journal of Environmental Quality 26:1240–1247.

Robertson, G. P. 1997. Nitrogen use efficiency in row crop agriculture: crop nitrogen use and soil nitrogen loss. Pages 347–365 in L. Jackson, editor. Ecology in agriculture. Academic Press, New York, New York, USA.

Robertson, G. P., J. C. Broome, E. A. Chornesky, J. R. Frankenberger, P. Johnson, M. Lipson, J. A. Miranowski, E. D. Owens, D. Pimentel, and L. A. Thrupp. 2004. Rethinking the vision for environmental research in US agriculture. BioScience 54:61–65.

Robertson, G. P., L. W. Burger, C. L. Kling, R. Lowrance, and D. J. Mulla. 2007. New approaches to environmental management research at landscape and watershed scales. Pages 27–50 in M. Schnepf and C. Cox, editors. Managing agricultural landscapes for environmental quality. Soil and Water Conservation Society, Ankeny, Iowa, USA.

Robertson, G. P., V. H. Dale, O. C. Doering, S. P. Hamburg, J. M. Melillo, M. M. Wander, W. J. Parton, P. R. Adler, J. N. Barney, R. M. Cruse, C. S. Duke, P. M. Fearnside, R. F. Follett, H. K. Gibbs, J. Goldemberg, D. J. Mladenoff, D. Ojima, M. W. Palmer, A. Sharpley, L. Wallace, K. C. Weathers, J. A. Wiens, and W. W. Wilhelm. 2008. Sustainable biofuels redux. Science 322:49–50.

Robertson, G. P., and S. K. Hamilton. 2015. Long-term ecological research at the Kellogg Biological Station LTER Site: conceptual and experimental framework. Pages 1–32 in S. K. Hamilton, J. E. Doll, and G. P. Robertson, editors. The ecology of agricultural ecosystems: long-term research on the path to sustainability. Oxford University Press, New York, New York, USA.

Robertson, G. P., S. K. Hamilton, S. J. Del Grosso, and W. J. Parton. 2011. The biogeochemistry of bioenergy landscapes: carbon, nitrogen, and water considerations. Ecological Applications 21:1055–1067.

Robertson, G. P., and R. R. Harwood. 2013. Sustainable agriculture. Pages 111–118 in S. A. Levin, editor. Encyclopedia of biodiversity. Second edition, Volume 1. Academic Press, Waltham, Massachusetts, USA.

Robertson, G. P., and E. A. Paul. 2000. Decomposition and soil organic matter dynamics. Pages 104–116 in E. S. Osvaldo, R. B. Jackson, H. A. Mooney, and R. W. Howarth, editors. Methods in ecosystem science. Springer-Verlag, New York, New York, USA.

Robertson, G. P., E. A. Paul, and R. R. Harwood. 2000. Greenhouse gases in intensive agriculture: contributions of individual gases to the radiative forcing of the atmosphere. Science 289:1922–1925.

Robertson, G. P., and S. M. Swinton. 2005. Reconciling agricultural productivity and environmental integrity: a grand challenge for agriculture. Frontiers in Ecology and the Environment 3:38–46.

Robertson, G. P., and P. M. Vitousek. 2009. Nitrogen in agriculture: balancing the cost of an essential resource. Annual Review of Environment and Resources 34:97–125.

Rudy, A. P., C. K. Harris, B. J. Thomas, M. R. Worosz, S. C. Kaplan, and E. C. O'Donnell. 2008. The political ecology of Southwest Michigan Agriculture, 1837–2000. Pages 152–205 in C. L. Redman and D. R. Foster, editors. Agrarian landscapes in transition. Oxford University Press, New York, New York, USA.

Russell, A. E., C. A. Cambardella, D. A. Laird, D. B. Jaynes, and D. W. Meek. 2009. Nitrogen fertilizer effects on soil carbon balances in Midwestern U.S. agricultural systems. Ecological Applications 19:1102–1113.

Ryan, M. R., D. A. Mortensen, L. Bastiaans, J. R. Teasdale, S. B. Mirsky, W. S. Curran, R. Seidel, D. O. Wilson, and P. R. Hepperly. 2010. Elucidating the apparent maize tolerance to weed competition in long-term organically managed systems Weed Research 50:25–36.

Sánchez, J. E., R. R. Harwood, T. C. Willson, K. Kizilkaya, J. Smeenk, E. Parker, E. A. Paul, B. D. Knezek, and G. P. Robertson. 2004. Managing soil carbon and nitrogen for productivity and environmental quality. Agronomy Journal 96:769–775.

Sánchez, J. E., E. A. Paul, T. C. Willson, J. Smeenk, and R. R. Harwood. 2002. Corn root effects on the nitrogen-supplying capacity of a conditioned soil. Agronomy Journal 94:391–396.

Schoof, J. T., S. C. Pryor, and J. Suprenant. 2010. Development of daily precipitation projections for the United States based on probabilistic downscaling. Journal of Geophysical Research 115, D13106. doi:10.1029/2009JD013030

Smith, R. G., and K. L. Gross. 2006a. Rapid change in the germinable fraction of the weed seed bank in crop rotations. Weed Science 54:1094–1100.

Smith, R. G., and K. L. Gross. 2006b. Weed community and corn yield variability in diverse management systems. Weed Science 54:106–113.

Smith, R. G., K. L. Gross, and G. P. Robertson. 2008. Effects of crop diversity on agroecosystem function: crop yield response. Ecosystems 11:355–366.

Smith, R. G., F. D. Menalled, and G. P. Robertson. 2007. Temporal yield variability under conventional and alternative management systems. Agronomy Journal 99:1629–1634.

Snapp, S. 2008. Agroecology: principles and practice. Pages 53–88 in S. Snapp and B. Pound, editors. Agricultural systems: agroecology & rural innovation for development. Academic Press, Burlington, Massachusetts, USA.

Snapp, S. S., M. J. Blackie, R. A. Gilbert, R. Bezner-Kerr, and G. Y. Kanyama-Phiri. 2010b. Biodiversity can support a greener revolution in Africa. Proceedings of the National Academy of Sciences USA 107:20840–20845.

Snapp, S. S., L. E. Gentry, and R. R. Harwood. 2010a. Management intensity—not biodiversity—the driver of ecosystem services in a long-term row crop experiment. Agriculture, Ecosystems and Environment 138:242–248.

Snapp, S. S., S. M. Swinton, R. Labarta, D. Mutch, J. R. Black, R. Leep, J. Nyiraneza, and K. O'Neil. 2005. Evaluating cover crops for benefits, costs and performance within cropping system niches. Agronomy Journal 97:322–332.

Studdert, G. A., and H. E. Echeverria. 2000. Crop rotations and nitrogen fertilization to manage soil organic carbon dynamics. Soil Science Society of America Journal 64:496–503.

Swinton, S. M., K. L. Gross, D. A. Landis, and W. Zhang. 2015a. The economic value of ecosystem services from agriculture. Pages 54–76 in S. K. Hamilton, J. E. Doll, and G. P. Robertson, editors. The ecology of agricultural ecosystems: long-term research on the path to sustainability. Oxford University Press, New York, New York, USA.

Swinton, S. M., N. Rector, G. P. Robertson, C. B. Jolejole, and F. Lupi. 2015b. Farmer decisions about adopting environmentally beneficial practices. Pages 340–359 in S. K. Hamilton, J. E. Doll, and G. P. Robertson, editors. The ecology of agricultural ecosystems: long-term research on the path to sustainability. Oxford University Press, New York, New York, USA.

Sylvester, K. M., and M. P. Gutmann. 2008. Changing agrarian landscapes across America: a comparative perspective. Pages 16–43 in C. L. Redman and D. R. Foster, editors. Agrarian landscapes in transition. Oxford University Press, New York, New York, USA.

Syswerda, S. P., B. Basso, S. K. Hamilton, J. B. Tausig, and G. P. Robertson. 2012. Long-term nitrate loss along an agricultural intensity gradient in the Upper Midwest USA. Agriculture, Ecosystems and Environment 149:10–19.

Syswerda, S. P., A. T. Corbin, D. L. Mokma, A. N. Kravchenko, and G. P. Robertson. 2011. Agricultural management and soil carbon storage in surface vs. deep layers. Soil Science Society of America Journal 75:92–101.

Syswerda, S. P., and G. P. Robertson. 2014. Ecosystem services along a management intensity gradient in Michigan (USA) cropping systems. Agriculture, Ecosystems, and Environment 189:28–35.

Tilman, D., K. G. Cassman, P. A. Matson, and R. L. Naylor. 2002. Agricultural sustainability and intensive production practices. Nature 418:671–677.

Tilman, D., P. B. Reich, J. Knops, D. A. Wedin, T. Mielke, and C. Lehman. 2001. Diversity and productivity in a long-term grassland experiment. Science 294:843–845.

West, T. O., and W. M. Post. 2002. Soil organic carbon sequestration rates by tillage and crop rotation: a global data analysis. Soil Science Society of America Journal 66:1930–1946.

Index